W9-CUK-233

Recent advances in materials and processing techniques have led to a revolution in the generation of ultrashort laser pulses. From novel fiber lasers to short pulse and high power diode lasers, development in this field has been very rapid. This comprehensive volume provides a survey of these innovations, and reviews the state of the art in compact, modelocked laser systems, discussing both their operational principles and potential applications.

The theory of short optical pulse generation by modelocking is covered in the first chapter, after which specific systems are discussed. These include passively modelocked solid state lasers, modelocked diode-pumped lasers, modelocked fiber lasers, nonlinear polarization evolution, modelocked surface emitting semiconductor lasers, ultrafast pulse generation by means of external cavity semiconductor lasers, hybrid soliton pulse sources, and monolithic colliding pulse modelocked diode lasers.

Presenting both theoretical and experimental aspects throughout, this book will be invaluable to anyone interested in short pulse laser systems, and particularly to researchers involved in high speed communications or the investigation of ultrafast phenomena.

CAMBRIDGE STUDIES IN MODERN OPTICS

Compact sources of ultrashort pulses

TITLES IN PRINT IN THIS SERIES

Optical Holography – Principles, Techniques and Applications
P. Hariharan

Fabry–Perot Interferometers
G. Hernandez

Holographic and Speckle Interferometry (second edition)
R. Jones and C. Wykes

Laser Chemical Processing for Microelectronics
edited by K. C. Ibbs and R. M. Osgood

The Elements of Nonlinear Optics
P. N. Butcher and D. Cotter

Optical Solitons – Theory and Experiment
edited by J. R. Taylor

Particle Field Holography
C. S. Vikram

Ultrafast Fiber Switching Devices and Systems
M. N. Islam

Optical Effects of Ion Implantation
P. D. Townsend, P. J. Chandler and L. Zhang

Diode–Laser Arrays
edited by D. Botez and D. R. Scifres

The Ray and Wave Theory of Lenses
A. Walther

Design Issues in Optical Processing
edited by J. N. Lee

Atom–Field Interactions and Dressed Atoms
G. Compagno, R. Passante and F. Persico

Compact Sources of Ultrashort Pulses
edited by I. Duling

Compact sources of ultrashort pulses

Edited by

Irl N. Duling, III

US Naval Research Laboratory,
Washington, DC

Published by the Press Syndicate of the University of Cambridge
The Pitt Building, Trumpington Street, Cambridge CB2 1RP
40 West 20th Street, New York, NY 10011-4211, USA
10 Stamford Road, Oakleigh, Melbourne 3166, Australia

First published 1995

Printed in Great Britain at the University Press, Cambridge

A catalogue record for this book is available from the British Library

Library of Congress cataloguing in publication data
Compact sources of ultrashort pulses / edited by Irl N. Duling.
p. cm. – (Cambridge studies in modern optics)
Includes bibliographical references.
ISBN 0–521–46192–8
1. Laser pulses, Ultrashort. 2. Lasers. I. Duling, Irl N.
II. Series.
QC689.5.L37C66 1995
621.36′6–dc20 94-41871 CIP

ISBN 0 521 46192 8

KW

Contents

List of contributors *x*
Acronyms and abbreviations *xii*
Preface *xv*

1 **Short pulse generation** **1**

H. A. HAUS

1.1 Active modelocking in the frequency domain 4
1.2 Active modelocking in the time domain and passive modelocking 7
1.3 Group velocity dispersion, self-phase modulation and the master equation 11
1.4 The nonlinear Schrödinger equation and solitons 18
1.5 The soliton laser 22
1.6 The coupled cavity modelocking and/or the additive pulse modelocking principle 23
1.7 Additive pulse modelocking with nonlinear interferometer 27
1.8 Kerr lens modelocking 32
1.9 All-fiber ring laser 35
1.10 Performance of some APM and KLM systems 42
1.11 Third-order dispersion and its effect on pulse width 44
1.12 Discussion 48
Acknowledgments 50
Appendix I 50
Appendix II 53
References 55

2 **Passive modelocking in solid state lasers** **57**

THOMAS BRABEC, STEPHEN M. J. KELLY AND FERENC KRAUSZ

2.1 Initial modelocked pulse formation 59
2.2 Steady-state pulse shaping dynamics 72
2.3 Conclusion 89

Acknowledgments 90
References 90

3 **Compact modelocked solid state lasers pumped by laser diodes** **93**
JOHN R. M. BARR
3.1 Introduction 93
3.2 Active modelocking of laser diode pumped solid state lasers 96
3.3 Passive modelocking 115
3.4 Alternative modelocking techniques 126
3.5 Amplification and tunability 129
3.6 Conclusion and future prospects 131
References 134

4 **Modelocking of all-fiber lasers** **140**
IRL N. DULING, III AND MICHAEL L. DENNIS
4.1 Methods of modelocking fiber lasers 140
4.2 The figure eight laser 146
4.3 Other modelocked fiber sources 168
4.4 Summary 174
References 175

5 **Nonlinear polarization evolution in passively modelocked fiber lasers** **179**
MARTIN E. FERMANN
5.1 Introduction 179
5.2 Linear polarization evolution in fiber lasers 180
5.3 Nonlinear polarization evolution in fiber lasers 184
5.4 Fiber laser cavities 187
5.5 Experiments 191
5.6 Summary 204
Acknowledgments 205
References 205

6 **Ultrafast vertical cavity semiconductor lasers** **208**
WENBIN JIANG AND JOHN BOWERS
6.1 Introduction 208
6.2 Optically pumped modelocked vertical cavity lasers 216
6.3 Analysis of laser pulse chirping in modelocked VCSELs 229
6.4 Carrier transport effect on modelocked VCSELs 246
6.5 Electrically pumped modelocked semiconductor lasers 255
6.6 Conclusions 260
References 261

7 **High power ultrafast semiconductor injection diode lasers** **274**

PETER J. DELFYETT

7.1 Introduction 274
7.2 Active modelocking 276
7.3 Passive and hybrid modelocking with multiple quantum well saturable absorbers 284
7.4 Cubic phase compensation 293
7.5 Intracavity dynamics 296
7.6 Amplification characteristics/dynamics 300
7.7 Applications of modelocked semiconductor laser diodes in synchronous optical networks 308
7.8 Conclusion and future directions 323
Acknowledgments 325
References 325

8 **The hybrid soliton pulse source** **329**

PAUL A. MORTON

8.1 Introduction 329
8.2 Pulse source requirements for soliton transmission systems 331
8.3 Hybrid soliton pulse source with a silicon optical bench reflector 333
8.4 Hybrid soliton pulse source with a fiber Bragg reflector 339
8.5 Spectral instabilities: cause and solution 342
8.6 Wide operating frequency range using a chirped Bragg reflector 348
8.7 CW operation with a chirped Bragg reflector 354
8.8 Packaged HSPS characteristics and soliton transmission results 373
8.9 Outlook 380
Acknowledgments 380
References 380

9 **Monolithic colliding pulse modelocked diode lasers** **383**

MING C. WU AND YOUNG-KAI CHEN

9.1 Introduction 383
9.2 Monolithic modelocked semiconductor lasers 384
9.3 Monolithic colliding pulse modelocked semiconductor lasers 387
9.4 Applications of monolithic CPM lasers 415
9.5 Other monolithic modelocked semiconductor lasers 416
9.6 Conclusion and future direction 420
Acknowledgments 421
References 421

Index *425*

Contributors

Chapter 1
Short pulse generation

H. A. HAUS
Department of Electrical Engineering and Computational Science, Massachusetts Institute of Technology, 79 Massachusetts Avenue, Cambridge, MA 02139, USA

Chapter 2
Passive modelocking in solid state lasers

THOMAS BRABEC, STEPHEN M. J. KELLY AND FERENC KRAUSZ
Abteilung für Quantenelektronik und Lasertechnik, Technische Universität Wien, Gußhausstraße 27, A-1040 Wien, Austria

Chapter 3
Compact modelocked solid state lasers pumped by laser diodes

JOHN R. M. BARR
Optoelectronics Research Centre, University of Southampton, Southampton, SO9 5NH, UK
now with
Pilkington Optronics, Barr & Stroud Limited, 1 Linthouse Road, Glasgow, G51 4BZ, UK

Chapter 4
Modelocking of all-fiber lasers

IRL N. DULING, III AND MICHAEL L. DENNIS
Naval Research Laboratory, Code 5670, Washington, DC 20375-5338, USA

Chapter 5
Nonlinear polarization evolution in passively modelocked fiber lasers

MARTIN E. FERMANN
IMRA America, 1044 Woodridge Avenue, Ann Arbor, MI 48105-9774, USA

Chapter 6
Ultrafast vertical cavity semiconductor lasers

WENBIN JIANG AND JOHN BOWERS
Department of Electrical and Computer Engineering, University of California at Santa Barbara, Santa Barbara, CA 93106-9560, USA

Chapter 7
High power ultrafast semiconductor injection diode lasers

PETER J. DELFYETT
Bellcore, 331 Newman Springs Road, Red Bank, NJ 07701, USA

Chapter 8
The hybrid soliton pulse source

PAUL A. MORTON
AT&T Bell Laboratories, 600 Mountain Avenue, Murray Hill, NJ 07974, USA

Chapter 9
Monolithic colliding pulse modelocked diode lasers

MING C. WU AND YOUNG-KAI CHEN
Department of Electrical Engineering, University of California at Los Angeles, Los Angeles, CA 90024, USA
AT&T Bell Laboratories, 600 Mountain Avenue, Murray Hill, NJ 07974, USA

Acronyms and abbreviations

A-FPSA	antiresonant Fabry–Perot saturable absorber
AFSA	artificial fast saturable absorber
AM	amplitude modulated
AOM	acousto-optic modulator
APM	additive pulse modelocking
AR	antireflection
ASE	amplified spontaneous emission
BER	bit error rate
BH	buried heterostructure
BITS	building integrated timing supply
BW	bandwidth
CCM	coupled cavity modelocking
CPM	colliding pulse modelocking
CTM	carrier-type modelocking
CW	continuous wave
DBR	distributed Bragg reflector
DC	direct current
DDF	dispersion decreasing fiber
DDL	dispersive delay line
DEMUX	demultiplexer
DFB	distributed feedback
DH	double heterojunction
ECL	emitter coupled logic
EDFA	erbium-doped fiber amplifier
e-hh	electron-heavy hole
EOT	electro-optic tuner
F8L	figure eight laser
FM	frequency modulated
FP	Fabry–Perot
FR	Faraday rotator
FRM	Faraday rotator mirror
FSF	frequency shifted feedback
FWHM	full width at half-maximum

GDD	group delay dispersion
GRIN-SCH	graded index separate confinement heterostructure
GVD	group velocity dispersion
HR	high reflector
HSPS	hybrid soliton pulse source
IMPATT	impact ionization avalanche transit time
IR	infrared
I-V	current versus voltage
KLM	Kerr lens modelocking
KSM	Kerr shift modelocking
LBO	lithium triborate
LD	laser diode
LDPSSL	laser diode pumped solid state laser
L-I	light versus current
ML	modelocked
MQW	multiple quantum well
MUX	multiplex, multiplexer
MZ	Mach–Zehnder
NALM	nonlinear amplifying loop mirror
NLSE	nonlinear Schrödinger equation
NOLM	nonlinear optical loop mirror
OC	output coupler
OE	opto-electronic
OEIC	opto-electronic integrated circuit
OMVPE	organometallic vapor phase epitaxy
OPO	optical parametric oscillator
OSA	optical spectrum analyzer
PM	polarization maintaining
Pr	praesedymium
PZT	piezo-electric transducer
RBW	resolution bandwidth
RF	radio frequency
RIN	relative intensity noise
RSS	Raman self-scattering
RWA	rotating wave approximation
SAM	self-amplitude modulation
SDA	saturable diode amplifier
SF10, SF18, SF57	Corning glass types
SH	second harmonic
SHG	second harmonic generation
SIOB	silicon optical bench
SMSR	side-mode suppression ratio
SP	synchronous pumped
SP-APM	stretched pulse-additive pulse modelocking
SPM	self-phase modulation
SQW	single quantum well
SRD	step recovery diode

SWP	sweep time
TDM	time division multiplexed
TE	transverse electric
TEM_{00}	lowest order transverse electric mode
TM	transverse magnetic
TOD	third-order dispersion
TWA	traveling-wave (laser) amplifier
UV	ultraviolet
VBW	video bandwidth
VCSEL	vertical cavity surface emitting laser
WDM	wavelength division multiplexing
WP	wave plate
WPS	weak pulse shaping
XPM	cross-phase modulation
YAG	yttrium aluminum garnet
YLF	yttrium lithium fluoride

Preface

Since the development of the first diode lasers and the recent proliferation of fiber amplifiers, the dream of researchers for a compact, efficient, turn-key source of ultrashort pulses has come closer to reality. A number of candidates for this ultimate source have been proposed and researched and a large body of information has been produced. To date there has not been a compilation of this valuable information in one place. It is the intent of this work to present the state of the art in the development of these sources. It is the added intention to provide a basis for the future research of others attempting to enter this still very active field.

In 1983, when the technique of electro-optic sampling was developed, the dream was that the sampler could be combined with a compact ultrashort pulse source to produce a subpicosecond resolution oscilloscope. This technology has languished due to the missing source. In 1988 the development of high efficiency low temperature GaAs photoconductive switches has produced similar high promise, but there is still no suitable low cost compact ultrashort pulse source. It is hoped that this book will show that much progress has been made since the early 80s and the groundwork has been laid for a new class of instrumentation, where a laser is included for optical processing and may not ever leave the instrument.

By producing compact and hopefully low cost sources of ultrashort pulses, probing of materials for characterization becomes more practical and optical ranging becomes more precise. Techniques such as surface harmonic generation and two photon absorption, both of which rely on the intensity of the light, can benefit from these sources. In fact the shortness of the pulse is not the issue any more, but it may be a distinct advantage in that the thermal loading on the sample is less while the peak intensity is the same. This may allow biological samples to be probed without damage.

A number of sensors have been developed which utilize short pulse sources. Modelocked gyroscopes may prove superior to their CW counterparts. Short pulses can be used to distinguish the signals from distributed sensors in detector arrays or smart structures. Even the modelocking process itself can be used for this purpose, turning the entire sensor into a modelocked laser.

As the applications multiply for compact short pulse sources it is necessary for the researchers developing these systems to have a full knowledge of the state of the art and a clear understanding of the limitations of the different modelocking techniques. It is the intent of this book to bring together the leading edge research in this field presented in such a way that it is useful to both the experienced researcher and to that person wishing to enter the field.

It is also important that the engineer who is designing applications based on these sources be presented with the strong points of each so that he can choose the optimal source for the given application. He should also get a clear understanding of the required conditions to make the laser operate as necessary for the given application. It is our hope that this book can fulfill all of these goals. Some of the criteria which must be addressed and which will vary from source to source are the output power, the shortest duration necessary for the application, the repetition rate required, the wavelength at which the system must operate, amplitude and phase noise, and allowed cost of the system. At a workshop on 'Real World Sources' at the 1994 Ultrafast Phenomena Conference in Dana Point, CA, the consensus was that the market applications for ultrashort pulse sources would open up if the cost could be brought below \$10 000. At this point none of the sources can meet that criterion, but it is expected that in the next few years there will be more than one technology meeting that threshold.

With the current emphasis being placed on the national information infrastructure and the best way to increase the bandwidth of the current installed fiber, it hardly needs to be said that a major motivation for working in this area is to produce viable sources for high speed telecommunications. A large portion of the work being done on optical fiber sources operates in the anomalous dispersion regime where solitons can be formed. Soliton communications systems are already planned for installation in the transoceanic links. Time division multiplexed communication has been carried out at 100 Gb/s and will soon reach 160 Gb/s and beyond. The importance of reliable sources of clean stable solitons cannot be avoided.

Chapter 1 treats the theory of the generation of short optical pulses by modelocking. This discussion will cover the effects of dispersion and nonlinearity within the fiber laser for a generalized switching mechanism, concentrating on the APM type of nonlinear switching. The analysis will present the master equation for the laser dynamics and its consequences for various laser configurations.

The pulse shaping dynamics of solitary solid state lasers utilizing modelocking techniques exploiting the Kerr nonlinearity of solids is the subject of Chapter 2. Specifically the pulse buildup in passively modelocked solid state lasers and the new phenomena associated with solitary modelocking (modelocking when soliton-like pulse shaping is occurring in the cavity) is examined.

The recent developments in the modelocking of diode pumped solid state lasers is covered in Chapter 3. It includes a practical review of standard modelocking and pulse measurement techniques with a strong emphasis on active modelocking.

Chapter 4 is a review of the modelocking of fiber lasers. With the exception of nonlinear polarization rotation modelocking, which is treated in Chapter 5, the passive, active and hybrid techniques of modelocking fiber lasers is covered with a comparison between them. A detailed discussion of cavity dispersion is presented, with its effect on pulse formation.

The phenomenon of nonlinear polarization evolution and its use for modelocking is covered in Chapter 5. The implementation of this technique in both bulk and fiber lasers is addressed, and analysis with and without dispersion in the cavity based on the soliton and solitary laser models is included.

After a review of the modelocking in edge-emitting diode lasers, Chapter 6 covers the modelocking of surface emitting semiconductor lasers. The analysis includes chirp analysis and carrier transport effects and their relationship to pulse formation.

Ultrafast optical pulse generation techniques utilizing external cavity semiconductor lasers are described in Chapter 7. Active modelocking, passive modelocking, hybrid modelocking and chirp compensation techniques are examined. The pulses from these lasers are used to study the ultrafast amplification characteristics of semiconductor lasers. The chapter covers the nature of the effects which dominate the pulse shaping mechanisms in external cavity hybrid modelocked diode lasers. These systems are applied to providing synchronous timing signals for clock distribution, photonic network synchronization, and all-optical clock recovery.

An alternative solution to the extended cavity semiconductor laser is to integrate the extended cavity in semiconductor material or fiber. These options are treated in Chapter 8. The development of this source, elimination of instabilities and optimization of pulse length and chirp are covered. The final results on a packaged device are also presented. With the use of a two-section laser diode high power pulses are generated.

The final chapter of the book covers what are probably the most compact of the lasers presented here. The monolithic colliding pulse modelocked diode lasers incorporate the modelocking mechanism into the diode laser cavity providing a high repetition rate source of ultrashort optical pulses. The stability criterion and the different modelocking techniques for these unique lasers are addressed.

1

Short pulse generation

H. A. HAUS

Introduction

Optical frequencies are so high that a small relative bandwidth is a very large bandwidth in absolute terms. This fact is behind the success of optical short pulse generation. The first successful generation of short optical pulses via modelocking started in the 60s with Nd:glass. Figure 1.1 gives the history of the advances in short pulse generation. Steady progress in generating shorter and shorter pulses was achieved with dye lasers, principally because these were CW lasers that had much more predictable modelocking behavior than Q-switched lasers. After subpicosecond pulse generation was achieved with dye lasers, it was soon realized that the short pulses produced by the lasers were much shorter than the relaxation times of the dyes used for the gain and absorption media. It was recognized that the pulses were generated through combined action of the saturable absorber dye that, by saturating, opened the 'shutter' for the transmission of the pulse, and the gain, by saturating, closed the 'shutter.' This was an important discovery since there exist very few saturable absorbers with subpicosecond relaxation times appropriate for intracavity operation. The shortest pulses achieved were 27 fs in duration, and after spectral spreading in a fiber and subsequent compression 6 fs was achieved[1], still the record today.

In 1978 the first modelocking of semiconductor lasers was achieved[2]. They are of prime importance in communications applications and are limited to about 100 fs pulse duration, a pulse width adequate for communications for many years to come. Color center lasers operating around 1.5 micron wavelength were pioneered by Mollenauer and

Based on a tutorial given at QELS'93 Baltimore, MD, May 1993

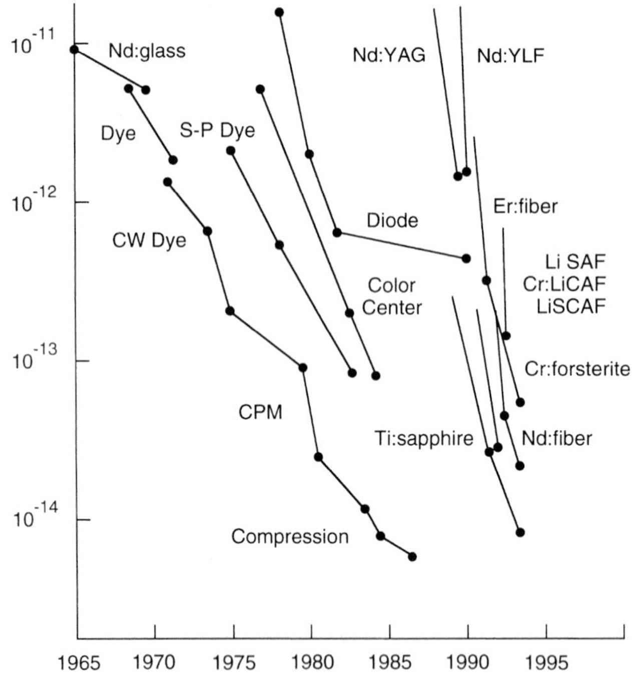

Figure 1.1. Progress in short pulse generation. CPM: colliding pulse modelocking, S-P: synchronously pumped. The acronyms for laser types are spelled out in Tables 1.1 and 1.2 (e.g. LiSAF = $LiS_r\ AlF_6$).

Stolen in their 'soliton' laser[3]. A major breakthrough occurred when it was recognized that the soliton principle employed in the color center laser was not necessary, and that a more general mechanism can be made to work: the construction of an artificial fast saturable absorber (AFSA) using the Kerr nonlinearity[4–7]. Since the Kerr effect extends over all near IR and optical frequencies, AFSAs could be constructed over this entire range of frequencies. Further, since these AFSAs have very short response times, they do not need the cooperative action of the gain medium; they can operate with gain media of long relaxation times, such as solid state lasers. This led to the breakthrough in short pulse generation with Ti:sapphire lasers which have achieved 11 fs pulse generation[8], the shortest pulses generated directly from a laser.

In this chapter, we shall start with the theory of active modelocking with a driven modulator inside the cavity. This is done mainly because the theory illustrates the 'locking' of the cavity modes, the origin of the term 'modelocking.' The theory is best cast into the time domain when treating passive modelocking. We shall study modelocking in this context through the remainder of the chapter. We shall develop a master equation that treats systems within which the change of the pulse upon passage through any element in the cavity is small. This theory yields analytic results that can be used as a guide in the understanding of modelocking operation. There are successful modelocking systems that do not satisfy this criterion and cannot be described by the simple master equation. However, for modelocking to lead to clean pulse shapes and mode spectra it is necessary that the changes wrought by the nonlinear effects in the resonator be small, whereas no such restriction exists for the linear processes such as filtering, group velocity dispersion, linear loss and gain. Such systems can be handled with a generalization of the master equation as we shall show when we treat some specific systems of importance.

A simplified form of the master equation is the nonlinear Schrödinger equation (NLSE) which describes soliton behavior. We shall study it in anticipation for an understanding of the 'soliton' laser of Mollenauer and Stolen. Also, soliton behavior is often of importance in the modelocking of all-fiber lasers. Thus it is useful to derive the pulse shape of the fundamental soliton from the NLSE. Since the soliton laser operates through injection of a second order soliton into the laser cavity, we derive the expression for the second order soliton as well. With this background one can gain an understanding of the 'soliton' laser.

The master equation describes other passive modelocking mechanisms which will be discussed in detail. It covers the simple saturable absorber case, but also the cases of coupled cavity modelocking[4,5] (CCM) and of additive pulse modelocking (APM)[6,7], and of Kerr lens modelocking (KLM)[9,10], all under the assumption of small change per pass. CCM and APM are different acronyms for systems that construct an AFSA using an auxiliary cavity interferometrically coupled to the laser cavity. APM also describes interferometric pulse interaction inside the main cavity either via an equivalent Mach–Zehnder interferometer inside the laser cavity, or by interference of two polarizations in a nonlinear medium. KLM produces nonlinear net gain variation by self-focusing effects. As mentioned earlier, the assumption of small change per pass is often not valid, but can be relaxed in most cases of interest as will also be shown.

The all-fiber ring laser operates via APM produced by polarization-ellipse rotation. If the dispersions of the fiber segments in the ring are all negative, the operation can be described as that of a perturbed soliton.

Recently it has been found that such an operation is not desirable, and that the shortest pulses are obtained if positive dispersion of one fiber segment is canceled by negative dispersion of another segment. This kind of operation has led to the shortest pulses obtained from an all-fiber laser. This operation will also be discussed in some detail.

1.1 Active modelocking in the frequency domain

Figure 1.2 summarizes the concepts involved in active modelocking[11]. A resonator with two mirrors contains a gain medium and a modulator that changes the loss in the cavity, cosinusoidally as a function of time. The modulation frequency ω_M is synchronized with the round-trip time T_R of a pulse through the cavity, or, in other words, the modulation frequency coincides with the frequency separation of the axial modes of the resonator.

$$\omega_M = \frac{2\pi}{T_R} = \frac{\pi c}{n_{eff} L_o} \tag{1.1}$$

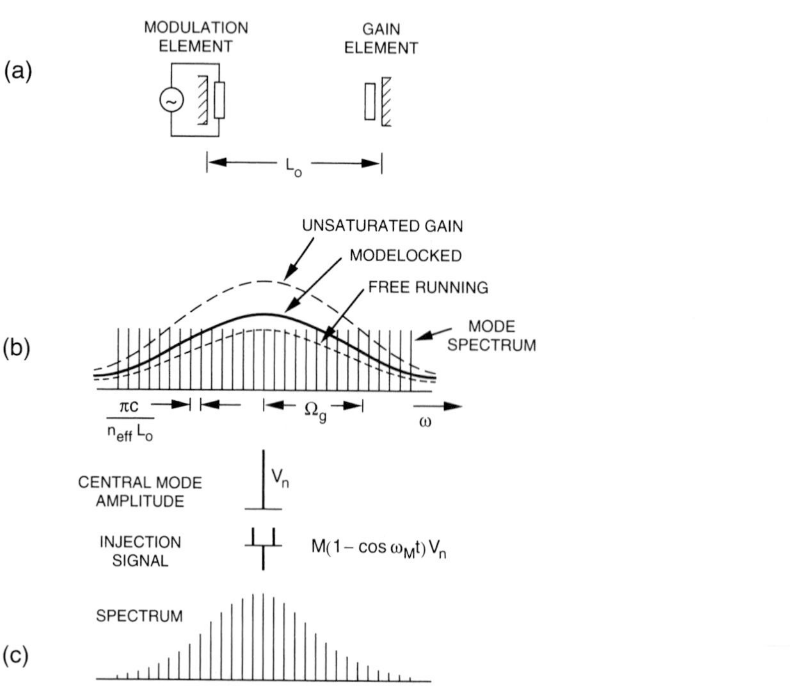

Figure 1.2. Active modelocking in the frequency domain.

where c is the speed of light, n_{eff} the effective index and L_o the cavity length. No higher-order transverse modes are assumed to exist. This can be accomplished by optical wave-guiding and/or aperturing.

Figure 1.2 shows a resonator with two reflecting mirrors with modes that are standing waves. A ring resonator has traveling waves as eigenmodes. In this case the frequency separation of the axial modes is $2\pi c/(n_{\text{eff}}L_o)$. Everything that follows relating to the former modes is equally applicable to ring resonator modes. There are subtle differences, however. Traveling wave modes do not tend to saturate the gain medium by producing 'spatial hole-burning' in which a periodic spatial pattern is imposed upon the gain. Also, in standing wave resonators the pulses pass the modulating element twice and care has to be taken that the changes experienced by the pulse upon double passage are (approximately) additive. In this respect, ring resonators present some advantages. Let us now consider the effects of the different elements in the resonator as a Fourier component a_n passes through it.

The loss

There are various sources of loss in the resonator, internal loss and the loss due to the output coupling. We lump them into one single parameter ℓ so that the nth mode amplitude a_n is multiplied by $1 - \ell$ as it passes through the resonator once.

The gain

We assume that a_n is multiplied by $1 + g(n)$ when the Fourier component passes through the gain medium. If the gain has a spectral response, g is a function of n. A common gain profile is Lorentzian, so that the nth Fourier component a_n passing through the gain medium experiences the change:

$$G(n)a_n = \left[1 + \frac{g}{1 + \frac{(n\omega_M)^2}{\Omega_g^2}}\right]a_n \simeq \left\{1 + g\left[1 - \frac{(n\omega_M)^2}{\Omega_g^2}\right]\right\}a_n \qquad (1.2)$$

where Ω_g is the gain bandwidth. Here we have centered the spectrum at $n = 0$. We shall write the frequency of the nth spectral component as $\omega_n = \omega_o + n\omega_M$, where ω_o is the carrier frequency.

The modulator

The modulator is assumed to pass the radiation perfectly at periodically timed instants so that a Fourier component a_n with the time dependence $a_n \exp(j\omega_o t)\exp(jn\omega_M t)$ is multiplied by $M[1 - \cos(\omega_M t)]$, where M is the modulation depth.

$$M(1 - \cos\omega_M t)a_n e^{j(\omega_o + n\omega_M)t} = -Ma_n e^{j\omega_o t}\left[\frac{1}{2}e^{j(n-1)\omega_M t} - e^{jn\omega_M t} + \frac{1}{2}e^{j(n+1)\omega_M t}\right] \tag{1.3}$$

Each Fourier component acquires sidebands that are injected into the adjacent axial modes (Figure 1.2b). Considering each action of the three elements separately and additive, consistent with the assumption that the changes wrought by each element are small, one obtains for the Fourier component $a_n^{(k+1)}$ entering the resonator the $(k+1)$st time in terms of its value $a_n^{(k)}$ upon kth re-entry:

$$a_n^{(k+1)} = a_n^{(k)} + g\left(1 - \frac{(n\omega_M)^2}{\Omega_g^2}\right)a_n^{(k)} - \ell a_n^{(k)} + \frac{M}{2}(a_{n+1}^{(k)} - 2a_n^{(k)} + a_{n-1}^{(k)}) \tag{1.4}$$

In the steady state, $a_n^{(k+1)} - a_n^{(k)} = 0$, the superscript (k) can be dropped, and the above reduces to a second order difference equation. The problem can be simplified further, if we replace the second order difference appearing in (1.4) by a second order derivative. This is legitimate if the spectrum is broad, i.e. contains many spectral components. Since in practical cases thousands of spectral components make up a pulse, this assumption is always justified. One obtains in the steady state:

$$0 = \left[g\left(1 - \frac{\omega^2}{\Omega_g^2}\right) - \ell + \frac{M\omega_M^2}{2}\frac{\mathrm{d}^2}{\mathrm{d}\omega^2}\right]a(\omega) \tag{1.5}$$

where a_n has been replaced by the function $a(\omega)$ of the continuous frequency variable $\omega = n\omega_M$. This equation has a simple Hermite Gaussian solution

$$a(\omega) = H_\nu(\omega\tau)e^{-\omega^2\tau^2/2} \tag{1.6}$$

with the constraints

$$\frac{1}{\tau^4} = \frac{M\omega_M^2\Omega_g^2}{2g} \tag{1.7}$$

$$g - \ell = M\omega_M^2\tau^2(\nu + \frac{1}{2}) \tag{1.8}$$

Equation (1.8) shows that the peak gain rises above the loss to compensate for the dispersion of the spectral components by the modulation. This is shown in Figure 1.2b which indicates the free-running gain peak pulled down to the loss line of the cavity modes, and the peak modelocked gain above the loss line. Higher order Hermite Gaussians have higher loss. Thus the lowest order Hermite Gaussian, $\nu = 0$, the simple Gaussian, is stable against the growth of higher order Hermite Gaussians. The spectrum is Gaussian in shape. Since the pulse train is periodic, the spectrum is, in fact, discrete with the Gaussian envelope (see Figure 1.2c)

$$a(\omega) = Ae^{-\omega^2\tau^2/2} \tag{1.9}$$

Its Fourier transform is a Gaussian pulse:

$$a(t) = \frac{\sqrt{2\pi}}{\tau} Ae^{-t^2/2\tau^2} \tag{1.10}$$

In fact, the time function $a(t)$ consists of a sequence of pulses, spaced times T_R apart, a consequence of the discreteness of the spectrum. The inverse pulsewidth $1/\tau$ is, according to (1.7), proportional to the geometric mean of the modulation frequency and the gain bandwidth. If the modulation is at the Nth harmonic of the mode spacing, ω_M in (1.7) and (1.8) would have to be replaced by $N2\pi/T_R$. Harmonic modelocking is one way of producing shorter pulses by active modulation, for a fixed modulation depth M.

1.2 Active modelocking in the time domain and passive modelocking

The preceding formalism was particularly simple in the frequency domain, because of the assumed sinusoidal modulation. Passive modelocking does not produce sinusoidal modulation, and thus its formulation in the frequency domain is not a convenient one. The formalism is easily adapted to the time domain by Fourier transformation. When we Fourier transform (1.4), treating the spectrum as continuous and using the Fourier transform pairs:

$$\int d\omega a(\omega) e^{j\omega t} = a(t) \tag{1.11}$$

with the inverse Fourier transform:

$$\frac{1}{2\pi}\int dt a(t) e^{-j\omega t} = a(\omega) \tag{1.12}$$

we obtain for the transient case:

$$a^{(k+1)}(t) - a^{(k)}(t) = \left[g\left(1 + \frac{1}{\Omega_g^2}\frac{d^2}{dt^2}\right) - \ell - \frac{M}{2}\omega_M^2 t^2\right] a^{(k)}(t) \tag{1.13}$$

Of course, the steady state solutions of (1.13) are the Fourier transforms of (1.6).

$$a(t) = \frac{(j)^\nu}{\sqrt{2\pi}} H_\nu\left(\frac{t}{\tau}\right) e^{-t^2/2\tau^2} \tag{1.14}$$

Figure 1.3 shows the process in the time domain. The periodic modulation of the loss provides a temporal window of gain. The pulse occurs within this window. The transition from gain to loss is at the inversion points of the Gaussian envelope.

Equation (1.13) is easily generalized to modelocking with a fast saturable absorber. Indeed, one recognizes that the M-coefficient is of the form $(M/2)(\omega_M t)^2$ which is the expansion of $M[1 - \cos(\omega_M t)]$ to second order in t. The sinusoidal modulation has been approximated by a parabola in time, a legitimate approximation when the sinusoid has a period much longer than the pulse width. But it is also obvious that the term can

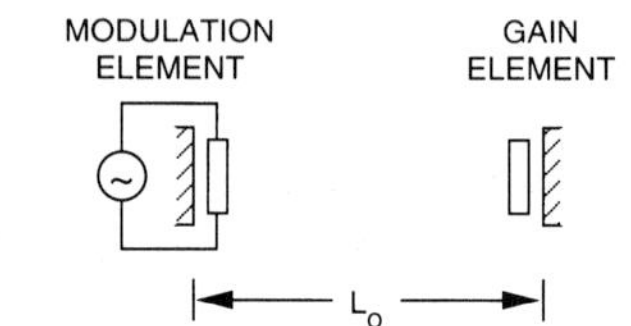

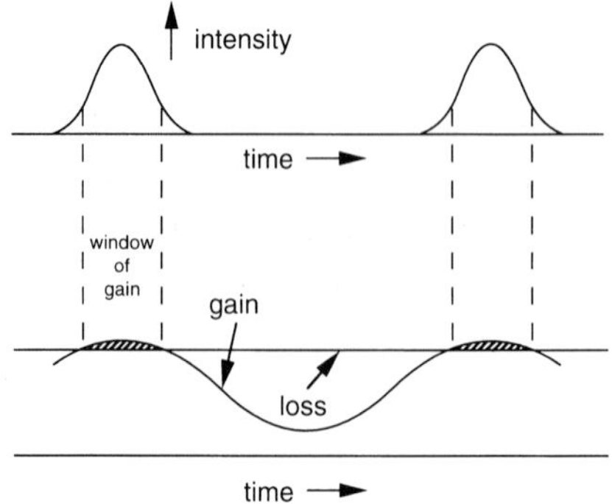

Figure 1.3. Active modelocking in the time domain.

be replaced in general by $f(t)a(t)$, where $f(t)$ is the temporal modulation of the loss. A fast saturable absorber of small signal absorption coefficient α_o saturates with intensity according to the law:

$$\alpha(|a(t)|^2) = \frac{\alpha_o}{1 + \frac{|a(t)|^2}{\mathcal{A}_{\text{eff}} I_a}} \tag{1.15}$$

Here we have assumed that $|a(t)|^2$ is equal to the power of the mode; $\mathcal{A}_{\text{eff}}$ is the effective cross sectional area of the mode in the gain medium, and I_a is the saturation intensity of the absorber. We have assumed that the relaxation time of the absorber is shorter than the pulse width τ, so that the saturation may be considered instantaneous. A saturable absorber medium of length L_a multiplies the amplitude $a(t)$ of the mode passing through it by:

$$e^{-\alpha L_a} \simeq 1 - \alpha L_a = 1 - \alpha_o L_a \left(1 - \frac{|a(t)|^2}{\mathcal{A}_{\text{eff}} I_a}\right) = 1 - \alpha_o L_a + \gamma |a|^2 \tag{1.16}$$

where

$$\gamma = \frac{\alpha_o L_a}{\mathcal{A}_{\text{eff}} I_a} \tag{1.17}$$

and we have expanded the exponential to first order in αL_a, and α to first order in $|a(t)|^2/(\mathcal{A}_{\text{eff}} I_a)$. The coefficient γ is called the self-amplitude modulation (SAM) coefficient.

A comment is in order with regard to the change of the spatial mode profile that is unavoidable when a beam of initially Gaussian profile passes through a bulk saturable absorber. Since the absorption depends on the local intensity the beam ceases to be a pure Gaussian after passage through the absorber. However, higher order Hermite Gaussians produced in this process have higher diffraction loss as they pass through the apertures in the resonator and thus get eliminated. In this way one may assume that the absorber provides a simple, intensity dependent loss for the fundamental Gaussian mode. The definition of effective mode area, $\mathcal{A}_{\text{eff}}$, is derived under this premise. Later on, when we introduce nonlinear phase modulation, the same considerations apply.

The linear loss $\alpha_o L_a$ can be lumped into the coefficient ℓ of the loss per pass. The master equation for the amplitude $a^{(k)}$ of fast saturable absorber modelocking is[12]:

$$a^{(k+1)} - a^{(k)} = \left\{ g\left(1 + \frac{1}{\Omega_g^2}\frac{d^2}{dt^2}\right) - \ell + \gamma |a^{(k)}|^2 \right\} a^{(k)} \tag{1.18}$$

We have assumed that the gain does not saturate, i.e. that the gain is approximately time independent during passage of the pulse. This assumption is justified in all systems with long gain relaxation times (compared with T_R) such as solid state lasers. It is not satisfied in semiconductor lasers or dye lasers.

In the steady state, $a^{(k+1)} - a^{(k)} = 0$, the superscript (k) can be dropped, and the solution of the master equation is:

$$a(t) = A_o \operatorname{sech}\left(\frac{t}{\tau}\right) \tag{1.19}$$

with

$$\gamma |A_o|^2 = \frac{2g}{\Omega_g^2 \tau^2} \tag{1.20}$$

$$g - \ell + \frac{g}{\Omega_g^2 \tau^2} = 0 \tag{1.21}$$

The loss ℓ is greater than the gain, providing stability of the system against growth of noise preceding and following the pulse. The pulse itself experiences net gain because the saturable absorber is bleached while the pulse lasts. Equation (1.20) can be rewritten:

$$\frac{1}{\tau^4} = \frac{\gamma |A_o|^2 \Omega_g^2}{2g\tau^2} \tag{1.22}$$

When we compare this expression with (1.7) we note that the role of $M\omega_M^2$ of the former is played by $\gamma |A_o|^2/\tau^2$ of the latter. In both cases these coefficients are proportional to the curvature of the loss modulation function plotted versus time. It is this curvature that provides pulse shortening which combats pulse lengthening due to the finite spectral width of the gain. In the saturable absorber case the pulse itself provides this curvature. Consequently, the curvature can be much larger than in the case of active modelocking, leading to much shorter pulses.

The preceding analysis covers modelocking with a fast saturable absorber. The processes described thus far are indicated in Figure 1.3 for a standing wave resonator. Figure 1.4 shows the processes schematically in a ring configuration. The gain is time independent, the loss decreases during the time of the pulse. This provides a net gain window that amplifies the pulse center, attenuates the wings, and shortens the pulse in the process. This pulse shortening combats the pulse lengthening due to the finite bandwidth of the gain medium.

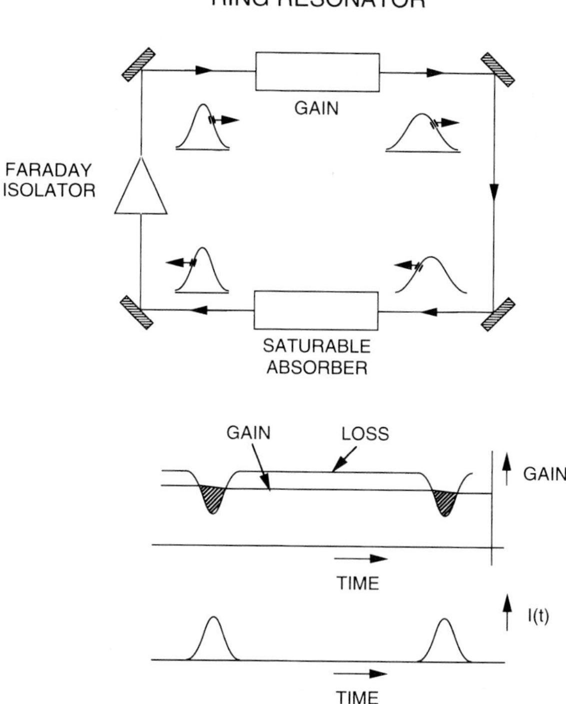

Figure 1.4. Action of a fast saturable absorber.

Thus far we have discussed fast saturable absorber action. Figure 1.5 shows schematically the operation of a system with a slow saturable absorber, such as a dye laser system. The loss saturates first and provides net gain to the pulse. Then the gain medium saturates and turns off the net gain. Both media have to have comparable relaxation times for such a mechanism to work. For this reason it is limited to systems like dye lasers and semiconductor lasers. The theory was presented first in the mid-seventies[13,14]. We shall not go into it in greater detail here.

1.3 Group velocity dispersion, self-phase modulation and the master equation

With shorter and shorter pulses, new effects show up. The larger peak intensities lead to nonlinear effects in the components of the cavity, the lenses, the gain crystal, etc. These tend to produce chirped pulses and

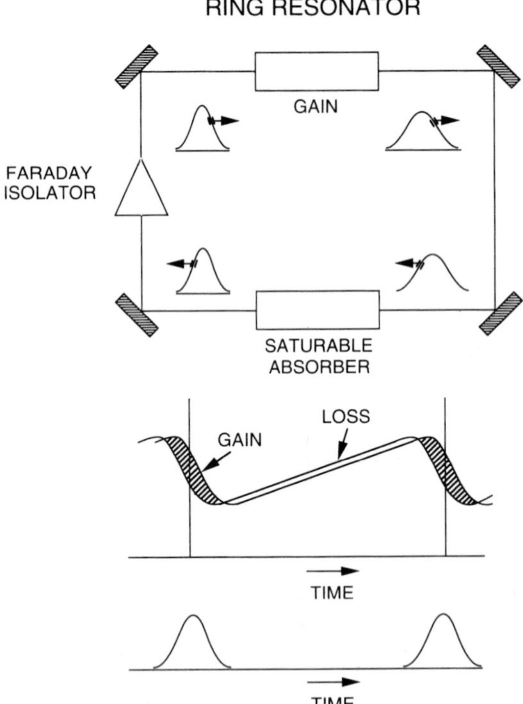

Figure 1.5. Combined actions of a slow saturable absorber and a gain medium.

prevent the generation of the shortest possible pulses. It was recognized by Fork *et al.*[15] that a prism pair introduced in the resonator can be constructed (See Figure 1.6) so that

(a) The beam axis is not changed in the resonator
(b) Both positive and negative group velocity dispersion can be produced. The dispersion is generated 'geometrically,' by deflecting different colors differently and having them travel different distances before recombining them.

The dispersion introduced by the prism is adjustable and provides another degree of freedom for the optimization of modelocking. The master equation can be generalized to include these two effects. The group velocity dispersion is due to the quadratic dependence on frequency of the propagation constant if $k(\omega)$ is expanded about ω_0.

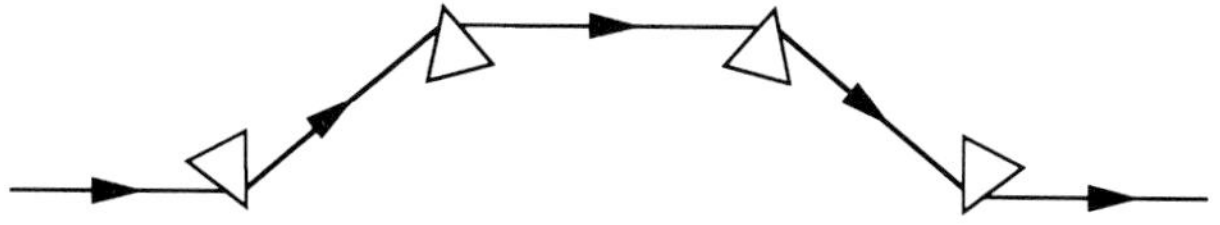

Figure 1.6. Prism pair for the adjustment of group velocity dispersion.

$$k(\omega) = k(\omega_o) + k'\Delta\omega + \frac{1}{2}k''\Delta\omega^2 \tag{1.23}$$

A wave passing through a dispersive medium of length L_d, experiences the net phase shift:

$$a(L_d, \omega) = \exp -\{j[k(\omega_o) + k'\Delta\omega + \frac{1}{2}k''\Delta\omega^2]L_d\}\ a(0, \omega) \tag{1.24}$$

The constant phase delay $k(\omega_o)L_d$ is not of interest. The phase proportional to frequency deviation $\Delta\omega$ is simply the group velocity delay. The phase term of interest is that which expresses the dependence of group velocity on frequency, the k'' term. If we expand the term to first order, we obtain

$$\exp\left\{-\frac{1}{2}jk''\Delta\omega^2 L_d\right\}a(L_d, \omega) \simeq \left(1 - \frac{1}{2}jk''\Delta\omega^2 L_d\right)a(0, \omega) \tag{1.25}$$

in the frequency domain, and in the time domain

$$a(L_d, t) = \left(1 + \frac{1}{2}jk''L_d\frac{d^2}{dt^2}\right)a(0, t) \equiv \left(1 + jD\frac{d^2}{dt^2}\right)a(0, t) \tag{1.26}$$

where

$$D \equiv \frac{1}{2}k''L_d \tag{1.27}$$

The operator representing group velocity dispersion (GVD) looks like the operator representing bandwidth limiting by gain, the derivative term on the right hand side of (1.13) except it is imaginary instead of real.

Next consider self phase modulation (SPM)[16]. The nonlinear phase shift is due to the optical Kerr effect: the index changes with intensity I.

$$n = n_o + n_2 I \tag{1.28}$$

n_2 is the nonlinear index coefficient and is usually positive (for glass and dyes). Thus, when a beam of modal cross section $\mathcal{A}_{\text{eff}}$ passes through a Kerr medium of length L_{K}, the phase shift is

$$\Phi = \frac{2\pi}{\lambda} n_2 \frac{L_{\text{K}}}{\mathcal{A}_{\text{eff}}} |a(t)|^2 \equiv \delta |a(t)|^2 \tag{1.29}$$

where δ is the self phase modulation (SPM) coefficient.

We can rewrite the master equation, assuming again that the Kerr effect and the group velocity dispersion are small per pass:

$$a^{(k+1)} - a^{(k)} = \left\{ g\left(1 + \frac{1}{\Omega_{\text{g}}^2} \frac{\text{d}^2}{\text{d}t^2}\right) + jD\frac{\text{d}^2}{\text{d}t^2} - \ell + (\gamma - j\delta)|a^{(k)}|^2 \right\} a^{(k)} \tag{1.30}$$

In the steady state, $a^{(k+1)} - a^{(k)} = j\psi a(t)$, where ψ is the phase shift per pass, the superscript (k) can be dropped and we obtain an equation for $a(t)$ in terms of the time variable t alone:

$$\left\{ -j\psi + (g - \ell) + \left(\frac{g}{\Omega_{\text{g}}^2} + jD\right) \frac{\text{d}^2}{\text{d}t^2} + (\gamma - j\delta)|a|^2 \right\} a = 0 \tag{1.31}$$

This equation is known as the Landau–Ginzburg equation. Its solution is

$$a(t) = A_{\text{o}} \left[\text{sech}\left(\frac{t}{\tau}\right) \right]^{(1+j\beta)} \tag{1.32}$$

with the constraints

$$-j\psi + g - \ell + \frac{(1+j\beta)}{\tau^2} \left(\frac{g}{\Omega_{\text{g}}^2} + jD\right) = 0 \tag{1.33}$$

$$\frac{1}{\tau^2} \left(\frac{g}{\Omega_{\text{g}}^2} + jD\right)(2 + 3j\beta - \beta^2) = (\gamma - j\delta) A_{\text{o}}^2 \tag{1.34}$$

If the gain medium relaxes slowly compared with one round-trip time T_{R}, as is the case for solid state lasers, g can be considered fixed at the level determined by the saturation formula

$$g = \frac{g_{\text{o}}}{1 + \frac{W}{T_{\text{R}} P_{\text{g}}}} \tag{1.35}$$

where W is the energy in the pulse, and P_{g} is the saturation power of the gain. One may use this equation to obtain the pulse energy by noting that, roughly, the loss ℓ is equal to the gain g. Then

$$\frac{W}{P_g T_R} = \frac{g_o}{\ell} - 1 \tag{1.36}$$

But the pulse envelope is a secant hyperbolic, and hence $|A_o|$ is fixed by the relation

$$2\tau |A_o|^2 = W \tag{1.37}$$

We can enter (1.34) and (1.35) with this result and obtain two equations for the pulse width and the chirp parameter.

$$\left(\frac{2g}{W\Omega_g^2\tau}\right)(2 - \beta^2 - 3\beta D_n) = \gamma \tag{1.38}$$

$$\frac{3\beta}{2 - \beta^2} = \frac{\delta + \gamma D_n}{\delta D_n - \gamma} \tag{1.39}$$

where

$$D_n = \frac{\Omega_g^2}{g} D \tag{1.40}$$

is the normalized group velocity dispersion.

Figure 1.7 shows the pulsewidth τ and the chirp parameter β for a fixed SAM coefficient γ, as a function of GVD and with the SPM coefficient δ as parameter. One sees that the chirp parameter vanishes when $\delta = 0$ and $D = 0$, and more generally, at one pair of values of δ and D, when $\delta > 0$ and $D < 0$. The pulse width is minimum at that point. The spectrum is not necessarily broadest at that point. Figure 1.8 shows a plot of the spectral width $(1 + \beta^2)^{0.5}/(\Omega_g\tau)$ for the same set of parameters. It is clear that a broader spectrum is obtained at positive D values. If means are provided to compress the pulse externally by sending it through a negatively dispersive medium, shorter pulse widths can be achieved.

The real part of (1.38) in its unnormalized form gives information on the pulse shortening and lengthening mechanisms:

$$\frac{g}{\Omega_g^2\tau^2}(2 - \beta^2) - \frac{3\beta D}{\tau^2} = \gamma A_o^2 \tag{1.41}$$

On the right-hand side is the pulse shortening action of the saturable absorber. On the left-hand side are the mechanisms responsible for the pulse lengthening. In the steady state the effects balance. The term $(g/\Omega_g^2\tau^2)(2 - \beta^2)$ expresses the pulse lengthening due to the finite gain bandwidth. It changes sign when $\beta^2 > 2$, the gain medium shortens the

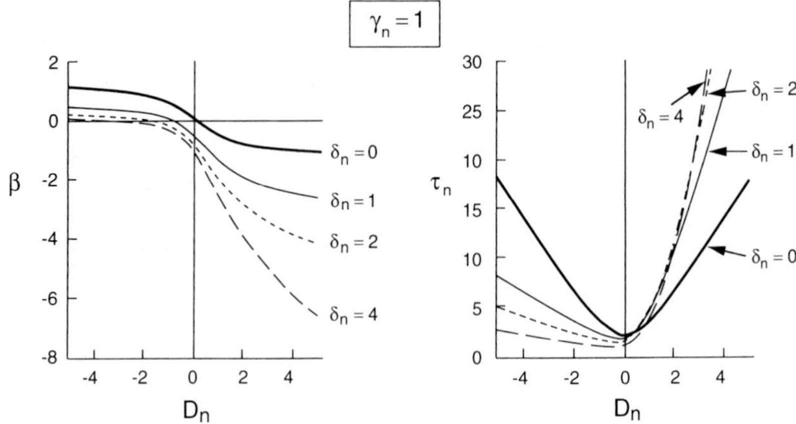

Figure 1.7. The plots of pulsewidth τ and chirp parameter β as functions of dispersion, with the SPM coefficient as parameter. β, chirp parameter; D_{n}, normalized GVD parameter; γ_{n}, normalized equivalent fast saturable absorber, $\gamma_{\mathrm{n}} = \gamma \frac{W}{2g} \Omega_{\mathrm{g}}$; δ_{n}, normalized self-phase modulation parameter, $\delta_{\mathrm{n}} = \delta \frac{W}{2g} \Omega_{\mathrm{g}}$; τ_{n} normalized pulse width $\tau_{\mathrm{n}} = \Omega_{\mathrm{g}} \tau$.

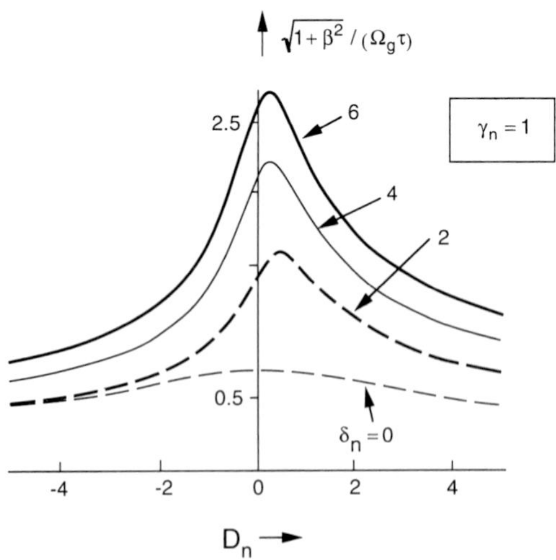

Figure 1.8. The spectral width as function of dispersion, with the SPM coefficient as parameter.

pulse if the chirp is large. This is the well known phenomenon of pulse shortening that occurs when a highly chirped pulse passes through a filter. The filter cuts off the wings that have large frequency deviation. When the gain medium shortens the pulse, there must be a mechanism that lengthens it, so as to balance the net pulse shortening action. This is now the dispersion. Positive β and negative GVD broaden the pulse, just as negative β and positive GVD do.

The existence of solutions does not guarantee stability of the solutions. One very simple stability criterion is the requirement that the gain be less than the loss before and after passage of the pulse. One may evaluate g and ℓ from the master equation. Figure 1.9 shows a plot of $\ell - g$ versus D for the parameters used previously. We find that for large δ, regions of instability are found near $D = 0$. Thus, if in a specific laser system the dispersion is varied by varying the prism pairs inserted

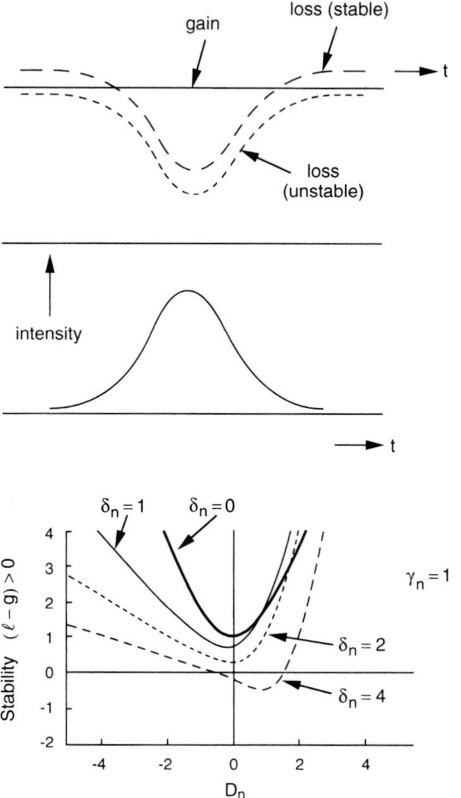

Figure 1.9. The stability requirement and the stability regime.

into the resonator, the system would cease to modelock when the instability region is approached, and resume modelocking after it is passed.

1.4 The nonlinear Schrödinger equation and solitons

As a special case, the master equation of modelocking contains the nonlinear Schrödinger equation which gives soliton solutions. We shall study these solutions for three reasons. First of all, in modelocked systems with very short pulses in which the optical Kerr effect is strong and compensated by negative dispersion as described above, the saturable absorber action and the gain bandwidth limitation can be considered as perturbations on the soliton of the nonlinear Schrödinger equation. Since this perturbation theory has been worked on extensively, useful analytic approaches are made available. Second, the breakthrough in modelocking techniques that led to the recent renaissance of modelocking and to new records of short pulse generation was initiated by Mollenauer and Stolen's 'soliton' laser which requires an understanding of soliton behavior. Third, some all-fiber modelocked lasers operate with pulses that are, on the average, solitons.

The master equation of modelocking describes the change effected on the pulse $a^{(k)}(t)$ to make it into the $k+1$ pulse. If the changes per pass are small, then the difference $a^{(k+1)} - a^{(k)}$ can be replaced by a differential. If the resonator is, in fact, unfolded so that the elements cascaded in the resonator are not identical with themselves, but simply repeated, one is confronted with single mode propagation along a medium (fiber) with gain, loss, gain bandwidth limiting, GVD, saturable absorber action and Kerr effect. The difference $a^{(k+1)} - a^{(k)}$ is interpreted as a spatial derivative, $\mathrm{d}a/\mathrm{d}z$, if all parameters in the equation are reinterpreted as parameters assigned to unit length of propagation distance. It should be noted further that, in the absence of all these actions, the change of any pulse is zero. This shows unequivocally that the spatial derivative has to be interpreted as taken in a frame co-moving with the pulse at the group velocity. With this reinterpretation (1.30) becomes:

$$\frac{\partial a}{\partial z} = \left[g\left(1 + \frac{1}{\Omega_g^2}\frac{\partial^2}{\partial t^2}\right) + jD\frac{\partial^2}{\partial t^2} - \ell + (\gamma - j\delta)|a|^2 \right] a \qquad (1.42)$$

The nonlinear Schrödinger equation (NLSE) results when all effects, except the Kerr effect and GVD, are turned off:

$$\frac{\partial a}{\partial z} = \left(jD\frac{\partial^2}{\partial t^2} - j\delta|a|^2 \right) a \tag{1.43}$$

The solution of (1.43) is:

$$a = A_o \operatorname{sech}\left(\frac{t - 2D\Delta\omega z}{\tau} \right) \exp(j\Delta\omega t)\ \exp -j\left(\frac{\delta|A_o|^2}{2} + D\Delta\omega^2 \right) z \tag{1.44}$$

where

$$\frac{1}{\tau^2} = -\frac{\delta|A_o|^2}{2D} \tag{1.45}$$

In (1.44), $\Delta\omega$ is an adjustable parameter, it is the choice of the carrier frequency offset from the nominal carrier frequency ω_o. Such a frequency offset changes the inverse group velocity by $2D\Delta\omega$ and the phase shift per unit distance by $D\Delta\omega^2$. The term $\delta|A_o|^2/2$ is the nonlinear phase shift per unit length, a phase shift corresponding to the average pulse intensity $|A_o|^2/2$. From (1.45) we see that solutions are obtained only when $\delta/D < 0$, when the two coefficients are of opposite sign. We shall assume henceforth that $D < 0$, since the Kerr coefficient, the SPM coefficient, is usually positive. One sees from (1.45) that the area of the pulse amplitude is an invariant:

$$\text{Area} = \int_{-\infty}^{\infty} |a(t)| \mathrm{d}t = \pi|A_o|\tau = \pi\sqrt{\frac{2|D|}{\delta}} \tag{1.46}$$

The pulse energy $W = 2|A_o|^2\tau = 4|D|/(\delta\tau)$ is inversely proportional to the pulse width. The name 'soliton' is applied to solutions of nonlinear partial differential equations which retain their shape upon collision with another soliton. In the present case, such a collision could occur when a pulse of higher carrier frequency catches up with one of lower frequency. Figure 1.10 shows such a collision. When the two pulses overlap one sees characteristic interference between the two pulses. After they have separated, they have completely recovered. If the system were linear, such a behavior would not be surprising. Here we are dealing with a nonlinear system and thus this behavior is truly remarkable. It indicates that wavelength division multiplexed (WDM) soliton streams could pass through each other on a nonlinear fiber without crosstalk. One must note, however, that the collisions lead to phase shifts and displacements. WDM

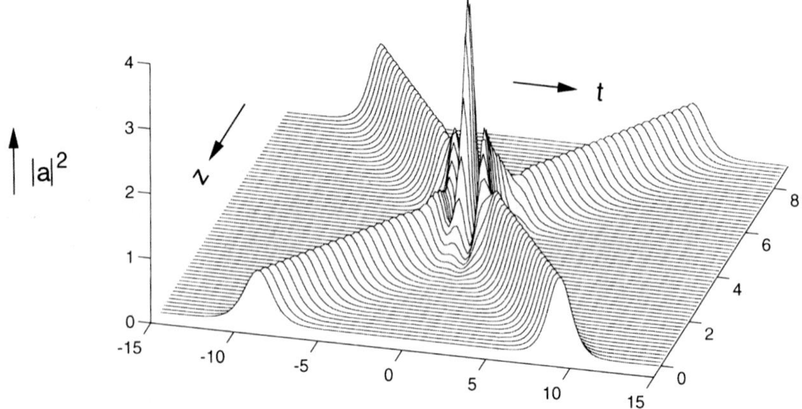

Figure 1.10. Collision of two solitons.

would work if, on the average, these effects are common to every pulse in the stream.

Another interesting property of solitons is shown in Figure 1.11. One may observe that an initial square pulse at $z = 0$ evolves into a soliton by shedding off radiation that does not fit into the final soliton shape. It should be mentioned that the soliton evolves only when the normalized area of the input pulse as defined in (1.45) is in the range

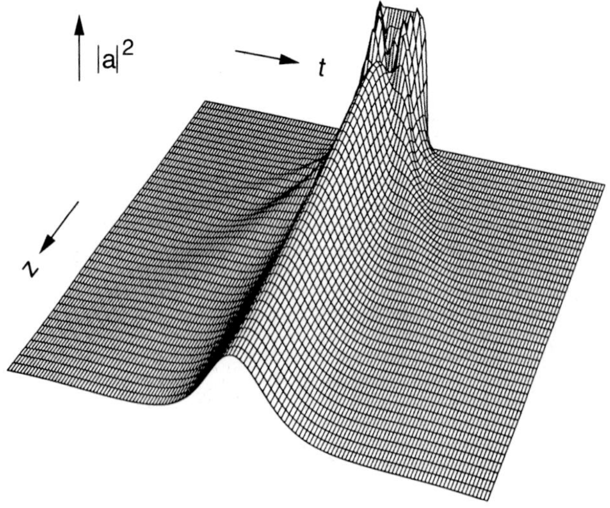

Figure 1.11. Evolution of a soliton from an initial square pulse.

$$\frac{\pi}{2} < \frac{\text{Area}}{\sqrt{2\,|D|/\delta}} < \frac{3\pi}{2} \tag{1.47}$$

In the range greater than $3\pi/2$ and less than $5\pi/2$, the pulse evolves into a second order soliton. Such a soliton is shown in Figure 1.12. It may be viewed as two 'bound' first order solitons that oscillate back and forth periodically. The spectrum alternately widens and narrows, and the pulse width does the opposite. An analytic expression for the symmetric second order soliton is[17]:

$$u(\xi, x) = \frac{\frac{4\eta_1(\eta_1+\eta_2)}{|\eta_2-\eta_1|} e^{j2\eta_1\xi}\{\cosh(2\eta_2 x) + \frac{\eta_2}{\eta_1}\cosh(2\eta_1 x)e^{2j[\eta_2^2-\eta_1^2]\xi}\}}{\cosh[2(\eta_1+\eta_2)x] + \left(\frac{\eta_1+\eta_2}{\eta_2-\eta_1}\right)^2 \cosh[2(\eta_2-\eta_1)x] + \frac{4\eta_1\eta_2}{(\eta_2-\eta_1)^2}\cos[2(\eta_2^2-\eta_1^2)\xi]} \tag{1.48}$$

where

$$\xi \equiv \delta z$$

$$x \equiv \sqrt{\frac{\delta}{2|D|}}t$$

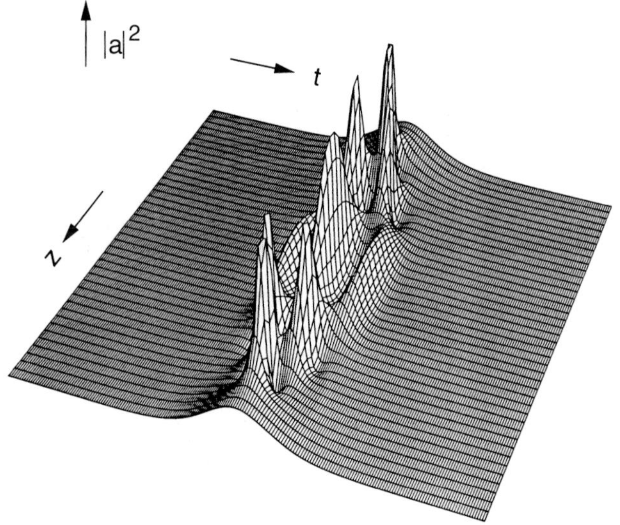

Figure 1.12. Second order soliton.

and where η_1 and η_2 are adjustable parameters. The plot of Figure 1.12 is carried out with the values $\eta_2 = 3/2$, $\eta_1 = 1/2$. The pulse is long initially, shortens and peaks as can be seen at half the propagation distance, and then returns to its original shape at the end of the so called 'soliton period' Z_o. With this background we are ready to discuss the 'soliton' laser.

1.5 The soliton laser

The historical significance of the soliton laser [3] is that it ushered in a new era of mode-locking systems. Figure 1.13 shows the schematic diagram. The main cavity contains the F-center crystal that is sync-pumped by a modelocked Nd:YAG laser. A partially transmitting mirror connects the cavity to an auxiliary resonator containing a fiber. The auxiliary resonator length is feedback controlled. When the auxiliary cavity is blocked, the system generates 10 ps pulses. When the auxiliary cavity is unblocked, 100 fs pulses are achieved. The feedback is adjusted to maintain a certain fixed power level.

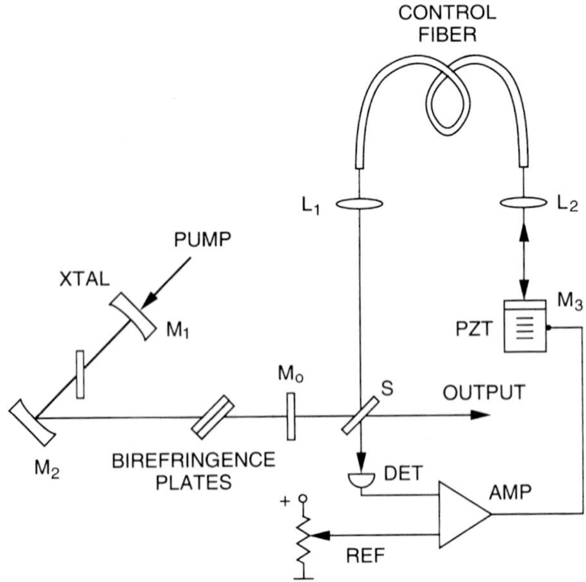

Figure 1.13. The soliton laser.

The explanation of the operation of the system is briefly as follows: The pulses fed into the fiber form second order solitons as they pass the fiber forward and back. The net distance is made equal or close to half the soliton period $Z_o/2$. This shortened pulse reinjected into the main cavity shortens the main pulse. This process continues until the pulse-shortening effects balance the pulse lengthening.

This explanation was valid for the laser originally constructed by Mollenauer and Stolen. However, computer simulations by K. J. Blow and D. Wood[4] showed that it was not necessary to have negative dispersion in the fiber. Short pulse operation was obtained also for positive dispersion. Experiments supported this prediction[5–7]. Clearly, a more general principle was at work in this system. This is how the coupled cavity modelocking principle or additive pulse modelocking principle was discovered.

1.6 The coupled cavity modelocking and/or the additive pulse modelocking principle

The coupled cavity and/or additive pulse modelocking principle (CCM and APM) was discovered when it was realized that soliton formation via negative dispersion in the auxiliary cavity as shown in Figure 1.13 was not necessary to achieve modelocking. In fact, it was soon observed experimentally that the pulse returned from the auxiliary cavity into the main cavity was in fact longer when the laser was tuned so as to operate in the positive dispersion regime of the fiber. Yet the net effect was a shortening of the pulse in the main cavity by the following process: if the returned pulse was chirped, so that its wings added in antiphase to the main pulse and its center added in phase, then a net shortening of the pulse could still be produced by superposition of the two pulses.

Let us look at the operation of a coupled cavity system analytically. Using Figure 1.14, note that the pulse $a(t)$ from the main cavity is impingent on the partially transmitting mirror from the left. Denote $a(t)$ by $a_1(t)$ to distinguish it from the returning pulse from the auxiliary cavity, $a_2(t)$. The scattering matrix of the mirror, with r equal to the amplitude reflectivity, establishes the relationship[18]:

$$b_1 = ra_1 + j\sqrt{1-r^2}a_2 \tag{1.49}$$

$$b_2 = j\sqrt{1-r^2}a_1 + ra_2 \tag{1.50}$$

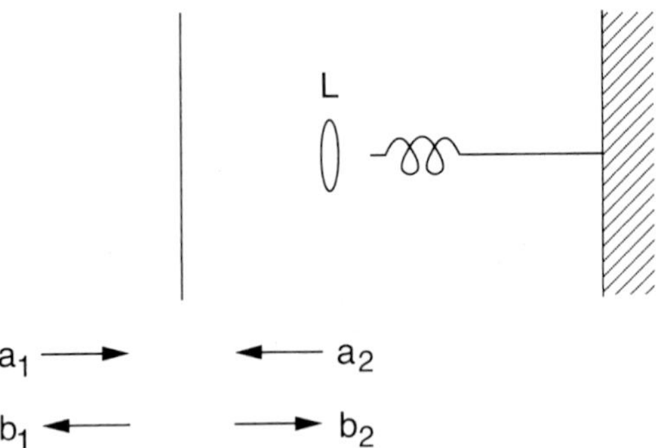

Figure 1.14. The operation of an auxiliary cavity.

The incident wave from the right, $a_2(t)$, is related to the outgoing wave $b_2(t)$ by a loss factor $L(< 1)$ and a net roundtrip phase delay ϕ:

$$a_2 = -Lb_2e^{-j\phi} \tag{1.51}$$

We use here L to denote a multiplier of the signal, in contrast to the use of the symbol ℓ indicating the change of the pulse passing through a loss section. For $\ell \ll 1$, we have $L = 1 - \ell$. When (1.51) is introduced into (1.50), this equation is solved for a_2 in terms of a_1, and the result is put into (1.49), one obtains:

$$b_1 = \frac{re^{j\phi} + L}{e^{j\phi} + rL}a_1 \tag{1.52}$$

The wave b_1 is reflected with the reflection coefficient Γ:

$$\Gamma = \frac{re^{j\phi} + L}{e^{j\phi} + rL} \tag{1.53}$$

Note that Γ depends on the intensity of the wave in the auxiliary cavity, because the phase delay ϕ depends on it.

The preceding analysis may appear to apply to a sinusoidal (CW) wave of single frequency only, since the auxiliary cavity is a resonator, and ϕ is a function of frequency. Note, however, that ϕ is a periodic function of frequency. If incident upon the structure is an excitation $a(t)$, which is composed of evenly spaced frequency components such that ϕ is the same for all of them, the reflection coefficient is legitimately applied to such an excitation as well. But, such an excitation is a (quasi) delta function of a carrier frequency ω_o close to the resonance frequency of one of the cavity modes, and repeating with a period equal to the round-trip time in the auxiliary resonator. An arbitrary pulse shape can be constructed of such trains of time-shifted delta functions. Hence the reflection coefficient (1.53) can be used in the master equation of modelocking for a steady state analysis. For a transient analysis, it is legitimate to use (1.53) if the evolution of $a^{(k+1)} - a^{(k)}$ is slow compared with the relaxation time of the auxiliary resonator.

We shall now make further simplifications. We shall assume that $L \ll 1$, so that we may expand (1.53) to first order in L. This means that we are retaining the contribution of one transit only in the auxiliary cavity. Then

$$\Gamma = r + (1 - r^2)Le^{-j\phi} \tag{1.54}$$

The phase shift ϕ can be separated into a bias phase ϕ_b and a nonlinear Kerr phase Φ:

$$\phi = \phi_b + \Phi \tag{1.55}$$

where

$$\Phi = \frac{2\pi}{\lambda} n_2 \frac{|a_f|^2}{\mathcal{A}_{\mathrm{eff}}} 2L_f = \frac{2\pi}{\lambda} n_2 \frac{2L_f}{\mathcal{A}_{\mathrm{eff}}} (1 - r^2)L^2 |a|^2 \equiv \kappa |a|^2 \tag{1.56}$$

and a_f is the field amplitude in the fiber, $\mathcal{A}_{\mathrm{eff}}$ is the mode cross section in the fiber, L_f is its length, n_2 is the Kerr coefficient of the fiber. When $\Phi \ll \phi_b$, one can expand (1.55) with the result

$$\Gamma = r + (1 - r^2)Le^{-j\phi_b} + \underbrace{(1 - r^2)Le^{-j\phi_b}(-j\Phi)}_{\text{intensity dependent}} \tag{1.57}$$

Relation (1.57) is illustrated in Figure 1.15. The reflection coefficient is r plus the contribution $(1 - r^2)L \exp -j\phi$ of one bounce in the auxiliary cavity. When $\Phi \neq 0$, the reflection coefficient can increase with increasing Φ, i.e. with increasing intensity, providing SAM action, if the bias phase lies in the proper range. There is a contribution to the SPM coefficient too.

$$\gamma = -\frac{2\pi}{\lambda} n_2 \frac{2L_f}{\mathcal{A}_{\mathrm{eff}}} (1 - r^2)L^2 (1 - r^2)L \sin \phi_b \tag{1.58}$$

$$\delta = \frac{2\pi}{\lambda} n_2 \frac{2L_f}{\mathcal{A}_{\mathrm{eff}}} (1 - r^2)L^2 (1 - r^2)L \cos \phi_b \tag{1.59}$$

Γ_o is the small signal reflection coefficient.

$$\Gamma_o = r + (1 - r^2)Le^{-j\phi_b} \tag{1.60}$$

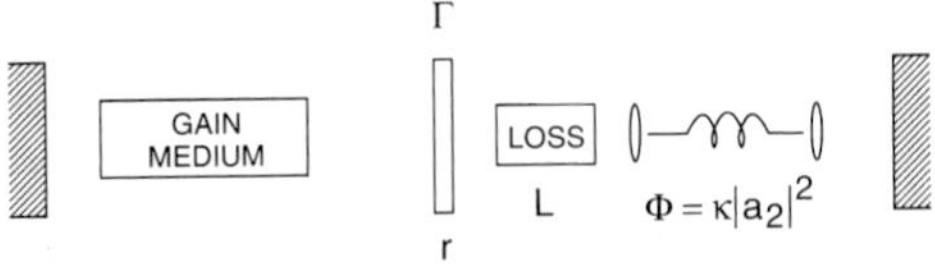

AMPLITUDE REFLECTIVITY

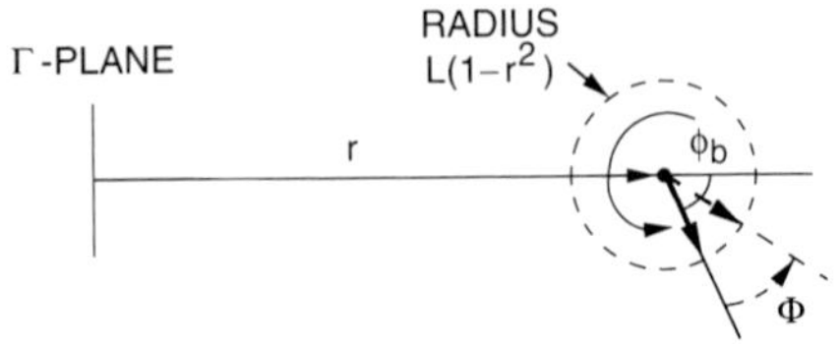

ϕb -BIAS PHASE; Φ -NONLINEAR (KERR) PHASE

Figure 1.15. Intensity dependence of reflection from an auxiliary cavity.

This analysis shows how the auxiliary cavity can provide a passive mode-locking mechanism, an artificial saturable absorber (AFSA). The advantage of the system is that the equivalent saturable absorber is fast, its relaxation time is that of the Kerr medium in the auxiliary cavity, and its linear and nonlinear absorption (Γ_o and γ) can be adjusted by choice of the mirror reflectivity, loss L and the length L_f of the fiber.

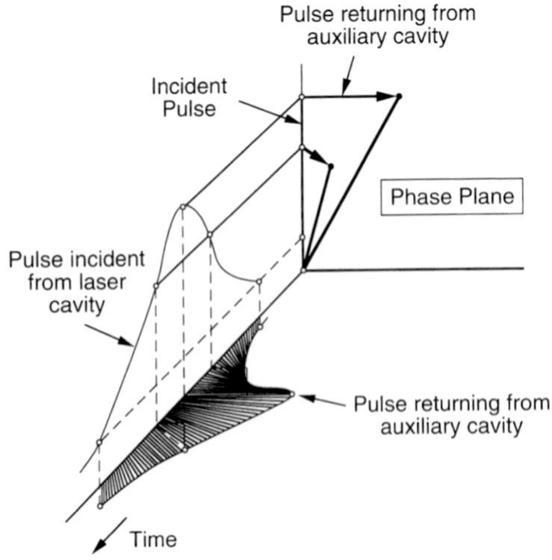

Figure 1.16. Addition of excitations at a mirror.

Figure 1.16 shows schematically the addition of the two pulses, one directly reflected at the mirror, the other emerging from the auxiliary cavity and transmitted through the mirror. For the purpose of the illustration, the former is assumed unchirped. The latter pulse has a bias phase of $-\pi$ and a nonlinear peak phase shift of $\pi/2$, so that, at the peak, the amplitudes add in quadrature, in the wings in antiphase. Superposition of the two pulses results in a shortened and slightly chirped pulse. The chirp can be compensated by GVD in the main cavity, so that the pulse on the next round trip can return unchirped. The phased additions are shown as projected onto the phasor plane. It is this picture that justifies the name 'additive pulse modelocking.'

The analysis suggests that the nonlinear phase shift in the auxiliary cavity must be much less than $\pi/2$. One should note that this, in general, is not necessary to obtain APM action. If the phase shift in the auxiliary cavity is comparable with or greater than $\pi/2$, but a filter structure is introduced either in the auxiliary cavity or in the main cavity, the pulse fed back from the auxiliary cavity can still lead to pulse shortening, and in fact to a small pulse shortening per pass. It has been observed experimentally that such a performance can be quite stable[19]. In fact, it has one great advantage: it lowers the value of the effective SAM coefficient with increasing intensity in the auxiliary cavity. The γ value is large initially to initiate self starting of modelocking. As the intensity of the pulse grows, and the phase shift in the auxiliary cavity becomes greater than $\pi/2$, this operation can be accomodated by filtering so that the total change of phase shift across the pulse returned by the auxiliary cavity does not exceed π.

The bias phase has to be just right. This means of course that CCM or APM operation with an auxiliary cavity calls for a feedback arrangement that stabilizes the bias phase against random drifts of the cavity lengths. Such feedback may not be necessary if the interferometeric structure that changes phase modulation into amplitude modulation is part of the main resonator. One such structure is a Mach–Zehnder resonator.

1.7 Additive pulse modelocking with nonlinear interferometer[18,20]

Consider the nonlinear interferometer of Figure 1.17. The partially transmitting mirrors have reflectivity r and Kerr media are introduced symmetrically into the two arms. A phase bias ϕ_b is introduced in

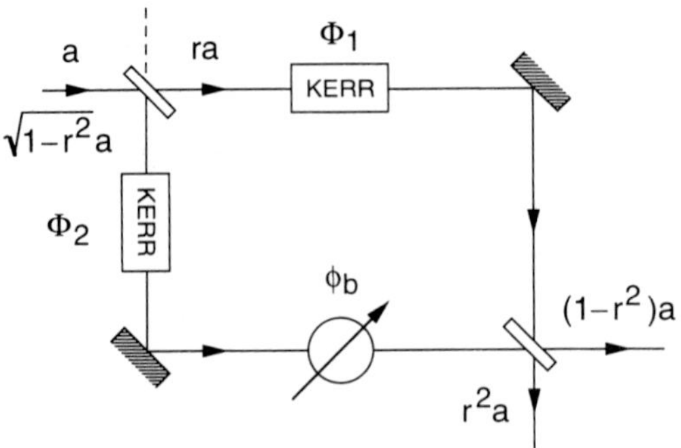

Figure 1.17. A nonlinear Mach–Zehnder interferometer.

arm (2). The nonlinear Kerr-induced phase shifts in the two arms are Φ_1 and Φ_2 respectively. The transmission of the interferometer is:

$$b = \left[Re^{-j\Phi_1} + (1 - R)e^{-j\phi_b}e^{-j\Phi_2}\right]a = Ta \tag{1.61}$$

where $R = r^2$ and T is the transmission coefficient. The nonlinear phase shifts are

$$\Phi_1 = \frac{2\pi}{\lambda} n_2 \frac{L_k}{A_{eff}} R|a|^2 \tag{1.62}$$

$$\Phi_2 = \frac{1 - R}{R} \Phi_1 \tag{1.63}$$

where L_k is the length of the Kerr media. Clearly, the transmission T through the interferometer depends on the intensity. When this interferometer is introduced into the laser resonator, its first effect is to introduce a linear loss and linear phase shift that may shift the frequency of resonance of the resonator. The nonlinear operation of the interferometer starts with this as given. We modify (1.61) by a linear phase multiplier $\exp j\phi$ chosen so that T is real under linear operation, $\Phi_1 = \Phi_2 = 0$. This phase shift incorporates the effect of introducing the interferometer into the resonator and shifting the resonator frequency by a change of the optical path length. By making T real, we have incorporated this linear effect of the interferometer. Denoting the transmission coefficient under small signal operation by T_o, we have for $1 - R \ll 1$,

$$T_o = [R + (1 - R)e^{-j\phi_b}]e^{j\phi} \simeq Re^{j\phi} + (1 - R)e^{-j\phi_b} \tag{1.64}$$

and thus

$$\phi = \frac{1 - R}{R} \sin \phi_b \tag{1.65}$$

Now turn on the nonlinearity, $\Phi_1 \neq 0$, and $\Phi_2 \neq 0$. Expand T to first order in Φ_1 and Φ_2 and set $\cos\phi \simeq 1$

$$\begin{aligned} T &= \cos(\phi - \Phi_1)R + (1-R)\cos(\phi_b + \Phi_2) + j[R\sin(\phi - \Phi_1) \\ &\quad - (1-R)\sin(\phi_b + \Phi_2)] \\ &\simeq T_o - (1-R)\sin\phi_b(\Phi_2 - \Phi_1) - j[(1-R)\cos\phi_b\Phi_2 + R\Phi_1] \end{aligned} \tag{1.66}$$

We find that the transmission is now a linear function of $|a|^2$. From this dependence we may evaluate the SAM and SPM coefficients of the nonlinear interferometer:

$$\gamma = -\frac{2\pi}{\lambda} n_2 \frac{L_k}{A_{eff}} (1 - 2R)(1 - R)\sin\phi_b \tag{1.67}$$

$$\delta = \frac{2\pi}{\lambda} n_2 \frac{L_k}{A_{eff}} [\cos\phi_b(1-R)^2 + R^2] \tag{1.68}$$

Hence a nonlinear Mach–Zehnder can provide equivalent fast saturable absorber action. When inserted into a resonator, it can produce modelocking. This operation is aptly described by the words 'additive pulse modelocking', because at the output of the interferometer the superposition of the pulses emerging from the two arms of the interferometer can produce a nonlinear dependence upon intensity of the throughput.

A nonlinear interferometer can be built to be self-stabilizing. One example of such a self-stabilized design is a nonlinear birefringent crystal with a polarizer and wave plate, as shown in Figure 1.18. The two arms of the interferometer are represented by the two polarizations. If the birefringence is sufficiently strong, the two polarizations change each

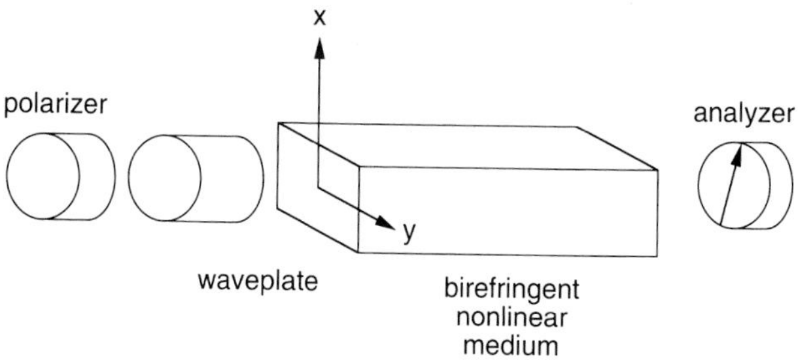

Figure 1.18. A self-stabilized nonlinear Mach–Zehnder interferometer.

other's index, without producing interference contributions. If the group velocity difference is so large that the envelopes of the two polarizations drift apart, the drift can be compensated by using two crystals with their fast and slow axes interchanged.

A similar kind of nonlinear interferometer is an isotropic Kerr medium operating via polarization rotation. Let us consider only forward propagation (Figure 1.19). Linear polarization at the input is transformed into elliptic polarization by the polarization transformer. The ellipse rotates due to the Kerr effect, the angle of rotation being proportional to intensity. The power transmitted through the polarizer at the output is a function of intensity. This mechanism is particularly suited for use in all-fiber APM systems.

We start with the column matrix representation of a linear polarization represented in $x - y$ coordinates:

$$\mathbf{a} = \begin{bmatrix} E_x \\ E_y \end{bmatrix} \tag{1.69}$$

The amplitudes are so normalized that their squares are equal to the power. Passage through a polarization transformer produces an output **b** from an input **a** via the polarization transformation matrix:

$$\mathbf{b} = \mathbf{T}\mathbf{a} \tag{1.70}$$

The most general polarization transformation matrix of a lossless structure is the unitary matrix:

$$\mathbf{T} = \begin{bmatrix} se^{-j\phi} & j\sqrt{1-s^2}e^{-j\theta} \\ j\sqrt{1-s^2}e^{j\theta} & se^{j\phi} \end{bmatrix} \tag{1.71}$$

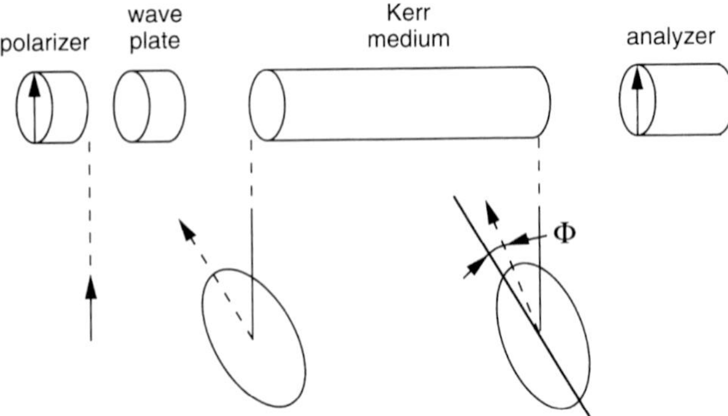

Figure 1.19. Additive pulse modelocking by polarization rotation.

A unitary matrix of second rank has four real parameters. Equation (1.71) has three parameters. The fourth is omitted since it represents solely a phase shift of both polarizations. The nonlinear phase shift in the fiber is most conveniently expressed in terms of circular polarization variables (see Appendix I):

$$\Phi_{\pm} = \frac{2\pi}{\lambda}\Delta n_{\pm} = \kappa_c\{|E_{\pm}|^2 + 2|E_{\mp}|^2\} \tag{1.72}$$

where κ_c is the Kerr coefficient for circular polarization (compare with (1.121) of Appendix I):

$$\kappa_c = \frac{4\pi}{3\lambda} n_2 \frac{L_k}{A_{\text{eff}}} \tag{1.73}$$

We are now ready to evaluate the SAM and SPM parameters of this structure. The incident radiation is linearly polarized by the polarizer and then transformed by the polarization transformer. The transformation into circular polarization basis is accomplished by the unitary transformation matrix U:

$$\mathbf{a}_c \equiv \begin{bmatrix} E_+ \\ E_- \end{bmatrix} = \mathbf{Ua} \tag{1.74}$$

where

$$\mathbf{U} = \frac{1}{\sqrt{2}}\begin{bmatrix} 1 & j \\ 1 & -j \end{bmatrix} \tag{1.75}$$

For a linear polarization along x of amplitude A transmitted through the polarization transformer (1.71), the Kerr phase shifts are found to be

$$\Phi_{\pm} = \frac{3}{2}\kappa_c|A|^2\left[1 \pm \frac{2}{3}s\sqrt{1-s^2}\cos(\phi+\theta)\right] \tag{1.76}$$

The output of the second polarizer is

$$\begin{aligned} \mathbf{b} &= [1 \quad 0]\mathbf{U}^{\dagger}\begin{bmatrix} e^{-j\Phi_+} & 0 \\ 0 & e^{-j\Phi_-} \end{bmatrix}\mathbf{UTa} \\ &= A\exp -j\left(\frac{\Phi_+ + \Phi_-}{2}\right)\left[s\cos\left(\frac{\Phi_+ - \Phi_-}{2}\right)e^{-j\phi}\right. \\ &\quad \left. + j\sqrt{1-s^2}\sin\left(\frac{\Phi_+ - \Phi_-}{2}\right)e^{j\theta}\right] \end{aligned} \tag{1.77}$$

The throughput power is

$$|b|^2 = \frac{1}{2}|A|^2[1 + (2s^2 - 1)\cos(\Phi_+ - \Phi_-) \\ - 2s\sqrt{1 - s^2}\sin(\Phi_+ - \Phi_-)\sin(\phi + \theta)] \tag{1.78}$$

from which one may evaluate the linear loss parameter ℓ and the SAM coefficient γ. Taking $1 - s \ll 1$ one finds:

$$\ell = 1 - s \tag{1.79}$$

$$\gamma|a|^2 = -\sqrt{1 - s^2}\sin(\phi + \theta)\left(\frac{\Phi_+ - \Phi_-}{2}\right) \tag{1.80}$$

From (1.77), for s close to unity, the SPM parameter is simply:

$$\delta = \frac{3}{2}\kappa_c \tag{1.81}$$

This completes the analysis of the self-stabilized Mach–Zehnder operating via polarization rotation. It is a particularly convenient system for the modelocking of all-fiber lasers, as will be described later on.

1.8 Kerr lens modelocking[20]

Kerr lens modelocking (KLM) was discovered experimentally[9], and only later identified as a nonlinear self amplitude modulation via self-focusing of the beam inside the laser resonator[10]. Changes of the beam radius can lead to changes of loss or gain when the beam is apertured by an iris, or by its overlap with the gain profile. Clearly, KLM cannot be employed in a system in which the beam profile is determined by waveguiding in an optical waveguide or fiber. In such systems changes of the beam profile induced by the Kerr effect would require exorbitantly high intensities. It can be employed in open resonators in which the self-focusing effects can produce relatively large changes of beam diameter.

The self-focusing is a nonlinear propagation phenomenon which can be treated accurately only by numerical methods. As always, it is useful to develop an analytic theory which can help our understanding of the processes involved and can serve as a guide for a rough layout of a projected system.

Consider first KLM in a monolithic cavity such as shown in Figure 1.20. Suppose that the cavity is filled with a medium of linear index n and Kerr index n_2. An aperture of radius w_a is placed at the output mirror. There is, of course, gain in the resonator to compensate for the loss, but

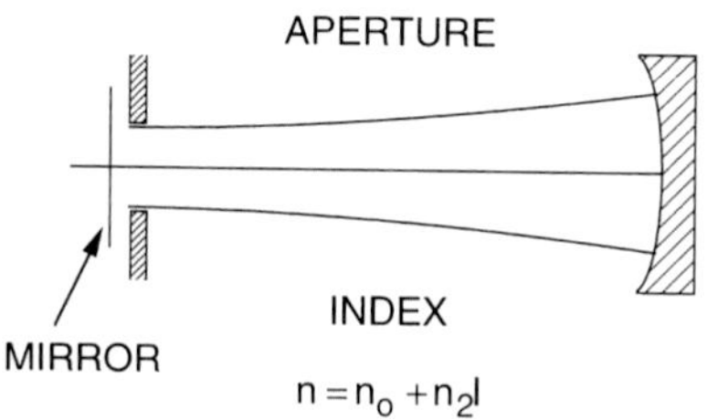

Figure 1.20. Kerr lens modelocking in a monolithic cavity.

this can be treated as a perturbation in the master equation. Our goal is to derive equivalent γ and δ parameters for the resonator. A Gaussian beam of amplitude A_o and beam radius $w(z)$ produces a nonlinear phase shift $\Delta\Phi$ within the distance Δz:

$$\Delta\Phi = \frac{2\pi}{\lambda} n_2 A_o^2 \exp[-2r^2/w^2]\Delta z \simeq \frac{2\pi}{\lambda} n_2 A_o^2 \left(1 - \frac{2r^2}{w^2}\right)\Delta z \qquad (1.82)$$

Here we have expanded the Gaussian profile to first order into a parabolic dependence upon radius. The parabolic phase profile preserves the Gaussian beam. In fact, the actual profile generates higher-order Hermite Gaussian modes. Yet the model adopted here is not unreasonable, since apertures in a laser resonator discriminate against higher-order modes. The phase profile modifies the beam propagation. Under the parabolic approximation of the phase profile it is possible to develop an analytic theory for the q parameter as shown in Appendix II. The renormalized q parameter q' propagates as in free space, where:

$$\frac{1}{q'} = \mathrm{Re}\left[\frac{1}{q}\right] + j\,\mathrm{Im}\left[\frac{1}{q}\right]\sqrt{1-2K} \qquad (1.83)$$

with

$$2K = \frac{8P}{\pi}\left(\frac{\pi}{\lambda}\right)^2 n_2 \qquad (1.84)$$

and P is the power. In a Kerr medium of length L_k and index n, the q' parameter transforms into

$$q_2' = q_1' + \frac{L_k}{n} \qquad (1.85)$$

In the cavity of Figure 1.20, the q parameter is pure imaginary at $z = 0$:

$$q_1 = jy_1 \tag{1.86}$$

where

$$y_1 = \frac{\pi w_1^2}{\lambda}$$

The renormalized q parameter is

$$q_1' = \frac{jy_1}{\sqrt{1-2K}} \simeq jy_1(1+K) \tag{1.87}$$

The transformation (1.85) gives the q' parameter at reference plane (2) at $z = L$

$$\frac{1}{q_2'} = \frac{1}{L_k/n + jy_1(1+K)} = \frac{L_k/n - jy_1(1+K)}{(L_k/n)^2 + y_1^2(1+K)^2} \tag{1.88}$$

The real part of $1/q'$ is equal to the real part of $1/q$, the inverse radius of curvature of the phase front. The radius of curvature of the phase front at the mirror of radius of curvature R must not change as a function of intensity. The change of the radius of curvature is prevented by a change δy_1. To first order in K:

$$\delta\left[\frac{L_k/n}{(L_k/n)^2 + y_1^2}\right] - \frac{2K(L_k/n)y_1^2}{[(L_k/n)^2 + y_1^2]^2} = 0 \tag{1.89}$$

The first term changes due to a change of y_1. One may evaluate it as

$$\frac{\delta y_1}{y_1} = -K \tag{1.90}$$

A decrease of the beam radius leads to a decrease of the loss. Denote the power loss per pass through the aperture by 2ℓ. It evaluates to:

$$2\ell P = \int_{R_o}^{\infty} 2\pi r \mathrm{d}r A_o^2 e^{-2r^2/w_a^2} = e^{-2R_o^2/w_a^2} P \tag{1.91}$$

Thus

$$2\ell = e^{-2R_o^2/w_a^2} = e^{-2\mathcal{R}/y_1} \tag{1.92}$$

where

$$\mathcal{R} \equiv \frac{\pi R_o^2}{\lambda} \tag{1.93}$$

The SAM parameter γ is evaluated from:

$$\gamma|a|^2 = -P\frac{d\ell}{dP} = -P\frac{d\ell}{dy_1}\frac{dy_1}{dP} = P\frac{\mathcal{R}}{y_1}e^{-2\mathcal{R}/y_1}\frac{K}{P} \tag{1.94}$$

With $|a|^2$ equal to the power P, and the value of K introduced from (8.3) one finds:

$$\gamma = \frac{\mathcal{R}}{y_1}\ell\frac{8\pi}{\lambda^2}n_2 \tag{1.95}$$

The factor $\mathcal{R}/y_1$ is of the order of unity. Using the value of n_2 for glass, $n_2 = 3.2 \times 10^{-16}$ cm^2/w and a wavelength of 1.5 μm, one finds for γ/ℓ the value 89.4[Gw]$^{-1}$. This shows that very large powers are required to achieve KLM action in a monolithic resonator.

The power requirement for KLM action can be reduced if one uses a composite cavity such as shown in Figure 1.21. If the resonator is so designed that it is near the mode–stability boundary, then slight changes of the lensing within the resonator can lead to much larger changes of beam radius. Figure 1.22 shows how much enhancement can be achieved by proper design of the composite cavity[20]. One should note, however, that the enhancement of the SAM coefficient is obtained at the expense of resonator stability. If stable operation is required, the enhancement is limited. Therefore, KLM is better suited for lasers with large internal power.

1.9 All-fiber ring laser

With the development of fiber amplifiers such as the erbium amplifier, all-fiber lasers are easily constructed. Because they need not have air gaps they can be made more stable under simple environmental control. All-fiber lasers can be passively modelocked using various APM principles. The ‘figure eight’ laser[21] uses the interference of two versions of the pulse at the coupler of a nonlinear Sagnac interferometer in the

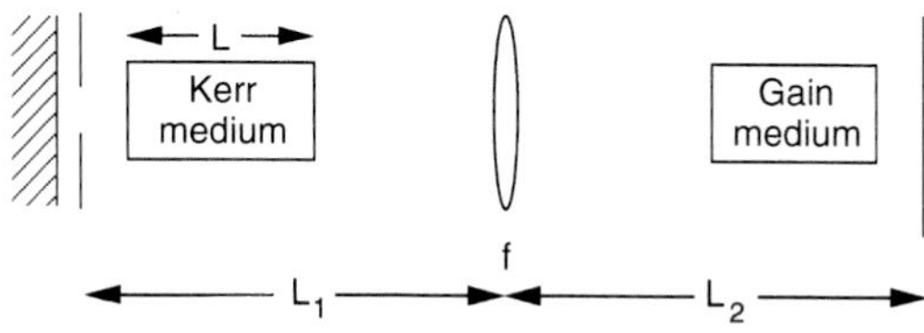

Figure 1.21. A composite resonator for KLM.

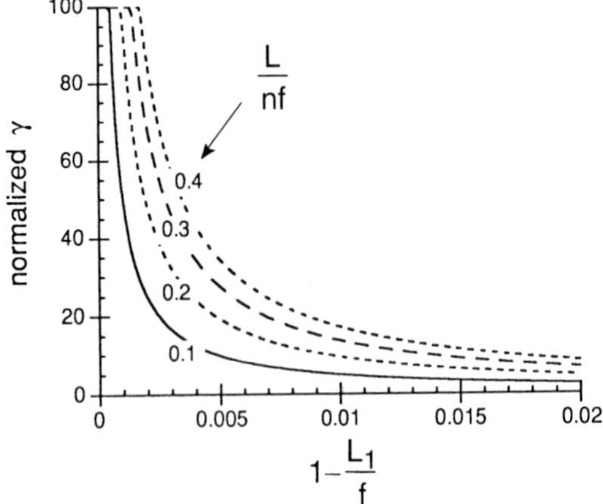

Figure 1.22. Enhancement of the SAM coefficient by a composite resonator.

cavity. Modelocked unidirectional all-fiber ring lasers were first constructed by L. F. Mollenauer[22]. The MIT unidirectional ring laser[23] uses polarization rotation as described in Section 1.7. Here we shall concentrate on the discussion of the ring laser.

Figure 1.23 shows a schematic diagram of the unidirectional ring laser. The isolator causes lasing to occur into a traveling wave in one direction. The polarizer and the polarization transformer following it realize the self-stabilized nonlinear Mach–Zehnder interferometer. The second polarization transformer is not needed in principle, but is required in practice to correct for polarization transformation in the fiber which, even though nominally nonbirefringent, causes polarization transformations over lengths of the order of 4 m, the length of the ring laser. The

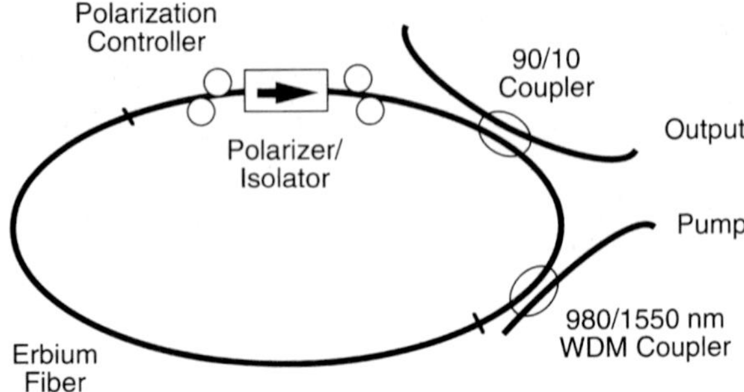

Figure 1.23. The MIT unidirectional ring laser.

overall length of the laser was made short, since one of the difficulties with passive modelocking of fiber lasers is the fact that the Kerr nonlinearity in the fiber is large on account of the small mode profile and the long distance of propagation, even though the powers of fiber lasers are much lower than those of bulk solid state lasers.

Typically, if an all-fiber laser is constructed in the 1.5 μ wavelength regime in which the fibers are usually negatively dispersive, multiple quantized pulses occur, as shown in Figure 1.24. This is due to processes that limit the peak intensity of the pulse. Since the pulses are soliton-like, their width is determined from (1.45), and thus their energy is fixed. The gain medium, whose relaxation time is much longer than the round-trip time in the fiber, delivers a certain amount of energy per transit time. If the single-pulse energy is smaller than this value, multiple pulses are generated. The timing of these multiple pulses is not controlled, they wander back and forth randomly, leading to unacceptable performance. This means that the length of the gain fiber must be kept short enough so that only one pulse occurs per transit, unless effective means are found to control the timing of multiple pulses.

Let us now discuss briefly the causes that limit the peak intensity of the pulses. One cause is the SAM action of polarization rotation. In the master equation we represented it by the term $\gamma|a|^2$. However, the APM process is periodic with regard to the polarization rotation angle

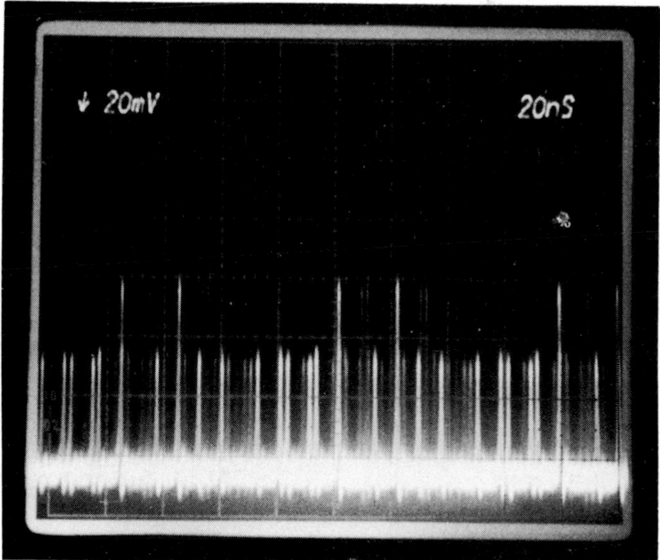

Figure 1.24. Multiple pulses in a fiber ring laser.

and is represented more strictly by $\sin(\gamma|a|^2)$. The process saturates. There is a peak intensity at which it gives minimum loss, and beyond that point the loss starts to increase. This value is found from maximization of the integral

$$\int dt \sin(\gamma|a|^2)|a(t)|^2 \tag{1.96}$$

for an assumed pulse shape $a(t) = A_o \mathrm{sech}(t/\tau)$. For $A_o\tau = \mathrm{const}$, i.e. for a soliton circulating in the ring, the maximum occurs when the argument $\gamma|A_o|^2 = 0.6\ \pi$. Beyond this value of $|A_o|^2$ the occurrence of two pulses is energetically more favorable.

Another possible mechanism of pulse intensity quantization is the generation of continuum radiation via the perturbations of the pulses as they circulate through the ring. The loss and the gain change the pulse periodically. When the distance within which the (quasi) solitons acquire a nonlinear phase shift of 2π approaches the resonator length, the perturbation is particularly strong, because it becomes 'phase-matched.' The pulses shed energy, their peak intensity tends to be limited and the continuum radiation coalesces into another pulse.

If care is taken so that the peak energy of the pulses is not less than the energy provided by the gain medium, single pulses occur that are very stable, as shown in Figure 1.25. The figure shows the intensity autocorrelation, which indicates a pulse width of 520 fs. Figure 1.26 shows two spectra obtained with slightly different polarization transformer adjustments. One sees the pulse spectrum and sharp spectral components. In

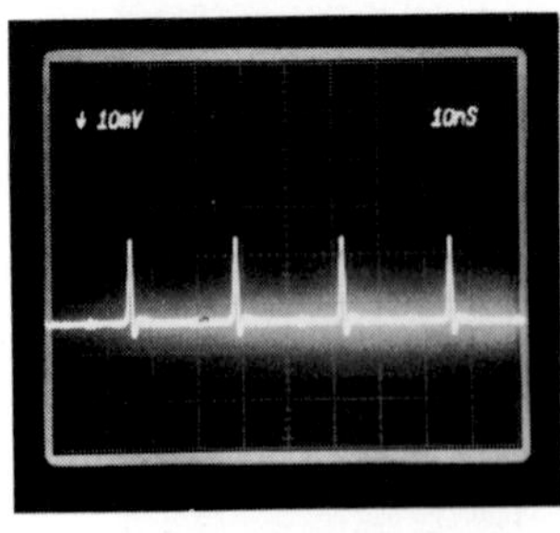

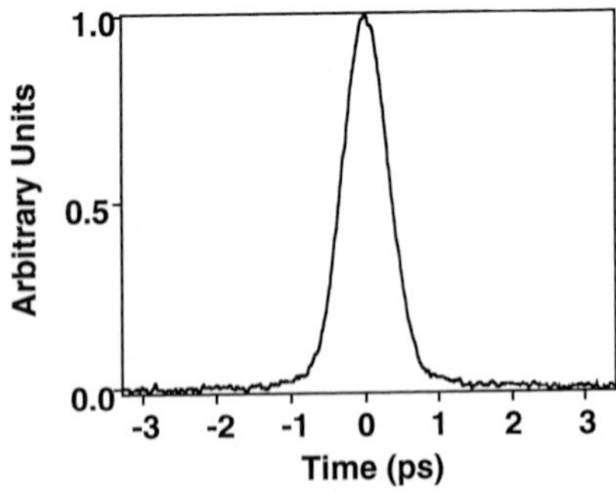

Figure 1.25. Characteristics of single pulse operation. Repetition period: 23.8 ns, pulse width: 452 fs, average power: 0.28 mW, peak power: 8.4 W.

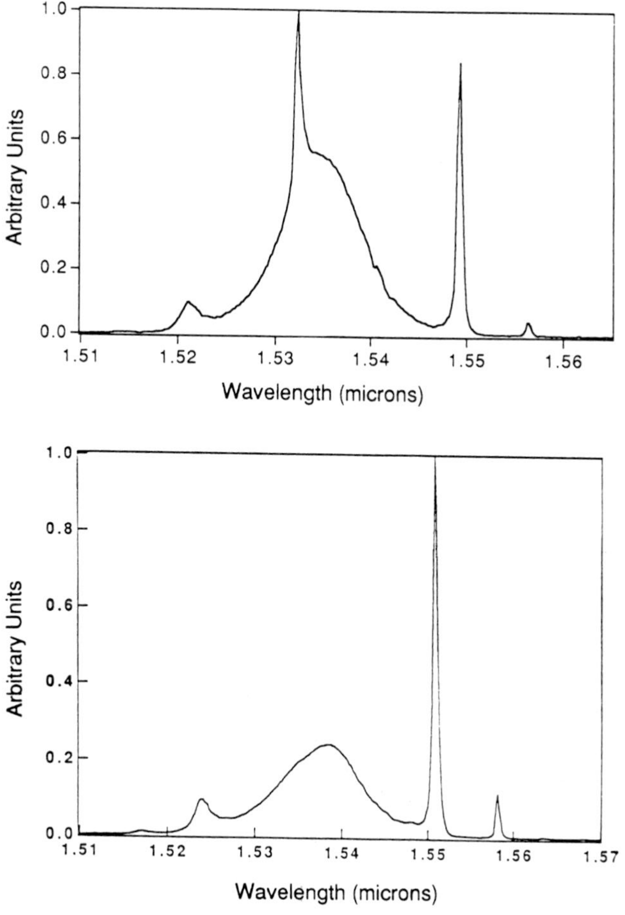

Figure 1.26. The spectra of pulses in a ring laser: (a) 520 fs pulsewidth, (b) 450 fs pulsewidth.

the spectrum (b), the component at 1.533 μ has been eliminated. It is believed that the spectrum of Figure 1.26b could be cleaned up if proper filters were introduced. However, it is not easy to introduce such filters into an all-fiber ring.

It should be mentioned that the fiber ring laser is automatically self-starting, in contrast to most APM and KLM systems that need some starting trigger. This can be explained, if self-starting is prevented by spurious reflections[24]. The spurious reflections cause nonuniform frequency-spacings of the resonator modes. To produce modelocking, sidebands produced by the SAM and SPM modulation must increase with increasing nonuniformity of mode spacing. In a standing wave cavity only one reflection is necessary to change the mode spacing. In a ring resonator two reflections are necessary, one backward and the other forward. To test this hypothesis, a Ti:sapphire laser was put into a ring resonator; it was found to be fully self-starting[25].

Recently a major advance was achieved in the APM modelocking of an all-fiber laser. The section of erbium doped fiber of Figure 1.23 was replaced with a doped fiber that had positive dispersion. The pulses from the resonator were as short as 75 fs, if the length of the negatively dispersive output fiber was properly adjusted. What is this breakthrough? It is a new principle of modelocking that can be dubbed stretched pulse-additive pulse modelocking (SP-APM).

Both experiment and theoretical prediction show that the pulse changes from a width as short as 75 fs to a width of the order of 1 ps as it travels around the loop. By spreading by more than a factor of 10, the nonlinear effects in the fiber are greatly reduced from those that would prevail if the pulse remained at constant width of 75 fs. As mentioned above, all-fiber lasers exhibit too much Kerr modulation. A mechanism that reduces this nonlinearity is generally beneficial.

Even though the pulse experiences large linear changes per pass, one may still develop a master equation for this mode of operation. Suppose at the start that the dispersions of the two fiber segments exactly cancel. Turn off the gain dispersion and the nonlinear effects and look for a self-consistent pulse solution circulating in the loop.

For an assumed minimum pulse width at a given position in the positively dispersive fiber, as shown in Figure 1.27, a minimum width pulse occurs symmetrically in the negatively dispersive fiber. The position of the minimum width is arbitrary at this point. Define $a(t)$ of the master equation as the pulse shape at the fiber cross-section where the pulse is of minimum width. Now turn on the various effects in the master equation separately. The gain dispersion is still approximated by an inverted parabola in the frequency domain. If the positive and negative dispersion of the two fiber segments are exactly cancelled, then the D parameter of the master equation is equal to zero. If they do not balance, the D of the master equation has to be interpreted as the effect on the pulse at the reference cross-section due to the difference between the two dispersions. The SAM and SPM coefficients per pass must be evaluated, taking the change of the pulse shape during one transit into account. If the effect of the nonlinearity is small, one may still define the γ and δ coefficients as perturbations on the pulse. They both depend on the position of the

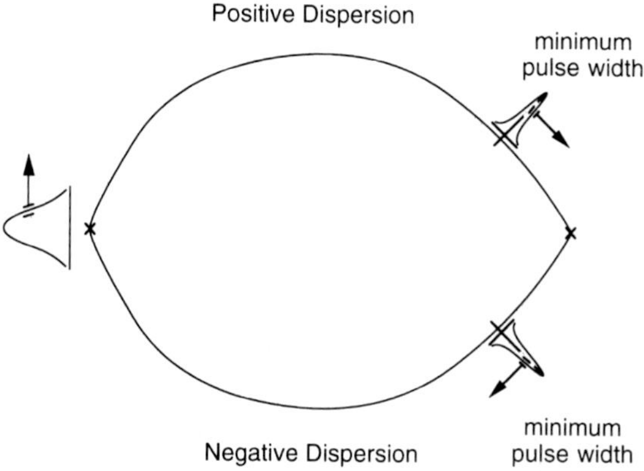

Figure 1.27. Location of minimum pulse width in a dispersion compensated ring.

minimum pulse width in the loop, because the average pulse profile depends upon it. The pulse polarization rotates across the pulse profile proportionally to the intensity averaged around the loop.

The greater the intensity dependent rotation at the polarizer, the larger is the APM effect. Hence the SAM coefficient is a maximum if the averaged pulse intensity is a maximum. This occurs when the pulse width is a minimum at the center of each of the fiber segments. Thus, this is the way the system will tend to adjust itself initially. As the pulse energy is increased by raising the gain, and the mechanisms that limit the peak pulse intensity come into play, the position of the minimum pulse width may well move to reduce the APM action.

It should be mentioned that the pulse spreading has a desirable side effect. Initially, with pulse width constant around the ring, a γ value is obtained, which can be large and thus is more likely to initiate self-starting. As the pulse intensity builds up, and the average pulse width in the cavity is larger than the minimum pulse width, the effective value of γ decreases. This is desirable, since it was pointed out earlier that a value of $\gamma|A_o|^2 > 0.6\,\pi$ leads to pulse break-up. Thus a system using SP-APM can both be self-starting, and can lead to higher pulse energies and shorter pulses by reducing γ as its pulse energy increases. Since the figure eight fiber laser also suffers from excess nonlinearity, the SP-APM principle could also help in providing shorter and more energetic pulses from that laser.

There was a tendency in the literature to explain the operation of all-fiber lasers in terms of solitons and soliton theory. Stretched pulse–additive pulse modelocking is achieved when the pulses are far from soliton-like as they pass around the system. They may behave like solitons only on the average[26]. The new thinking triggered by this approach should be helpful in designing new short-pulse all-fiber lasers.

1.10 Performance of some APM and KLM systems

Rapid progress has been made in recent years in the generation of even shorter pulses using the APM and KLM principles. Tables 1.1 and 1.2 give some recent advances in the generation of short pulses. Consider first Table 1.1 which describes results with APM systems. The field was led by the short pulse generation in titanium sapphire lasers, first at St. Andrews, with some significant results also obtained at MIT. The relatively narrowband Nd:YAG and Nd:YLF systems have been modelocked to give pulses with spectra that are as broad as the bandwidth of the gain medium. The Vienna results in Nd:glass were first achieved in bulk, and then continued by Fermann at Bellcore resulting in 100 fs pulses in an erbium-doped fiber laser amplifier with external

Table 1.1. *APM in solid state lasers*

Laser	Wavelength	Performance	Group
Ti:Al_2O_3	670–1000 nm	60 fs	MIT, St. Andrews
Nd:YAG			
(lamp pumped)	1.06 μm	6 ps	MIT
	1.32 μm	6 pf	MIT
(diode pumped)	1.06 μm	1.7 ps	MIT
Nd:YLF			
(lamp pumped)	1.053 μm	3.7 ps (7W)	UCLA
(diode pumped)	1.053 μm	1.5 ps	Strathclyde, MIT
Cr:forsterite	1.23 μm	150 fs	Cornell
NaCl:OH	1.5–1.6 μm	75 fs	Cornell, AT&T
KCl:Tl	1.45–1.55 μm	80 fs	St. Andrews, MIT
Nd:glass	1.054 μm	42 fs	T.U. Vienna
Er:glass	1.56 μm	300 fs	Southampton
		100 fs	NTT
		90 fs	NRL
		84 fs	Bellcore
		75 fs	MIT
Nd:LMA	1.054 μm	600 fs	Southampton

Table 1.2. *KLM in solid state lasers*

Laser	Wavelength	Performance	Group
Ti:Al_2O_3	670–1000 nm	60 fs	St. Andrews (1991)
			Coherent (1991)
		9 fs	U. Washington (1993)
Nd:YAG			
diode pumped	1.053 μm	6 ps	Strathclyde (1991)
Cr:$LiCaAlF_6$	790–830 nm	170 fs	U. Florida (1992)
	(720–840) nm		
Cr:$LiSrAlF_6$	800–920 nm	50 fs	St. Andrews (1992)
			Imperial Coll. (1992)
Cr:Mg_2SiO_4	1230–1280 nm	60 fs	City College NY (1992)
			Cornell (1992)

bulk elements[27]. The record in an all-fiber laser is held, at the moment, by MIT at 75 fs with an all-fiber unidirectional ring laser at 1.5 μm.

Pulses much shorter than those obtained by APM have been obtained by KLM. The work was initiated by Sibbett's group at St. Andrews, perfected by Coherent, and recently, at the Washington State University and at the Technical University of Vienna, has culminated in the shortest pulses ever obtained from a laser, 11 fs. The KLM system does not need stabilization, is much more rugged, and this is the main reason for the ability of the experimenter to fine-tune the system to peak performance.

One compact system producing KLM is the microdot mirror laser developed by Fujimoto's group at MIT (see Figure 1.28). The end mirror backs up against a Kerr medium. The mirror has reflecting dots of different radii. The KLM action takes place in the Kerr medium and is adjustable by a proper choice of the reflecting dot. Half of the mirror

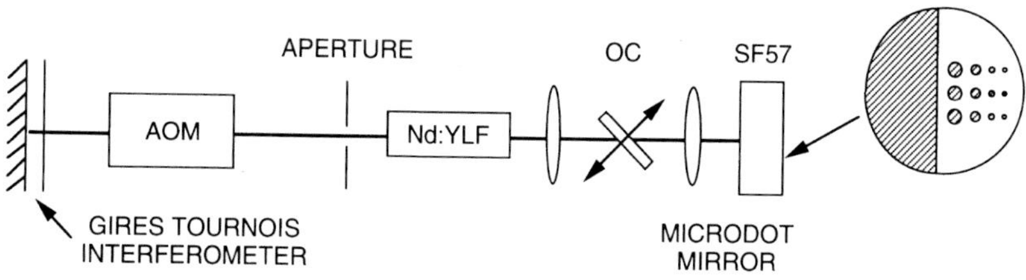

Figure 1.28. The microdot-mirror laser. Microdot mirror: saturable absorber + SPM; Gires Tournois: negative dispersion; AOM: starting mechanism.

surface is fully coated. The laser is aligned and started using this surface, until the mirror is translated into the position of the appropriate reflecting dot.

1.11 Third-order dispersion and its effect on pulse width

Recent experiments with very short pulses have run into the problem of third order dispersion, the effects of the third derivative of the propagation constant, k''', in the components of the system. The prism pairs can also be a source of third order dispersion. We start with (1.42) and introduce a third order dispersion term:

$$\frac{\partial a}{\partial z} = \left[g\left(1 + \frac{1}{\Omega_g^2}\frac{\partial^2}{\partial t^2}\right) + jD\frac{\partial^2}{\partial t^2} - \ell + (\gamma - j\delta)|a|^2 \right] a \\ + \frac{1}{3!}k''' L_d \frac{\partial^3}{\partial t^3} a = 0 \tag{1.97}$$

where L_d is the length of the medium with third order dispersion. This equation can be used as the starting point of a numerical analysis[28].

It is found in the numerical analysis that the pulses can be unstable if the gain bandwidth limiting is not strong enough. To allow for greater variation of the gain bandwidth, the pulses are stabilized in the numerical treatment by introducing saturation of the self-amplitude modulation $\gamma|a|^2$. This term is replaced by $\gamma_3|a|^2 - \gamma_5|a|^4$. Physically, this is justified, because any mechanism producing saturation limits in its own right. It is not easy to find analytic solutions for this generalized equation. In the analysis presented here, the term $\gamma_5|a|^4$ is treated as a perturbation.

One may make some progress analytically[28]. One assumes that the pulse is soliton-like with zero chirp, in the absence of third order dispersion. This is the state an experimenter would want to realize for the generation of chirp-free pulses. This happens when:

$$\frac{2g}{\Omega_g^2 \tau^2} = \gamma|A_o|^2 \tag{1.98}$$

and the pulse is of the form:

$$a(t) = A_o \operatorname{sech}\left(\frac{t}{\tau}\right) \tag{1.99}$$

Next we introduce the third order dispersion as a perturbation and suppose that its most important effect is to introduce an additional chirp of time dependent phase $\phi(t)$, and a time shift δT. If one retains only first order derivatives of $\phi(t)$, one obtains an equation for δT and the derivative of $\phi(t)$:

$$\frac{2|D|}{\tau^2}\tau\dot{\phi} + \frac{k''' L_\mathrm{d}}{3!\tau^3}\left[1 - 6\,\mathrm{sech}^2\left(\frac{t}{\tau}\right)\right] + \frac{\delta T}{\tau} = 0 \tag{1.100}$$

Integration of (1.100) gives a phase that is a superposition of t and $\tanh(t/\tau)$; the phase is antisymmetric, as opposed to the symmetric phase versus time produced by second order dispersion. The equation does not give a value for δT; δT simply changes the chirp. In order to pin down the timing shift one makes the reasonable assumption that δT adjusts itself so as to minimize the loss of the pulse as it passes through the elements in the resonator. For this purpose one considers the energy balance equation. It follows from the steady state version of (1.97) by multiplying it by a^*, adding the complex conjugate and integrating over the pulse. The third order dispersion term and the time shift do not contribute directly to the energy balance. Using the solution of (1.99) one finds:

$$(g-\ell)\int |a|^2 \mathrm{d}t - \frac{g}{\Omega_\mathrm{g}^2}\int \left|\frac{\mathrm{d}a}{\mathrm{d}t}\right|^2 \mathrm{d}t + \int (\gamma_3|a|^4 - \gamma_5|a|^6)\mathrm{d}t = 0 \tag{1.101}$$

One chooses δT to minimize the loss due to filtering. Then δT is related to $k''' L_\mathrm{d}/\tau^2$ and one finds:

$$\frac{g}{\Omega_\mathrm{g}^2}\int \left|\frac{\mathrm{d}a}{\mathrm{d}t}\right|^2 \mathrm{d}t = \frac{g}{\Omega_\mathrm{g}^2\tau^2} W \frac{1}{3}\left[1 + \frac{1}{15}\frac{gB^2}{\Omega_\mathrm{g}^2\tau^2}\right] \tag{1.102}$$

where

$$B = \frac{k''' L_\mathrm{d}\Omega_\mathrm{g}}{|D|\sqrt{g}} \tag{1.103}$$

and W is the energy of the pulse, $W = 2\tau|A_\mathrm{o}|^2$. The B parameter increases the loss because the time dependent phase $\phi(t)$ caused by third-order dispersion broadens the spectrum. Finally, introducing (1.102) into (1.101), one obtains an equation for the pulse width:

$$(g-\ell) - \frac{g}{\Omega_\mathrm{g}\tau^2}\frac{1}{3}\left[1 + \frac{1}{15}\frac{gB^2}{\Omega_\mathrm{g}\tau^2}\right] + \frac{1}{3}\gamma_3\frac{W}{\tau} - \frac{2}{15}\gamma_5\frac{W^2}{\tau^2} = 0 \tag{1.104}$$

The net gain is reduced by the presence of the phase factor $\exp[j\phi(t)]$, because it broadens the spectrum of a pulse of given pulse width τ and is filtered by bandwidth limiting. Hence, third order dispersion is harmful and will tend to produce longer pulses by lowering the peak intensity of the pulses passing the saturable absorber. Figure 1.29 shows a plot labeled 'analytic' which is based on equation (1.104). In plotting it we have used the gain and pulse energy obtained from a computer simulation for every value of $k'''L_d$. The simulation gives results that are close to the analytic results, when the gain bandwidth is narrow. With widening bandwidth the analytic theory and the simulation start deviating from each other. This is due to the fact that the pulses shed radiation into the continuum, somewhat analogously to the continuum produced during soliton evolution, as shown in Figure 1.11. This radiation into the continuum is harmful.

The evidence of radiation into the continuum is shown in Figure 1.30, which shows a computer simulation in the case when the analytic theory does not predict the pulse width correctly. One sees the emergence of a sideband. The sideband is located at a frequency at which the phase velocity of the continuum is matched to the phase velocity of the (nonlinear) soliton-like pulse.

The spectral sideband manifests itself in the 'wiggles' in the temporal envelope of the pulse, Figure 1.31. These effects would escape observation, since the autocorrelation function does not show anomalies, as shown in Figure 1.32.

Given the importance attached to third order dispersion, it is important to be able to determine it in any particular system. W. Knox[29]

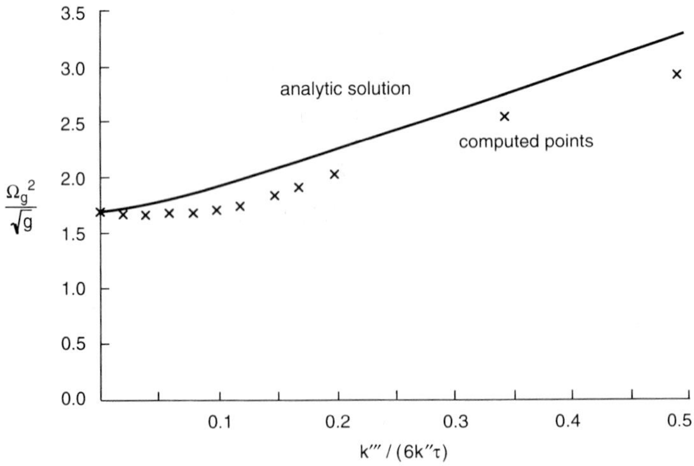

Figure 1.29. The effect of third order dispersion.

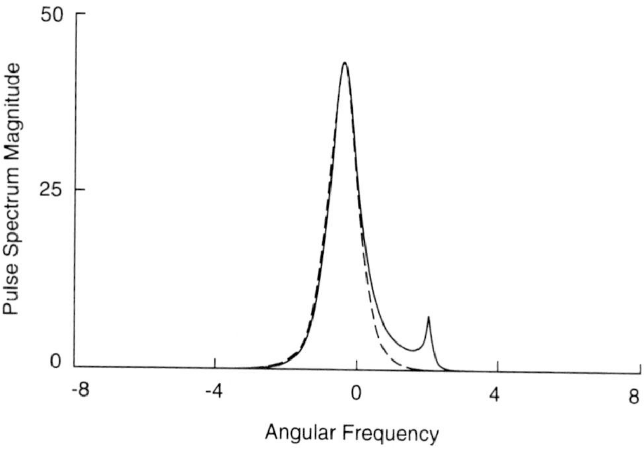

Figure 1.30. Spectrum of a pulse in the presence of third order dispersion.

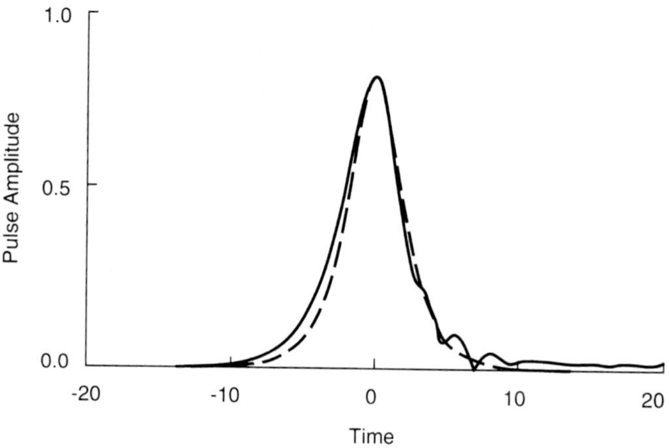

Figure 1.31. The temporal profile of the pulse.

developed an ingenious way to measure it 'in situ,' by tuning a mode-locked Ti:sapphire laser and recording the round-trip frequency of the pulses. The pulse widths were in the subpicosecond or picosecond range so that their spectrum was still narrow compared with the total band-width covered. The resulting curve can be used to plot group velocity versus frequency and to determine the dispersion characteristic. The master equation shows that the round trip is unaffected by the

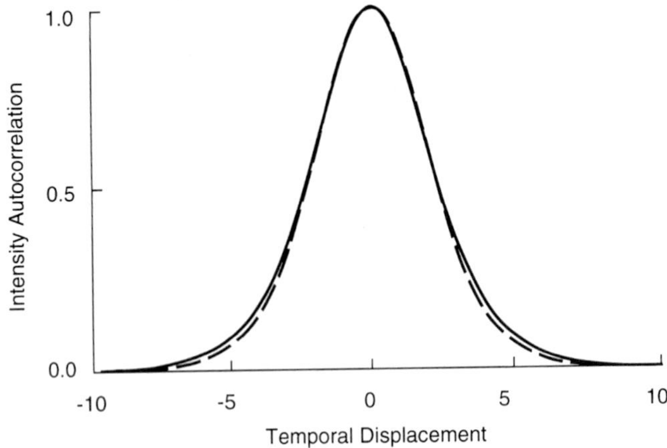

Figure 1.32. The autocorrelation function of the pulse.

(effective) fast saturable absorber action and by the self-phase modulation. Hence, the group velocity versus frequency curve obtained in this way contains only the effects of the linear resonator components in the resonator. It is surprising that this method had not been used before, but there is good reason for it. When conventional saturable absorbers were used to modelock, the laser frequency could not be tuned over a wide range, given the spectral response of the saturable absorber. KLM and APM have an equivalent saturable absorber action that is broadband, and hence modelocking can be realized over practically the full laser bandwidth.

1.12 Discussion

We have reviewed in some detail the generation of short optical pulses by modelocking. The word 'modelocking' implies the locking of modes, and this is indeed the picture formed in the frequency domain. Passive modelocking, which produces the shortest pulses, is viewed more conveniently in the time domain in which the modelocking mechanism does not show up clearly.

We have developed a master equation of modelocking which, eventually, incorporated group velocity dispersion and Kerr phase shift. The solution of the master equation gives chirped secant hyperbolic pulses. The chirp can be removed by a proper balance of the Kerr effect and negative group velocity dispersion. In open resonators, the negative group velocity dispersion can be achieved by prism pairs.

A special form of the master equation is the nonlinear Schrödinger equation whose solutions are solitons. The operation of the so-called soliton laser can be explained by formation of a higher order soliton in the auxiliary cavity, which when returned foreshortened, can reduce the pulse width of the pulse in the main cavity. This is not the most general pulse shortening mechanism, because soliton formation is not required for the generalized passive modelocking described by the acronyms CCM and APM. Once this was recognized, systems were modelocked for which negative dispersion in the Kerr medium was not readily available. Solid state lasers like the Ti:sapphire laser are modelocked more effectively with Kerr lens modelocking, which does not require interferometric stabilization of two cavities. However, the APM principle can also be realized in single cavity systems with self-stabilized nonlinear interferometric schemes. All-fiber lasers are good candidates for this modelocking scheme.

The master equation which provided the theme for this chapter was developed under the assumption of a small change per pass of the pulse. In so far as nonlinear changes (Kerr phase shifts) of the pulse in the main cavity have to be kept small, if spectral distortions are to be avoided, this is an acceptable assumption. Linear changes via dispersion, gain or loss need not be small for the master equation to be valid. The nonlinear modulation coefficients that enter the master equation have to be evaluated properly by taking into account the pulse shape changes through the resonator.

The generation of ultrashort pulses encounters several difficulties. If their energy is to be maintained at a reasonable level, the nonlinearities in the resonator can become excessive. A large nonlinear phase shift per pass leads to shedding of the pulse energy into the 'continuum' and hence to deterioration of the performance. Further, for a SAM coefficient sufficiently large to have the system start to modelock, the SAM coefficient 'saturates' when the peak intensity becomes excessive. Higher order dispersion starts to affect very short pulses. The third order dispersion effect can be countered to some extent by direct compensation, e.g. a Gires–Tournois interferometer[30]. When the pulses are ultrashort, the excessive peak intensity in the resonator can be combatted in part by the stretched pulse technique. This technique is particularly important in fiber lasers, in which the Kerr nonlinearity is large. However, it should prove useful also in bulk-component KLM systems.

Acknowledgments

This work was supported in part by the Joint Services Electronics Program Contract DAAH04-95-1-0038, the Air Force Office of Scientific Research Contract F49620-95-1-0221, and the National Science Foundation Grant ECS-94-23737.

Appendix I: The polarization transformation by isotropic Kerr medium

In this chapter we use extensively the constitutive relations for the Kerr effect in isotropic media. The Kerr tensor is a fourth rank tensor that has to obey certain symmetry relations related to the symmetry group of the crystal lattice. The elasticity tensor of an isotropic medium, which relates strain and stress, has two free parameters: Young's modulus and the Poisson ratio. On the other hand, the Kerr tensor relates three vectors, three factors of the electric field, to a vector, the polarization. Hence it is not *a priori* obvious whether it has the same number of degrees of freedom as the elasticity tensor. It seems appropriate to derive the relation between polarization and electric field for the special case of an isotropic Kerr medium with instantaneous response, in particular because the derivation presented here is a simple one. We limit the analysis to elliptic polarization of the electric field in the $x - y$ plane, because this is the important case within the paraxial wave propagation.

If the Kerr medium is isotropic and responds instantaneously, then the polarization density $P(t)$ is related to the electric field $E(t)$ by

$$P(t) = \epsilon_o \chi E^3(t) \tag{1.105}$$

and points in the direction of E at every instant of time; χ is related to the third order susceptibility.

Linear polarization basis

Consider first a linear polarization of the electric field $E(t)$, with sinusoidal time dependence, of amplitude A.

$$E_x(t) = \frac{1}{2}[Ae^{j\omega t} + A^* e^{-j\omega t}] \tag{1.106}$$

Then $P(t)$ has both a fundamental and third harmonic contribution. The complex amplitude of $P(t), P_x$, of the response at the fundamental frequency is:

$$P_x = \frac{3}{4}\epsilon_o \chi |A|^2 A \tag{1.107}$$

Consider next a pure circular polarization with $|E_x| = |E_y| = |A|$. In an isotropic medium, the polarization vector points along the electric field vector and rotates with it at the fundamental frequency. No third harmonic is generated. The magnitude of the complex polarization amplitude is:

$$|P| = \epsilon_o \chi |A|^3 \tag{1.108}$$

Circular polarization basis of general polarization

Next, we turn to a circular polarization basis and write down the general constitutive law for elliptic polarization of the electric field in the $x - y$ plane, made up of two counterrotating circular polarizations E_+ and E_-. We shall first prove that the relation between the polarization and the electric field at the fundamental frequency, expressed in the circular polarization basis, must be of the restricted form:

$$P_+ = \epsilon_o \{\alpha |E_+|^2 + \beta |E_-|^2\} E_+ \tag{1.109}$$

$$P_- = \epsilon_o \{\alpha |E_-|^2 + \beta |E_+|^2\} E_- \tag{1.110}$$

This form of the equation is dictated in part by the fact that the complex field amplitudes must appear squared, and multiplied by a complex conjugate, so as to lead to the proper time dependence $\exp(j\omega t)$. If this were the only constraint, then (1.109) and (1.110) would also contain a 'coherence' term, e.g. of the form $E_-^2 E_+^*$. No such coherence term has been added, because it is ruled out by the following reasoning: consider as a special case a linear polarization for which $|E_+| = |E_-|$ and for which $\arg[E_-/E_+]$ gives the orientation of the E-field in the $x - y$ plane. If a coherence term appeared in (1.109), then the polarization rotation would depend upon the orientation of the E-field. The medium being isotropic, no such dependence can occur. Hence, we have proven that, in the circular polarization basis, the constitutive law relating the fundamental frequency components of E and P must be of the form (1.109) and (1.110).

Next, let us relate the coefficients α and β to χ by considering the special cases of linear and circular polarization of the E-field and by comparing the answers with (1.107) and (1.108). For linear polarization, $E_+ = E_- = A/\sqrt{2}$, and $P_+ = P_- = P/\sqrt{2}$. Using this fact we find:

$$|P_\pm| = \epsilon_o \frac{\alpha + \beta}{2} \frac{1}{\sqrt{2}} |A|^3 = \frac{1}{\sqrt{2}} |P_x| \tag{1.111}$$

Comparing (1.111) with (1.107) we find:

$$\frac{\alpha + \beta}{2} = \frac{3\chi}{4} \tag{1.112}$$

Next consider circular polarization. If the magnitude of the field is $|A|$, then the magnitude of E_+ is:

$$|E_+| = \sqrt{2}A \tag{1.113}$$

and from (1.109) the magnitude $|P|$ of P_+ is:

$$|P| = \frac{1}{\sqrt{2}} |P_+| = \epsilon_o \chi |A|^3 = 2\epsilon_o \alpha |A|^3 \tag{1.114}$$

Comparison with (1.108) gives

$$\alpha = \frac{\chi}{2} \tag{1.115}$$

and with (1.112)

$$\beta = \chi \tag{1.116}$$

Finally, let us relate χ to the Kerr coefficient n_2. A linear polarization produces the index change

$$\Delta n = n_2 I \tag{1.117}$$

which, written in terms of ϵ, is:

$$\frac{P_x}{E_x} = \frac{\Delta\epsilon}{\epsilon_o} = 2n\Delta n = \frac{3}{4}\chi|E|^2 = 2nn_2 I \tag{1.118}$$

If $|E|^2$ is normalized to equal intensity, then

$$\chi = \frac{8}{3} nn_2 \tag{1.119}$$

The polarization $P_\pm$ of (1.109) and (1.110) can be written

$$P_\pm = \epsilon_o \frac{4}{3} nn_2 \{|E_+|^2 + 2|E_-|^2\} \tag{1.120}$$

and the index change is accordingly:

$$\Delta n_{\pm} = \frac{2}{3} n_2 \{|E_+|^2 + 2|E_-|^2\} \tag{1.121}$$

Appendix II: The γ-parameter analysis of the Kerr medium

Consider the propagation of a beam through a short section of space of length $\mathrm{d}z$, containing a weak lens of (differential) strength $\mathrm{d}(1/f)$. The ABCD matrix for this section is:

$$\begin{bmatrix} A & B \\ C & D \end{bmatrix} = \begin{bmatrix} 1 & \mathrm{d}z \\ -\mathrm{d}(\frac{1}{f}) & 1 \end{bmatrix} \tag{1.122}$$

The transformation of the q parameter by this matrix is:

$$q + \mathrm{d}q = \frac{q + \mathrm{d}z}{1 - \mathrm{d}(\frac{1}{f})q} = q + \mathrm{d}z + q^2 \mathrm{d}\left(\frac{1}{f}\right) \tag{1.123}$$

leading to the differential equation for $1/q$:

$$-\frac{\mathrm{d}}{\mathrm{d}z}\left(\frac{1}{q}\right) = \frac{1}{q^2} + \frac{\mathrm{d}}{\mathrm{d}z}\left(\frac{1}{f}\right) \tag{1.124}$$

We are now ready to introduce the Kerr medium and compute the focal strength of the medium of length $\mathrm{d}z$. The intensity distribution of the beam, $A_o^2 e^{-2r^2/w^2}$, is so normalized that the power P is equal to the integral of the square of the intensity:

$$P = \int 2\pi r \mathrm{d}r A_o^2 e^{-2r^2/w^2} = \frac{\pi w^2}{2} A_o^2 \tag{1.125}$$

The Kerr medium causes a phase delay via the nonlinear optical coefficient n_2 that is radial coordinate dependent

$$\Phi = \frac{2\pi}{\lambda} n_2 A_o^2 e^{-2r^2/w^2} \mathrm{d}z \simeq \frac{2\pi}{\lambda} n_2 \frac{2P}{\pi w^2}\left(1 - \frac{2r^2}{w^2}\right)\mathrm{d}z \tag{1.126}$$

The relative phase advance is

$$\Delta\Phi_a = \frac{2\pi}{\lambda} n_2 \frac{2P}{\pi} \frac{2r^2}{w^4} \mathrm{d}z = \frac{8P}{\lambda} n_2 \left(\frac{\pi}{\lambda}\right)^2 \mathrm{d}z \left[\mathrm{Im}\left(\frac{1}{q}\right)\right]^2 r^2 \tag{1.127}$$

where we have introduced the relation between the beam radius squared and $\mathrm{Im}(1/q)$. Since the inverse radius of curvature of the phase front, $1/R$, is related to the parabolic dependence on radius of the phase:

$$\frac{kr^2}{2R} = \Delta\Phi_{\rm a} \tag{1.128}$$

with $k = 2\pi/\lambda$, we find for the focal strength of the Kerr medium of length dz:

$$\frac{1}{R} = \mathrm{d}\left(\frac{1}{f}\right) = \frac{8P}{\pi}\left(\frac{\pi}{\lambda}\right)^2 n_2 \mathrm{d}z\left[\mathrm{Im}\left(\frac{1}{q}\right)\right]^2 \tag{1.129}$$

We may now enter the dependence of the focal strength d$(1/f)$ upon the q parameter into (1.124) and obtain:

$$-\frac{\mathrm{d}}{\mathrm{d}z}\left(\frac{1}{q}\right) = \frac{1}{q^2} + 2K\left[\mathrm{Im}\left(\frac{1}{q}\right)\right]^2 \tag{1.130}$$

where we have defined the dimensionless Kerr parameter

$$2K = \frac{8P}{\pi}\left(\frac{\pi}{\lambda}\right)^2 n_2 \tag{1.131}$$

By separating (1.130) into real and imaginary parts one obtains:

$$-\frac{\mathrm{d}}{\mathrm{d}z}\mathrm{Re}\left(\frac{1}{q}\right) = \left[\mathrm{Re}\left(\frac{1}{q}\right)\right]^2 - \left[\mathrm{Im}\left(\frac{1}{q}\right)\right]^2(1-2K) \tag{1.132}$$

$$-\frac{\mathrm{d}}{\mathrm{d}z}\mathrm{Im}\left(\frac{1}{q}\right) = 2\mathrm{Re}\left(\frac{1}{q}\right)\mathrm{Im}\left(\frac{1}{q}\right) \tag{1.133}$$

We introduce the renormalized q parameter q':

$$\frac{1}{q'} = \mathrm{Re}\left(\frac{1}{q}\right) + j\mathrm{Im}\left(\frac{1}{q}\right)(1-2K)^{1/2} \tag{1.134}$$

In terms of this parameter, (1.132) and (1.133) can be recombined into one single complex equation

$$-\frac{\mathrm{d}}{\mathrm{d}z}\left(\frac{1}{q'}\right) = \frac{1}{q'^2} \tag{1.135}$$

which is identical with the free space propagation equation. Thus, a problem involving self-focusing can be analyzed like free space propagation; the effect of the nonlinearity is fully contained in the renormalizations at the input and output ends.

References

[1] R. L. Fork, C. H. Brito Cruz, P. C. Becker and C. V. Shank, 'Compression of optical pulses to six femtoseconds by using cubic phase compensation,' *Opt. Lett.* **12**, 483, July 1987.

[2] P. T. Ho, L. A. Glasser, E. P. Ippen and H. A. Haus, 'Picosecond pulse generation with a cw GaAlAs laser diode,' *Appl. Phys. Lett.* **33**, 241–2, 1978.

[3] L. F. Mollenauer and R. H. Stolen, 'The soliton laser,' *Opt. Lett.* **9**, 13, 1984.

[4] K. J. Blow and D. Wood, 'Mode-locked lasers with nonlinear external cavities,' *J. Opt. Soc. Am. B* **5**, 629, 1988.

[5] K. J. Blow and B. P. Nelson, 'Improved mode locking of an F-center laser with a nonlinear nonsoliton external cavity,' *Opt. Lett.* **13**, 1026, 1988; P. N. Kean, X. Zhu, D. W. Crust, R. S. Grant, N. Langford and W. Sibbett, 'Enhanced mode locking of color-center lasers,' *Opt. Lett.* **14**, 39, 1989.

[6] J. Mark, L. Y. Liu, K. L. Hall, H. A. Haus and E. P. Ippen, 'Femtosecond pulse generation in a laser with a nonlinear external resonator,' *Opt. Lett.* **14**, 48, 1989.

[7] E. P. Ippen, H. A. Haus and L. Y. Liu, 'Additive pulse mode locking,' *J. Opt. Soc. Am. B* **6**, 1736, 1989.

[8] M. T. Asaki, C. P. Huang, D. Garvey, J. Zhou, H. C. Kapteyn and M. M. Murnane, 'Generation of 11 femtosecond pulses from a self mode-locked Ti:sapphire laser,' *Opt. Lett.* **18**, 977, 1993.

[9] U. Keller, G. W. 't Hooft, W. H. Knox and J. E. Cunningham, 'Femtosecond pulses from a continuously self-starting passively mode-locked Ti:sapphire laser,' *Opt. Lett.* **16**, 42, 1991; D. K. Negus, L. Spinelli, N. Boldblatt and G. Feuget, 'Sub-100 fs pulse generation by Kerr lens modelocking in Ti:Al_2O_3,' in *Tech. Dig. Topic. Meet. Advan. Solid State Lasers*, Washington, DC, Optical Society of America, 1991, postdeadline paper.

[10] F. Salin, J. Squier and M. Piché, 'Mode locking of Ti:Al_2O_3 lasers and self-focusing: a Gaussian approximation,' *Opt. Lett.* **16**, 1674–6, November 1991.

[11] H. A. Haus, 'A theory of forced mode locking,' *IEEE J. Quant. Electron.* **QE-11**, 323–30, July 1975.

[12] H. A. Haus, 'Theory of mode locking with a fast saturable absorber,' *J. Appl. Phys.* **46**, 3049–3058, July 1975.

[13] H. A. Haus, 'Theory of modelocking with a slow saturable absorber,' *IEEE J. Quant. Electron.* **QE-10**, 736–46, Sept. 1975.

[14] G. H. C. New, 'Pulse evolution in mode-locked quasi-continuous lasers,' *IEEE J. Quant. Electron.* **QE-10**, 115-24, Feb. 1974.

[15] R. L. Fork, O. E. Martinez and J. P. Gordon, 'Negative dispersion using pairs of prisms,' *Opt. Lett.* **9**, 150, 1984.

[16] R. H. Stolen and C. Lin, 'Self-phase-modulation in silica optical fibers,' *Phys. Rev. A* **17**, 1448-53, April 1978.

[17] H. A. Haus and M. N. Islam, 'Theory of the soliton laser,' *IEEE J. Quant. Electron.* **QE-21**, 1172–88, Aug. 1985.

[18] H. A. Haus, J. G. Fujimoto and E. P. Ippen, 'Structures for additive pulse mode locking,' *J. Opt. Soc. Am. B* **8**, 2068–76, Oct. 1991.

[19] F. I. Khatri, G. Lenz, J. D. Moores, H. A. Haus and E. P. Ippen, 'Extension of coupled-cavity additive pulse mode-locked laser theory,' *Opt. Commun.* **110**, 131–6, 1994.

[20] H. A. Haus, J. G. Fujimoto and E. P. Ippen, 'Analytic theory of additive pulse and Kerr lens mode locking,' *IEEE J. Quant. Electron.* **28**, 2086–96, Oct. 1992.

[21] I. N. Duling, 'All-fiber modelocked figure eight laser,' in *Optical Society of America 1990 Annual Meeting*, Vol. 5 of OSA 1990 Technical Digest Series (Optical Society of America, Washington, DC, 1990), pp. 306–10.

[22] L. F. Mollenauer, in *Digest of Conference on Optical Fiber Communication* (Optical Society of America, Washington, DC, 1991), p. 200.

[23] K. Tamura, H. A. Haus and E. P. Ippen, 'Self-starting additive pulse mode-locked erbium fibre ring laser,' *Electron. Lett.* **28**, 2226–7, Nov. 1992.

[24] H. A. Haus and E. P. Ippen, 'Self-starting of passively mode-locked lasers,' *Opt. Lett.* **16**, 1331–3, Sept. 1991.

[25] K. Tamura, J. Jacobson, E. P. Ippen, H. A. Haus and J. G. Fujimoto, 'Unidirectional ring resonators for self-starting passively mode-locked lasers,' *Opt. Lett.* **18**, 220–2, Feb. 1993.

[26] L. F. Mollenauer, S. G. Evangelides and H. A. Haus, 'Long-distance soliton propagation using lumped amplifiers and dispersion shifted fiber,' *J. Lightwave Tech.* **9**, 194v–7, Feb. 1991.

[27] M. E. Fermann, M. J. Andrejco, M. L. Stock, Y. Silberberg and A. M. Weiner, 'Passive mode locking in erbium fiber lasers with negative group delay,' *Appl. Phys. Lett.* **62**, 910, Mar. 1993.

[28] H. A. Haus, J. D. Moores and L. E. Nelson, 'Effect of third-order dispersion on passive mode locking,' *Opt. Lett.* **18**, 51–3, Jan. 1993.

[29] W. H. Knox, '*In situ* measurement of complete intracavity dispersion in an operating Ti:sapphire femtosecond laser,' *Opt. Lett.* **17**, 514–16, Apr. 1992.

[30] J. M. Jacobson, K. Naganuma, H. A. Haus and J. G. Fujimoto, 'Femtosecond pulse generation in a Ti:Al_2O_3 laser using second- and third-order intracavity dispersion,' *Opt. Lett.* **17**, 139, 1992.

2

Passive modelocking in solid state lasers

THOMAS BRABEC, STEPHEN M. J. KELLY
AND FERENC KRAUSZ

Introduction

The appearance of broadband solid state laser media has, within the last few years, generated considerable interest in the field of ultrashort pulse generation. The broad fluorescence emission (Moulton, 1986; Petricevic *et al.*, 1989; Borodin *et al.*, 1992; Smith *et al.*, 1992) of novel transition-metal-doped laser crystals (Figure 2.1) can potentially support light pulses of a few femtoseconds in duration. To utilize these bandwidths for femtosecond pulse generation it has proved essential to develop new modelocking techniques. The traditional approach to passive modelocking of solid state lasers was to use organic dye saturable absorbers just as in dye lasers. Slow saturable absorbers work very efficiently in dye systems owing to the contribution of gain saturation to the pulse formation (New, 1974). However, this scheme imposes a severe constraint on the resonator design since the cavity round-trip time is

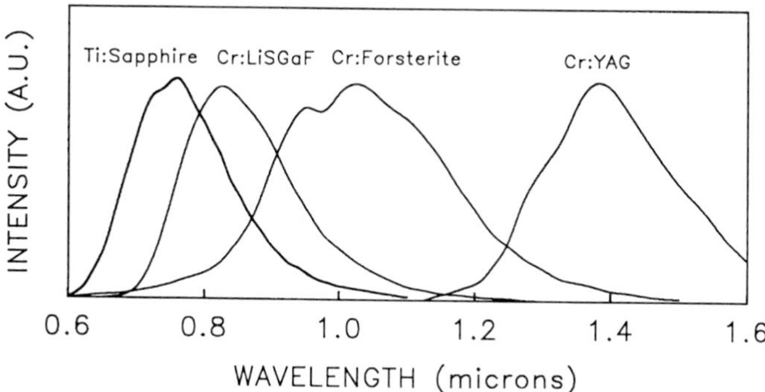

Figure 2.1. Fluorescence emission spectra of some transition metal-doped broadly tunable near infrared solid state materials.

required to be comparable with the upper state lifetime of the laser medium (Haus, 1975a). The μs relaxation time of current solid state gain media does not permit practical systems to meet this requirement, and hence resonant nonlinearities are not capable of generating femtosecond pulses (< 100 fs) in solid state systems.

A breakthrough in solid state ultrafast technology was achieved by coupling a nonlinear auxiliary resonator containing an optical fiber to the main laser cavity. In contrast with dye saturable absorbers this concept allows the realisation of an artificial absorber with an extremely fast response time and broad bandwidth due to the properties of the fiber nonlinearity. The first realisation of this concept was Mollenauer and Stolen's soliton laser (Mollenauer and Stolen, 1984). In several theoretical works (Ouelette and Piché, 1986; Blow and Wood, 1988; Ippen *et al.*, 1989) it was then clarified that subpicosecond pulse formation achieved by this scheme is based on the coherent addition of a weak nonlinear phase-shifted pulse returning from the coupled cavity to the main cavity pulse. The technique has been called additive pulse modelocking (APM) (Ippen *et al.*, 1989; Mark *et al.*, 1989) coupled-cavity modelocking (Kean *et al.*, 1989) or interferential modelocking (Morin and Piché, 1990) and has also been utilized for modelocking in fiber lasers in a modified form (Hofer *et al.*, 1991).

Soon after the appearance of APM another saturable absorber scheme using the optical Kerr nonlinearity was developed. Whereas APM relies on the time-dependent phase shift obtained in the fiber, this modelocking technique is based on the transverse intensity variation of the beam profile due to a bulk Kerr nonlinearity. When combined with spatial aperturing, this approach provides an extremely simple means of ultrafast all-optical modulation. Spence *et al.* (1991) were the first to utilize this effect for modelocking, and Piché (1991) suggested self-focusing as the responsible modelocking mechanism. This novel scheme has been referred to as self-modelocking (Spence *et al.* 1991) or Kerr lens modelocking (KLM) (Spinelli *et al.*, 1991). Several authors have studied KLM numerically (Piché, 1991; Salin *et al.*, 1991; Chen and Wang, 1991) and analytically (Brabec *et al.*, 1992b, 1993; Haus *et al.*, 1992; Georgiev *et al.*, 1992).

In addition to fast saturable absorber-like pulse shaping another key feature distinguishes femtosecond solid state lasers from their dye forerunners: the modelocked pulses always experience strong self-phase modulation (SPM) arising from the intensity-dependent refractive index associated with the host material of the gain medium. Self-phase modulation can be converted into efficient pulse shortening in the presence of a net negative group delay dispersion (GDD) in the cavity. As the passive amplitude modulation introduced by APM or KLM is comparatively weak, the steady state pulse parameters in these systems are primarily determined by the interplay between GDD and SPM, often referred to as soliton-like shaping. However, the discrete interaction of SPM and negative GDD on its own is unstable (Brabec *et al.*, 1991), and self-amplitude modulation (SAM) as supplied by APM or KLM provides the necessary stabilizing action for the femtosecond pulse.

In this chapter transient and steady-state pulse shaping dynamics in passively modelocked femtosecond solid state lasers are addressed. In the first part, the buildup of a single pulse from mode-beating fluctuations in a free running laser is investigated (Krausz *et al.*, 1991, 1993) and possible mechanisms responsible for the existence of a self-starting threshold are discussed. This is of particular interest for solid state systems which are, in contrast with dye femtosecond lasers, in general not self-starting due to the relatively weak SAM action of APM or KLM. In the second part of this chapter steady state operation of femtosecond solitary lasers is analysed drawing on work in Brabec *et al.* (1991, 1992a); Brabec and Kelly (1993); Krausz *et al.* (1992); Kelly (1992); Spielmann *et al.* (1994). As indicated above, the steady-state in solid state femtosecond lasers is dominated by soliton-like pulse shaping. Solitary effects are taken into account using a generalized formalism which allows for a detailed analysis of the stationary state in femtosecond solid state lasers. In particular, steady-state pulse parameters are determined and stability issues of solitary systems are discussed. Our analysis provides insight into the basic processes and mechanisms which impose a limit on the shortest pulse widths obtainable from passively modelocked solid state oscillators.

2.1 Initial modelocked pulse formation

2.1.1 *General considerations*

In continuously-pumped laser oscillators the introduction of a suitable nonlinearity into the resonator can transform the randomly occurring variations in intensity and phase into a train of intense well-defined single short pulses. The method by which this is achieved is commonly referred to as modelocking and describes the situation where the individual modes of the free-running laser are locked. There are in principle two ways to obtain modelocking, active and passive.

Active modelocking implies the use of an external drive signal, such as from an acousto-optic or phase modulator, whereas in passive laser modelocking the amplitude and/or phase modulation is self-induced by the interaction of the circulating radiation with a suitable intracavity nonlinearity. In this section conditions for passive modelocking to start spontaneously from mode beating fluctuations, which have (recently) been found to occur with a free-running laser, are investigated.

An initial requirement of the nonlinear element is that it must create an amplitude instability so that an intense fluctuation experiences an increased net gain with respect to the less intensive parts of the intracavity radiation. A further requirement is this induced gain window should, ideally, be as short as the fluctuation itself so as to lock all initially oscillating modes. In lasers with an upper-state lifetime comparable with the cavity round-trip time (e.g. dye lasers), this requirement can be easily fulfilled by employing a 'slow' saturable absorber. The combined action of saturated absorption and saturated gain forms a gain window which automatically follows the intensity pulse shape and thereby provides efficient pulse shortening, until the pulse duration becomes comparable with the dephasing time (transverse relaxation time) of the absorber and/or amplifier, see Figure 2.2. The transverse relaxation times of excited states are usually well below 100 fs for condensed media (Weiner and Ippen, 1984), which allows femtosecond pulse generation in broadband lasers with short upper-state lifetimes (e.g. dye lasers (Fork *et al.*, 1981; Valdmanis *et al.*, 1985), and some color center lasers (Langford *et al.*, 1987)).

The situation is quite different in lasers having long relaxation times (e.g. most solid state lasers). Here, passive modelocking relies on a 'fast' saturable absorber mechanism to create a short gain window by itself. This is because dynamic gain saturation in these lasers is very weak, which implies a gain window limited by the finite excited-state lifetime (longitudinal relaxation time) of the absorber (dashed line in Figure 2.2(c)). Although a partial recovery in the loss may occur on a subpicosecond time scale in novel semiconductor absorbers (Cesar *et al.*, 1990; Soccolich *et al.*, 1990; Keller and Chiu, 1992), the relaxation times are still too long for sub-100 fs pulse generation and, as an additional disadvantage, the absorption cross-sections exhibit a wavelength-dependence. All optical modulators (or self-amplitude modulators) based on the nonresonant optical Kerr nonlinearity are free from these shortcomings. The response of these devices is almost instantaneous (< 10 fs) (Grudinin *et al.*, 1987) and independent of the carrier frequency over a

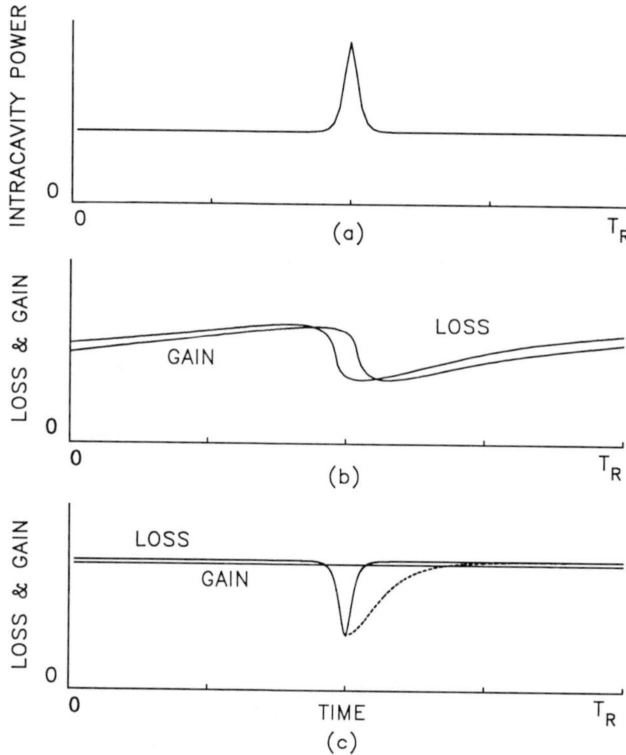

Figure 2.2. (a) CW intracavity radiation with a short pulse-like fluctuation. (b) Response of a fast-relaxation-time gain medium and slow saturable absorber to the fluctuation under quasi-steady-state conditions. (c) The same as in (b) with a long-relaxation-time gain medium and a saturable absorber of instantaneous response time; (dashed line, absorber with finite relaxation time).

wide wavelength range (solid line in Figure 2.2(c)). Consequently, efficient pulse shaping down to the sub-100 fs regime becomes possible for a wide range of lasers with long upper-state lifetimes.

With the fulfilment of the above requirements, the laser can favor pulsed (modelocked) operation over CW oscillation, however, this still does not ensure the free-running laser actually becomes modelocked. This can be understood by noting the initial mode-beating fluctuation(s), from which the modelocked pulse could evolve, have a finite 'lifetime' due to the finite mutual mode coherence time and uneven mode spacings in the free-running laser (Krausz *et al.*, 1991). This suggests the buildup of an intense pulse must be completed within the lifetime of mode-beating fluctuations.

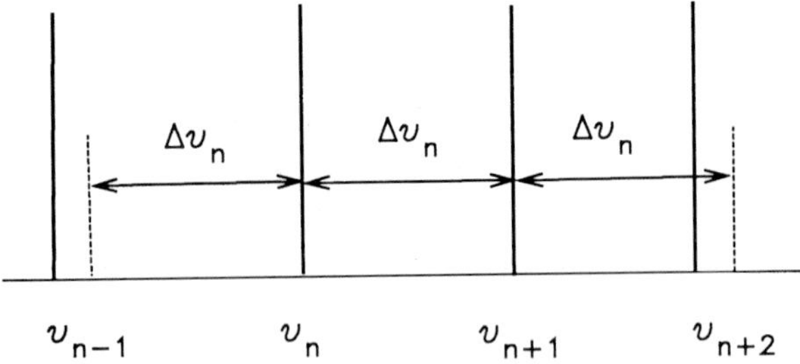

Figure 2.3. Start of the transient modelocking process in the frequency domain. The sidebands (dashed bars) are generated by a saturable absorber-like nonlinearity in the resonator. See text for further explanation.

Additional insight can be gained by a frequency domain analysis. Here the nonlinearity mixes adjacent axial cavity modes, e.g., n and $n+1$ in Figure 2.3, to generate a signal at the difference frequency $\Delta\nu_n = \nu_{n+1} - \nu_n$. This, in turn, modulates the modes themselves, creating sideband signals at $\nu_{s,n-1} = \nu_n - \Delta\nu_n$ and $\nu_{s,n+2} = \nu_{n+1} + \Delta\nu_n$ (dashed bars in Figure 2.3) which can injection lock the ν_{n-1} and ν_{n+2} modes, respectively. However, in general neither are the ν_i equal to $\nu_{s,i}$, nor are the corresponding spectral lines infinitely narrow, hence the sideband intensity is required to reach a minimum (threshold) level in order to injection lock the corresponding cavity mode (see e.g. Siegman, 1986). The sideband intensity is in turn proportional to the strength of self-amplitude modulation in the cavity and the deviation of $\nu_{s,i}$ from ν_i relates to the lifetime of mode-beating fluctuations and thus provides a link between the frequency and time domain. These frequency-domain considerations support the conclusions from the time-domain analysis. To formulate the start-up condition more precisely, the transient modelocking process is now investigated in more detail.

2.1.2 *Self-starting condition for passive modelocking*

We assume the electromagnetic field $E(t)$ circulating inside the resonator can be described by

$$E_k(t) = a_k(t)e^{i\omega_0 t}, \tag{2.1}$$

where the evolution of a_k is governed by the equation (Krausz *et al.*, 1991)

$$|a_{k+1}|^2 - |a_k|^2 = (g - l + q)|a_k|^2, \tag{2.2}$$

where $a_k = a_k(t)$ is the electric field amplitude of the electromagnetic wave at a fixed position in the cavity after the kth round trip. Here, the time parameter t takes values between 0 and T_r, where T_r stands for the cavity round-trip time. The electric field is normalized so that $|a_k|^2$ gives the power traveling in one direction. The parameters l and g are the round-trip power loss and gain, respectively, and q describes the effect of the nonlinearity. The parameters g and q may, in general, also be dependent on k and t. Because $q \ll l$ for the initial stage of the transient modelocking process, the change in g is relatively small and the above equation can be used irrespective of the origin of the longitudinal modes in the free-running laser (spatial or spectral hole-burning). As a consequence, Equation (2.2) allows a unified treatment of transient modelocking in homogeneously and inhomogeneously broadened lasers as long as the number of modes during the initial stage of buildup remains unchanged. During this stage, pulse evolution is not affected by gain dispersion or any other kind of bandwidth limitation since all the initially oscillating modes experience the same round-trip gain. As SAM is the principal driving force for the evolution of a single modelocked pulse out of a free-running (CW) oscillation, phase-affecting nonlinearities (e.g. Kerr-induced self-phase modulation) have been neglected in Equation (2.2).

The laser pulse will evolve from the most intense mode-beating fluctuation present in the multimode free-running laser. Neglecting the smaller fluctuations the intracavity power can be subdivided into two parts: $|a_k|^2 = P_k + \Delta P_k(t)$, where P_k is the power of the slowly varying average background radiation and $\Delta P_k(t)$ is the most intense mode-beating fluctuation in the cavity. We assume moderate intracavity powers and a fast saturable absorber action with an instantaneous response

$$q_k(t) = \gamma |a_k(t)|^2, \tag{2.3}$$

where γ is a characteristic of the nonlinear device giving the change in round-trip power gain per unit intracavity power.[1] This is always a good approximation for the initial stage of the modelocking process, where the mode-beating fluctuations have relatively low powers and long durations. For a fast saturable absorber and a gain medium with a long relaxation

[1] Note that in our previous work κ instead of γ has been chosen as a characteristic for SAM.

time, the evolution of the peak power p_k of the mode-beating perturbation $\Delta P_k(t)$ is governed by the simple equation

$$p_{k+1} - p_k = B_k p_k^2 - \frac{T_r}{T_c} p_k, \tag{2.4}$$

where

$$B_k = \gamma - \alpha_k \frac{g_{0,k}}{2P_s} \frac{\tau_k}{T_g}. \tag{2.5}$$

Here, T_r is the cavity round-trip time, $\alpha_k = e_k/(p_k \tau_k)$ is a pulse-shape-dependent numerical factor of order unity, e_k is the energy of the fluctuation, $g_{0,k}$ represents the gain before the arrival of the pulse, τ_k is the duration of the fluctuation, and T_g is the lifetime of the upper laser level. The saturation power of the gain medium is given by $P_s = \hbar\omega A_g/(\sigma T_g)$, where $\hbar\omega$ is the photon energy, A_g denotes the average beam cross-section in the gain medium, and σ is the stimulated emission cross-section. The last term on the right-hand side of Equation (2.4) accounts for the finite lifetime T_c of the initial fluctuation (Krausz *et al.* 1991). In Equation (2.4) $l - g_0 \approx 0$ has been assumed owing to the approximation $e_k \ll P_k T_r$, which applies for the initial transient mode-locking process. The existence of a gain window is ensured by $B > 0$, i.e:

$$\gamma > \alpha \frac{g_0}{2P_s} \frac{\tau_0}{T_g}, \tag{2.6}$$

where $\tau_0 = \tau_{k=0}$ is the width of the initial fluctuation. This condition was also obtained by Ippen *et al.* (1990) by requiring energy growth of the initial fluctuation. Since the saturation power varies less than an order of magnitude around $P_s \approx 1$ W in a wide range of laser-pumped CW lasers including both solid state and dye systems, the magnitude of the right-hand side accounting for dynamic gain saturation changes primarily due to the variation of τ_0/T_g. In solid state lasers with microsecond gain relaxation times the right-hand side of Equation (2.6) takes values between $10^{-8}\,\mathrm{W}^{-1}$ and $10^{-6}\,\mathrm{W}^{-1}$, for which condition (2.6) can be easily satisfied in practice. In dye/diode lasers with nanosecond gain relaxation times, gain saturation may be as strong as $10^{-3}\,\mathrm{W}^{-1}$ which, however, is overcome by strong SAM due to real physical absorbers. The finite relaxation time of the absorbers does not limit pulse shortening in this case because the gain window is closed by gain saturation rather than loss recovery in the absorber.

The existence of this gain window still does not guarantee the initial fluctuation will grow. An initial fluctuation will monotonically increase until steady state modelocking is reached if and only if

$$B_0 p_0 > \frac{T_r}{T_c}. \tag{2.7}$$

For solid state lasers with weak gain saturation this relationship simplifies to:

$$\gamma P_0 > \frac{1}{\ln N} \frac{T_r}{T_c}, \tag{2.8}$$

where N represents the number of initially oscillating modes, and we have assumed $p_0 \approx P_0 \ln N$ (Kryukov and Letokhov, 1972). The lifetime of the initial mode-beating fluctuations T_c has been postulated to be equal to the effective mode correlation time defined by the inverse 3dB full-width $\Delta\nu_{3dB}$ of the first beat note of the free-running laser (Krausz *et al.*, 1991):

$$T_c = \frac{1}{\pi} \frac{1}{\Delta\nu_{3dB}}. \tag{2.9}$$

This relationship provides an important link between the time- and frequency-domain characterization of multimode free-running oscillation in a laser cavity. Expressions (2.8) and (2.9) can now be combined into a simple condition for self-starting passive modelocking of CW-pumped lasers:

$$\gamma P_0 > \frac{\pi}{\ln N} \frac{\Delta\nu_{3dB}}{\Delta\nu_{ax}}, \tag{2.10}$$

where $\Delta\nu_{ax}$ stands for the axial mode spacing at the center of the laser spectrum. Using this condition, simple spectral measurements of the free-running laser output (in the optical and radio-frequency domain for N and $\Delta\nu_{3dB}$, respectively) allow a prediction of the necessary nonlinearity and/or intracavity power if spontaneously starting passive modelocking operation is to be accomplished. The significant parameter that determines the self-starting threshold is the beat-note linewidth of the free-running laser.

Typical values of the beat-note linewidths in laser-pumped homogeneously-broadened lasers, e.g. Ti:sapphire, Nd:glass, and Nd:YLF lasers range from 2 kHz to 10 kHz, depending on the saturated gain and the intracavity power, which imply mode correlation times of $\sim< 100\,\mu$s. Substituting this into (2.10) gives the minimum dynamic increase in round-trip gain required for startup to be of the order of 10^{-4} in typical laser resonators with $\Delta\nu_{ax} \approx 100\,\text{MHz}$, in close agreement with experimental results (Krausz *et al.*, 1992). Although this simple phenomenological theory is useful for the design and development of CW passively modelocked solid state lasers, one important question still remains to be answered: what are the basic physical origins of a finite beat-note linewidth? In what follows, the principal mechanisms that can contribute to the broadening of the beat-note line, and thus impair the self-starting passive modelocking performance of practical laser oscillators, are addressed.

2.1.3 *Beat-note line broadening*

In the preceding sections it has been pointed out that for modelocking to prevail, the excess round-trip gain experienced by the initial fluctuation must be large enough to complete the modelocking process within the lifetime of the fluctuation. This lifetime relates to the linewidth of the first beat-note in the RF power spectrum of the free-running laser, which can be attributed to two distinct physical effects: (i) the finite mutual coherence time of the longitudinal modes, and (ii) mode pulling which is nonlinear in the mode frequency. A primary consequence of the latter effect is uneven frequency spacings between adjacent longitudinal modes. Since the mode spacing is given by the ratio of phase velocity of light to the round-trip cavity length, the effect of uneven mode spacings can be described in terms of a phase velocity which is slightly different for different axial modes. Clearly, a frequency-dependent phase velocity leads to a decay of the initial pulse-like fluctuation just as the finite mutual coherence of the axial modes.

The limited mutual coherence of co-oscillating axial cavity modes manifests itself in a broadening of the beat-note signal of any single pair of modes, therefore we refer to this type of broadening as homogeneous. If more than two modes are oscillating, the overall beat-note signal is subject to an additional broadening caused by the slightly different center frequencies of the beat-note signals arising from different pairs of adjacent axial modes. This inhomogeneous broadening comes from a slight variation of the mode spacing over the oscillation spectrum. The different broadening mechanisms are illustrated in Figure 2.4. Since homogeneous broadening arising from spontaneous emission

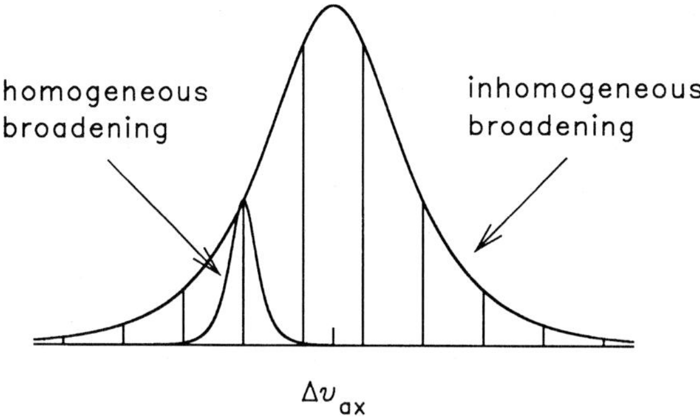

Figure 2.4. Schematic diagram of the radio-frequency spectrum of the free-running laser output around $\Delta\nu_{ax}$, the mode spacing at the laser oscillation spectrum. The individual bars represent beat-note signals of different pairs of adjacent axial cavity modes. Each of these is broadened to some extent due to the finite mutual coherence time of the modes (homogeneous broadening) and shifted with respect to each other owing to nonlinear mode pulling effects.

noise of the amplifier is of quantum nature and thus very small,[2] we concentrate on inhomogeneous beat-note line broadening mechanisms. The following sources of uneven shifts of the resonator eigenfrequencies have been identified: (i) intracavity dispersion (Yan *et al.*, 1989); (ii) spurious reflections inside or outside the cavity (Haus and Ippen, 1991); and (iii) complex refractive index grating in the gain medium (Krausz and Brabec, 1993).

The most apparent origin of uneven mode spacings is intracavity dispersion. Elementary calculations for the dispersion-induced broadening yield (Krausz *et al.*, 1992)

$$\Delta\nu_{3\mathrm{dB,disp}} \approx \pi|D|\Delta\nu_{\mathrm{osc}}\Delta\nu_{\mathrm{ax}}^2 \tag{2.11}$$

where D is the round-trip group delay dispersion (GDD) at the center frequency,[3] and $\Delta\nu_{osc}$ stands for the free-running laser oscillation bandwidth. Since $\Delta\nu_{osc}$ is typically of the order of 10^2 GHz in homogeneously-broadened lasers, and $|D|$ is usually smaller than $10^4\,\mathrm{fs}^2$, in a $\Delta\nu_{ax} = 100\,\mathrm{MHz}$ cavity the dispersion-induced broadening is of the

[2] The major classical source of line broadening of single cavity modes, namely resonator length fluctuations, introduces correlated perturbations to the different cavity modes, and hence does not affect the mutual coherence.

[3] The round-trip cavity GDD is defined by $D = (\mathrm{d}^2\Phi/\mathrm{d}\omega^2)_{\omega=\omega_0}$, where $\Phi(\omega)$ is the frequency-dependent round-trip phase delay and ω_0 is the center laser frequency. This gives a factor of (1/2) difference from the notation chosen by H. A. Haus.

order of 0.1 kHz or less. Thus, we may conclude that in typical modelocked solid state lasers cavity dispersion is not the dominant effect opposing self-starting passive modelocking. Nevertheless, under specific circumstances (e.g. large oscillation bandwidth in an inhomogeneously-broadened laser and/or short cavity length) the effect of dispersion may require reconsideration.

The adverse effect of spurious reflections on passive modelocking has recently been observed experimentally (Krausz *et al.*, 1990) and studied theoretically (Haus and Ippen, 1991). The broadening of the beat-note linewidth in the presence of spurious reflections may be calculated using the formalism developed by Haus and Ippen (1991). Here the frequency shifts $\delta\nu_n$ suffered by the individual resonator modes are calculated, and the associated broadening of the RF linewidth is approximately given by (Krausz *et al.*, 1992)

$$\Delta\nu_{3\mathrm{dB,refl}} \approx 2\left\langle \delta\nu_n^2 \right\rangle^{1/2} \approx \frac{r_{\mathrm{ext}}}{\pi\sqrt{2}} \frac{1 - r_{\mathrm{oc}}^2}{r_{\mathrm{oc}}} \Delta\nu_{\mathrm{ax}}, \tag{2.12}$$

where r_{ext}, and r_{oc} are the amplitude reflectivities of the external reflector and output coupler, respectively.[4] This result demonstrates that a very small fraction ($r_{\mathrm{ext}}^2 \approx 10^{-5}$) of the output power fed back into the resonator can lead to a substantial broadening of the beat-note linewidth and thus seriously impair modelocking, as is often observed in experiments. Note that even the back face ($r_{\mathrm{ext}} \approx 0.2$) of a high reflectivity mirror with $1 - r_{\mathrm{oc}}^2 \approx 10^{-3}$ may adversely affect modelocking, in agreement with the experimental results reported in Krausz *et al.* (1990). For an intracavity reflection one has to omit the factor $(1 - r_{\mathrm{oc}}^2)/r_{\mathrm{oc}}$ and replace r_{ext} by r_{int}, the reflectivity of the intracavity reflector. In either case very small reflectivities can disturb ultrashort pulse formation, and therefore special care has to be taken in the design of femtosecond systems to avoid spurious reflections.

After elimination of all identifiable spurious intracavity and extracavity reflections, the beat-note linewidths of optimized systems appeared to exhibit a strong dependence on both the intracavity power and the total internal cavity losses (Krausz *et al.*, 1992). This behavior suggests that the beat-note linewidth is ultimately affected by optical nonlinearities in the cavity. One nonlinear element present in every laser is the gain medium. In standing-wave laser oscillators the field energy density periodically varies along the resonator axis and creates a complex

[4] We note that the proportionality factor in Equation (14) in Krausz *et al.* (1992) is incorrect, and Equation (2.12) in this text gives the correct result.

refractive index grating in the gain medium, a phenomenon often referred to as spatial hole-burning (Tang *et al.*, 1963). This grating couples the originally independent counterpropagating waves of the individual axial cavity modes. The coupling induces a mode frequency shift that is a nonlinear function of the unperturbed mode frequency. To quantify the effect of spatial hole-burning on the start-up performance of passively modelocked lasers, the following assumptions are used: (a) the laser medium is much shorter than the resonator length and is positioned approximately in the middle of the cavity, (b) the number of free-running modes N is much higher than 1, and (c) N does not significantly exceed L_r/L_g, the ratio of the resonator length to that of the gain medium. With these approximations, then (Krausz and Brabec, 1993)

$$\Delta\nu_{3\text{dB,shb}} \approx \frac{g_s}{2\pi\sqrt{2}}|\alpha_1|L_g\Delta\nu_{ax} \tag{2.13}$$

for the spatial-hole-burning-induced broadening of the beat-note line. Here g_s is the round-trip saturated power gain and α_1 is the magnitude of the spatial modulation of the gain coefficient along the resonator axis, which is given by

$$\alpha_1 = \frac{g_s}{2L_g}\frac{P_0}{P_s}\left(1+\frac{2P_0}{P_s}\right)^{-1}\Delta\rho. \tag{2.14}$$

Here, P_s denotes the saturation power of the gain medium, and $\Delta\rho$ represents the effective spatial modulation amplitude of the field energy density relative to its average value in the gain medium.[5] The explicit dependence of α_1 on the intracavity power vanishes for $P_0/P_s \gg 1$. Figure 2.5 plots $\Delta\rho$ versus N for a representative case characterized in the figure caption. With $\Delta\nu_{ax} = 100\,\text{MHz}$, $g_s = 0.2$, $P_0/P_s = 2$, and $N \approx 10^2$ implying $\Delta\rho \approx 10^{-2}$ (see Figure 2.5), substitution of Equation (2.14) in Equation (2.13) yields $\Delta\nu_{3\text{dB,shb}} \approx 1\,\text{kHz}$ for a typical homogeneously-broadened CW solid-state laser. Considering the fact that $\Delta\nu_{3\text{dB,shb}}$ is expected to be broader for an asymmetrically located gain medium because of more pronounced spatial hole-burning (larger $\Delta\rho$), this prediction is in reasonable agreement with recent experimental results (Krausz *et al.*, 1992). An important implication of the theory is the explicit quadratic dependence of $\Delta\nu_{3\text{dB,shb}}$ on g_s, i.e. on the net

[5] Here we assumed a long-relaxation-time gain medium and ignored the time-varying part of $\rho(z)$, because this sinusoidal modulation is so rapid that the atomic population cannot respond to it.

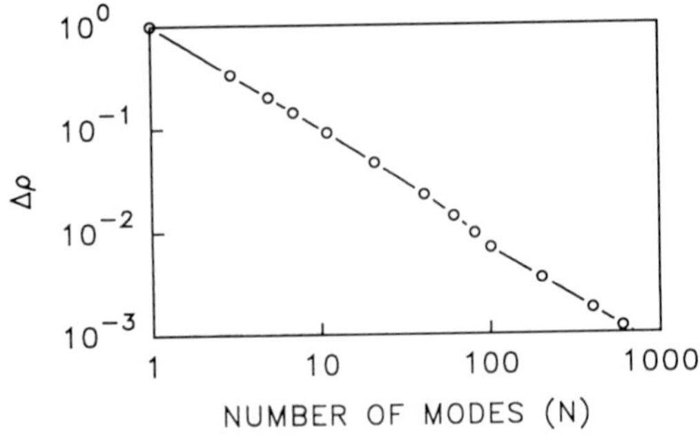

Figure 2.5. Effective modulation amplitude of the energy density relative to the average energy density in the gain medium as a function of the number of oscillating modes (N) for a resonator length $L_r = 1$ m, center laser wavelength $\lambda_o = 1\,\mu$m, and an optical length of the gain medium $nL_g = 10$ mm. The field energy was assumed to be evenly distributed among the oscillating modes.

intracavity loss.[6] The rapid increase in the beat-note linewidth with increasing cavity loss has been experimentally verified in Nd:YLF, Nd:glass, and Ti:sapphire lasers (Krausz *et al.*, 1992) and more recently also in a Nd:fiber laser (Ober *et al.*, 1993).

Beyond this quantitative comparison, some qualitative experimental observations can also be interpreted in terms of the above theoretical results. Since it takes a finite time for the population grating to build up, a fast enough displacement of the standing-wave pattern of the field energy density in the gain medium can lead to a reduction of α_1. Hence, a sudden perturbation (Spence *et al.*, 1991) or periodic variation (Liu *et al.*, 1992) of the cavity length implies not only increased mode-beating fluctuations but also a partially erased population grating in the gain medium. Previous observations (Flood *et al.*, 1990) in an actively modelocked system support this hypothesis. The complex index grating in the gain medium can be efficiently eliminated by constructing a ring cavity in which only one propagation direction is allowed. Recently, it has been demonstrated experimentally that unidirectional ring cavities improve the start-up performance of both actively (Flood *et al.*, 1990) and passively (Tamura *et al.*, 1993) modelocked lasers. These findings and the reasonable quantitative predictions of our theory suggest that the beat-note line broadening originating from spatial hole-burning in the gain medium is likely to

[6] The explicit quadratic dependence may be somewhat modified by $N = N(g_s)$, which makes $\Delta\rho$ dependent on g_s.

be the dominant effect opposing self-starting passive modelocking in practical standing-wave laser oscillators.

2.1.4 *The role of self-phase modulation during the transient modelocking process*

In the simple theoretical model presented above self-phase modulation (SPM) arising from the Kerr effect has not been taken into account. Nevertheless, Kerr-induced SPM may influence the transient modelocking process under some circumstances. Specifically, dispersion-controlled systems with an overall negative GDD ($D < 0$) may exhibit an instability that can lead to an 'amplification' of the initial fluctuations in the free-running laser. This phenomenon, referred to as modulation instability, results from an interplay between self-phase modulation and negative dispersion. Even though this effect cannot give rise to modelocked pulse formation by itself, it may help the amplitude nonlinearity in building up the modelocked state. Loosely speaking, the modulation instability increases the peak power of the initial fluctuation and thus reduces the minimum value of γP_0 necessitated for startup. To quantify the conditions for optimum exploitation of this phenomenon, recall that the round-trip (power) gain experienced by a weak perturbation of frequency Ω which is superimposed on a CW radiation is given by (Agrawal, 1989):

$$g(\Omega) = |D|\Omega(\Omega_c^2 - \Omega^2)^{1/2}. \tag{2.15}$$

The gain is maximum at $\Omega_{max} = \Omega_c/\sqrt{2} = (2P_0\delta/|D|)^{1/2}$ and has a value of $g_{max} = 2P_0\delta$, where δ accounts for the ultrafast Kerr nonlinearity and is defined such that $P_0\delta$ gives the round-trip nonlinear phase shift.[7] Consequently, for the modulation instability to noticeably support the buildup of modelocking, δ should be greater than or comparable with γ, a condition which is almost always met in practical systems. Also, the frequency region experiencing maximum gain should coincide with the spectrum of the fluctuations, i.e. $\Omega_{max} \approx \Delta\nu_{osc}$. From the latter requirement one can obtain the optimum value of D for self-starting:

$$|D_{opt}| \approx \frac{2P_0\delta}{\Delta\nu_{osc}^2}. \tag{2.16}$$

[7] Note that in our previous work ϕ instead of δ was used to account for the Kerr nonlinearity.

For practical solid state systems the right-hand side of Equation (2.16) often exceeds $10^4\,\mathrm{fs}^2$, a value which is usually greater than that of $|D|$ providing optimum CW modelocked performance. Still, if the difference between $|D|$ and the right-hand side of (2.16) is not excessively large, and $g_{\max}$ is high enough, modulation instability may have a noticable effect on the self-starting performance of practical femtosecond solid state laser oscillators (Spielmann *et al.*, 1991).

2.2 Steady-state pulse shaping dynamics

The introduction of this chapter indicated that the formation of short intense femtosecond pulses in dye lasers is primarily based on SAM and gain saturation. In such lasers, small fractional changes in the phase and amplitude of the pulse occur when passing through each of the cavity elements. Under these conditions the modelocking dynamics are essentially described by a weak pulse shaping (WPS) approximation (Krausz *et al.*, 1992). A consequence of applying the WPS approximation is that the individual pulse shaping elements situated at different resonator positions are treated as though they were continuously distributed throughout the laser cavity. This approach has enabled the development of a number of analytical theories on passively modelocked lasers (Haus, 1975a,b; Haus *et al.*, 1991; Martinez *et al.*, 1985) which have been successful in explaining the dependence of the principal pulse parameters, such as pulse width, energy and intensity, as a function of the laser gain, bandwidth and SAM parameters.

In contrast to dye lasers, femtosecond modelocked solid state lasers tend to rely on a large nonlinear phase shift in order to provide sufficiently strong pulse shaping, the contribution from SAM being comparatively weak. For example, on one cavity round trip the nonlinear phase shift is typically in excess of unity (Krausz *et al.*, 1992) whereas SAM is much less than unity. In this case pulse shaping is dominated by the interplay between SPM and negative GDD. In a visible or near-infrared solid state laser, SPM and GDD arise from different cavity elements and therefore occur at different positions within the resonator. This leads to additional contributions to the pulse dynamics which are not accounted for in the WPS approximation. Therefore, to include such effects a more general theoretical description of the pulse shaping process should be used. In what follows, steady state modelocking of practical femtosecond solid state lasers is discussed, where the WPS approximation is usually not applicable. Although exact closed form solutions are not readily obtained from the extended model, our treatment does allow access to a qualitative description of a number of new and hitherto unforseen phenomena in this class of novel femtosecond source.

2.2.1 *Principal pulse shaping mechanisms*

Consider the situation inside the laser resonator after a significant time has elapsed such that conditions suitable for the propagation of a single intense pulse exist. We assume this radiation can be described by a continuous function, $E(t, z)$, written as

$$E(t, z) = a(t, z)e^{i\omega_o t}, \tag{2.17}$$

where ω_0 is the center frequency, $a(t, z)$ is a slowly varying (complex) envelope, t is measured in a frame of reference moving at the group velocity of the pulse, and z is the direction of propagation normalized to the length of the cavity. Now consider a laser resonator which contains a number of discrete pulse shaping elements which relate to a specific operation. It is assumed that propagation of the field envelope, $a(t, z)$, through any one of these cavity elements is described by an equation of the form

$$\frac{\partial a}{\partial z} = \hat{F}_i a, \tag{2.18}$$

where $\hat{F}_i$ is an operator which describes the action of the ith cavity element. Repeated application of Equation (2.18) for each cavity element leads to a description of pulse propagation throughout the whole laser cavity. Considerable insight into this description can be obtained by noting that Equation (2.18) can be written as

$$a_{k+1}(t, z) = \prod_i e^{\hat{F}_i} a_k(t, z) = \hat{T}(z) a_k(t, z), \tag{2.19}$$

where the transfer operator $\hat{T}$ describes a full cavity round trip and k indicates the number of roundtrips. The exponential representation in Equation (2.19) is exact when $\hat{F}_i$ represents a phase operation. For operators acting on the pulse amplitude Equation (2.19) may be regarded as a first-order approximation which is applicable provided the change brought about by $\hat{F}_i$ is small. The operator $\hat{T}(z)$ can be rearranged into a more appropriate form by employing the Campbell–Baker–Hausdorff theorem (Agrawal, 1989),

$$\hat{T}(z) = \exp\left\{ \sum_i \hat{F}_i + \sum_{ij} c_{ij}(z)\left[\hat{F}_i, \hat{F}_j\right] + \sum_{ijk} c_{ijk}(z)\left[\hat{F}_i, \left[\hat{F}_j, \hat{F}_k\right]\right] + \ldots \right\}. \tag{2.20}$$

In Equation (2.20) the roles played by the discrete pulse shaping elements have been separated into two parts. The first summation denotes the simultaneous action of the cavity elements, whereas the second and higher summation terms reflect the physical situation that each cavity element is in fact a discrete pulse shaping operation. This information is contained in the z dependence of the coefficients $c_{ij}, c_{ijk}, \ldots$ and through the commutator of the operators $\hat{F}_i$ and $\hat{F}_j$.

The above introduced formalism can now be applied to investigate steady-state pulse shaping dynamics. For stable modelocked operation the radiation propagating inside the resonator has to fulfill a steady-state condition such that the pulse envelope $a_{k+1}(t, z)$ remains unchanged between one cavity round trip and the next

$$a_{k+1}(t, z) = e^{i\Psi} a_k(t, z), \tag{2.21}$$

where Ψ is a constant phase shift per cavity roundtrip. The substitution of Equation (2.21) into Equation (2.19) along with the consideration of specific pulse shaping operations for $\hat{T}$, Equation (2.20), yields then an equation which determines steady-state modelocked operation.

Consider the situation when $\hat{T}$ represents the principal pulse shaping effects appropriate for a solid-state laser oscillator, as illustrated in Figure 2.6. The principle operators governing modelocked operation in this system are as follows: $\hat{D} = i(D/2)\partial^2/\partial t^2$ and $\hat{N} = -i\delta|a|^2$ are operators which describe the action of negative GDD and SPM, respectively, and $\hat{A} = (1/2)(g - l - \gamma|a|^2)$ describes the contributions from gain, loss

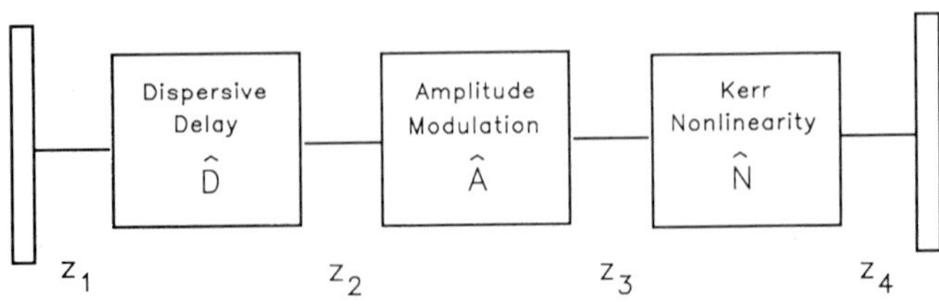

Figure 2.6. Schematic diagram of a solitary laser.

and SAM, respectively.[8] As defined in the preceding section, the parameter D denotes the round-trip GDD of the dispersive delay line, and δ is the round-trip nonlinear phase shift per unit power in the Kerr medium (unit: W^{-1}). In terms of the parameters introduced above, a solitary system can be defined by the relationships $D/\delta < 0$, and $\gamma/\delta \ll 1$. Physically, the condition $D/\delta < 0$ implies negative GDD and positive SPM, which is necessary for soliton-like pulse shaping, whereas $\gamma/\delta \ll 1$ ensures weak SAM action as compared with SPM.

On passage through any of the cavity elements, if the fractional change in the pulse parameters is small, the laser operates under conditions which are suitably described by the WPS approximation. This implies that the coefficient c_{ij} and higher order coefficients, as contained within the transfer operator in Equation (2.20), can be neglected and therefore $\hat{T}(z)$ reduces to

$$\hat{T}(z) = \exp(\hat{A} + \hat{N} + \hat{D}). \tag{2.22}$$

This simplified description for $\hat{T}$ models the modelocking dynamics for uniformly distributed pulse shaping action throughout the laser cavity. Under these conditions the substitution of Equation (2.22) into Equation (2.19) along with Equation (2.21) and, by taking the limit $\gamma/\delta \to 0$, i.e., ignoring the contribution from SAM, allows the derivation of the following stationary solution

$$a(t, z) = \left(\frac{W}{2\tau_s}\right)^{\frac{1}{2}} \operatorname{sech}\left(\frac{t}{\tau_s}\right) e^{i\Psi z} \tag{2.23}$$

$$\Psi = \frac{D}{2\tau_s^2} \tag{2.24}$$

$$\tau_s = \frac{2D}{W\delta}. \tag{2.25}$$

Here W is the pulse energy, τ_s is the soliton pulse width and Ψ is the phase shift per resonator round trip. Equations (2.23)–(2.25) represent a pulse identical to the fundamental ($N = 1$) soliton propagating without change of shape in a nonlinear optical fibre having anomalous dispersion (Agrawal, 1989). This description is clearly an idealization of the pulse shaping process which takes place inside a real laser, and may be

[8] Note that in our previous work ϕ and κ instead of δ and γ have been chosen as characteristics for SPM and SAM, respectively. Furthermore, the round-trip cavity GDD (D) is defined by a factor 1/2 different from the notation of H. A. Haus.

considered as a first order approximation to actual modelocked operation. However, a notable failing with this description is the behavior of the pulse width as the negative GDD is decreased. For example, from Equation (2.25), as $D \to 0$ the pulse width $\tau_s \to 0$, which is clearly physically unacceptable. In real laser systems additional effects are present which prevent this behavior. In the following sections, 2.2.2–2.2.4, the influences of such effects are considered and related to recent experimental studies.

2.2.2 *Periodic and higher-order dispersive perturbations*

There are a number of important effects which contribute to the femtosecond pulse shaping dynamics in solitary lasers. In many cases of practical interest these effects can be regarded as perturbations to the basic soliton pulse described above which can be grouped into contributions arising from higher-order dispersion and/or those due to periodical perturbations. These additional perturbations can be included in the transfer operator given in Equation (2.22), which leads to the following description

$$\hat{T}(z) = \exp(\hat{A} + \hat{N} + \hat{D} + \hat{P}(z, \hat{N}, \hat{D}) + \hat{D}_{\mathrm{h}}). \tag{2.26}$$

Here the operators $\hat{A}, \hat{D}$, and $\hat{N}$ have been defined in the preceding section and describe fundamental processes responsible for solitary pulse shaping. The new operators $\hat{P}$ and $\hat{D}_{\mathrm{h}}$ represent additional complicating physical processes which disturb the simple picture of ideal soliton shaping. These operators become particularly relevant when attempting to generate sub-100 fs pulses and will be considered in the remaining part of this section.

In the sub-100 fs region pulse shaping contributions from the separate action of SPM and negative GDD tend to become comparable with the commutation of such operations. This behavior is accounted for by $\hat{P}(z, \hat{N}, \hat{D})$, a position-dependent perturbation operator which can be expanded in commutators of increasing order in $\hat{N}$ and $\hat{D}$ according to Equation (2.20). The fractional change of the pulse envelope due to either SPM or GDD scales with the ratio of the resonator round-trip length L_{r} to that of the soliton period L_{s} (Brabec *et al.*, 1991)

$$r = \frac{L_{\mathrm{r}}}{L_{\mathrm{s}}} = \frac{1}{2\pi} \frac{(W\delta)^2}{|D|}. \tag{2.27}$$

Hence, as L_s becomes comparable with L_r, the parameter r in Equation (2.27) approaches unity. This implies that the effects of the first few terms in the expansion of $\hat{P}$ become of the order of the principle pulse shaping operators $\hat{D}$ and $\hat{N}$ (Krausz *et al.*, 1992). Consequently, the phase changes incurred by the pulse upon propagation through a nonlinear and/or dispersive element per cavity round trip become large enough so that the discreteness or inhomogeneity of the laser cavity becomes evident. To estimate the significance of this type of perturbation in femtosecond solid state lasers note that $r \approx 3$ for a Ti:sapphire system (Spielmann *et al.*, 1994), supporting pulses down to 10 fs ($|D| \approx 80\,\mathrm{fs}^2$, $W \approx 40\,\mathrm{nJ}$, $\delta \approx 1 \times 10^{-6}\,\mathrm{W}^{-1}$, and $W\delta \approx 40\,\mathrm{fs}$). In high power lasers (output power $\approx 1\mathrm{W}$) generating pulses in the 50 fs range, r can also take values exceeding unity. Consequently, a strong influence of $\hat{P}$ on femtosecond pulse formation is expected, which is analysed in more detail in Section 2.2.3.

Besides the contribution from inhomogeneous perturbations, as above, dispersive effects originating from the finite bandwidth of the oscillator (gain dispersion) and frequency dependence of the cavity GDD (higher-order dispersion) are also present. Such effects are described by the operator $\hat{D}_h = D_g\partial^2/\partial t^2 - (D_3/6)\partial^3/\partial t^3 + \ldots$ Here D_3 defines the magnitude of the round-trip third-order dispersion (TOD). The magnitude of the gain dispersion is determined by $D_g = g/\Omega_g^2$, where Ω_g is a spectral width associated with the gain medium and additional bandwidth limiting elements. The operator $\hat{D}_h$ increasingly affects steady-state modelocking performance as the negative GDD is reduced. To estimate the significance of this perturbation note that the pulse shaping effect of gain dispersion, D_g, scales with the parameter (Brabec *et al.*, 1992a)

$$\epsilon_g = \frac{D_g}{|D|}. \tag{2.28}$$

In solid state lasers with fluorescence linewidths of the order of 100 THz (see Figure 2.1), D_g typically takes a value of $< 5\,\mathrm{fs}^2$, unless tuning or other additional bandwidth-limiting elements are introduced. This low value of D_g implies $\epsilon_g \ll 1$ so that modelocking is essentially unaffected by bandwidth limitations when $|D| > 50\,\mathrm{fs}^2$, and thus for pulse durations as short as 10 fs.

In addition to the finite fluorescence linewidth, higher-order contributions to $\hat{D}_h$ arise from the frequency-dependence of the intracavity GDD. In particular, third-order dispersion (TOD) provides a linear and usually dominant contribution. The fractional change of the intracavity GDD over the modelocked spectrum due to TOD can be estimated from the ratio

$$\epsilon_t = \frac{1}{3}\frac{|D_3|\omega_s}{|D|}, \tag{2.29}$$

where $\Delta\omega_s = 2/\pi\tau$ is the spectral width of the modelocked laser spectrum. In a dispersion optimised system $|D_3|$ can be quite low, typically $500 - 1000\,\text{fs}^3$. Thus a similar GDD, i.e. $|D| > 50\,\text{fs}^2$ and pulse duration $\approx 10\,\text{fs}$, implies $\epsilon_t \approx 0.3$. In comparison with ϵ_g it is clear that TOD provides a much stronger influence on the modelocking. Indeed, a number of experimental studies (Lemoff and Barty, 1992; Curley *et al.*, 1993; Asaki *et al.*, 1993; Proctor and Wise, 1993; Spielmann *et al.*, 1994) indicate that in the sub-100 fs domain the rapid increase of TOD with decreasing negative GDD can become a major limitation to further pulse shortening. The nature in which this occurs is discussed in more detail in Section 2.2.4.

To conclude this section the reader is referred to Figure 2.7, which summarises, schematically, the principal pulse shaping effects in femtosecond solid state lasers. For sufficiently large values of negative intracavity GDD the magnitude of the parameters, as defined by Equations (2.28) and (2.29), are typically much less than unity. However, as the GDD is reduced, the influence of dispersive ($\hat{D}_h$) and discrete pulse shaping ($\hat{P}$) effects increases. If one of the scaling parameters r, ϵ_g and ϵ_t approaches unity, the soliton-like modelocked pulse is subjected to

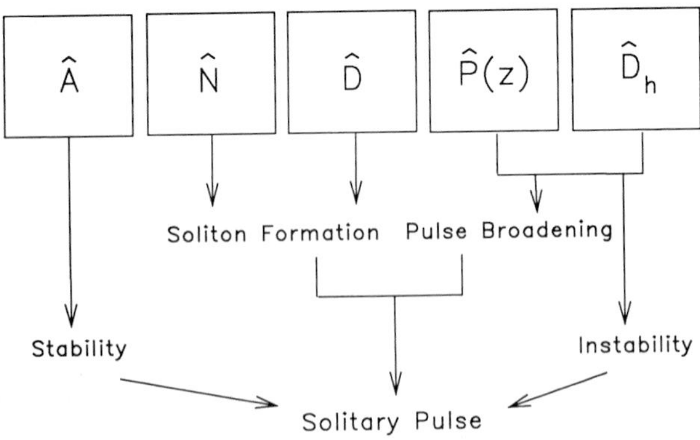

Figure 2.7. Summary of the various processes which influence the generation and evolution of femtosecond pulses in a solitary laser. $\hat{A}, \hat{N}, \hat{D}, \hat{P}(z)$ and $\hat{D}_h$ denote the operators for SAM, SPM, negative GDD, periodic and dispersive perturbations, respectively.

modifications which may limit further pulse shortening. The principal mechanism opposing the emergence of instabilities is SAM. The presence of SAM appears to be crucial to keep the influence of these perturbations at a low level in solitary modelocked lasers, and thus allows the generation of pulses down to 10 fs. The primary role played by SAM is to introduce a differential gain across the modelocked pulse. In the time domain this leads to a suppression of the pulse wings compared with the pulse center.

In the remainder of this chapter this qualitative consideration of solitary modelocking is extended to a quantitative analysis. In particular, in Section 2.2.3 attention is paid to steady-state pulse shaping dynamics, and in Section 2.2.4 considerations about the role of SAM as a stabilizing mechanism are presented.

2.2.3 *Steady-state pulse parameters*

In the last section the r parameter, see Equation (2.27), was introduced. This parameter provides a measure of the deviation of solitary modelocking from ideal soliton pulse propagation. When r is comparable with or greater than unity the discrete nature of the soliton-like pulse shaping is expected to significantly affect the modelocking dynamics. Mathematically, this manifests in the appearance of the non-commuting parts associated with the operators $\hat{N}$ and $\hat{D}$ in $\hat{T}(z)$, thus

$$\hat{P}(z, \hat{N}, \hat{D}) = \underbrace{c_1(z)[\hat{N}, \hat{D}]}_{\hat{O}(r^2)} + \underbrace{c_2(z)\big[\hat{N}, [\hat{N}, \hat{D}]\big] + c_3(z)\big[\hat{D}, [\hat{N}, \hat{D}]\big]}_{\hat{O}(r^3)} + \ldots \tag{2.30}$$

Here the perturbation operators $\hat{O}(r^2)$ and $\hat{O}(r^3)$ contain commutators $[\hat{N}, \hat{D}]$ and double commutators $[\hat{N}, [\hat{N}, \hat{D}]], [\hat{D}, [\hat{N}, \hat{D}]]$ (see Equation (2.20)), respectively. The expansion, Equation (2.30), shows the commutator $[\hat{N}, \hat{D}]$, is of second order in r, whilst further commutation products are third-order or higher in r. Using this terminology, the position-independent transfer operator (2.22), derived in the WPS approximation, and its solution are therefore accurate to order r. From (2.30) the coefficients $c_i(z)$ associated with the commutators in $\hat{O}(r^2)$ and $\hat{O}(r^3)$ are readily obtained for different positions in the cavity by using the Campbell–Baker–Hausdorff theorem (Agrawal, 1989). Specifically, it can be shown (Krausz *et al.*, 1992) that $\hat{O}(r^2)$ vanishes at the resonator ends z_1 and z_4 (Figure (2.8a)), and the leading perturbation term in the exponent is

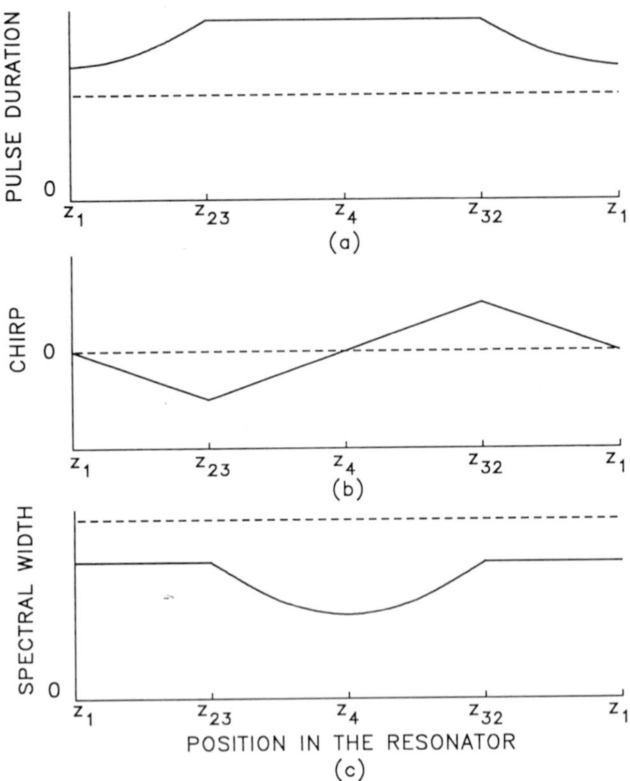

Figure 2.8. Periodic evolution of the (a) pulse duration, (b) chirp, and (c) bandwidth of the steady-state femtosecond pulse in discrete solitary systems with $r \approx 1$–3. The full lines depict the pulse dynamics of the solitary system whereas the dotted lines indicate the exact $N = 1$ soliton solution. The chirp is defined by $(\partial^2\phi_{\rm nl}/\partial t^2)_{t=0}$, where $\phi_{\rm nl}$ is the nonlinear phase shift imposed by the Kerr medium.

$\hat{O}(r^3)$ at these positions. As the double commutators in $\hat{O}(r^3)$ are imaginary they primarily affect the amplitude rather than the phase of the pulse, implying a modelocked pulse which is nearly bandwidth-limited at z_1 and z_4, a result which is accurate to order r^3.

Let us consider steady-state operation in a femtosecond broadband solid state laser under the assumption that sufficient SAM is provided to guarantee stable modelocked operation. The periodic pulse shaping dynamics is depicted in Figures 2.8(a–c), where full and dashed lines show solitary and exact $N = 1$ soliton solutions, respectively. As pointed out above, the pulse is nearly bandwidth-limited at the resonator positions z_1 and z_4. Further, the bandwidth-limited pulse at z_4 becomes compressed after propagating from z_4 to z_1 due to the combined action of SPM and negative GDD. Conversely, propagation from z_1 to z_4 is accompanied by broadening as a consequence of the reversed order of dispersion and nonlinearity (Figure 2.8(a)). Clearly, chirps of opposite signs are imposed on the pulse when passing through z_{23} and z_{32}, respectively (Figure 2.8(b)). This is consistent with the real second-order operator $\hat{O}(r^2)$ nonvanishing at these positions. Finally, bandwidth-limited pulses of different duration at z_1 and z_4 also imply a position-dependent pulse spectrum (Figure 2.8(c)).

These qualitative considerations enable us to make a prediction of the duration of the solitary pulse as a function of position within the laser resonator. Assume the ratio of the solitary pulse width to that of the fundamental soliton can be a set equal to $f(r)$, i.e. a function of only r (Brabec *et al.*, 1991). On physical grounds $f(r)$ smoothly approaches unity as $r \to 0$. Thus a Taylor expansion is possible, and to first order in r yields $f(r) \approx 1 + \alpha' r$. Now using Equations (2.25) and (2.27) allows us to calculate the FWHM (full width at half-maximum) duration of the solitary pulse. After some algebra it is found that this takes the form

$$\tau = \frac{3.53|D|}{W\delta} + \alpha W\delta, \tag{2.31}$$

where $\alpha = 0.56\alpha'$. The second (perturbation) term on the right-hand side of Equation (2.31) is simply proportional to the pulse energy and the optical Kerr nonlinearity. Computer studies (Brabec *et al.*, 1991) have verified the applicability of Equation (2.31) over a remarkably wide range of parameters ($r < 10, \gamma/\delta < 0.2$), the coefficient α depending only on z. The numerical simulations have yielded $\alpha(z_1) = 0.10$ and $\alpha(z_4) = 0.25$, i.e. $\alpha(z_1) < \alpha(z_4)$, this behavior is in accordance with the conclusions of the qualitative analysis. It should be mentioned that although Equation (2.31) is reasonable for $r < 10$, Fourier limited pulses with time–bandwidth products of ≈ 0.3 at the resonator ends can be expected only up to $r \approx 1 - 3$ in practical systems, depending on the efficiency of SAM (Brabec *et al.*, 1991).

The first part of Equation (2.31) denotes the ideal soliton pulse width, whereas the second part comes from the separate action of SPM and negative GDD. A typical range of parameters for $W\delta$ in experimental femtosecond bulk systems is $50 - 200$. This gives, according to (2.31) for $D \to 0$, a residual pulse width of $5 - 20$ fs, which shows that the discrete action of $\hat{N}$ and $\hat{D}$ sets a constraint on the shortest obtainable pulse widths. However, this is not the only limitation to pulse shortening in femtosecond solitary lasers. Reduction of D increases the magnitude of the parameters ϵ_t and r, scaling the influence of discrete and TOD perturbations. These effects tend also to induce unstable operation in real laser systems owing to an additional source of instabilities which is not included in Equation (2.31). This issue and the analysis of such effects form the basis of the next section.

2.2.4 *Dispersive waves and related stability considerations*

This final section summarises more recent investigations into a new type of instability associated with lasers whose principal mode of operation is based on soliton-like pulse shaping. Of late, considerable experimental and theoretical attention has been devoted to understanding this instability, particularly with regard to stable modelocking and in limiting the achievable pulse duration in femtosecond solid state oscillators. From the previous sections, 2.2.3 and 2.2.2, the leading terms in the expansion of the perturbation operators $\hat{P}$ and $\hat{D}_h$ were shown to induce distortions in the duration, energy and shape of the modelocked pulse from that associated with an ideal soliton. The aim of this section is to show that such disturbances can give rise to an energy growth in the pulse wings which ultimately can oppose stable modelocked operation.

A characteristic feature of the instability is the appearance of one or more discrete spectral lines in the wings of the soliton spectrum. These features have been related to a phase matching instability, an intuitive understanding of which, regardless of the actual source of perturbation, can be obtained from the following argument. The reshaping or disturbance of a soliton pulse is always accompanied by the emission of some energy which is lost to a weak radiative field. This low power field normally disperses according to a linear dispersive equation and, in general, will be phase mismatched from the soliton. Perturbations which provide the wavevector difference between the soliton and the dispersive field can lead to a resonant coupling so that an enhanced resonant energy exchange can occur. Note that phase matching can only occur at discrete frequencies where the wave numbers are equal in sign and magnitude. In femtosecond solid-state lasers resonantly enhanced energy loss of the solitary pulse can arise from two distinct physical effects. First, the spatially varying action of SPM and GDD (Gordon, 1992; Kelly, 1992; Elgin and Kelly, 1993; Spielmann *et al.*, 1994) introduces a periodic perturbation, which induces a continuous energy loss associated with resonant sidebands in the modelocked spectrum. Second, in soliton-like lasers some residual TOD is always present in the cavity which can also lead to the appearance of a spectral resonance line (Elgin, 1992, 1993; Curley *et al.*, 1993). To appreciate the difference between these two sources the schematic diagrams, Figures 2.9(a) and (b), depict the dispersion characteristics and the associated spectral signature of the modelocked pulse.

In Figure 2.9(a) a periodic perturbation supplies the additional wavevectors $k_p = 2\pi n$, where n is a positive integer. Thus, resonant energy transfer occurs when $k_d = k_s - k_p$ is fulfilled. The wavevectors of the soliton and the dispersive wave are $k_s = D/(2\tau^2)$ and $k_d = (D/2)\Delta\omega^2$, respectively, where $\Delta\omega$ is the separation from the center frequency of the

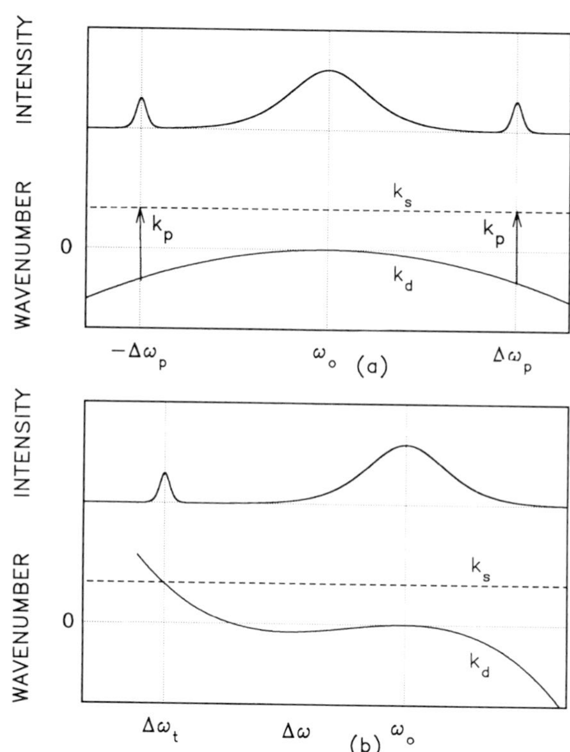

Figure 2.9. Wavenumber of the soliton pulse k_s and the dispersive wave k_d versus the angular frequency relative to the center laser oscillation frequency. The upper trace sketches qualitatively the expected laser spectrum. (a) Phase matching in the presence of periodic perturbations; is the wavenumber of the periodic perturbation. (b) Phase matching in the presence of TOD.

modelocked spectrum. Note that a normalized propagation parameter z is used so that the resonator length does not appear in the above expressions (D denotes the negative GDD per cavity round trip). From the phase matching condition and the assumption $k_s \ll$, which is met in practical laser systems, the frequencies $\Delta\omega_{\mathrm{p,n}}$ where resonances can appear are evaluated as

$$\Delta\omega_{\mathrm{p,n}} \approx \pm\sqrt{\frac{\pi n}{|D|}} \quad n = 1, 2, \ldots \tag{2.32}$$

This result Equation (2.32), illustrates that the resonances evolve at fixed spectral positions, the scaling properties of which have been verified by several experimental groups investigating soliton-like pulse shaping dynamics in both bulk and fiber solid state lasers (Noske *et al.*, 1992; Dennis and Duling, 1993, 1994; Spielmann *et al.*, 1994). It is interesting to note that the appearance of periodic sidebands in solitary lasers can be due to two different mechanisms, although the position of the sidebands is in both cases determined by Equation (2.32): (i) high intracavity gain/loss and therewith strong variation in the pulse energy per cavity round trip; (ii) spatially separate action of SPM and negative GDD. According to this, two different classes of solitary lasers can be distinguished: erbium-doped fiber systems with continuously distributed pulse shaping, where the high intracavity gain is responsible for sideband generation; and contrary to this, bulk lasers with low round-trip gain/loss, where the resonances can be attributed to the discrete operation of SPM and negative GDD.

In contrast to periodic perturbations, the presence of TOD modifies the dispersion characteristic by an additional term cubic in frequency, see Figure 2.9(b). Then, phase matching between the soliton and the dispersive radition occurs for a frequency in the positive GDD region where $k_{\mathrm{d}} = k_{\mathrm{s}}$, with the wavenumber of the linear wave $k_{\mathrm{d}} = (D/2)\Delta\omega^2 + (D_3/6)\Delta\omega^3$. This indicates the position of the TOD-induced resonant sideband as

$$\Delta\omega_{\mathrm{t}} \approx -\frac{3D}{D_3}, \tag{2.33}$$

where its position scales in proportion to D in contrast to the $1/\sqrt{|D|}$ dependence of the periodical case, Equation (2.32).

Experimentally, the existence of one or more of these resonances is found to severely compromise modelocked laser stability (Curley *et al.*, 1993; Dennis and Duling, 1993, 1994). To relate the appearance of the resonances, $\Delta\omega_r(r = p, t)$, to stability in solitary lasers, the amount of energy scattered into the dispersive radiation has to be determined. This can be done using a perturbation expansion of the inverse scattering transform (Gordon, 1992; Elgin, 1993; Brabec and Kelly, 1993). The most important result for these calculations is that the energy loss rate F can be written in the form

$$-F = \frac{dW}{dz} \propto \text{sech}^2\left(\frac{\Delta\omega_r}{\Delta\omega_s}\right), \tag{2.34}$$

where $\Delta\omega_s$ is the spectral width. Equation (2.34) shows that the resonant energy transfer rate is proportional to the spectral intensity of the soliton at the resonance frequency, and hence scales exponentially with the position of the sideband relative to the center of the modelocked spectrum. Therefore a rapid increase of the scattering process is expected as the sideband moves towards the center of the modelocked spectrum. This strong transfer of energy from the modelocked pulse into dispersive radiation, if allowed to continue, can induce unstable operation. Consequently, stable operation requires a mechanism which compensates for the energy loss. The action of a bandwidth limiting element or the fast saturable absorber action of SAM are potential mechanisms which are able to compensate for this loss. Both cases result in a higher gain coefficient for the soliton than for the dispersive radiation. The following analysis is based on the assumption that energy lost from the modelocked pulse can be supplied by the action of SAM.

Since the action of SAM relies on an intensity dependent mechanism, there exists a gain differential across the modelocked pulse which can suppress the dispersive (low power) component. This can be seen from the energy rate equations for the modelocked pulse energy W and resonant wave energy W_r (Brabec and Kelly, 1993)

$$\frac{dW}{dz} = (g - l)W + S - F \tag{2.35}$$

$$\frac{dW_r}{dz} = (g - l)W_r + F, \tag{2.36}$$

where the parameter

$$S \propto \gamma W^2/\tau \tag{2.37}$$

describes an effective gain coefficient associated with SAM. Equations (2.35) and (2.36) are coupled by the saturated gain parameter $g = g_o/(1 + W_o/(T_r P_s))$ where g_o is the unsaturated gain, W_o is the total energy in the resonator, and P_s and T_r are the saturation power and the cavity round-trip time, respectively. For small perturbations, the total energy in the resonator $W_o = W(z) + W_r(z)$ can be assumed to be a constant, and after some algebra an equilibrium between W and W_r is evaluated, where (Brabec and Kelly, 1993)

$$\frac{W_r}{W_o} \propto \frac{\delta}{\gamma} \operatorname{sech}^2\left(\frac{\omega_r}{\omega_s}\right) \tag{2.38}$$

In Figure 2.10 the behaviour of W_r/W_o, as described by Equation (2.38), is plotted as a function of ω_s/ω_r using $\gamma/\delta = 5 \times 10^{-2}$ and 5×10^{-3}. The values of γ/δ are considered to be representative of experimental conditions for strong and weak SAM, respectively.

According to Figure 2.10 the threshold-like response of a solitary laser to a resonance can be divided into three regions: (i) below threshold, where SAM is effective and nearly all the resonator energy is contained by the modelocked pulse; (ii) around threshold, where most of the resonator energy is still modelocked but the energy balance between SAM and resonant energy loss becomes highly sensitive to small changes in the position of the sideband; (iii) above threshold, where a rapid increase in energy associated with the dispersive wave can occur. This behavior, and in particular the existence of a threshold-

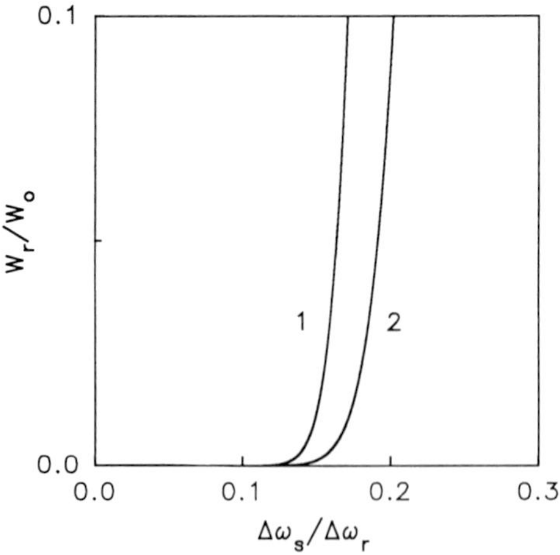

Figure 2.10. Ratio of the dispersive energy to the total intracavity energy W_r/W_o versus the ratio of spectral width to resonance position $\Delta\omega_s/\Delta\omega_r$. Graphs 1 and 2 are for SAM to SPM $\gamma/\delta = 5 \times 10^{-3}$ and 5×10^{-2}, respectively.

like region (ii), where stability is compromised, is consistent with experimental observations (Curley *et al.*, 1993; Dennis and Duling, 1993). In support of this we note that the magnitude of the energy loss described by Equation (2.38) is primarily determined by the exponential-like dependence on the ratio $\Delta\omega_r/\Delta\omega_s$. In other words the position of the resonance with respect to the peak of the modelocked pulse spectrum is the dominant parameter determining the range of stable operation, rather than the nature of the source, i.e. TOD or periodic perturbation. With respect to this point, evaluation of a critical value $\Delta\omega_r/\Delta\omega_s$ from experiments in the presence of TOD (Curley *et al.*, 1993) and periodical perturbations (Dennis and Duling, 1993) yields for both cases $\Delta\omega_r/\Delta\omega_s \approx 3$.

The above discussion suggests that region (ii) imposes a stability limit to solitary modelocking and hence the generation of femtosecond pulses. Experimentally, the shortest femtosecond pulses tend to be generated in lasers by reducing the intracavity GDD. In this case Equation (2.33) indicates the position of the TOD resonance then moves towards the modelocked spectrum in proportion to D, whereas for a periodical perturbation, Equation (2.32), the $1/\sqrt{|D|}$ dependence implies resonances which move away from the pulse spectrum, see Figure 2.11. This argument shows the TOD resonance as being more influential in destabilising modelocking, e.g. Ti:sapphire lasers (Curley *et al.*, 1993). In Er:fiber

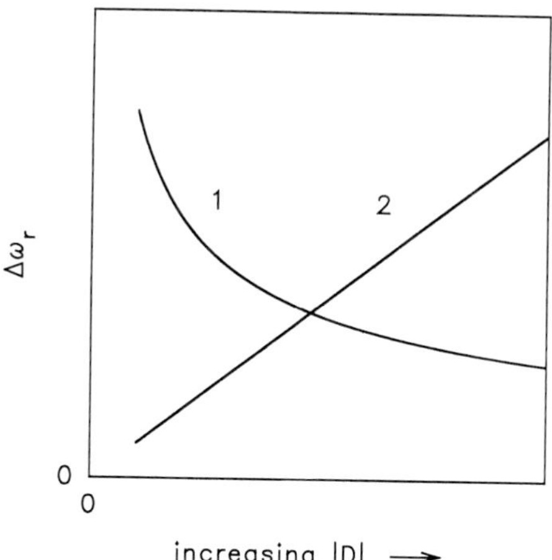

Figure 2.11. Position of the resonant frequency $\Delta\omega_r$ versus negative GDD. Graphs 1 and 2 denote the perturbations due to periodic and TOD perturbations, respectively.

lasers however, where $\epsilon_t = \Delta\omega_s/\Delta\omega_t \approx 0.01$, experiments (Dennis and Duling, 1993, 1994) indicate a dominant influence of the periodic resonances on the stability. This can be qualitatively explained by the fact that the spectral width due to the soliton relation in Equation (2.25) increases ($\propto 1/D$) faster than the sideband moves away from the soliton ($\propto 1/\sqrt{D}$). Therefore, for decreasing negative GDD the parameter $\Delta\omega_p/\Delta\omega_s$ can approach values (see graph 1 in Figure 2.12) for which the sideband generation is very strong, thus leading to instability and limiting further pulse compression.

In the case of strong periodic pulse shaping, modifications to the soliton pulse width have to be included in the stability considerations. A qualitative argument indicates that, for laser systems where SPM and negative GDD act separately, the maximum achievable spectral width for $D \to 0$ is limited to $\Delta\omega_s \propto 1/(W\delta)$, see Equation 2.31. Evaluating the ratio $\Delta\omega_p/\Delta\omega_s$ using Equation 2.31, a dependence depicted by graph 2 in Figure 2.12 is obtained, where the minimum in graph 2 depends on the magnitude of $W\delta$. This shows stability and shortest achievable pulse width can be strongly influenced by the magnitude of $W\delta$, as was first reported in the numerical investigations of Brabec *et al.* (1991).

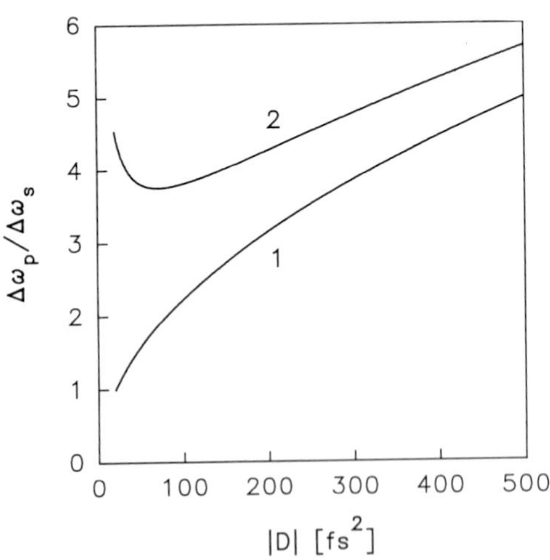

Figure 2.12. Ratio of resonance position to width of the modelocked spectrum $\Delta\omega_p/\Delta\omega_s$ versus negative GDD; in graphs 1 and 2 $\Delta\omega_s$ is determined by the soliton pulse width (2.25) and the solitary pulse width (2.31) respectively. See text for further information.

2.3 Conclusion

Recent research on femtosecond pulse generation in solid state lasers has been stimulated by the emergence of new efficient modelocking techniques exploiting the Kerr nonlinearity of solid state media. In this chapter we have summarized theoretical investigations on new aspects raised by this rapid progress in femtosecond solid state laser technology: (i) pulse buildup in passively modelocked lasers; this is of particular importance in solid state oscillators which are in general not self-starting, due to the low self-amplitude modulation achievable by nonresonant nonlinearities; (ii) solitary modelocking; due to the strong self-phase modulation in solid state Kerr nonlinearities, theoretical models developed for steady state analysis in femtosecond dye lasers have been found to be only partially applicable and had to be extended to account for a number of new phenomena.

In Section 2.1 an equation for the pulse buildup starting from an initial fluctuation has been presented. Solution of the equation has yielded a self-starting threshold condition as a function of a few easily measurable laser parameters. In solid state lasers the beat-note linewidth of the free running (CW) laser is the dominant parameter determining the threshold. Possible sources for beat-note linewidth broadening have been identified and their influence on the self-starting threshold has been evaluated. As a result, spatial hole-burning in the gain medium has been found to be the most dominant effect opposing self-starting in standing wave laser oscillators, which is supported by a reasonable agreement of theoretical predictions with experiments.

In Section 2.2 soliton-like pulse formation in femtosecond solid state lasers has been discussed. We have shown that solitary systems can be regarded as soliton systems subject to various types of perturbations, where the main perturbations are third-order dispersion and periodic perturbations. Scaling parameters have been introduced to estimate the influence of dispersive and periodic perturbations. Investigation of steady state in the presence of these perturbations has revealed a modification of the pulse parameters and additional resonant effects adversely affecting the stability of modelocked operation. It has been pointed out that a stabilizing mechanism suppressing the instabilities is indispensable for femtosecond pulse formation. Self-amplitude modulation is an essential mechanism to support stability of soliton-like pulse shaping. For decreasing pulse widths, however, it has been demonstrated that the destabilizing influence of the perturbations prevails, thus setting a limit to the shortest obtainable pulses. Simple relationships for pulse parameters and stability limits in the presence of perturbations have been obtained, which can be used as a help for design and optimisation of femtosecond solid state oscillators.

Acknowledgments

The authors are deeply indebted to Professors A. J. Schmidt and E. Wintner for their support. T. Brabec is supported by the Österreichische Akademie der Wissenschaften, APART grant. S. M. J. Kelly acknowledges the support of the Royal Society, in the form of a European Exchange Fellowship. This research was sponsored by the Jubiläumsfonds der Österreichischen Nationalbank, grant 4418, and the Fonds zur Förderung der Wissenschaftlichen Forschung in Österreich, P8566 and P09710.

References

G. P. Agrawal (1989) *Nonlinear fiber optics*, Academic Press, New York

M. Asaki, C. Huang, D. Garvey, J. Zhou, H. C. Kapteyn and M. M. Murnane (1993) *Opt. Lett.* **18**, 977

N. L. Borodin, A. G. Okrimchuk and A. V. Shestakov (1992) *OSA Proceedings on Advanced Solid State Lasers*, Vol. 13 (Optical Society of America, Washington, DC), 42

K. J. Blow and D. Wood (1988) *J. Opt. Soc. Am. B* **5**, 629

T. Brabec and S. M. J. Kelly (1993) *Opt. Lett.* **18**, 2002

T. Brabec, Ch. Spielmann and F. Krausz (1991) *Opt. Lett.* **16**, 1961

T. Brabec, Ch. Spielmann and F. Krausz (1992a) *Opt. Lett.* **17**, 748

T. Brabec, Ch. Spielmann, P. F. Curley and F. Krausz (1992b) *Opt. Lett.* **17**, 1292

T. Brabec, P. F. Curley, Ch. Spielmann, E. Wintner and A. J. Schmidt (1993) *J. Opt. Soc. Am. B* **10**, 1029

C. L. Cesar, M. N. Islam, C. E. Soccolich, R. D. Feldman, R. F. Austin and K. R. German (1990) *Opt. Lett.* **15**, 1147

S. Chen and J. Wang (1991) *Opt. Lett.* **16**, 1689

P. F. Curley, Ch. Spielmann, T. Brabec, F. Krausz, E. Wintner and A. J. Schmidt (1993) *Opt. Lett.* **18**, 54

M. L. Dennis and I. N. Duling (1993) *Appl. Phys. Lett.* **62**, 2911

M. L. Dennis and I. N. Duling (1994) to be published in *IEEE J. Quantum Electron*

J. N. Elgin (1992) *Opt. Lett.* **17**, 1409

J. N. Elgin (1993) *Phys. Rev. A*, **47**, 4331

J. N. Elgin and S. M. J. Kelly (1993) *Opt. Lett.* **18**, 787

C. J. Flood, G. Giuliani and H. M. van Driel (1990) *Opt. Lett.* **15**, 218
R. L. Fork, B. I. Greene and C. V. Shank (1981) *Appl. Phys. Lett.* **38**, 671
D. Georgiev, J. Hermann and U. Stamm (1992) *Opt. Commun.* **92**, 368
J. P. Gordon (1992) *J. Opt. Soc. Am. B* **9**, 91
A. B. Grudinin, E. M. Dianov, D. V. Korobkin, A. M. Prokhorov, V. N. Serkin and D. V. Khaidarov (1987) *JETP Lett.* **46**, 221
H. A. Haus (1975a) *IEEE J. Quantum Electron.* **QE-11**, 736
H. A. Haus (1975b) *J. Appl. Phys.* **46**, 3049
H. A. Haus and E. P. Ippen (1991) *Opt. Lett.* **16**, 1331
H. A. Haus, J. G. Fujimoto and E. P. Ippen (1991) *J. Opt. Soc. Am. B* **8**, 2068
H. A. Haus, J. G. Fujimoto and E. P. Ippen (1992) *IEEE. J. Quantum Electron.* **QE-28**, 2086
M. Hofer, M. E. Fermann, F. Haberl, M. H. Ober and A. J. Schmidt (1991) *Opt. Lett.* **16**, 502
E. P. Ippen, H. A. Haus and L. Y. Liu (1989) *J. Opt. Soc. Am.* B **6**, 1736
E. P. Ippen, L. Y. Liu and H. A. Haus (1990) *Opt. Lett.* **6**, 183
P. N. Kean, X. Zhu, D. W. Crust, R. S. Grant, N. Langford and W. Sibbett (1989) *Opt. Lett.* **14**, 39
U. Keller and T. H. Chiu (1992) *IEEE J. Quantum Electron.* **QE-28**, 1710
S. M. J. Kelly (1992) *Electron. Lett.* **28**, 806
F. Krausz and T. Brabec (1993) *Opt. Lett.* **18**, 888
F. Krausz, Ch. Spielmann, T. Brabec, E. Wintner and A. J. Schmidt (1990) *Opt. Lett.* **15**, 1082
F. Krausz, T. Brabec and Ch. Spielmann (1991) *Opt. Lett.* **16**, 235
F. Krausz, M. E. Fermann, T. Brabec, P. F. Curley, M. Hofer, M. H. Ober, Ch. Spielmann, E. Wintner and A. J. Schmidt (1992) *IEEE J. Quantum Electron.* **QE-28**, 2097
P. G. Kryukov and V. S. Letokhov (1972) *IEEE J. Quantum Electron.* **QE-8**, 766
N. Langford, K. Smith and W. Sibbett (1987) *Opt. Lett.* **12**, 903
B. E. Lemoff and C. P. J. Barty (1992) *Opt. Lett.* **17**, 1367
Y. M. Liu, K. W. Sun, P. R. Prucnal and S. A. Lyon (1992) *Opt. Lett.* **17**, 1219
J. Mark, L. Y. Liu, K. L. Hall, H. A. Haus and E. P. Ippen (1989) *Opt. Lett.* **14**, 48
O. E. Martinez, R. L. Fork and J. P. Gordon (1985) *J. Opt. Soc. Amer. B*, **2**, 753
L. F. Mollenauer and R. H. Stolen (1984) *Opt. Lett.* **9**, 13
M. Morin and M. Piché (1990) *Opt. Lett.* **14**, 1119
P. F. Moulton (1986) *J. Opt. Soc. Am. B*, **3**, 125
G. H. C. New (1974) *IEEE J. Quantum Electron.* **QE-10**, 115
D. U. Noske, N. Pandit and J. R. Taylor (1992) *Opt. Lett.* **17**, 1515
M. H. Ober, M. Hofer and M. E. Fermann (1993) *Opt. Lett.* **18**, 367
F. Ouellette and M. Piché (1986) *Opt. Commun.* **60**, 99
V. Petricevic, S. K. Gayen and R. R. Alfano (1989) *Opt. Lett.* **14**, 612
M. Piché (1991) *Opt. Commun.* **86**, 156
B. Proctor and F. Wise (1993) *Appl. Phys. Lett.* **62**, 470

F. Salin, J. Squier and M. Piché (1991) *Opt. Lett.* **16**, 1674

A. E. Siegman (1986), *Lasers*, Mill Valley, CA: University Science

L. K. Smith, S. A. Payne, W. L. Kway, L. L. Chase and B. H. T. Chai (1992) *IEEE J. Quantum Electron.* **QE-28**, 2612

C. E. Soccolich, M. N. Islam, M. G. Young and B. I. Miller (1990) *Appl. Phys. Lett.* **56**, 2177

D. E. Spence, P. N. Kean and W. Sibbett (1991) *Opt. Lett.* **16**, 42

Ch. Spielmann, F. Krausz, T. Brabec, E. Wintner and A. J. Schmidt (1991) *Appl. Phys. Lett.* **58**, 2470

Ch. Spielmann, P. F. Curley, T. Brabec and F. Krausz (to be published 1994) *IEEE J. Quantum Electron.*

L. Spinelli, B. Couillaud, N. Goldblatt and D. K. Negus (1991) *Digest of Conference on Lasers and Electro-Optics* (Optical Society of America, Washington, DC), paper CPDP7

K. Tamura, J. Jacobson, E. P. Ippen, H. A. Haus and J. G. Fujimoto (1993) *Opt. Lett.* **18**, 220

C. L. Tang, H. Statz and G. deMars (1963) *J. Appl. Phys.* **34**, 2289

J. A. Valdmanis, R. L. Fork and J. P. Gordon (1985) *Opt. Lett.* **10**, 131

A. M. Weiner and E. P. Ippen (1984) *Opt. Lett.* **9**, 53

L. Yan, P.-T. Ho, C. H. Lee and G. L. Burdge (1989) *IEEE J. Quantum Electron.* **QE-25**, 2431

3

Compact modelocked solid state lasers pumped by laser diodes

JOHN R. M. BARR

3.1 Introduction

The development of commercially available laser diodes (LDs) has revolutionised solid state lasers so much so that today the term *solid state lasers* normally implies laser diode pumping. The change started in the mid 1980s with the commercial release of GaAs LD arrays capable of generating between 50 mW and 250 mW around 800 nm and capable of optically pumping Nd doped laser hosts. Subsequently, the range of available wavelengths and the power levels from LDs has increased dramatically, particularly around 980 nm, 800 nm and 670 nm, so that almost every possible dopant and host combination is capable of being LD pumped. In parallel with the improving technology for the pump sources, a new understanding of techniques for modelocking solid state lasers, that allowed the pulse durations directly from the laser to approach the limit set by the gain bandwidth of the laser, has dramatically changed the pulse duration expectations from solid state lasers. Consequently the design of a modern modelocked solid state laser is quite different from previous designs. It is the purpose of this chapter to summarise the developments that have been made over the past five to ten years in developing compact modelocked sources based on LD pumped solid state lasers.

3.1.1 *Laser diode pumping*

Pumping using LDs has been a major factor in allowing the phrases *compact* or *miniature* to be applied to LD pumped solid state lasers (LDPSSLs). This claim will be justified by quoting typical values for various parameters that should be achievable. It cannot be emphasised enough that the developments in this field, particularly in the performance of LDs and in the growth of new laser hosts, will make these Figures somewhat prone to revision. LD arrays are capable of producing up to 20 W continuous wave output from a total device volume (LD chip, heatsink and package) of only a few cm^3. Due to the high electrical to optical conversion efficiency of about 40% (device dependent), very little cooling of the pump source is required (no expensive water cooling system) and the power supply can be attached to a normal laboratory (or domestic) single phase supply. The directional nature and monochromaticity of the LD output allow efficient coupling between the pump source and the laser host. Around 80% of the LD output can be delivered to, and absorbed within, a single absorption feature of the laser host. The LDPSSL can reach optical to optical conversion efficiencies of around 50%. The net result is a laser system with a conversion from electrical power to optical power of around $0.4 \times 0.8 \times 0.5 = 16\%$. System efficiencies of this magnitude have indeed been obtained.

In addition to providing a compact, easily utilised pump source, the LD has a number of other advantages:

(a) It can have a very long lifetime, approaching 10 000–50 000 hours of operation or 1.1 to 5.7 years of **continuous** operation. This feature makes LDs very attractive for applications that make frequent servicing impractical, e.g. systems installed in satellites. For comparison, a standard arc pumped CW Nd:YAG laser might require a lamp change every 200 hours.

(b) The monochromatic pumping mechanism allows selective and hence efficient pumping of a single absorption band. Compared with broadband pumping mechanisms, e.g. flashlamps or arc lamps, the thermal loading for a given output is reduced so that thermal problems, for example thermal lensing, thermal birefringence, and thermal damage such as fracture or melting, can be reduced. This allows successes such as the CW operation of thermally sensitive materials like Nd:glass when optically pumped by an LD. However, thermal effects are likely to be one of the main limits to the high power scaling of LDPSSLs.

(c) The utilisation of the LD as a pump source removes some of the major problems associated with water cooled solid state lasers in the past. The long fluorescence lifetimes, in the region of 200 μs to 10 ms of most (but not all) solid state lasers, makes them prone to developing relaxation oscillations that reduce the amplitude stability of the laser. The mechanical vibration caused by cooling water flow coupled to the optical resonator drives low amplitude relaxation oscillations and degrades the laser stability. LDs avoid the need for cooling water and provide a quieter environment for the solid state laser. Further, the LD output power stability can also be controlled by stabilising the drive current allowing the construction of a very quiet laser so that small effects, that may be masked in other lasers by mechanical noise, may be exploited in LDPSSLs.

(d) A less obvious benefit of laser pumping of solid state lasers, for example using a LD, is that a wide variety of laser hosts can be used. Flashlamp pumping requires that the laser material can be produced in long rods of high optical quality to match the length of the flashlamps and sufficient thickness to allow efficient absorption of the pump light. Examples of materials that can be produced with these dimensions include Nd:YAG, Nd:glass, ruby, and Nd:YLF. Laser pumping requires a sample only slightly larger than the diameter of the pump beam and roughly one absorption length long. Many more materials can be grown with these dimensions than in the larger sizes required for flashlamp pumping.

The general historical development of LDPSSLs has been described in two reviews. The first one, by Fan and Byer (1988), covers the literature until 1987, while the second, by Hughes and Barr (1992a), is current up to 1991. It is important to remember that the LDPSSL is not a new development. In fact the first semiconductor laser pumped solid state laser ($U^{3+}:CaF_2$, $\lambda = 2.613\,\mu m$) dates from 1963 (Keynes and Quist, 1964). The recent development of LDs is described in a review by Streifer *et al.* (1988).

3.1.2 *Modelocked solid state lasers*

A good review of the status of modelocking techniques suitable for use with various laser hosts and covering the period up to 1982 is that of New (1983). Prior to the development of the new passive modelocking schemes now commonly used with LDPSSLs (see Chapters 1 and 2 in this book), only a few techniques were suitable for use with solid state lasers. These include active modelocking techniques based on either amplitude or phase modulation via intracavity modulators, and traditionally described by the analysis of Kuizenga and Siegman (1970). Normally these lasers are continuous wave (CW) or operated with a prelase and

Q-switch, because of the many round trips it takes for the pulse to develop within the cavity (Kuizenga, 1981). As will be noted in Section 3.2, the drawback of active modelocking in solid state lasers is that the resulting pulse duration does not make full use of the available gain bandwidth, and in particular, is unlikely to result in pulse durations less than 1 ps. Alternatively, passive modelocking using nonlinear techniques based on the saturation of dyes (Mocker and Collins, 1965), nonlinear (intensity dependent) modulators (Dahlstrom, 1973), or perhaps including an active modulator in a hybrid active–passive scheme to avoid the random nature of true passive modelocking (Lewis and Knudtson, 1982), can result in subpicosecond pulses. However, the output of these lasers tended to be in the form of a short Q-switched train of pulses where the pulse energy and pulse duration evolve during the train. Clearly these types of lasers are not ideal for use in precision measurements where repeatability of the pulse characteristics that can be achieved from CW modelocked lasers is essential. The experimental results reported using the new passive modelocking schemes will be summarised in Section 3.3.

3.2 Active modelocking of laser diode pumped solid state lasers

Active modelocking of solid state lasers provides one of the more robust modelocking techniques presently available. The pulse formation and quality is largely independent of the laser performance resulting in a dependable device. The relatively long pulses that it generates does mean that it is not always the technique of choice, nonetheless whenever a new laser material is modelocked for the first time, it is likely to be an active technique. We will now briefly describe the principles of active modelocking and then review the results that have been achieved with LDPSSLs.

A LDPSSL that includes an active modulator is shown in Figure 3.1. The modulator changes either the amplitude (AM) or phase (FM) of the signal that passes through it and is driven by an external RF signal. Each element of the laser shown in Figure 3.1 plays a role in determining the final pulse duration. The analysis of active modelocking of homogeneously broadened solid state lasers described here is based on that of Kuizenga and Siegman (1970) but uses the notation of Siegman (1986). A solution in the form of a Gaussian pulse will be sought. Under steady state conditions, when the pulse is fully formed, the modifications made by passage through the gain material and the modulator will result in the pulse duration and amplitude remaining unchanged after each round trip.

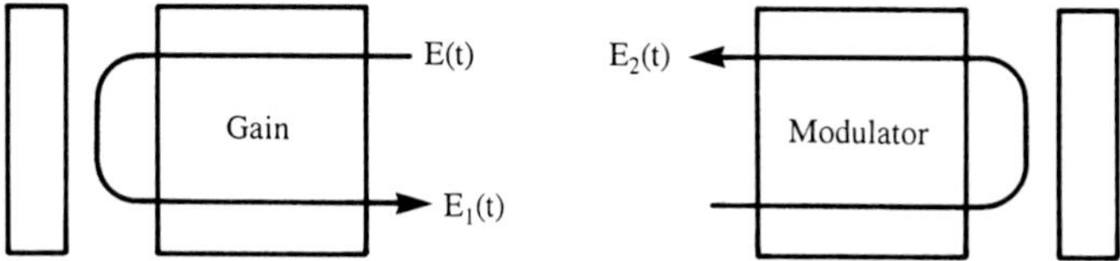

Figure 3.1. A schematic diagram of a laser containing an active modulator. The input pulse shape $E(t)$ passes through the gain medium and then through the modulator before returning to the starting plane.

The circulating pulse will be assumed to be Gaussian of the form:

$$E(t) = E_0 \exp(-\Gamma t^2 + i\omega_0 t), \tag{3.1}$$

where the pulse has a center frequency of ω_0 and $\Gamma = \alpha - i\beta$ is the Gaussian pulse parameter, so the pulse duration, defined as the full width at half-maximum (FWHM) of the intensity, is:

$$\tau_\mathrm{p} = \left(\frac{2\ln 2}{\alpha}\right)^{1/2}. \tag{3.2}$$

The pulse, in general, also has a linear chirp, so the instantaneous frequency at any point on the pulse envelope is:

$$\omega(t) = \omega_0 + 2\beta t. \tag{3.3}$$

The Fourier transform of the pulse described by Equation 3.2 has a Gaussian shape:

$$E(\omega) \propto \exp - \frac{(\omega - \omega_0)^2}{4\Gamma}. \tag{3.4}$$

The bandwidth (FWHM) of the spectral intensity, $\Delta\nu_\mathrm{p}$, and the time-bandwidth product, $\tau_\mathrm{p}\Delta\nu_\mathrm{p}$, of the pulse are given by:

$$\begin{aligned} \Delta\nu_\mathrm{p} &= \left(\frac{2\ln 2}{\pi^2}\right)^{1/2}\left(\frac{\alpha^2+\beta^2}{\alpha}\right)^{1/2}, \\ \tau_\mathrm{p}\Delta\nu_\mathrm{p} &= \frac{2\ln 2}{\pi}\left(1+\frac{\beta^2}{\alpha^2}\right)^{1/2}. \end{aligned} \tag{3.5}$$

For the special case of a chirp-free Gaussian pulse the time–bandwidth product is $\tau_\mathrm{p}\Delta\nu_\mathrm{p} = 0.441$. Other pulse shapes will have different time–bandwidth products, although the magnitude is close to unity, as shown in Table 3.1.

Table 3.1.

	$I(t); (x = t/T)$	$\tau_p \Delta\nu_p$	τ_p/T	$G_0^2(\tau); (y = \tau/T)$	τ_G/T	τ_p/τ_G
1. Square	$I(t) = \begin{cases} 1; & \lvert t\rvert \le T/2 \\ 0; & \lvert t\rvert > T \end{cases}$	0.8859	1	$\begin{cases} 1 - \lvert y\rvert; & \lvert\tau\rvert \le T \\ 0 \quad ; & \lvert\tau\rvert > T \end{cases}$	1	1
2. Parabolic	$I(t) = \begin{cases} 1 - x^2; & \lvert t\rvert \le T \\ 0 \quad ; & \lvert t\rvert > T \end{cases}$	0.7276	$\sqrt{2}$	$\begin{cases} 1 - \frac{5}{4}y^2 + \frac{5}{8}\lvert y\rvert^3 - \frac{1}{32}\lvert y\rvert^5; & \lvert\tau\rvert \le 2T \\ 0 \quad ; & \lvert\tau\rvert > 2T \end{cases}$	1.6226	0.8716
3. Diffraction function	$I(t) = \dfrac{\sin^2 x}{x^2}$	0.8859	2.7831	$\dfrac{3}{2y^2}\left[1 - \dfrac{\sin 2y}{2y}\right]$	3.7055	0.7511
4. Gaussian	$I(t) = e^{-x^2}$	0.4413	$2\sqrt{\ln 2}$	$e^{-y^2/2}$	$2\sqrt{2\ln 2}$	0.7071
5. Triangular	$I(t) = \begin{cases} 1 - \lvert x\rvert; & \lvert t\rvert \le T \\ 0 \quad ; & \lvert t\rvert > T \end{cases}$	0.5401	1	$\begin{cases} 1 - \frac{3}{2}y^2 + \frac{3}{4}\lvert y\rvert^3 \quad ; & \lvert\tau\rvert \le T \\ 2 - 3\lvert y\rvert + \frac{3}{2}y^2 - \frac{1}{4}\lvert y\rvert^3; & T < \lvert\tau\rvert \le 2T \\ 0 \quad ; & \lvert\tau\rvert > 2T \end{cases}$	1.445	0.6922
6. Hyperbolic sech	$I(t) = \operatorname{sech}^2 x$	0.3148	1.7627	$\dfrac{3}{\sinh^2 y}[y \coth y - 1]$	2.7196	0.6482
7. Lorentzian	$I(t) = \dfrac{1}{1 + x^2}$	0.2206	2	$\dfrac{1}{1 + (y/2)^2}$	4	0.5000
8. One-sided exp	$I(t) = \begin{cases} e^{-x}; & t \ge 0 \\ 0 \; ; & t < 0 \end{cases}$	0.1103	$\ln 2$	$e^{-\lvert y\rvert}$	$2\ln 2$	0.5000
9. Symmetric two-sided exp	$I(t) = e^{-2\lvert x\rvert}$	0.1420	$\ln 2$	$(1 + 2\lvert y\rvert)e^{-2\lvert y\rvert}$	1.6783	0.4130

A listing of commonly assumed pulse shapes $I(t)$, time bandwidth products $\tau_p\Delta_p$, autocorrelation shapes $G_0^2(\tau)$, and deconvolution factors τ_p/τ_G, after Sala *et al.*, (1980), τ_p and $\Delta\nu_p$ are the FWHM of the temporal and spectral intensities, respectively.

3.2.1 *The gain medium*

The gain bandwidth of the lasing material is finite and it, ultimately, limits the bandwidth of any intracavity signal. Using the time–bandwidth product from Equation 3.5 it is easy to see that the shortest possible pulse directly from a laser will be of the order of the inverse gain bandwidth. In most solid state lasers the gain is not depleted by the amplification of a single pulse (the pulse energy density is small compared with the saturation energy density), but only by the repeated passage of many pulses. Hence gain saturation, which is important in dye lasers and other lasers with large stimulated emission cross-sections, does not play a significant role in pulse shaping in solid state lasers. The time dependence of the gain may be neglected and only the frequency selectivity of the gain taken into account. The gain of a homogeneously broadened material is represented by:

$$g(\omega) = \exp\left[\frac{g_0}{1 + 2i\frac{(\omega-\omega_0)}{\Delta\omega_a}}\right], \tag{3.6}$$

where g_0 is the round trip saturated amplitude gain and $\Delta\omega_a$ is the linewidth of the transmission. If the pulse bandwidth is small compared with $\Delta\omega_a$, i.e. $\tau_p > 2\pi/\Delta\omega_a$, then the gain may be expanded about the line center, ω_0, to give:

$$g(\omega) = \exp[g_0]\exp\left[-2ig_0\left(\frac{\omega-\omega_a}{\Delta\omega_a}\right)\right]\exp\left[-4g_0\left(\frac{\omega-\omega_0}{\Delta\omega_a}\right)^2\right]. \tag{3.7}$$

The passage of the pulse through the gain medium modifies the amplitude, phase and bandwidth of the input signal. The Gaussian pulse parameter, Γ_1, of the amplified signal $E_1(\omega) = g(\omega)E(\omega)$ is:

$$\frac{1}{\Gamma_1} = \frac{1}{\Gamma} + \frac{16g_0}{\Delta\omega_a^2}. \tag{3.8}$$

This equation expresses the effect amplification has on the Gaussian pulse parameter. The following conclusions may be drawn:

(a) An unchirped pulse ($\beta = 0$ so Γ is real) remains unchirped, providing that the background dispersion of the gain medium may be ignored, and is always increased in duration by amplification since the finite bandwidth of the gain medium, $\Delta\omega_a$, effectively reduces the pulse bandwidth and hence increases its duration. In fact Equation 3.8 may be used to estimate the effect of gain narrowing on the effective bandwidth of an amplifier.

(b) Normally the change in pulse parameter per pass is small, so that Equation 3.8 may be re-expressed as:

$$\begin{aligned}\Gamma_1 - \Gamma &= -\Gamma^2 \frac{16g_0}{\Delta\omega_a^2} \\ &= -(\alpha^2 - \beta^2 - 2i\alpha\beta)\frac{16g_0}{\Delta\omega_a^2}.\end{aligned} \tag{3.9}$$

Amplification of a chirped pulse may be divided into three regions, depending on the relative sizes of α and β. If $\alpha > \beta$, then the pulse length is still increased on amplification. If $\alpha = \beta$ the pulse duration remains unchanged although the chirp (β_1) is altered. Finally, if $\beta > \alpha$, which is the case for a strongly chirped pulse, then the amplified pulse duration is actually shorter than the input pulse duration. These observations lie at the heart of frequency modulation modelocking, as we shall see shortly in Section 3.5.

3.2.2 *Amplitude modulation*

Assume that the modulator changes the amplitude of the injected signal according to:

$$m(t) = 1 - \Delta_m(1 - \cos\omega_m t) \approx \exp - \Delta_m(1 - \cos\omega_m t). \tag{3.10}$$

The time dependent transmission of such a modulator is shown in Figure 3.2, where ω_m is the angular frequency of the loss modulation, for a standing wave acousto-optic modulator this is twice the frequency of the driving RF field. The depth of modulation for a double pass is Δ_m. The form of the modulator transmission function is not unique and other expressions have been used in the literature. Any optical pulse incident on the modulator at a time corresponding to maximum transmission, when $\omega_m t = 2\pi \times$ an integer, will emerge with a shorter duration. Expanding $\cos\omega_m t$ for small time changes, t, with respect to the maximum transmission yields:

$$m(t) = \exp - \frac{\Delta_m}{2}\omega_m^2 t^2. \tag{3.11}$$

Thus, when the transmission of the modulator is close to its maximum, it is approximately Gaussian in time. A Gaussian pulse incident on the modulator at maximum transmission will be shortened and, from

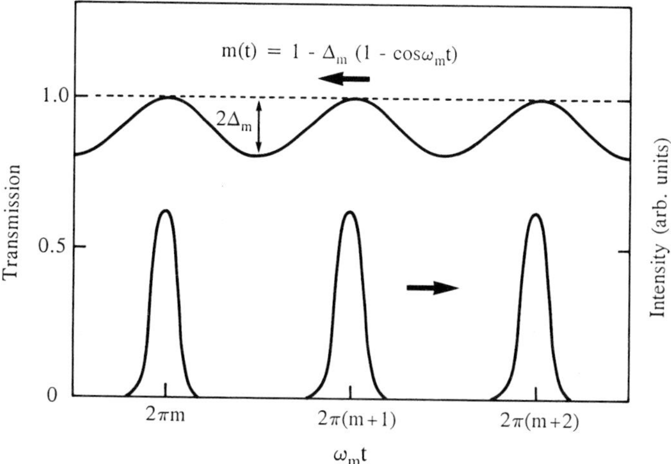

Figure 3.2. The time dependent electric field transmission of an amplitude modulator indicating where the pulse is formed.

$E_2(t) = m(t)E_1(t)$, the Gaussian pulse parameter can easily be shown to be modified to:

$$\Gamma_2 = \Gamma_1 + \frac{\Delta_m}{2}\omega_m^2. \tag{3.12}$$

Next we have to assume that the length of the laser cavity is carefully adjusted so that when the pulse repeats one round trip of the cavity, $\tau_c \approx 2L/c$, it returns to the modulator at a time corresponding to maximum transmission. The repetition rate of the laser must be:

$$\frac{1}{\tau_c} = \frac{\omega_m}{2\pi} = \nu_m.$$

Repeated passages of the pulse through the modulator will result in the pulse duration becoming unphysically short and its bandwidth becoming unphysically large. This limit will, of course, not be reached due to the spectral filtering by the gain medium.

3.2.3 *Steady state pulse duration*

With all the components of the model assembled, the complete change in Gaussian pulse parameter on one round trip may be found from Equations 3.9 and 3.12 to be:

$$\Gamma_2 - \Gamma = \frac{\Delta_m}{2}\omega_m^2 - \Gamma^2 \frac{16g_0}{\Delta\omega^2}. \tag{3.13}$$

In steady state, when $\Gamma = \Gamma_2$, we conclude that the pulse parameter is real for AM modelocking (i.e. no chirp) and that the pulse duration from Equations 3.2 and 3.13 is:

$$\tau_p = \left(\frac{2\ln 2}{\pi^2}\right)^{1/2} \left(\frac{2g_0}{\Delta_m}\right)^{1/4} \left(\frac{1}{\nu_m \Delta\nu_a}\right)^{1/2}. \tag{3.14}$$

The bandwidth for this unchirped pulse may be calculated from the time–bandwidth product for Gaussian pulses (Equation 3.5). The value for the saturated gain coefficient is calculated from $2g_0 = -\ln(1/R)$ where R is the effective output coupler including all other sources of loss within the cavity.

The important information to extract from Equation 3.14 is that the pulse duration drops slowly with increasing gain bandwidth $\Delta\nu_a$. As shown in Table 3.2, it will be very difficult to generate subpicosecond pulses from purely actively modelocked systems even if quite large modulation frequencies are used.

Table 3.2.

Material	Wavelength	Bandwidth	Modulation frequency 100 MHz	Modulation frequency 5 GHz	T_{ss} 100 MHz	T_{ss} 5 GHz
Nd:YAG	1064 nm	150 GHz	55 ps	7.8 ps	23.1 μs	9.2 ns
Nd:YLF	1047 nm	360 GHz	35.6 ps	5.0 ps	55.5 μs	22.2 ns
Nd:LMA	1054 nm	1.3 THz	18.7 ps	2.64 ps	200 μs	80.1 ns
	1082 nm	2.2 THz	14.4 ps	2.0 ps	339 μs	1.35 ns
Nd:glass	1054 nm	5.3 THz	9.3 ps	1.3 ps	816 μs	326 ns

The steady state pulse durations calculated for AM modelocking of a number of LD pumpable gain materials. The modulation depth was $\Delta_m = 1$, the modulation frequency was either 100 MHz or 5 GHz, the gain was calculated assuming that output coupling was $R = 90\%$ and the round-trip time for the resonator was taken to be $\tau_c = 1/\nu_m$. The final two columns display the value of T_{ss} calculated from Equation 3.18.

3.2.4 *Pulse formation time*

If the change in Gaussian pulse parameter on each round trip is sufficiently small, then Equation 3.13 may be recast as a differential equation in the form:

$$\begin{aligned}\frac{\mathrm{d}\Gamma}{\mathrm{d}t} &= \frac{1}{\tau_\mathrm{c}}\left[\frac{\Delta_\mathrm{m}}{2}\omega_\mathrm{m}^2 - \Gamma^2\frac{16g_0}{\Delta\omega_\mathrm{a}^2}\right] \\ &= \frac{\Gamma_\mathrm{ss}^2 - \Gamma^2}{\Gamma_\mathrm{ss}T_\mathrm{ss}}\end{aligned} \tag{3.15}$$

where the additional quantities Γ_ss and T_ss are given by:

$$\begin{aligned}\Gamma_\mathrm{ss} &= \left(\frac{\Delta_\mathrm{m}}{2g_0}\right)^{1/2}\frac{\omega_\mathrm{m}\Delta\omega_\mathrm{a}}{4}, \\ T_\mathrm{ss} &= \frac{\tau_\mathrm{c}\Delta\omega_\mathrm{a}^2}{16g_0\Gamma_\mathrm{ss}}.\end{aligned} \tag{3.16}$$

Here Γ_ss is the steady state pulse parameter and T_ss will now be shown to be a time constant that controls how long it takes for the pulse formation to occur. The solution to this equation, assuming that the pulse is initially infinitely long, $\Gamma(0) = 0$, is given by:

$$\Gamma(t) = \Gamma_\mathrm{ss}\tanh\left(\frac{t}{T_\mathrm{ss}}\right). \tag{3.17}$$

Clearly T_ss represents the time when $\Gamma(t)$ reaches 76% of its steady state value or when the pulse duration is only 1.146 times larger than its final value. Table 3.2 tabulates typical pulse durations and pulse formation times resulting from AM modelocking of a number of representative materials. The practical importance of T_ss is illustrated by writing it in a slightly more transparent form for the number of round trips required to form a pulse within 15% of its steady state value:

$$\frac{T_\mathrm{ss}}{\tau_\mathrm{c}} = \frac{1}{2}\frac{1}{\sqrt{2g_0\Delta_\mathrm{m}}}\frac{\Delta\nu_\mathrm{a}}{\nu_\mathrm{m}}. \tag{3.18}$$

In order to reduce the pulse formation time, high modulation frequencies and large modulation depths are necessary to maximise the pulse shaping on each pass through the modulator. Increasing g_0 and reducing $\Delta\nu_\mathrm{a}$, while having the same effect on T_ss, are obviously inappropriate if short pulses are required. The larger T_ss is, the more prone the laser will be to having the pulse formation disturbed by external perturbations, e.g. mechanical vibration, air turbulence or pump fluctuations; all of which make the realisation of the steady state pulse more difficult.

3.2.5 *Frequency modulation modelocking*

Instead of modulating the amplitude of the intracavity field it is possible to modulate the phase using a frequency modulator (Kuizenga and Siegman, 1970; Siegman, 1986; Harris and McDuff, 1965). Experimentally there may be advantages to this which will be noted later. From the point of view of developing a simple theory it is necessary to take the modulator transmission function as:

$$\begin{aligned} m(t) &= \exp i\Delta_{\mathrm{m}} \cos \omega_{\mathrm{m}} t \\ &\approx \exp \pm i\Delta_{\mathrm{m}}\left(1 - \frac{1}{2}\omega_{\mathrm{m}}^2 t^2\right). \end{aligned} \tag{3.19}$$

The modulator transmission has been expanded around the phase maxima (+) or minima (−) (where $\omega_{\mathrm{m}} t = \pi$ times an integer) to obtain a Gaussian-like function. Notice that the phase modulator does not, on its own, shorten the pulse, but simply modifies the phase (or frequency) of the pulse. Ignoring the constant term in Equation 3.19, and passing a Gaussian pulse through the modulator, yields:

$$\Gamma_2 = \Gamma_1 \pm i\frac{\Delta_{\mathrm{m}}}{2}\omega_{\mathrm{m}}^2, \tag{3.20}$$

for the change in pulse parameter. This results in a change to the frequency chirp β_1 and the pulse duration remains unchanged. Pulse shortening occurs when the chirped signal passes through the gain medium, as described by Equation 3.9 and pointed out in the subsequent discussion. The intracavity signal simply becomes increasingly frequency-chirped as it repeatedly passes through the modulator. Only when the chirp is large enough does any pulse shortening occur. As shown in Figure 3.3, the frequency of the signal passing through the modulator at the phase extrema does not change. Therefore these points see the maximum gain. At other times the signal can deviate markedly from line center, ω_0, resulting in a reduced gain and, hence, amplitude modulation and ultimately pulse formation. Combining Equations 3.9 and 3.20 enables the complete change in pulse parameter to be calculated on a single round trip and hence the pulse duration, in steady state, can be shown to be:

$$\tau_{\mathrm{p}} = \left(\frac{4\ln 2}{\pi^2}\right)^{1/2}\left(\frac{g_0}{\Delta_{\mathrm{m}}}\right)^{1/4}\left(\frac{1}{\nu_{\mathrm{m}}\Delta\nu_{\mathrm{a}}}\right)^{1/2}. \tag{3.21}$$

A further feature of FM modelocking that appears from this analysis is that the pulse is, in general, chirped with $\alpha = \beta$. Under these conditions its duration does not change as it shuttles round the cavity.

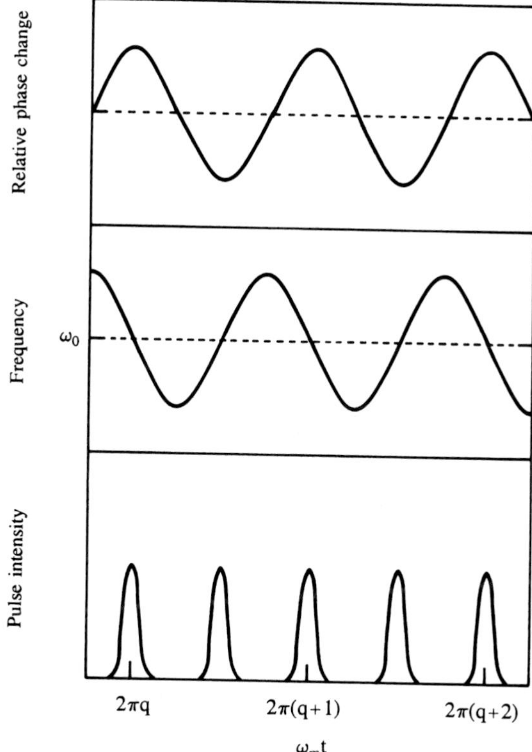

Figure 3.3. The transmission function of a phase modulator. The resulting frequency shift is also shown. The pulses form at the phase extrema where there is no frequency shift.

FM modelocking has a number of experimental advantages over AM modelocking:

(a) The RF drive frequency does not have to be accurately matched to the modulator resonance in a phase modulator. Typically the tolerance is about 1 MHz compared with about 20 kHz for FM and AM modulators respectively.

(b) Loss (AM) modulation always results in a drop in average power from the laser when it is modelocked compared with the free running performance. The reason is that some loss is essential in forming the pulse. In addition the standing wave acousto-optic modulator commonly adopted for AM modelocking may not be perfect and could have a small traveling acoustic wave component which gives a time independent loss.

(c) Finally, AM modulators require accurate adjustment of the modulator orientation to maximise the modulation depth. This is not necessary for FM modulators.

The analysis presented here and the information shown in Figure 3.3 imply that the repetition rate of an FM modelocked laser is $2\nu_m$. In practice most FM modelocked lasers oscillate with only one pulse in the cavity at a repetition rate ν_m. The pulse oscillates at one of the two phase extrema and can switch at random between the two states. It is also common to observe that the two pulses have different pulse durations, that the pulses are shorter than predicted by Equation 3.21, and this occurs when the drive frequency ν_m to the modulator is detuned slightly from an exact match to the cavity repetition rate (Kuizenga and Siegman, 1970). It is the interaction of the chirped pulses generated by FM modelocking with the dispersion within the laser cavity which causes this effect.

3.2.6 *Experimental results*

The robust nature of active modelocking has meant that there is a rich literature on the subject. It is more convenient to tabulate the majority of results by material rather than attempt to describe each experiment individually, and this has been done in Table 3.3. Not all materials have yet been LD pumped, however all the entries in the table have been laser pumped and can be potentially LD pumped.

The advantages of laser pumping of solid state materials became apparent when CW oscillation of an argon ion laser pumped Nd:phosphate glass laser was reported for the first time by Kishida *et al.* (1979). Phosphate glass is a low gain material due to its large gain linewidth and it has a low melting point and is prone to fracture under thermal stress as well as suffering from thermal lensing and thermal birefringence (Koechner, 1988a). However, it is commonly used for large Nd:glass amplifiers since large diameter rods and discs can be manufactured with excellent optical quality, so there is continuing interest in designing good quality, well characterised sources, including modelocked lasers, based on Nd:glass or other materials emitting at 1054 nm, e.g Nd:YLF or Nd:LMA. At present, this means CW sources. In addition to demonstrating CW operation, it was noted that end pumping also provides TEM_{00} operation with excellent discrimination against higher order modes without the use of an intracavity aperture. In essence, the

Table 3.3. *Experimental results on laser pumped actively modelocked solid state lasers*

Material	Wavelength	Bandwidth	Pump	AM or FM	Duration	Power	References
Nd:YAG	1064 nm	150 GHz	dye 590 nm	AM 247.7 MHz	240 ps	? mW	a
	1064 nm	150 GHz	LD ? nm	AM 174 MHz	105 ps	? mW	b
	1064 nm	150 GHz	LD 807 nm	AM 114 MHz	55 ps	45 mW	c
	532 nm	150 GHz	LD 807 nm	AM 250 MHz	520 ps	76 mW	d
	1064 nm	150 GHz	LD 807 nm	FM 350 MHz	12 ps	65 mW	e
	1064 nm	150 GHz	LD ? nm	AM 500 MHz	100 ps	60 mW	f
	1064 nm	150 GHz	LD ? nm	AM 1000 MHz	100 ps	60 mW	f
	1320 nm	? GHz	LD 807 nm	AM 114 MHz	< 50 ps	9 mW	c
	1320 nm	? GHz	LD 807 nm	FM 295 MHz	19 ps	15 mW	g
	1320 nm	? GHz	LD 807 nm	FM 1000 MHz	12 ps	162 mW	h
NdLaPO	1052 nm	750 GHz	dye 585 nm	FM 477 MHz	14 ps	8 mW	i
Nd:YLF	1053 nm	360 GHz	LD 806 nm	AM 114.8 MHz	18 ps	12 mW	j
	1053 nm	360 GHz	LD ? nm	FM 360 MHz	9 ps	49 mW	k
	1047 nm	360 GHz	LD ? nm	AM 500 MHz	9 ps	150 mW	l
	1047 nm	360 GHz	LD ? nm	AM 2000 GHz	7 ps	135 mW	m
	1047 nm	360 GHz	Ti:S 795 nm	AM 500 MHz	6.4 ps	40 mW	f
	1047 nm	360 GHz	Ti:S 795 nm	AM 1000 MHz	6.2 ps	20 mW	f
	1047 nm	360 GHz	LD 798 nm	AM 160 MHz	7 ps	160 mW	n
	1321 nm	? GHz	LD 792 nm	FM 1000 MHz	8 ps	240 mW	h
	1053 nm	360 GHz	Ti:S 797 nm	FM 5300 MHz	13 ps	350 mW	o
	?nm	360 GHz	LD ? nm	FM 100 MHz	21 ps	? mW	p
Nd:glass	1054 nm	5.3 THz	LD 802 nm	AM 174 MHz	10 ps	0.3 mW	b
	1054 nm	5.3 THz	Ar 514 nm	AM 100 MHz	9.2 ps	? mW	q
	1054 nm	5.3 THz	LD 800 nm	AM 76 MHz	7 ps	40 mW	r
	1054 nm	5.3 THz	AR 799 nm	AM 66 MHz	3.8 ps	? mW	s
	1054 nm	5.3 THz	LD 800 nm	AM 113.6 MHz	58 ps	15 mW	t
	1054 nm	5.3 THz	LD 800 nm	FM 235 MHz	9 ps	14 mW	u
	1054 nm	5.3 THz	LD 800 nm	AM 240 MHz	9 ps	30 mW	v
Nd:BEL	1070 nm	900 GHz	LD 810 nm	FM 250 MHz	7.5 ps	65 mW	w
	1070 nm	900 GHz	LD 810 nm	AM 500 MHz	7.5 ps	230 mW	x
	1070 nm	900 GHz	LD 810 nm	FM 20 GHz	3.9 ps	30 mW	y
	1070 nm	900 GHz	LD 810 nm	FM 238 MHz	2.9 ps	30 mW	y
Nd:LMA	1054 nm	1.3 THz	Ti:S 798 nm	FM 362 MHz	70 ps	50 mW	z
	1054 nm	1.3 THz	LD 802 nm	FM 230 MHz	14 ps	50 mW	aa
forsterite	1227 nm	36.5 THz	YAG 1064 nm	AM 76 MHz	31 ps	120 mW	ab
Cr:YAG	1440 nm	? THz	YAG 1064 nm	AM 125 MHz	26 ps	20 mW	ac
$(F_2^+)^*$:NaF	1070 nm	1 nm	LD 820 nm	SP 160 MHz	< 900 ps	? mW	ad

A tabulation of materials that have been, or have the potential of being, LD pumped and actively modelocked. The lasing wavelength, gain bandwidth, pump source (LD = laser diode, Ti:S = Titanium doped sapphire, Ar = argon ion laser, Kr = krypton ion laser), modelocking technique (AM = amplitude modulation, FM = frequency modulation, SP = synchronous pumping), repetition rate, pulse duration and average output power are listed. The references are (a) Alcock and Ferguson, 1986; (b) Basu and Byer, 1988; (c) Maker *et al.*, 1988; (d) Dimmick, 1989; (e) Maker and Ferguson, 1989a; (f) Walker *et al.*, 1990; (g) Keen and Ferguson, 1989; (h) Zhou *et al.*, 1991; (i) Chinn and Zwicker, 1979; (j) Maker and Ferguson, 1989b; (k) Maker and Ferguson 1989c; (l) Keller *et al.*, 1990; (m) Weingarten *et al.*, 1990; (n) Juhasz *et al.*, 1990 (o) Schulz and Henion, 1991; (p) Sweetser *et al.*, 1993; (q) Yan *et al.*, 1989; (r) Krausz *et al.*, 1989; (s) Krausz *et al.*, 1990a; (t) Dimmick, 1990; (u) Hughes *et al.*, 1991; (v) Hughes *et al.*, 1992b; (w) Godil *et al.*, 1991a; (x) Li *et al.*, 1991; (y) Godil *et al.*, 1991b; (z) Shi *et al.*, 1993; (aa) Hughes *et al.*, 1993a; (ab) Seas *et al.*, 1991; (ac) French *et al.*, 1993; (ad) Mazighi *et al.*, 1991.

finite transverse extent of the pumped region acts as the spatial mode selecting aperture.

Subsequently, a CW dye laser at $\lambda = 585\,\text{nm}$ was used by Chinn and Zwicker (1979) to end-pump a $Nd_{0.5}La_{0.5}P_5O_{14}$ laser which was mode-locked using a phase modulator. The gain material is interesting because it contains a high doping level of Nd and also exhibits a large linewidth of 750 GHz centered at 1052 nm. There are a number of notable points contained within this paper and the CW Nd:glass report which, as they have been repeated many times in subsequent work, are worth emphasising now.

The resonator design. The laser resonator adopted was similar to the three mirror astigmatically compensated cavity used for CW dye lasers and shown in Figure 3.4(a), together with a commonly used variant, Figure 3.4(b). An analysis of its properties was presented by Kogelnik *et al.* (1972). This type of resonator was chosen because it allows a tightly focused beam waist at the gain medium, thus helping to ensure a low threshold. In addition a quasi-collimated arm is also provided, where components such as modulators and tuning elements can be sited. Further, the gain material is inserted at Brewster's angle, which minimises the cavity losses and ensures a linear polarised output without the addition of another component. The problem is that an inclined interface causes astigmatism to an optical beam passing through it (Hanna, 1969). Unless remedial action is taken then, at best, the laser output will be astigmatic; at worst, the resonator could be unstable since the stability criteria might not be met in both planes of the resonator simultaneously. Stability and the production of a circular beam are assured by using an off-axis curved mirror to collimate the beam from the gain medium. The astigmatism caused by the Brewster surface(s) of the gain medium can be canceled by the astigmatism of the curved mirror if the correct angle of incidence on the mirror is chosen.

For completeness the equations describing the stability requirements of a three-mirror resonator will be stated. For the case shown in Figure 3.4(a) where the gain medium, in the form of a disk of thickness t (measured normal to the disk surface) and of refractive index n, is separate from the end mirror, astigmatism compensation occurs when θ, the angle of incidence onto the fold mirror of radius of curvature R, satisfies:

$$R \sin\theta \tan\theta = 2t\frac{n^2 - 1}{n^4}\sqrt{1 + n^2}. \tag{3.22}$$

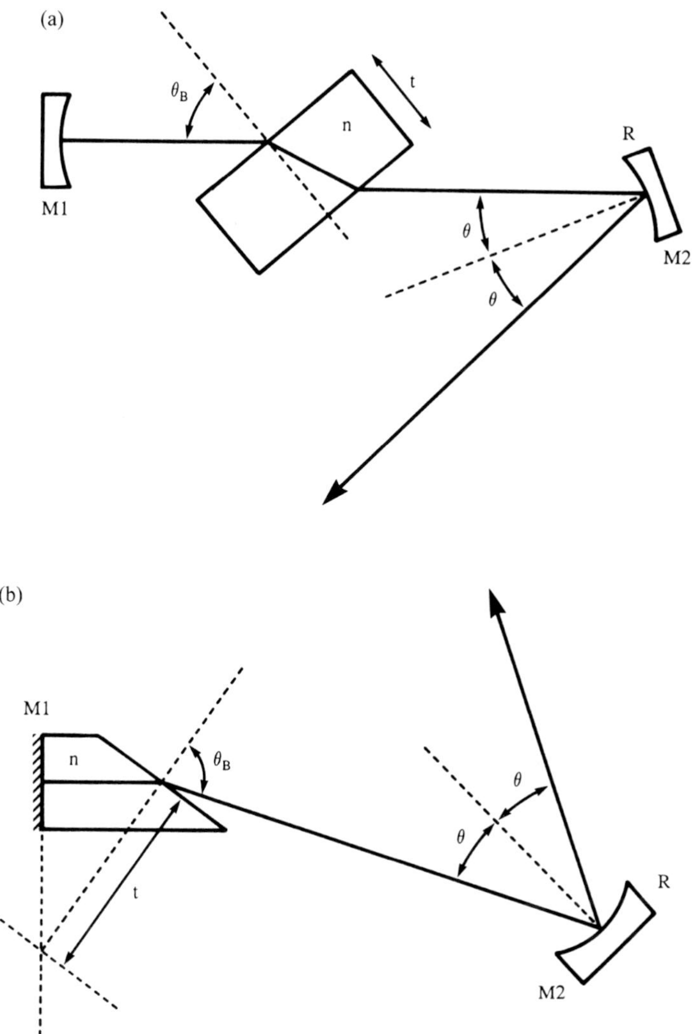

Figure 3.4. The two configurations of gain medium commonly adopted for laser pumped solid state lasers. In (a) the gain medium is separate from the cavity mirrors and has thickness t and refractive index n. Mirror M2, often known as the fold mirror, has a radius of curvature R and the beam is incident at an angle θ. In (b) the gain medium has a dielectric mirror M1 coated directly onto it. The thickness of the material is measured along the normal to the Brewster interface to the projection of M1. Mirror M2 again has a radius of curvature R and an angle of incidence θ.

The same equation may be used to calculate the angle of incidence required to compensate for the Brewster surface astigmatism for the case illustrated in Figure 3.4(b), where t is now the distance between the normal to the Brewster surface and the back surface of the gain medium. In addition, Ferguson (1989) showed that the waist size, ω_0, at the tight focus is:

$$\omega_0^2 \approx \frac{\lambda}{2\pi} \frac{R^2}{4(d_2 - R/2)} \tag{3.23}$$

and that the spot size at the fold mirror, and (approximately) throughout the quasi-collimated region, is given by:

$$\omega_{\text{mirror}} \approx \sqrt{\frac{\lambda_2 d_2}{\pi}}. \tag{3.24}$$

A final point in favor of a Brewster surface is that it avoids the need to use antireflection (AR) coatings on the normal incidence surface of the gain medium. AR coatings are prone to being optically damaged by high power pump beams focused to spot sizes of the order of $\approx 50\,\mu$m. Also, as AR coatings are never perfect, the gain material could act as an etalon and restrict the linewidth of the laser resulting in longer than expected pulses when modelocked.

End pumping. This type of resonator is ideal for end pumping by another laser and has a number of advantages over the alternative scheme, known as side or transverse pumping. The reason is that end pumping generally provides a higher gain than side pumping for a given absorbed power. With reference to Figure 3.5, the single pass small signal gain for end pumping with a pump beam of area A and an absorbed power P in a material of stimulated emission cross-section σ and lifetime τ is:

$$G = \exp \frac{P}{h\nu} \frac{\sigma\tau}{A} \tag{3.25}$$

The equivalent expression for side pumping is:

$$G = \exp \frac{P}{h\nu} \frac{\sigma\tau}{h l_{\text{a}}}, \tag{3.26}$$

where h is the height of the pump beam while l_a is the absorption length of the pump light and is typically in the range of $1 - 20$ mm depending on the material. Clearly end pumping can give rise to the larger gains, since

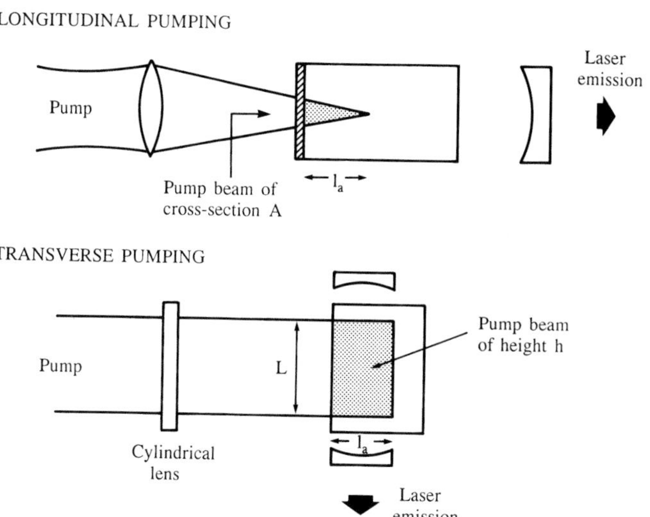

Figure 3.5. A schematic diagram of end pumping and side pumping.

the beam area can be made very small, and hence to lower laser thresholds than side pumping. Additionally, due to the small gain aperture caused by the pump beam in end pumping, it is far simpler to achieve single transverse mode operation.

The modulation frequency. Chinn and Zwicker (1979) chose to use a high modulation frequency of 476.95 MHz resulting in a compact laser design with an overall length of 31.4 cm. As noted before, a high modulation frequency is a good way of reducing the pulse duration. The measured pulse duration of 14 ps was in good agreement with the theoretical values calculated using the Kuizenga–Siegman analysis.

Frequency modulated operation. Operation of a phase modulator with the round-trip time of the resonator detuned from resonance with the RF drive frequency results in frequency modulated operation where the output power is constant (no pulses), but the frequency is sinusoidally changing with time about a central value. FM operation over a bandwidth of 240 GHz was obtained by Chinn and Zwicker. This is much larger than the bandwidth for the modelocked case.

A number of other reports of modelocked laser pumped solid state lasers (Alcock and Ferguson, 1986; Yan *et al.*, 1989) appeared during the 1980s. However, with the availability of LDs, the first reports of modelocked operation of LD pumped Nd:glass and LD pumped Nd:YAG appeared (Basu and Byer, 1988; Maker *et al.*, 1988). As might be expected from the previous discussion, the lasers were end pumped and the resonators were astigmatically compensated three mirror cavities. Basu and Byer reported AM modelocking of Nd:glass pumped by a single stripe low power LD at 800 nm. While this system worked, the output power was too low to measure the pulse duration directly using an autocorrelator, so additional pump power was supplied from a dye laser. Under these conditions the pulse duration was estimated to be in the range 5.8 ps to 10 ps depending on the pump power, the pulse duration decreased with higher pump powers, which is not predicted by the Kuizenga–Siegman analysis. Replacing the Nd:glass disk with a piece of Nd:YAG, Basu and Byer obtained a 105 ps pulse with LD pumping. Again increased pump power from the dye laser yielded shorter pulses. Maker *et al.* used a high power LD array providing 500 mW at 807 nm to pump a Nd:YAG laser. In order to obtain the shortest pulses they noted that all etalon effects within the cavity had to be avoided and used wedged optics throughout. The shortest pulse durations were 55 ps. Q-switched operation, using a low level lasing or prelase to allow the pulses enough round trips to form before switching the cavity Q, allowed peak pulse powers of 7 kW to be reported.

LD pumping and modelocking of Nd:YLF was reported by Maker and Ferguson (1989b). The larger gain linewidth and longer upper state lifetime of Nd:YLF compared with Nd:YAG allows shorter pulse durations and higher Q-switched pulse energies. AM modelocking of this system yielded 18ps pulses, assuming a sech^2 pulse shape, although the autocorrelation did show large wings. Table 3.1 shows a list of commonly assumed pulse shapes, the appropriate deconvolution factors and time–bandwidth products. The first report of FM modelocking of a LD pumped Nd:YAG laser and a Nd:YLF laser by Maker and Ferguson (1989a, 1989c) demonstrated pulse durations of 12 ps at 1064 nm and 9 ps at 1054 nm respectively. It was noted that the measured pulse durations tended to be significantly shorter than those calculated from the Kuizenga–Siegman theory.

Intracavity frequency doubling of a modelocked Nd:YAG laser was examined by Dimmick (1989) as a way of increasing the conversion efficiency to 532 nm and of avoiding the fluctuations in conversion efficiency observed in intracavity doubling of multimode CW lasers. In this way pulse durations of 520 ps at an average power of 76 mW were reported.

FM modelocking of a LD pumped Nd:YAG laser operating at 1.32 μm enabled pulses as short as 19 ps to be generated by Keen and Ferguson (1989). Q-switched operation allowed 2.3 μJ to be generated in a 350 ns long train of 30 ps pulses. Subpicosecond pulses were obtained by using stimulated Raman scattering in a 1 km length of single mode optical fiber (Gouvei-Neto *et al.*, 1988).

Nd:glass was successfully pumped using a high power 500 mW LD array at 800 nm and modelocked using an amplitude modulator by Krausz *et al.* (1989). In this case the shortest pulse durations of 7 ps were observed just above threshold, increasing to 12 – 20 ps at higher pump levels. Again, the pulse durations were claimed to be shorter than those derived from the Kuizenga–Siegman theory by a factor of two, the difference being attributed to self-phase modulation at the tight focus within the gain material. The output power from this material remained low at 40 mW compared with Nd:YLF and Nd:YAG. One important point was noted, the shortest pulse durations using active modelocking require precise cavity length stabilisation to the modulation frequency. For the case of Nd:glass, the allowed length fluctuations were estimated to be less than 1 μm. This is difficult to obtain passively, so some kind of active stabilisation technique was recommended. A scheme that does this has been recently demonstrated by Turi and Krausz (1991) using a Nd:glass laser, although other schemes may be found in the literature (Koechner 1988b). The length of the laser is monitored by measuring the first intermode beat frequency at $c/2L$ present in the laser output, frequency dividing by a factor of two, and using this frequency to drive the amplitude modulator. Any length fluctuations are automatically tracked and the laser remains optimally modelocked. The shortest pulse durations yet generated from Nd:glass used active modelocking with a special high modulation depth amplitude modulator to generate 3.8 ps pulses (Krausz *et al.*, 1990a). FM modelocking and FM operation of a LD pumped Nd:glass laser have been demonstrated to yield pulse durations of 9 ps and FM bandwidths of 850 GHz (Hughes *et al.*, 1991).

The quest for shorter pulses led to the development of high frequency modulators, since, as shown in Table 3.2, this is an effective way of reducing the pulse length. For the purpose of this article, high frequency is arbitrarily defined as those modulators operating at frequencies higher than 500 MHz. Additionally, high modulation frequencies result in a more compact laser, for a 1 GHz modulation frequency the cavity length is only 15 cm. These modelocked lasers are truly compact. Various modulator materials can be used for this purpose provided that the acoustic absorption is not too severe. As a rule the acoustic absorption increases with frequency and reduces the standing wave Q and hence the diffraction efficiency. AM modulators have been constructed from sapphire by Weingarten *et al.* (1990) which has a low optical absorption, a moderate acoustic figure of merit and a low acoustic loss combining to give reasonable diffraction efficiency and good laser performance at these high frequencies. GaP AM devices have also been demonstrated at 500 MHz and 1000 GHz, although GaP tends to have a higher insertion loss than other optical materials (Sizer II, 1989). High frequency FM devices have also been built. A 2 cm long Nd:YLF laser cavity was operated at 5300 MHz and the pulse duration was 13 ps (Schulz and Henion, 1991). The highest modulation frequency so far reported seems to be 20 GHz where pulses of 3.9 ps were generated in a Nd:BEL laser by Godil *et al.* (1991b). This report also features the shortest pulse yet generated from an actively modelocked system of 2.9 ps at 238 MHz. FM operation was chosen for these high frequencies due to the problem of constructing high diffraction efficiency AM modulators.

Wide bandwidth tunable solid state laser materials are also LD pumpable and promise short pulses from active modelocking. So far, however, the pulse durations have been disappointing, probably because these materials can easily and efficiently be passively modelocked. For the record, Cr^{4+}:forsterite (Seas *et al.*, 1991) and Cr^{4+}:YAG (French *et al.*, 1993) have produced pulses of 31 ps and 26 ps respectively by AM modelocking. These materials are presently pumped by conventional CW Nd:YAG lasers but could be pumped either directly by a LDPSSL or by a LD itself since the Cr ion has a wide absorption band around 650 nm.

Finally, since LDs can be easily power modulated, it is possible to synchronously pump suitable materials in a similar manner to a technique commonly used with dye lasers (New, 1983). At present the reported performance by Mazighi *et al.* (1991) is disappointing, nonetheless it is worth bearing in mind should high power ultrashort pulses from LDs fulfil their potential and become a viable source of pulses.

The Kuizenga–Siegman theory provides a simple expression for calculating the pulse duration in a homogeneously broadened solid state laser. However, it is found in practice that theory and experiment can differ by up to a factor of two. The main reason for the disagreement is that the theory neglects spatial hole-burning which has a dramatic effect on the free-running linewidths of these lasers. A true homogeneously broadened laser would have only a single longitudinal mode oscillating, and the Kuizenga–Siegman theory of modelocking then corresponds to the modulator creating a number of other longitudinal modes. A real laser will exhibit a degree of spatial hole-burning that allows more than one, and frequently many, longitudinal modes to oscillate. Consequently modelocking does not correspond to the mode creation picture but is more akin to phaselocking of existing longitudinal modes, as is expected in inhomogeneously broadened lasers. A detailed comparison of the theory and experiment of FM operation of a Nd:YAG laser by Adams *et al.* (1990) showed that closest agreement was obtained when spatial hole-burning effects were eliminated completely using a unidirectional ring laser, and similar conclusions are expected to apply to modelocked operation. A further problem that has recently been observed in an FM modelocked LD pumped Nd:YLF laser concerns the pulse shape. In this experiment Sweetser *et al.* (1993) passively modelocked a dye laser to produce 50 fs pulses and synchronised it with the Nd:YLF laser with a relative jitter of 0.5 ps. The cross-correlation between these two sources provides a direct measurement of the Nd:YLF pulse duration *and* shape. The results indicate that close to threshold reasonable agreement with theory is obtained both for the pulse shape, Gaussian as expected, and for the durations, 20 – 30 ps. Above threshold the pulse becomes asymmetric and shortens, which is not expected from the theory. Again spatial holeburning is suggested as a cause, although this view may be an oversimplification. Calculations that do not assume a Gaussian lineshape for the gain material but attempt to treat it exactly, indicate that the resulting pulse durations are not only shorter than those of the Kuizenga–Siegman theory, but are also asymmetric in time (Petrov and Stankov, 1990).

Clearly, the Kuizenga–Siegman theory does not describe the fine details of the modelocking process particulary, and not surprisingly, when the laser deviates from the assumptions of theory. However, the simplicity of this approach makes it a very good place to start from, especially when analysing more complicated modelocking techniques. The formalism of this theory will be used in the discussion of passive modelocking of solid state lasers in Section 3.3.

3.3 Passive modelocking

The shortest pulses from solid state lasers are generated using one of the new passive modelocking techniques. In this section a brief review of the main processes that allow the generation of these ultrashort pulses will be presented. A discussion of the experimental results follows in Section 3.3.1. A more detailed discussion of the techniques appears in Chapters 1 and 2.

The main problem with active modelocking is that the modulation frequency is fixed. Thus, as the pulse duration is decreased, the fractional change in pulse duration on each pass also decreases, as is shown by Equation 3.12. What is really required is a fictitious amplitude modulator that can increase its modulation frequency as the pulse gets shorter. In other words, the modulation frequency depends on the pulse duration in the form: $\omega_m \propto 1/\tau_p$ so that the pulse duration is a constant fraction of the modulation period. In this case it is straightforward to show that a modulator of this type will alter the Gaussian pulse parameter so that Equation 3.12 becomes:

$$\Gamma_2 = \Gamma_1 + \eta\Gamma_1, \tag{3.27}$$

where η is a constant related in some way to the effectiveness of the modulator. Notice that the fractional change in the Gaussian pulse parameter is now independent of the pulse duration, so that the modulator remains effective for all pulse durations. Proceeding, in a similar fashion as in Section 3.2, to calculate the change in pulse parameter for a round trip of the cavity yields:

$$\Gamma_2 - \Gamma = \eta\Gamma - \Gamma^2 \frac{16 g_0}{\Delta\omega_a^2}(1 + \eta). \tag{3.28}$$

Hence the steady state pulse duration is found to be either $\tau_p = \infty$ or

$$\tau_p = \frac{2}{\pi}\sqrt{\frac{2 g_0 (1 + \eta)\ln 2}{\eta}}\frac{1}{\Delta\nu_a}. \tag{3.29}$$

There are three points worth making now. Firstly, the steady state pulse duration is proportional to the inverse of the gain bandwidth, so it should be possible for the pulse duration to approach $1/\Delta\nu_a$. Many solid state gain materials have a sufficiently large bandwidth to allow the production of subpicosecond pulses. Secondly, the existence of two steady state solutions is profoundly important since it implies that a mechanism has to be found to ensure that the CW solution ($\tau_p = \infty$) is avoided and the modelocked solution preferred. Hence, the emphasis in the literature on starting mechanisms and ‘self-starting’ modelocking. Finally it should be noted that devices with roughly the properties of the fictitious modulator invoked here can be constructed.

All the nonlinear modulators work by incorporating a device that can be regarded as a nonlinear mirror into the laser cavity. This mirror has the property that, at least in the early stages of pulse formation, its reflectivity increases with intensity so that it may be regarded as an amplitude modulator. The laser can still operate CW, yet any intensity fluctuations that arise either naturally from inter-mode beats or mechanical vibration, or are created using an active modelocker, see an increased reflectivity from the nonlinear mirror and grow into short pulses and dominate the laser output at the expense of the CW component. The CW component is suppressed via multipass gain saturation by the short pulse, since the CW reflectivity is lower from the nonlinear mirror. Since there is very little dynamic gain saturation upon a single pass of the gain medium, the nonlinear mirror has to control both the leading and falling edge of the pulse. The effective nonlinear mirrors have been based either on fast nonresonant nonlinearities, e.g. self-phase modulation, self-focusing, and second harmonic generation, or on rapidly recovering saturation in multiple quantum well saturable absorbers.

A variety of laser hosts have been modelocked using different forms of the nonlinear mirror, and examples which have been LD pumped, or are capable of being LD pumped, are tabulated in Table 3.4. The sections that follow are discussions of each of the main passive techniques.

3.3.1 *Additive pulse modelocking*

Additive pulse modelocking (APM) or coupled cavity modelocking was the first general passive modelocking scheme devised for solid state lasers capable of generating CW subpicosecond pulses. It grew from an analysis of the soliton laser which used short, high order, solitons generated externally to the laser in a single mode fiber arm, in the wavelength region where the fiber exhibited negative group velocity dispersion, and then reinjected into the main laser cavity to compress the output from the main laser. It was theoretically predicted that the soliton laser was an example of a general modelocking scheme where various nonlinear components were placed in an external cavity. The nonlinearities could be either nonlinear phase modulation, e.g. self-phase modulation using the optical Kerr effect, or nonlinear amplitude modulation using a fast saturable absorber or amplifier (Blow and Wood, 1988). An example of this modelocking scheme had in fact already been demonstrated by Ouellette and Piche (1986) in an experiment where a CO_2 laser had been modelocked using self-phase modulation in a Kerr medium.

Table 3.4

Material	Wavelength	Pump	Technique	Nonlinearity	Pulse duration	Reference
Tl:KCl	1500 nm	YAG 1064 nm	APM	Fiber	3.0 ps	a
Tl:KCl	1500 nm	YAG 1064 nm	APM	Fiber	260 fs	b
Tl:KCl	1500 nm	YAG 1064 nm	APM	SDA	250 fs	b
F_2^+:LiF	900 nm	?	APM	Fiber	1 ps	b
Tl:KCl	1500 nm	YAG 1064 nm	APM	Fiber	127 fs	c
OH^-:NaCl	1550 nm	YAG 1064 nm	APM	Fiber	75 fs	d
Nd:YAG	1064 nm	Flash	APM	SHG	30 ps	e
Ti:S	790 nm	Ar	APM	Fiber	? ps	f
Nd:YAG	1064 nm	Flash	APM	Fiber	6 ps	g
Nd:YAG	1320 nm	Flash	APM	Fiber	< 10 ps	g
Nd:YAG	1064 nm	LD 809 nm	APM	Fiber	1.7 ps	h
Nd:glass	1054 nm	Kr 800 nm	APM	Fiber	900 fs	i
Nd:glass	1054 nm	Kr 800 nm	APM	Fiber	380 fs	j
Nd:YLF	1047 nm	LD 791	APM	Fiber	1.5 ps	k
Nd:glass	1054 nm	Kr 752–799 nm	APM	Fiber	88 fs	l
Ti:S	870 nm	Ar all lines	KLM	Gain	60 fs	m
Nd:YLF	1047 nm	Ti:S 798 nm	APM	MQW	4 ps	n
Nd:YLF	1047 nm	LD 792 nm	KLM	SF57	6 ps	o
Nd:YLF	1047 nm	Ti:S 798 nm	APM	MQW	3.3 ps	p
Nd:phosphate	1054 nm	Kr 752–799 nm	APM	Fiber	1.4 ps	q
Nd:silicate	1061 nm	Kr 752–799 nm	APM	Fiber	1.4 ps	q
forsterite	1260 nm	YAG 1064 nm	KLM	Gain	60 fs	r
Nd:YAG	1064 nm	LD 808 nm	KLM	Gain	8.5 ps	s
Nd:GSAG:YSAG	1064 nm	Kr 752–799 nm	APM	Fiber	260 fs	t
Cr:LiCaAlF	800 nm	Kr 647–676 nm	KLM	Gain	170 fs	u
Cr:LiSrAlF	860 nm	Ar 476 nm	KLM	Gain	50 fs	v
Nd:LMA	1054 nm	Ti:S 802 nm	APM	Fiber	600 fs	w
Nd:LMA	1054 nm	LD 800 nm	APM	Fiber	420 fs	x
Nd:YLF	? nm	Ti:S ? nm	APM	MQW	2.8 ps	y
Nd:YAG	1064 nm	Ti:S ? nm	APM	MQW	7 ps	y
Nd:YAG	1064 nm	LD ? nm	APM	MQW	8.3 ps	z
Nd:YLF	1047 nm	LD ? nm	APM	MQW	5.1 ps	z

Tabulation of potentially LD pumpable materials that have been passively modelocked. The pump sources are: YAG = Nd:YAG; Flash = flashlamp or arc lamp pumping; LD = laser diode; Ar = argon ion laser; Kr = krypton ion laser; Ti:S = titanium doped sapphire laser. The modelocking technique is either APM = additive pulse modelocking or KLM = Kerr lens modelocking. The source of the nonlinearity is: Fiber = self-phase modulation in a fiber; SDA = saturable diode amplifier; SHG = nonlinear mirror using second harmonic generation; Gain indicates that KLM used self-phase modulation in the gain material; MQW = saturable reflector based on a multiple quantum well absorber; SF57 = KLM uses self-phase modulation in SF57 glass. The references are: (a) Blow and Nelson, 1988; (b) Kean *et al.*, 1989; (c) Mark *et al.*, 1989; (d) Yakymyshyn *et al.*, 1989; (e) Barr and Hughes, 1989; (f) Goodberlet *et al.*, 1989; (g) Liu *et al.*, 1990; (h) Goodberlet *et al.*, 1990; (i) Krausz *et al.*, 1990b; (j) Krausz *et al.*, 1990c; (k) Malcolm *et al.*, 1990; (l) Spielmann *et al.*, 1991; (m) Spence *et al.*, 1991; (n) Keller *et al.*, 1991; (o) Malcolm and Ferguson, 1991; (p) Keller *et al.*, 1992; (q) Jehetner *et al.*, 1992; (r) Seas *et al.*, 1992; (s) Liu *et al.*, 1992; (t) Ober *et al.*, 1992; (u) LiKam Wa *et al.*, 1992; (v) Evans *et al.*, 1992; (w) Phillips *et al.*, 1992; (x) Hughes *et al.*, 1993b; (y) Keller *et al.*, 1993a; (z) Weingarten *et al.*, 1993.

Very rapidly a number of other authors, as shown in Table 3.4, demonstrated that this was a widely applicable technique, both for a range of solid state lasers and using a number of quite different nonlinearities. The initial experiments by Blow and Nelson (1988), Kean *et al.* (1989), Mark *et al.* (1989), and Yakymyshyn *et al.*, (1989) used color center lasers and demonstrated both very short pulses and the fact that the technique was more general than the soliton laser, since it could be applied to optical fibers that possessed positive group velocity dispersion. The first non-Kerr nonlinearity was a saturable laser diode amplifier which allowed Kean *et al.* (1989) to generate 250 fs pulses.

The next important development occurred when *self-starting* APM operation was demonstrated. All the previous experiments strictly only demonstrated additional pulse compression using the nonlinear arm, the main laser was crudely modelocked using an active technique. In addition these self-starting experiments demonstrated that the technique could be applied both to Nd:YAG (Barr and Hughes, 1989) and to Ti:sapphire (Goodberlet *et al.*, 1989). Since the pulses must be developing from an intensity fluctuation or prototype pulse it was pointed out by Ippen *et al.* (1990) that self-starting would be easier in laser hosts with small stimulated-emission cross-sections, like Nd:YAG and Ti:sapphire, so that the pulse duration would not be changed by single pass gain saturation. Later, it was noted by Krausz *et al.* (1990c) that an additional condition for self starting is that the prototype pulse should exist for a sufficiently large number of round trips so that it has time to be shortened by the nonlinear mirror.

A very simple picture of coupled cavity modelocking will now be presented. The mathematical details for an APM laser using self phase modulation in a fiber are presented in Chapters 1 and 2 of this book. This analysis will allow a basic understanding of how pulse shortening occurs for a variety of nonlinear or linear modulators. The laser system shown in Figure 3.6 comprises a main cavity coupled to a nonlinear arm by a mirror with electric field reflectivity r_1. Inside the external cavity is a modulator with a transmission function $m(t)$ for the electric field, and the external cavity is terminated by a mirror with reflectivity r_2. Assuming that r_2 is sufficiently small, the reflection from the coupled cavity is easily found to be:

$$r(t) = r_1 - r_2(1 - r_1^2)m(t)\exp[i\phi], \tag{3.30}$$

where $\phi = (2\pi/\lambda)2nL$ is the linear phase change for one round trip of the cavity of length L and refractive index n (Barr and Hughes, 1990). This

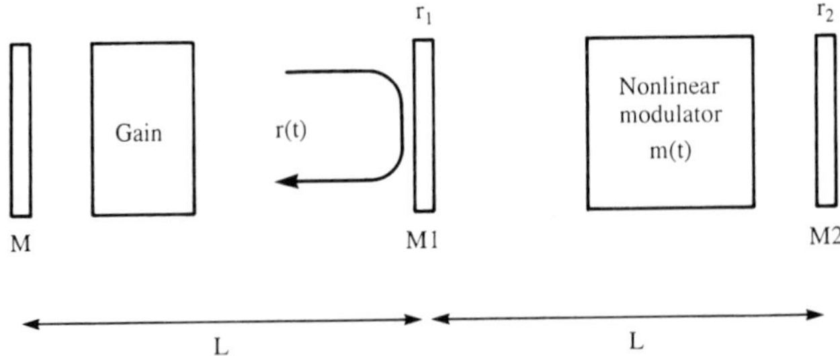

Figure 3.6. A schematic diagram of the coupled cavity arrangement analysed in the text.

equation was derived using the same scattering matrix as in Chapter 1. The definition of the linear phase change, however, differs by a factor π. For the special cases of a fast saturable absorber a suitable transmission function has the form:

$$\begin{aligned} m_{\text{abs}}(t) &= \exp\left[-\frac{\alpha}{1+\frac{I(t)}{I_{\text{sat}}}}\right] \\ &\approx \exp[-\alpha]\exp\left[\alpha\frac{I_{\max}}{I_{\text{sat}}}(1-2\Gamma t^2)\right], \end{aligned} \tag{3.31}$$

where $\exp[-\alpha]$ is the small signal transmission for the electric field, I_{sat} is the saturation intensity, and $I(t) = I_{\max}\exp(-2\Gamma t^2)$ is the intensity of the pulse incident on the saturable absorber. The second line corresponds to $m_{\text{abs}}(t)$ being expanded around the peak of the pulse where the transmission is a maximum. Neglecting the time independent factors, this expression has precisely the form required for the fictitious modulator introduced in Section 3.3. In a similar manner, self phase modulation in a fiber of length L leads to a transmission function:

$$m_{\text{sp}}(t) \approx \exp\left[i\frac{2\pi}{\lambda}\gamma L I_{\max}(1-2\Gamma t^2)\right], \tag{3.32}$$

where $I_{\max}$ is the peak intensity within the fiber and the contribution from the second pass has been neglected due to the small size of r_2.

Optimal pulse shortening will occur when a pulse incident on the external cavity sees the largest reduction in reflectivity as t changes from the peak of the pulse. This condition can be maximised by choosing special values for the linear phase change ϕ. These are illustrated in Figure 3.7, where the best operation points for a phase modulator and for an amplitude modulator are shown. For a fast saturable absorber it

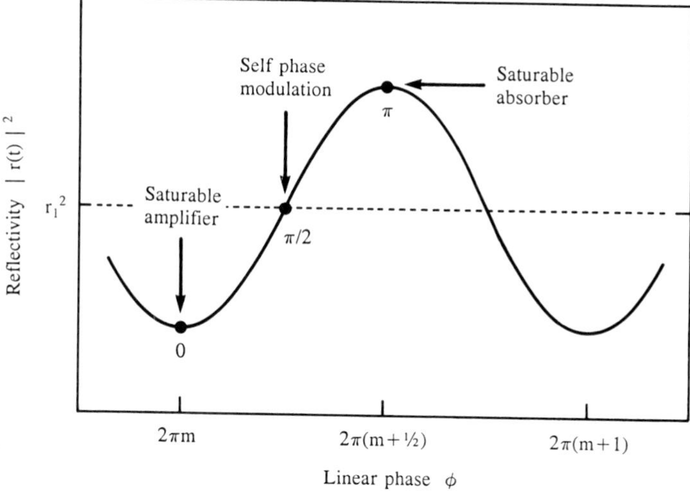

Figure 3.7. The optimum linear phase for various types of nonlinear modulator in an external cavity.

turns out that this is at antiresonance when $\phi = \pi$. At this point the intensity reflection coefficient from the coupled cavity is (neglecting terms of order r_2^2 or higher):

$$|r(t)|^2 = r_1^2 - 2r_2(1 - r_1^2)m_{\text{abs}}(t)\cos\phi = r_1^2 + 2r_2(1 - r_1^2)m(t)_{\text{abs}}, \tag{3.33}$$

and actually decreases with t around the pulse maximum. The reason is, of course, that as the intensity decreases, the absorption of the fast saturable absorber increases, i.e. $m_{\text{abs}}(t)$ decreases, causing a net effect similar to a decrease in r_2. Consequently the reflectivity of the coupled cavity decreases when t is away from the pulse maximum, and the pulse is shortened. A similar analysis for a fast saturable amplifier shows that the required operation point is on resonance, $\phi = 0$. Self-phase modulation can produce a coupled cavity that shortens pulses if the linear phase change is biased anywhere between $\phi = 0$ and π, and it can be shown that the maximum pulse shortening occurs at $\phi = \pi/2$. Normally self-phase modulation does not change a pulse duration, only its spectrum. Placing the nonlinear phase modulator within a Fabry–Perot resonator allows the change in phase to be converted into an amplitude change.

Two distinct resonator designs have been studied for APM operation. They are known as the Fabry–Perot and the Michelson by analogy with the well known interferometers and are shown in Figure 3.8. Experimentally the Fabry–Perot arrangement is easier to set up, particularly for low gain lasers, however, the Michelson scheme provides higher useful output power and tends to be more stable. The pulse durations do not differ for the two resonator designs.

These considerations point to one of the main problems with APM schemes, namely that the length of the external cavity has to be stabilised accurately so that the correct phase bias can be achieved and maintained. In practice, this is not too difficult for a laboratory-based instrument, see for example Mitschke and Mollenauer (1986), but it probably explains why this technique has not been implemented in a commercial laser. The most practical APM lasers are based on self-phase modulation in optical fibers. The alignment required to maintain good coupling into the fiber is difficult to realise for long periods of time. A further point is that the use of a separate nonlinear cavity uses a certain amount of output power that

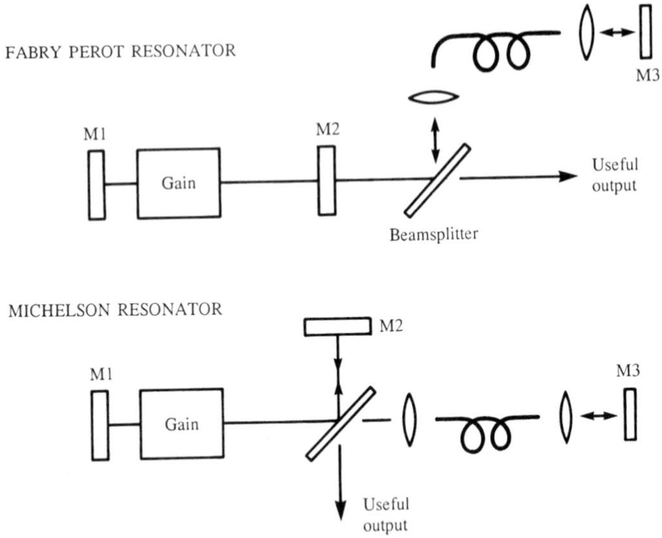

Figure 3.8. The two types of resonator, based on Fabry–Perot or Michelson interferometers, commonly used with APM schemes. In both cases the length of the resonators (not shown to scale here) are chosen so that pulses return simultaneously to the beamsplitter, where the pulse shaping takes place. Optional dispersion compensation may be inserted into these resonators to achieve the shortest pulses. It is also usual to actively stabilise the lengths of the linear and nonlinear parts of the resonators by mounting mirror M3, for example, onto a piezoelectric crystal.

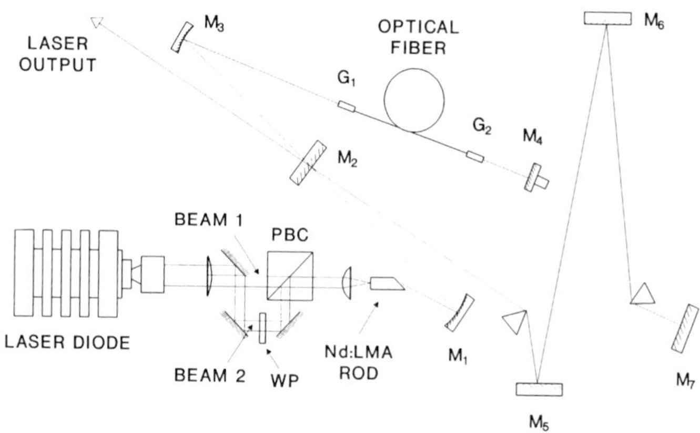

Figure 3.9. A LD pumped APM Nd:LMA laser that has generated pulses as short as 420 fs. The resonator is an example of a Michelson arrangement and M2 is the beamsplitter. The linear cavity contains two prisms for dispersion compensation. Light is coupled into the optical fiber via graded index lenses G_1 and G_2. The beam coupling from the LD into the gain medium is complicated by an arrangement that increases the effective brightness of the LD (Hughes *et al.*, 1993b).

would normally be available for use. This overhead makes APM lasers less efficient than they should be.

To date there are very few LD pumped APM systems reported (see Table 3.4) and these are Nd doped lasers operating around 1 μm. APM has achieved the shortest pulses yet produced from these lasers, including the shortest pulse yet produced from a LD pumped laser. The experimental configuration is shown in Figure 3.9 (Hughes *et al.*, 1993b). The laser is pumped by a 3 W LD at 800 nm. Over 300 mW can be obtained during optimised CW operation yet only 130 mW of useful output power is available from an optimised Michelson scheme. Due to the short pulse duration, dispersion compensation for the pulse stretching caused by the fiber is achieved using a prism pair within the laser cavity.

3.3.2 *Kerr lens modelocking*

An alternative passive modelocking scheme suitable for use with solid state lasers, which avoids the need for a complicated system of coupled resonators and stabilisation electronics, is known as Kerr lens modelocking (KLM). This mechanism for initiating short pulses was first observed in Ti:sapphire and subsequentially demonstrated in a wide range of vibronically tuned laser materials. It has also been demonstrated in a restricted range of LD pumped materials: Nd:YLF by Malcolm and Ferguson (1991b); Nd:YAG by Liu *et al.* (1992).

KLM uses the intensity dependent refractive index, also called the optical Kerr effect, to produce a lens with an intensity dependent focal length in a component within the laser cavity. The spatial variation in intensity across the beam profile causes a spatial change in refractive index that modifies the phase of the beam in a lens-like fashion and is often referred to as self-focusing. This modifies the spatial mode so that the resonator mode size becomes intensity dependent and can increase or decrease at various points around the resonator. An aperture is placed at a plane where the mode size decreases with increasing intracavity intensity and arranged so that the CW laser, with its low intensity, sees significant loss. A high intensity perturbation produces a weak positive lens in the Kerr material and modifies its beam profile through self-focusing, so that its mode size at the aperture is reduced and it experiences higher transmission. In this fashion the perturbation sees lower loss around the cavity and grows and develops into a pulse. The aperture coupled with the self focusing is capable of controlling both the leading and falling edges of the pulse and thus acts as a fast saturable absorber. The aperture can either be a physical slit within the cavity or it can be the finite pumped volume within the gain medium. In the latter case the optical mode size changes to improve the overlap with the available gain.

In an experiment with a LD pumped Nd:YLF laser, Malcolm and Ferguson (1991b) achieved pulses as short as 6 ps. In order to achieve a sufficiently large self-focusing effect a 1 cm long piece of SF57 glass, which has a large intensity dependent refractive index, was placed at an intracavity focus. In common with other KLM systems, the laser did not usually self-start, but when aligned correctly, a mechanical perturbation such as tapping the mirrors provided a sufficiently large intensity perturbation to initiate pulse formation. A later experiment extended the technique to LD pumped Nd:YAG where 8.5 ps pulses were obtained by Liu *et al.* (1992). They used the nonlinearity of Nd:YAG to provide KLM and also used an aperture caused by the aberrations of the thermal lens created in the gain material by the LD pump beam. Any thermal lens departs from an ideal spherical lens if it is sampled too far from the center of the pump source, and can introduce optical aberrations. Hence a thermal lens can assist KLM operation by reducing the aberrations of the laser beam if the spot size at the thermal lens decreases with intensity, as happens in this case.

KLM works very well in tunable solid state lasers. In fact the first observation was basically an accident in Ti:S by Spence *et al.* (1991). While the technique has been demonstrated in Nd:YAG and Nd:YLF, the pulse durations are not yet as short as can be obtained using APM.

A related technique, dubbed Kerr shift modelocking or KSM, has recently been demonstrated by Keller *et al.* (1993b). Instead of using self-focusing, which is a spatial effect, self phase modulation has been used to passively modelock the laser. An intracavity filter introduces a frequency dependent loss into the resonator that has a high loss for the CW laser frequency but a low loss nearby. An intensity fluctuation undergoes self-phase modulation and its spectrum increases in width. Those components seeing a low loss frequency nearby are favored and increase, ultimately modelocking the laser at a different center frequency relative to its CW operation. This technique has been demonstrated in Nd:glass where pulse as short as 130 fs were obtained.

3.3.3 *Multiple quantum well saturable absorbers*

Fast saturable absorbers should encourage passive modelocking in solid state lasers without the complicated cavity designs that are required for both KLM and APM. Multiple quantum well materials provide the right range of properties and have been used to passively modelock solid state lasers. The combination of very fast thermalisation of electrons in the conduction band, which allows a recovery of the absorption on timescales of $< 300\,\text{fs}$, and a controllable electron-hole recombination time of $\approx 10\,\text{ps}$ to $1\,\text{ns}$ depending on the growth conditions (Keller *et al.*, 1993a) leads to CW passive modelocking and pulse durations in the range $2 - 10\,\text{ps}$ for Nd:YAG and Nd:YLF. The MQW saturable absorber has been used in a coupled cavity arrangement where the external cavity length, was matched to the main laser cavity length, and also in a compact arrangement where the end mirror, with a reflectivity in the range $R = 95 - 98\%$, and the saturable absorber/HR mirror are in intimate contact, thus forming a coupled cavity arrangement where the external cavity is only a few μm thick (Keller *et al.*, 1992). In agreement with the simple theory presented in Section 3.3.1, the cavity has to be operated in antiresonance, where $\phi = \pi$, in order to achieve pulse shortening. This all solid state saturable absorber is known as an antiresonant Fabry–Perot saturable absorber (A-FPSA).

3.4 Alternative modelocking techniques

The modelocking techniques that have so far been discussed are well known and have been applied to a wide variety of laser hosts. In this section a number of less well-known techniques that have the potential to be applied to miniature LD pumped lasers will be described.

3.4.1 *Frequency shifted feedback lasers*

A laser resonator design that is increasingly attracting attention is one that incorporates frequency shifted feedback (Taylor *et al.*, 1971; Hale and Kowalski, 1990; Littler *et al.*, 1992; Phillips *et al.* 1993; Barr *et al.*, 1993). A schematic diagram of a typical linear cavity, where the diffracted and frequency shifted beam is retroreflected, is shown in Figure 3.10. Ring resonators can also be used. Unlike a conventional laser, the feedback is frequency shifted which prevents the build up of longitudinal mode structure. Under these conditions the frequency shifted feedback (FSF) laser output grows from broadband spontaneous emission that is repeatedly frequency shifted and amplified across the gain curve.

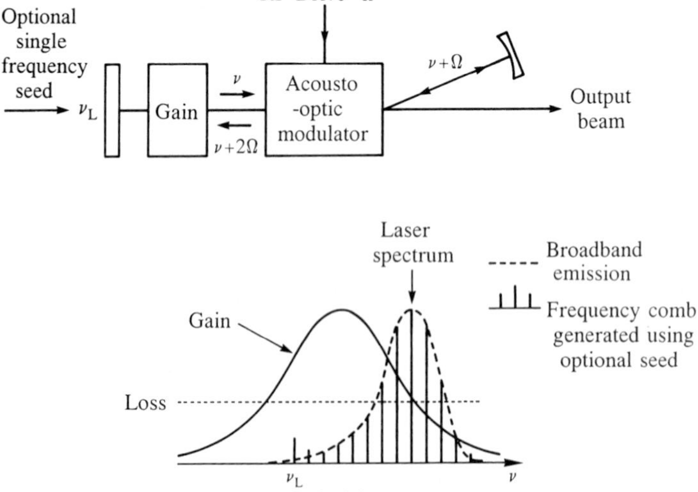

Figure 3.10. A schematic diagram of a frequency shifted feedback laser.

When the round-trip time of the resonator is matched to the round-trip frequency shift the laser can produce a train of pulses at a repetition rate equal to the frequency shift, as first shown by Hale and Kowalski (1990). Pulses of the order of 10 ps were obtained from a dye laser. A similar experiment was carried out on a Ti:sapphire pumped Nd:YLF laser by Phillips *et al.* (1993) but generated relatively long ns pulses. The reason for the different behavior is believed to be the additional pulse shaping that occurs in the dye laser from gain saturation. However, there is an alternative way of generating short pulses using an FSF laser. If a single frequency laser is injected into the FSF laser then it, rather than the spontaneous emission, dominates the laser output, and the laser spectrum consists of a discrete spectrum of frequencies separated by the frequency shift and often called a frequency comb. An example of a frequency comb generated using this technique is shown in Figure 3.11. The advantage of producing a frequency comb in this fashion is that the overall bandwidth of the comb does not depend on the laser length, unlike alternative comb generation techniques. Again, when the frequency shift is matched to the

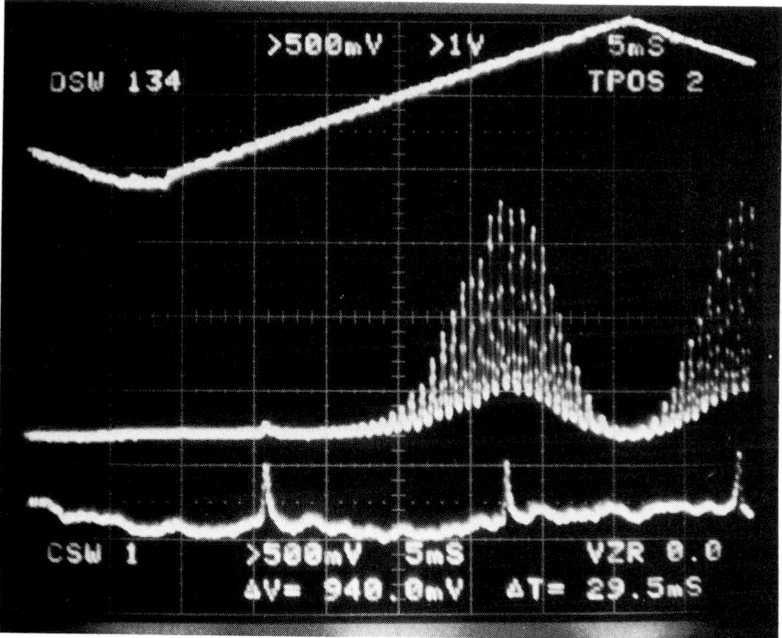

Figure 3.11. A frequency comb generated using a narrow linewidth FSF Nd:YLF laser, after Barr *et al.* (1993). The separation between the teeth is 160 MHz. The bottom trace shows the seed laser was operating on two longitudinal modes but only one component seeded the FSF laser. Similar frequency combs have been generated over 140 GHz.

repetition rate short pulses are generated with a duration approximately equal to the inverse of the frequency comb bandwidth. In this way 30 ps pulses have been generated from a Nd:YLF laser by Phillips *et al.* (1993), limited solely by the frequency stability of the injection laser. Pulse durations of a few ps are feasible from Nd:YLF and subpicosecond pulses from a variety of broader linewidth lasers. This type of modelocking is similar to active schemes in that it is robust and is not sensitive to the intracavity power of the laser, unlike all the passive schemes described in Section 3.3. It is at an early stage in its development and still needs to be examined with regard to pulse quality, stability and overall reliability.

3.4.2 *Planar waveguide lasers*

Planar waveguide technology is an emerging area that is distinct from both bulk lasers and fiber lasers. The guides are produced in crystalline hosts, with the possibilities of combining the gain and various nonlinearities such as second harmonic generation and electro-optic modulation. The waveguide also offers small spot sizes, giving rise to a low laser threshold and high intensities over reasonable lengths. From a device point of view these lasers could be easily combined with integrated optics technology.

There have been only a few reports of modelocked planar waveguide lasers, but they are very encouraging. A Nd-doped $MgO:LiNbO_3$ substrate had a 11 mm long stripe waveguide fabricated onto its surface that was single mode at both the pump wavelength, 814 nm, and the laser wavelength, 1085 nm (Lallier *et al.*, 1991). Electrodes were deposited parallel to the guide and appropriate mirrors directly deposited onto the ends of the guides. An RF signal applied to the electrodes modulated the phase of the light in the guide, allowing it to act as a phase modulator. The very short length requires a high frequency drive; in this case it was 6.2 GHz producing 7 ps pulses. More recently an FM modelocked Er-doped $Ti:LiNbO_3$ laser was demonstrated to give < 96 ps pulses at 1532 nm (Suche *et al.*, 1993). The pulse duration measurement was limited by their detection system. This laser had an overall length of 48 mm and was modulated at 1.441 GHz. Finally, AM modelocking of waveguide lasers has been achieved using a bulk modulator in an extended cavity arrangement. In this case the guides were made in Nd:glass and pulse durations of 80 ps were recorded at 1057 nm (Sanford *et al.*, 1991). A bulk modulator had to be used since Nd:glass does not exhibit the Pockels effect.

An alternative way of producing short pulses from short cavity waveguide lasers is simply to Q-switch them, as shown by Lallier *et al.* (1993). The short round-trip times and large gains combine to produce pulses of 300 ps and respectable peak powers of 350 W at 1085 nm.

3.5 Amplification and tunability

It is almost a truism that a laser that produces a desired pulse duration does so at the wrong part of the spectrum or with too low a power. The following sections cover only a small number of the experiments that have demonstrated ways of transforming the output of compact modelocked sources either in power or in frequency.

3.5.1 *Tunability*

Optical parametric oscillators (OPOs) provide the possibility of generating light (almost) anywhere within the transmission window of a nonlinear crystal. The reason is that the gain is provided by a nonresonant nonlinearity. Recently synchronously pumped OPOs have attracted a great deal of attention. The reasons for this are that short pulses are a good way of producing the intense pump required for high gain. Repeated pumping by a train of short pulses allows the laser to operate in steady state so that each pump pulse need only provide sufficient gain to overcome the round-trip losses that are typically small. A conventional single pulse OPO, on the other hand, grows from noise in a single pulse and requires many times the pump intensity to reach threshold. A synchronously pumped OPO can generate a stable train of transform limited pulses with excellent spatial characteristics that are ideal to use in sensitive experiments.

Synchronously pumped CW all solid state OPOs have been demonstrated using KTP and LBO as the parametric gain material by Ebrahimzadeh *et al.* (1992), McCarthy and Hanna (1992), and Butterworth *et al.* (1993). LBO is a very attractive material for pumping by a frequency doubled LD pumped passively modelocked 1 μm laser, since tunability over the range 650–2650 nm can be obtained simply by changing the crystal temperature (and the mirror set of course). Butterworth *et al.* (1993) used a LD pumped APM Nd:YLF laser generating 2 ps pulses at an average power of 500 mW at 1047 nm after resonant doubling to 523 nm using a non-critically phasematched temperature tuned $MgO:LiNbO_3$ crystal to pump an LBO OPO. The external build-up cavity increased the 1 μm intensity and produced 260 mW at the harmonic, a conversion efficiency of 66%. The OPO cavity design is similar to those used for the optically pumped laser materials described in Section 3.2, and is shown in Figure 3.12. When the length of the OPO resonator was matched to the Nd:YLF laser, synchronous pumped operation occurred providing outputs of up to 78 mW from the non-resonated beam. The pulse duration of the OPO output was slightly shorter than the pump at around 1.5 ps.

An alternative way of generating a tunable output is to synchronously pump a dye laser, as shown by Malcolm and Ferguson (1991a). The LD pumped FM modelocked Nd:YLF laser, after frequency doubling in an external ring enhancement cavity, produced 250 mW average power in 9 ps pulses at 523 nm. A R6G dye laser was matched in length to the pump laser and generated 1 ps pulses at 560 nm with a useful output power of 40 mW.

3.5.2 *Amplification*

Most solid state laser materials make excellent amplifiers of short pulses since the long fluorescence lifetime allows significant energy storage in the gain medium. The small stimulated emission cross-section,

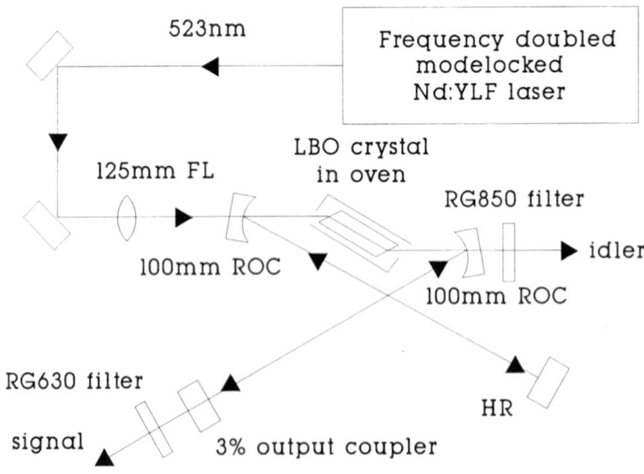

Figure 3.12. A schematic diagram of a singly resonant synchronously pumped optical parametric oscillator pumped by a passively modelocked Nd:YLF laser.

compared with dye or excimer lasers, means that the net gain can be small. This point is illustrated by a recent experiment, where a modulated laser diode providing average powers of mWs was amplified by a factor of 190 to powers of over 1 W in a CW LD pumped Nd:YLF amplifier operating at 1047 nm (Yu *et al.*, 1993). This represents an interesting combination of a simple oscillator that is cheap and easy to modulate with a solid state laser capable of providing multi-watt power levels.

The low gain that can be achieved in a solid state amplifier may be overcome by the use of regenerative amplification (Lowdermilk and Murray, 1980). A short pulse is injected into a resonator containing an amplifier and undergoes repeated passes and hence repeated gain. Even a small net gain per pass can build up to a respectable value in this fashion. Two examples of LD pumped regenerative amplifiers have been published by Dimmick (1990) and by Gifford and Weingarten (1992), the former used Nd:glass to achieve pulse levels of μJ s while the latter used Nd:YLF to amplify 8 ps pulses at 1047 nm to 88 μJ. The double-pass gain was estimated to be 70% and the total energy gain after 50 round trips reached 22×10^6 (73 dB) indicating that the net gain per pass including losses was more like 40%, or there was significant gain saturation. Due to the filtering effect of multiple passes of the gain medium (see Equation 3.8) the pulse duration of the amplified pulse increased slightly to 11 ps.

3.6 Conclusion and future prospects

Compact LD pumped modelocked sources are predominantly based on rare earth doped gain materials, which have absorption bands that match the high power LDs at 800 nm and 980 nm. There is a range of other materials, notably those based on Cr-doped materials, which are potentially LD pumpable and are simply awaiting the development of sufficiently high power sources so that the new passive modelocking techniques can be demonstrated.

Among the vibronically tunable solid state laser hosts the following CW LD pumped results have been obtained.

Ti:sapphire, which has a broad absorption band centered on 490 nm and can lase from 700 nm to 1000 nm, is capable of producing the shortest pulses obtained directly from a laser of 11 fs, see for example Asaki *et al.* (1993). As yet there are no LDs capable of pumping this host directly. However, by careful design and reduction of the laser threshold, Harrison *et al.* (1991) pumped Ti:sapphire using a frequency doubled LDPSSL based on Nd:YAG and generated 175 mW at 532 nm. The Ti:sapphire had a threshold as low as 119 mW. Usable modelocked Ti:sapphire systems require pump powers in the range 6 – 8 W, which should be available from commercial frequency doubled LDPSSLs in the next 2–5 years. In the longer term it is possible to anticipate that the recently demonstrated blue LDs may allow this important material to be pumped directly by an LD.

Direct LD pumping of tunable materials has been demonstrated for some of the Cr doped hosts: alexandrite (Scheps *et al.*, 1990); Cr:LiSrAlF_6 (Scheps *et al.*, 1991); and Cr:LiCaAlF_6 (Scheps, 1991). This series of results used AlGaInP LDs operating around 670 nm, which closely matches the Cr absorption bands. At present the power level commercially available from these LDs is limited to about 500 mW, but is likely to increase in the near future. Forsterite is conventionally pumped by 1064 nm directly from Nd:YAG lasers, which could at present be provided by commercially available LDPSSLs.

All these materials offer the potential for tunable femtosecond sources of pulses in an all solid state package. The impact of such devices is likely to be limited until the cost of the LD pump source reduces to compete with conventional pumps. This is a reasonable expectation given the historical reduction in price already observed in LDs.

Tunability is likely to come about with the use of all-solid state lasers to directly pump OPOs. The development of synchronously pumped OPOs in particular has demonstrated the importance and usefulness of these devices. Very large tuning ranges have been demonstrated using a single nonlinear crystal rather than a series of different laser hosts. Future work in this area is likely to be in the area of new nonlinear materials. Quasi-phasematching in materials like LiNbO_3, either in the bulk or in a waveguide, potentially allows the conversion efficiency to increase by about an order of magnitude compared with conventional birefringent phasematching, due to the large d_{33} nonlinear coefficient that may be accessed (see for example Ito *et al.*, 1991). Lower thresholds can be achieved by constructing a planar guide in a crystal, since high intensities may be maintained over long lengths. The results for modelocked waveguide lasers referred to earlier clearly illustrate the potential in this area, particularly when coupled with the fact that the laser may be integrated with an electro-optic modulator and designed for self-frequency doubling.

The passive modelocking schemes described here have been most successful when accessing the optical Kerr nonlinearity in materials. This is not the only way of constructing a nonlinear mirror. An intriguing possibility is to construct a Kerr-like nonlinearity by cascading two χ_2 processes, for example two frequency doubling crystals. An analogous scheme has been used to construct the SHG nonlinear mirror by Stankov (1988) and Barr and Hughes (1989), which gives rise to nonlinear amplitude modulation. The cascaded χ_2 processes can provide nonlinear phase modulation but with a nonlinearity many times larger than that obtained in silica (see for example Stegeman *et al.*, 1993). These schemes are of particular interest when combined with planar guides in integrated optics since the required conversion efficiencies can be reached at a reasonable power level.

The question of alternative modelocking schemes which do not rely on nonlinear processes yet can generate reasonably short pulses is technically important. Active modelocking is robust and insensitive to the laser performance and can be achieved in a controllable fashion in very low power lasers at an enormous range of repetition rates. As has been noted, the pulse duration is is unlikely to approach the bandwidth limit of the gain material due to the relatively small modulation frequencies that have been achieved or are likely to be realised. An alternative that has been recently demonstrated is to generate a frequency comb using a frequency shifted feedback laser. In one sense this is synthesising a pulse from a discrete frequency spectrum by altering the phase of the comb frequencies. The properties of this technique still need to be fully explored, particularly with regard to the pulse quality and stability. The cavity length stability controls the relative phase of the frequency components and will play a role in affecting the pulse quality.

In conclusion, compact modelocked sources can be devised. The small size of the laser diode is matched by the compact solid state laser cavity it pumps, particulary for high modulation frequency actively modelocked lasers. Passive modelocking has also been demonstrated in all-solid state systems and is being used for specific applications, for pumping OPOs for tunability and for generating subpicosecond pulses for seeding large Nd:glass amplifiers. In the future these devices are liable to be scaled to higher powers, but progress in this direction is driven mainly by the development in LD technology. The increased wavelength diversity of LDs will also allow a new range of materials to become LD pumpable. A wide variety of modelocking techniques have been developed that allow pulse durations to be reached that are fundamentally limited by the gain bandwidth of the laser material. As robust modelocking techniques are adopted this type of device is likely to emerge from the dedicated research laboratory and become a useable tool.

References

Adams, C. S., Maker, G. T. and Ferguson, A. I. (1990) 'FM operation of Nd:YAG lasers with standing wave and ring cavity configurations', *Opt. Comm.* **76**, 127.

Alcock, I. P. and Ferguson, A. I. (1986) 'Modelocking and Q-switching of an optically pumped miniature Nd:YAG laser', *Opt. Comm.* **58**, 417.

Asaki, M., Huang, C., Garvey, D., Zhou, J., Kapteyn, H. C. and Murnane, M. M. (1993) 'Generation of 11 fs pulses from a self modelocked Ti:sapphire laser', *Opt. Lett.* **11**, 977.

Barr, J. R. M. and Hughes, D. W. (1989) 'Coupled cavity modelocking of a Nd:YAG laser using second harmonic generation', *Appl. Phys. B.* **49**, 323.

Barr, J. R. M. and Hughes, D. W. (1990) 'Coupled cavity modelocking using nonlinear amplitude modulators', *J. Mod. Optics.* **37**, 447.

Barr, J. R. M., Liang, G. Y. and Phillips, M. W. (1993) 'Accurate frequency interval measurements based on non-resonant frequency comb generation', *Opt. Lett.* **18**, 1010.

Basu, S. and Byer, R. L. (1988) 'Continuous wave modelocked Nd:glass laser pumped by a laser diode', *Opt. Lett.* **13**, 458.

Blow, K. J. and Nelson, B. P. (1988) 'Improved modelocking of an F-centre laser with a nonlinear nonsoliton external cavity', *Opt. Lett.* **13**, 1026.

Blow, K. J. and Wood, D. (1988) 'Modelocked lasers with nonlinear external cavities', *J. Opt. Soc. Am. B* **5**, 629.

Butterworth, S. D., McCarthy, M. J. and Hanna, D. C. (1993) 'Widely tunable synchronously pumped optical parametric oscillator', *Opt. Lett.* **18**, to be published.

Chinn, S. R. and Zwicker, W. K. (1979) 'FM modelocked $Nd_{0.5}La_{0.5}P_5O_{14}$ laser', *Appl. Phys. Lett.* **34**, 847.

Dahlstrom, L. (1973) 'Modelocking of high power lasers by a combination of intensity and time dependent Q-switching', *Opt. Comm.* **7**, 89.

Dimmick, T. E. (1989) 'Semiconductor laser pumped, modelocked, and frequency doubled Nd:YAG laser', *Opt. Lett.* **14**, 677.

Dimmick, T. E. (1990) 'Semiconductor laser pumped, CW modelocked Nd:phosphate glass oscillator and regenerative amplifier', *Opt. Lett.* **15**, 177.

Ebrahimzadeh, M., Malcolm, G. P. A. and Ferguson, A. I. (1992) 'Continuous wave modelocked optical parametric oscillator synchronously pumped by a diode laser pumped solid state laser', *Opt. Lett.* **17**, 183.

Evans, J. M., Spence, D. E., Sibbett, W. and Chai, B. H. T. (1992) '50 fs pulse generation from a self modelocked $Cr:LiSrAlF_6$ laser', *Opt. Lett.* **17**, 1447.

Fan, T. Y. and Byer, R. L. (1988) 'Diode-laser pumped solid state lasers', *IEEE J. Quant. Electr.* **24**, 895.

Ferguson, A. I. (1989) in *The physics and technology of laser resonators,* Chap. 15, D. R. Hall and P. E. Jackson (eds.), Adam Hilger.

French, P. M. W., Rizvi, N. H. and Taylor, J. R. (1993) ' Continuous wave modelocked Cr^{4+}:YAG laser', *Opt. Lett.* **18**, 39.

Gifford, M. and Weingarten, K. J. (1992) 'Diode pumped Nd:YLF regenerative amplifier', *Opt. Lett.* **17**, 1788.

Godil, A. A., Li, K. D. and Bloom, D. M. (1991a) 'Pulsed FM modelocking of a Nd:BEL laser', *Opt. Lett.* **16**, 1243.

Godil, A. A., Hou, A. S., Auld, B. A. and Bloom, D. M. (1991b) 'Harmonic modelocking of a Nd:BEL laser using a 20 GHz dielectric resonator/optical modulator', *Opt. Lett.* **16**, 1765.

Goodberlet, J., Wang, J., Fujimoto, J. G. and Schultz, P. A. (1989) 'Femtosecond passively modelocked Ti:Al_2O_3 laser with a nonlinear cavity', *Opt. Lett.* **14**, 1125.

Goodberlet, J., Jacobsen, J., Fujimoto, J. G., Schultz, P. A. and Fan, T. Y. (1990) 'Self starting additive pulse modelocked diode pumped Nd:YAG laser', *Opt. Lett.* **15**, 504.

Gouvei-Neto, A. S., Gomes, A. S. L. and Taylor, J. R. (1988) 'High efficiency single pass soliton like compression of Raman radiation in an optical fibre around 1.4 μm', *Opt. Lett.* **12**, 1035.

Hale, P. D. and Kowalski, F. V. (1990) 'Output characterisation of a frequency shifted feedback laser: theory and experiment', *IEEE J. Quant. Electr.* **26**, 1845, and references therein.

Hanna, D. C. (1969) 'Astigmatic Gaussian beams produced by axially asymmetric laser cavities', *IEEE J. Quant. Electr.* **5**, 483.

Harris, S. E. and McDuff, O. P. (1965) 'Theory of FM laser oscillation', *IEEE J. Quant. Electr.* **1**, 245.

Harrison, J., Finch, A., Rines, D. M., Rines, G. A. and Moulton, P. F. (1991) 'Low threshold, CW all solid state Ti:Al_2O_3 laser', *Opt. Lett.* **16**, 581.

Hughes, D. W., Barr, J. R. M. and Hanna, D. C. (1991) 'Modelocking of a diode laser pumped Nd:glass laser by frequency modulation', *Opt. Lett.* **16**, 147.

Hughes, D. W. and Barr, J. R. M. (1992a) 'Laser diode pumped solid state lasers', *J. Phys. D:Appl. Phys.* **25**, 563.

Hughes, D. W., Phillips, M. W., Barr, J. R. M. and Hanna, D. C. (1992b) 'A laser diode pumped Nd:glass laser: Modelocked, high power, and single frequency performance', *IEEE J. Quant. Electr.* **28**, 1010.

Hughes, D. W., Majdabadi, A., Barr, J. R. M. and Hanna, D. C. (1993a) 'An FM modelocked laser diode pumped $La_{1-x}Nd_xMgAl_{11}O_{19}$ laser', *Appl. Opt.* **32**, 5958.

Hughes, D. W., Majdabadi, A., Barr, J. R. M. and Hanna, D. C. (1993b) 'Subpicosecond pulse generation from a laser diode pumped, self starting additive pulse modelocked Nd:LMA laser', paper AMC3, *OSA Topical meeting on Advanced Solid State Lasers, New Orleans.*

Ippen, E. P., Liu, L. Y. and Haus, H. A. (1990) 'Self starting condition for additive pulse modelocked lasers', *Opt, Lett.* **15**, 183.

Ito, H., Takyu, C. and Inaba, H. (1991) 'Fabrication of periodic domain grating in $LiNbO_3$ by electron beam writing for application of nonlinear processes', *Electr. Lett.* **27**, 1221.

Jehetner, J., Spielmann, Ch. and Krausz, F. (1992) 'Passive modelocking of homogeneously and inhomogeneously broadened lasers', *Opt. Lett.* **17**, 871.

Juhasz, T., Lai, S. T. and Pessot, M. A. (1990) 'Efficient short pulse generation from a diode pumped Nd:YLF laser with a piezoelectrically induced diffraction modulator', *Opt. Lett.* **15**, 1458.

Kean, P. N., Zhu, X., Crust, D. W., Grant, R. S., Langford, N. and Sibbett, W. (1989) 'Enhanced modelocking of colour centre lasers', *Opt. Lett.* **14**, 39.

Keen, S. J. and Ferguson, A. I. (1989) 'Subpicosecond pulse generation from an all solid state laser', *Appl. Phys. Lett.* **55**, 2165.

Keller, U., Li, K. D., Khuri-Yakub, B. T., Bloom, D. M., Weingarten, K. J. and Gersternberger, D. C. (1990) 'High frequency acousto-optic modelocker for picosecond pulse generation', *Opt. Lett.* **15**, 45.

Keller, U., Woodward, T. K., Sivco, D. L. and Cho, A. Y (1991) 'Coupled cavity resonant passive modelocked Nd:yttrium lithium fluoride laser', *Opt. Lett.* **16**, 390.

Keller, U., Miller, D. A. B., Boyd, G. D., Chiu, T. H., Ferguson, J. F. and Asom, M. T. (1992) 'Solid state, low-loss intracavity saturable absorber for Nd:YLF lasers: an antiresonant semiconductor Fabry–Perot saturable absorber', *Opt. Lett.* **17**, 505.

Keller, U., Chiu, T. H. and Ferguson, J. F. (1993a) 'Self starting and self Q-switching dynamics of modelocked Nd:YLF and Nd:YAG lasers', *Opt. Lett.* **18**, 217.

Keller, U., Chiu, T. H. and Ferguson, J. F. (1993b) 'Femtosecond Kerr shift modelocked Nd:glass laser using an A-FPSA as a continuous starting mechanism', paper JTuC4, *Technical digest series, vol. 11, CLEO'93, Baltimore*. (OSA, Washington, DC).

Keynes, R. J. and Quist, T. M. (1964) 'Injection luminescent pumping of $CaF_2 : U^{3+}$ with GaAs diode lasers', *App. Phys. Lett.* **4**, 50.

Kishida, S., Washio, K., Yoshikawa, S. and Kato, Y. (1979) 'Cw oscillation in a Nd:phosphate glass laser'. *Appl. Phys. Lett.* **34**, 273.

Koechner, W. (1988a) *Solid state lasers*, Chap. 7, 2nd edn. Springer-Verlag.

Koechner, W. (1988b) *Solid state lasers*, Chap. 9, 2nd edn. Springer-Verlag.

Kogelnik, H. W., Ippen, E. P., Dienes, A., and Shank, C. V. (1972) 'Astigmatically compensated cavities for CW dye lasers', *IEEE J. Quant. Electr.* **8**, 373.

Krausz, F., Brabec, T., Wintner, E., and Schmidt, A. J. (1989) 'Modelocking of a continuous wave Nd:glass laser pumped by a multistripe diode laser', *App. Phys. Lett.* **55**, 2386.

Krausz, F., Turi, L., Kuti, C. and Schmidt, A. J. (1990a) 'Active modelocking of lasers by piezoelectrically induced diffraction modulation', *Appl. Phys. Lett.* **56**, 1415.

Krausz, F., Spielmann, Ch., Brabec, T., Wintner, E. and Schmidt, A. J. (1990b) 'Subpicosecond pulse generation from a Nd:glass laser using a nonlinear external cavity', *Opt. Lett.* **15**, 737.

Krausz, F., Spielmann, Ch., Brabec, T., Wintner, E. and Schmidt, A. J. (1990c) 'Self starting additive pulse modelocking of a Nd:glass laser', *Opt. Lett.* **15**, 1082.

Kuizenga, D. J. (1981) 'Short pulse oscillator development for the Nd:glass laser-fusion systems', *IEEE J. Quant. Electr.* **17**, 1694.

Kuizenga, K. J. and Siegman, A. E., (1970) 'FM and AM modelocking of the homogeneous laser – part I: theory', *IEEE J. Quant. Electr.* **6**, 694.

Lallier, E., Pocholle, J. P., Papuchon, M., He, Q., De Micheli, M., Ostrowsky, D. B., Grezes-Besset, C. and Pelletier, E. (1991) ' Integrated Nd:MgO:$LiNbO_3$ FM modelocked waveguide laser', *Opt. Lett.* **27**, 936.

Lallier, E., Papillon, D., Pocholle, J. P., Papuchon, M., De Micheli, M. and Ostrowsky, D. B. (1993) 'Short pulse, high power Q-switched Nd:MgO:$LiNbO_3$ laser', *Electron. Lett.* **29**, 175.

Lewis, M. A. and Knudtson, J. T. (1982) 'Active passive modelocked Nd:YAG oscillator', *Appl. Opt.* **21**, 2897.

Li, K. D., Sheridan, J. A. and Bloom, D. M. (1991) 'Picosecond pulse generation in Nd:BEL with a high frequency acousto-optic modelocker', *Opt. Lett.* **16**, 1505.

LiKamWa, P., Chai, B. H. T. and Miller, A. (1992) 'Self modelocked Cr^{3+}:$LiCaAlF_6$ laser', *Opt. Lett.* **17**, 1438.

Littler, I. C. M., Balle, S. and Bergmann, K. (1992) 'The cw modeless laser: spectral control, performance data and build up dynamics', *Opt. Comm.* **88**, 514, and references therein.

Liu, K. X., Flood, C. J., Walker, D. R. and van Driel, H. M. (1992) 'Kerr lens modelocking of a diode pumped Nd:YAG laser', *Opt. Lett.* **17**, 1361.

Liu, L. Y., Huxley, J. M., Ippen, E. P. and Haus, H. A (1990) 'Self starting additive pulse modelocking of a Nd:YAG laser', *Opt. Lett.* **15**, 553.

Lowdermilk, W. H. and Murray, J. E. (1980) 'The multipass amplifier: theory and numerical analysis', *J. Appl. Phys.* **51**, 2436.

McCarthy, M. J. and Hanna, D. C. (1992) 'Continuous wave modelocked singly resonant optical parametric oscillator synchronously pumped by a laser diode pumped Nd:YLF laser', *Opt. Lett.* **17**, 402.

Maker, G. T. and Ferguson, A. I. (1989a) 'Frequency modulated modelocking of a diode pumped Nd:YAG laser', *Opt. Lett.* **14**, 788.

Maker, G. T. and Ferguson, A. I. (1989b) 'Modelocking and Q switching of a diode laser pumped neodymium doped yttrium lithium fluoride laser', *Appl. Phys. Lett.* **54**, 403.

Maker, G. T. and Ferguson, A. I. (1989c) 'Frequency modulation modelocking and Q-switching of a diode laser pumped Nd:YLF laser', *Electr. Lett.* **25**, 1025.

Maker, G. T., Keen, S. J. and Ferguson, A.I. (1988) 'Modelocked and Q-switched operation of a laser diode pumped Nd:YAG at 1.064 μm', *Appl. Phys. Lett.* **53**, 1675.

Malcolm, G. P. A., Curley, P. F. and Ferguson, A. I. (1990) 'Additive pulse modelocking of a diode pumped Nd:YLF laser', *Opt. Lett.* **15**, 1303.

Malcolm, G. P. A. and Ferguson A. I. (1991a) 'Synchronously pumped continuous wave dye laser pumped by a modelocked frequency doubled diode pumped Nd:YLF laser', *Opt. Lett.* **16**, 814.

Malcolm, G. P. A. and Ferguson, A. I. (1991b) 'Self modelocking of a diode pumped Nd:YLF laser', *Opt. Lett.* **16**, 1967.

Mark, J., Liu, L. Y., Hall, K. L., Haus, H. A. and Ippen, E. P. (1989) 'Femtosecond pulse generation in a laser with a nonlinear external resonator', *Opt. Lett.* **14**, 48.

Mazighi, K., Doualan, J. L., Margerie, J., Mounier, D. and Ostrovsky, A. (1991) 'Active modelocked operation of a diode pumped colour centre laser', *Opt. Comm.* **85**, 232.

Mitschke, F. M. and Mollenauer, L. F. (1986) 'Stabilising the soliton laser', *IEEE J. Quant. Electr.* **22**, 2242.

Mocker, H. W. and Collins, R. J. (1965) 'Mode competition and self locking effects in a Q-switched ruby laser', *Appl. Phys. Lett.* **7**, 270.

New, G. H. C., (1983) 'The generation of ultrashort pulses', *Rep. Prog. Phys.* **46**, 877.

Ober, M. H., Sorokin, S., Sorokina, I., Krausz, F., Wintner, E. and Shcherbakov, I. A. (1992) 'Subpicosecond modelocking of a Nd^{3+} doped garnet laser', *Opt. Lett.* **17**, 1364.

Ouellette, F. and Piche, M. (1986) 'Pulse shaping and passive modelocking with a nonlinear Michelson interferometer', *Opt. Comm.* **60**, 99.

Petrov, V. P. and Stankov, K. A. (1990) 'Theory of active passive modelocking of solid state lasers using a nonlinear mirror', *Appl. Phys. B.* **50**, 409.

Phillips, M. W., Chang, Z., Barr, J. R. M., Hughes, D. W., Danson, C. N., Edwards, C. B., and Hanna, D. C. (1992) 'Self starting additive pulse modelocking of a Nd:LMA laser', *Opt. Lett.* **17**, 1453.

Phillips, M. W., Liang, G. Y. and Barr, J. R. M. (1993) 'Frequency comb generation in a Nd:YLF laser using frequency shifted feedback', *Opt. Comm.* **100**, 473.

Sala, K. L., Kenney-Wallace, A. and Hall, G. L. (1980) 'CW autocorrelation measurements of picosecond laser pulses', *IEEE J. Quant. Electr.* **16**, 990.

Sanford, N. A., Malone, K. J. and Larson, D. R. (1991) 'Extended cavity operation of rare earth doped glass waveguide lasers', *Opt. Lett.* **16**, 1095.

Schepes, R. (1991) 'Cr:$LiCaAlF_6$ laser pumped by visible laser diodes', *IEEE J. Quant. Electr.* **27**, 1968.

Scheps, R., Gately, B. M., Myers, J. F., Krasinski, J. S. and Heller, D. F. (1990) 'Alexandrite pumped by semiconductor lasers', *Appl. Phys. Lett.* **56**, 2288.

Scheps, R., Myers, J. F., Serreze, H. B., Rosenberg, A., Morris, R. C. and Long, M. (1991) 'Diode pumped Cr:$LiSrAlF_6$ laser', *Opt. Lett.* **16**, 820.

Schulz, P. A. and Henion, S. R. (1991) '5-GHz modelocking of a Nd:YLF laser', *Opt. Lett.* **16**, 1502.

Seas, A., Petricevic, V. and Alfano, R. R. (1991) 'Continuous wave modelocked operation of a chromium doped forsterite laser', *Opt. Lett.* **16**, 1668.

Seas, A., Petricevic, V. and Alfano, A. A. (1992) 'Generation of sub-100fs pulses from a cw modelocked chromium doped forsterite laser', *Opt. Lett.* **17**, 937.

Shi, Y., Ragey, J. P. and Haugen, H. K. (1993) 'Continuous wave, Q-switched and modelocked operation of a laser pumped Nd:LMA laser', *IEEE J. Quant. Electr.* **29**, 435.

Siegman, A. E. (1986) *Lasers*, Chap. 27, University Science Books.

Sizer II, T. (1989) 'Modelocking of high power neodymium:yttrium aluminium garnet lasers at ultrahigh repetition rates', *App. Phys. Lett.* **55**, 2694.

Spence, D. E., Kean, P. N. and Sibbett, W. (1991) '60 fs pulse generation from a self modelocked Ti:sapphire laser', *Opt. Lett.* **16**, 42.

Spielmann, Ch., Krausz, F., Brabec, T., Wintner, E. and Schmidt, A. J. (1991) 'Femtosecond passive modelocking of a solid state laser by dispersively balanced nonlinear interferometer', *App. Phys. Lett.* **58**, 2470.

Stankov, K. (1988) 'A mirror with an intensity dependent reflection coefficient', *Appl. Phys. B.* **45**, 191.

Stegeman, G. I., Sheik-Bahae, M., van Stryland, E. and Assanto, G (1993) *Opt. Lett.* **18**, 13.

Streifer, W., Scifres, D. R., Harnagel, G. L., Welch, D. F., Berger, J. and Sakamoto, M. (1988) 'Advances in diode laser pumps', *IEEE J. Quant. Electr.* **24**, 883.

Suche, H., Baumann, L., Hiller, D. and Sohler, W. (1993) 'Modelocked Er:Ti:$LiNb_3$ waveguide laser', *Electr. Lett.* **29**, 1111.

Sweetser, J., Dunn, T. J., Walmsey, I. A., Radzewicz, C., Palese, S. and Miller, R. J. D. (1993) 'Characterisation of an FM modelocked Nd:YLF laser synchronised with a passively modelocked dye laser', *Opt. Comm.* **97**, 379.

Taylor, D. J., Harris, S. E., Nieh, S. T. K. and Hansch, T. W. (1971) 'Electronic tuning of a dye laser using the acousto-optic filter ', *Appl. Phys. Lett.* **19**, 269.

Turi, L. and Krausz, F. (1991) 'Amplitude modulation modelocking of lasers by regenerative feedback', *Appl. Phys. Lett.* **58**, 810.

Walker, S. J., Avramopoulos, H. and Sizer II, T. (1990) ' Compact modelocked solid state lasers at 0.5 and 1 GHz repetition rates', *Opt. Lett.* **15**, 1070.

Weingarten, K. J., Shannon, D. C., Wallace, R. W. and Keller, U. (1990) 'Two gigaHertz repetition rate diode pumped modelocked Nd:YLF laser', *Opt. Lett.* **15**, 962.

Weingarten, K. J., Keller, U., Chiu, T. H., and Ferguson, J. F. (1993) 'Passively modelocked diode pumped solid state lasers that use an antiresonant Fabry Perot saturable absorber', *Opt. Lett.* **18**, 640.

Yakymyshyn, C. P., Pinto, J. F. and Pollock, C. R. (1989) 'Additive pulse modelocked NaCl:OH^- laser', *Opt. Lett.* **14**, 621.

Yan, L., Ho, P. T., Lee, C. H. and Burdge, G. (1989) 'Generation of ultrashort pulses from a neodymium glass laser system', *IEEE J. Quant. Electr.* **25**, 2431.

Yu, A. W., Krainak, M. A. and Unger, G. L. (1993) '1047 nm laser diode master oscillator Nd:YLF power amplifier laser system', *Electr. Lett.* **29**, 678.

Zhou, F., Malcolm, G. P. A. and Ferguson, A.I. (1991) '1-GHz repetition rate frequency modulation modelocked neodymium lasers at 1.3 μm', *Opt. Lett.* **16**, 1101.

4

Modelocking of all-fiber lasers

IRL N. DULING, III AND MICHAEL L. DENNIS

4.1 Methods of modelocking fiber lasers

It has long been imagined that the unique characteristics of the fiber laser cavity might allow for the development of versatile, compact, efficient sources of coherent light. Particularly in the area of ultrashort pulse production, the fiber geometry provides access to the nonlinear index of the cavity providing additional degrees of freedom for pulse formation mechanisms. The first modelocked fiber laser, reported in 1983, actually used a material saturable absorber to provide the passive loss mechanism necessary for modelocking. The UV radiation from a flashlamp produced transient color centers in the glass matrix which in turn modelocked the laser, producing Q-switched trains of ultrashort pulses (Dzhibladze *et al.*, 1983).

Since that time nearly every method of modelocking that has been used in a bulk laser has been translated into the fiber geometry. In this chapter we will attempt to provide an overview of the more successful attempts and provide the insight necessary to choose the best technique for a given application.

The chapter is organized according to modelocking technique, starting with active modelocking, both amplitude and phase modulation, and then proceeding with the newer techniques of passive modelocking. Of the methods that deal with artificial fast saturable absorbers, this chapter will concentrate on the use of the nonlinear loop mirror (Sagnac interferometer) in the figure eight laser (F8L) configuration (4.2), rather than polarization rotation which will be treated in Chapter 5. Included in this section will be an extensive discussion of the generation and consequences of the sidebands present in the spectra. Although this discussion is in the section on the F8L, the conclusions are applicable to any soliton based fiber laser source. We will also review the work that has been done with alternative modelocking methods (4.3) and include a discussion of some methods of producing short pulses for which, although they are not strictly fiber lasers, the optical fiber plays a crucial role in the construction of the source (4.3.7).

4.1.1 *Active modelocking*

It is not the intent of this section to describe in detail the theory or practice of active modelocking in the general sense. This topic has been covered in some detail elsewhere (with a review in Chapters 1 and 3). Instead, we will present the state of the art in compact fiber sources and the significant experimental techniques involved. A complete list of references is included for the reader who desires further information. One of the major advantages of active modelocking is the control which it gives the researcher over the characteristics of the laser. The idea of active modelocking is that there is some form of external drive (either optical or electrical) which modulates the parameters of the cavity causing phase locking of the laser longitudinal modes. The synchronization of the pulse train to an electrical drive allows the laser to be referenced to external phenomena. This has particular advantage when the laser is being used to study an electrical device, as in electro-optic sampling, or when the modelocked train must be modulated by an electrical signal, as in digital communications.

The two main methods of active modelocking are amplitude and phase modulation. The principal electrically driven modulators that have been used are the acousto-optic modulator, due to its ease of operation and ready availability, and the lithium niobate integrated optic modulator, which if pigtailed is closely compatible with the fiber nature of the laser and provides a completely integrated package.

The first modelocked fiber lasers with amplitude modulators used Nd:fiber as the gain medium. Neodymium is readily incorporated into silica fiber, it is a 4-level laser, and the optics necessary for operation are widely available due to the popularity of the Nd:YAG and Nd:YLF bulk lasers. In addition, neodymium has a pump absorption band around 810 nm which can be pumped by readily available AlGaAs laser diodes. The Nd:silica lasing transition is at 1.08 μm, but this can be shifted by glass composition to as short as 1.054 μm. In using a bulk modulator in a fiber laser care must be taken to minimize back reflections into the lasing mode from all internal surfaces. These include the surfaces of the modulator, lenses and the fiber end itself. The surfaces are most often angled to send the reflection out of the cavity. The additional loss which is incurred is not a problem for most fiber lasers due to the high gain of the transition.

The high gain in doped fibers provides a unique situation among lasers. Bulk lasers generally fall into two categories which are roughly delineated by the magnitude of the emission cross-section of the lasing transition. Lasers with a large emission cross-section are generally high gain per pass and show a large amount of gain saturation with the passage of a single pulse. The dye laser and the semiconductor diode laser fall into this category. On the other hand lasers with a small emission cross-section, like the Nd:YAG laser, show little gain saturation from a single pulse and low gain per pass. In contrast to both of these situations the fiber laser shows little single pulse saturation but a high gain due to the length of the gain medium. In effect it is like having a Nd:YAG rod that is 10 m long and where the lasing mode is able to remain at focus through the entire length of the rod. The laser therefore can tolerate rather large intracavity loss, but shows the characteristics of lasers with little or no gain saturation, as discussed in Chapter 2.

By straightforwardly placing an acousto-optic modulator into a fiber laser cavity, analogous to a Nd:YAG laser, pulses on the order of 70 ps can be generated. The pulse durations obtained generally agree with what would be calculated from a simple Kuizenga–Siegman theory (Siegman and Kuizenga, 1974). In this theory the modelocked pulse width τ_p is estimated as,

$$\tau_\mathrm{p} = \frac{(2\ln 2)^{\frac{1}{2}}}{\pi}\frac{g_0^{\frac{1}{4}}}{\delta^{\frac{1}{2}}}\left(\frac{1}{f_\mathrm{m}\Delta f_\mathrm{a}}\right)^{\frac{1}{2}} \tag{4.1}$$

where g_0 is the saturated single-pass gain, δ the modulation parameter, f_m the modulator drive frequency and Δf_a the lasing bandwidth. In a typical experiment, an acousto-optic modulator driven at 41 MHz provides a modulation depth of 46%. The modulation depth is easily measured by examining the ASE transmitted through the modulator with a fast photodiode. This gives a measurement at the wavelength and angle of incidence of the actual lasing mode. The predicted pulse width is 64 ps, which is in good agreement with the value derived from the experiment, which was 66 ps (Duling *et al.*, 1988).

Because of their higher modulation depth and higher bandwidth, Mach–Zehnder integrated optic modulators have been very successful at modelocking fiber lasers. Their wide bandwidth allows the modulator to be driven by a short pulse at the cavity round-trip time (usually generated from RF by a step-recovery diode) which allows the drive frequency in the Kuizenga–Siegman Equation to be replaced by an effective frequency close to the maximum response of the step-recovery diode. This allows shorter pulses to be produced than with sinusoidal modulation. The first example of this type of laser constructed from Er:doped fiber was reported by Kafka *et al.* (1989). This laser consisted of an erbium doped fiber amplifier pumped by a Ti:sapphire laser, a lithium niobate amplitude modulator, and a 50% output coupler which served also as the wavelength division multiplexer (WDM) for coupling the pump light into the Er:fiber. The laser produced 50 ps pulses with sufficient bandwidth for 2 ps pulses at a repetition rate of 90 MHz. In order to insure that the laser was operating in the anomalous dispersion regime, 2 km of standard telecommunications fiber was added to the cavity. By operating in the anomolous dispersion regime the pulse could access the additional pulse shortening mechanism that produces solitons. In this way 4 ps pulses were produced. When the modulator was driven by RF directly, the pulse formation followed closely the Kuizenga–Siegman theory. In other work this type of modulator has been used to produce pulses as short as 2 ps and at repetition rates as high as 30 GHz, although not simultaneously.

It has been realized by a number of researchers that phase modulation can produce shorter pulses than amplitude modulation, since the time varying frequency imposed by the modulator can be compensated by the dispersion of the cavity (see for example Krausz *et al.*, 1992). For this reason a number of systems have been produced which utilize either bulk or integrated phase modulators of various types. Bulk electro-optic modulators have produced pulses as short as .9 ps (Davey *et al.*, 1991c), and repetition rates as high as 480 MHz (Davey *et al.*, 1991b). These very short pulses do use passive polarization modelocking in addition to the active modelocking which starts the process. If we restrict ourselves to systems that do not use passive modelocking, then the shortest pulses produced with bulk components are 20 ps in duration (Phillips *et al.*, 1989a).

In order to eliminate the bulk components and reduce the size while removing the need for alignment of the laser, the bulk phase modulators have been replaced by various types of integrated optic modulators. For the ultimate in integration, a phase modulator can be integrated into the fiber itself. A radial or transverse acoustic field is generated by an acoustic horn applied to the fiber (Phillips *et al.*, 1989b), or an acoustic transducer can be coated right onto the fiber itself. The fiber is used as the resonant cavity and the resultant laser is extremely low loss. Lasers of this type have produced 200 ps pulses at 417 MHz (Phillips *et al.*, 1989b). The alternative integrated phase modulator is a $LiNbO_3$ waveguide modulator (Shan *et al.*, 1992). Due to the higher modulation depth these modulators generally outperform acoustic phase modulators. In Nd:fiber pulses as short as 35 ps and repetition rates of 5 GHz have been obtained (Davey *et al.*, 1991a).

An intermediate form of modelocking is accomplished by integrating a diode amplifier into a ring fiber laser (Burns and Sibbett, 1990). The modulation of the drive current provides both amplitude and phase modulation to the light in the fiber laser. Chirp induced by the drive current and the gain saturation can be compensated with a proper choice of the dispersion in the cavity. In this manner pulses have been produced of 3.7 ps duration at 612 MHz. An additional degree of complexity is added when the semiconductor amplifier is included inside an optical loop mirror. The interferometric nature of this mirror transforms the phase modulation of the amplifier into an amplitude modulation. Lasers of this type have produced pulses less than 25 ps duration in either semi-linear or figure eight cavities (Frisken *et al.*, 1991; Cochlain *et al.*, 1993). (The nonlinear loop mirror and its use inside a fiber laser will be treated more fully in the following sections.)

The advantage of active modelocking is that it provides an electrical reference that can be synchronized to an external process, whether that is an electrical circuit under study or a modulator encoding the modelocked pulse train with data. In addition, by actively modulating the cavity, interpulse noise is reduced and, in harmonic modelocking, the pulses will be evenly spaced within the cavity period.

Since environmental perturbations (thermal, mechanical and acoustic) can change either the length or the birefringence of the cavity as a function of time it is necessary, particularly in the case of active modelocking, to take steps to stabilize the cavity. The most straightforward way to control the polarization effects is to use polarization maintaining fiber in the cavity. The stabilization of the cavity length is a bit more complicated and usually requires active stabilization. A number of feedback techniques have been implemented (Shan *et al.*, 1992; Harvey and Mollenauer, 1993). The simplest technique to correct the cavity length drive frequency mismatch is to look at the difference in phase of the output bit stream relative to the drive to the active modelocker. This can be done in an RF mixer with the error signal controlling a PZT in the cavity and/or a thermal control on a section of the fiber. Harvey and Mollenauer (1993) have been able to obtain excellent results using a metal coated fiber where current is driven through the coating to resistively heat the fiber and adjust its length. Recently systems have been constructed which have obtained error free operation without feedback by constructing the cavity from polarization maintaining fiber. These lasers will obviously have some limitation of operating temperature range, but point out the fact that the length fluctuations observed in other fiber lasers may have more to do with shifts in the polarization state and sampling of the birefringence than they do with actual fiber length changes.

4.1.2 *Passive modelocking*

By far the shortest pulses to be produced by fiber lasers have been by those utilizing passive modelocking to some extent. The advantage of passive modelocking is that the strength of the modulation is constant or increases as the pulse width decreases. The two major categories of passive modelocking possible in a fiber laser are those based on the Kerr effect (e.g. the nonlinear index or the nonlinear birefringence) and those based on other materials such as semiconductors or organic dyes. The intrinsic response time of these materials dictates the ultimate speed of its modelocking as mentioned in Section 2.1.1. A theoretical treatment of modelocking, using the Kerr effect is presented in Chapters 1 and 2, and the use of a semiconductor saturable absorber is mentioned in Section 4.3.3.

It is worthwhile, however, to mention that there is a sub-class of passive modelocking in which more than one passive technique is used, or where there may be an active modulator included in the cavity as well. Generally described as hybrid modelocking, these methods attempt to combine the strong points of more than one technique. Some systems falling into this category have already been mentioned in the section on active modelocking. Systems where the pulses have sufficient energy to propagate as nonlinear pulses (Kafka *et al.*, 1989), with an active modelocker, or the work by Davey *et al.* (1991c) where an active phase modulator was combined with soliton propagation and nonlinear polarization rotation. Passive techniques have also been combined, where a semiconductor saturable absorber has been combined with nonlinear polarization rotation to provide self-starting and fundamental modelocking (Fermann *et al.*, 1994).

4.2 The figure eight laser

As presented in Chapter 1, very strong modelocking can be obtained using interferometric techniques. These systems usually require stabilization of the two arms of the interferometer to maintain the phase bias necessary for modelocking. By constraining the two arms of the interferometer to be in the same fiber, the interferometer becomes self-stabilizing. Modelocking mechanisms of this type are polarization switching, where each polarization state can be considered a different arm of the interferometer, and the nonlinear optical loop mirror type of modelocking where a Sagnac interferometer is used and the two arms are distinguished by their direction through the fiber. Lasers incorporating a fiber loop to produce modelocking will be treated in this section. The most prominent member of this family of lasers is the figure eight laser (F8L), where the nonlinear loop mirror is included as an element in a ring laser. An illustration of the F8L is presented in Figure 4.1. The right-hand half of the figure constitutes a nonlinear amplifying loop mirror (NALM) where the counterpropagating pulses in the loop mirror are different in amplitude due to the presence of the EDFA close to one end of the loop (Fermann *et al.*, 1990a). The left-hand half of the illustration is merely a feedback loop which takes the pulse exiting the loop mirror, extracts the usable output power through an output coupler, and reinjects the pulse into the loop mirror. By placing an isolator in this feedback loop unidirectional operation is assured and the NALM is forced to operate in

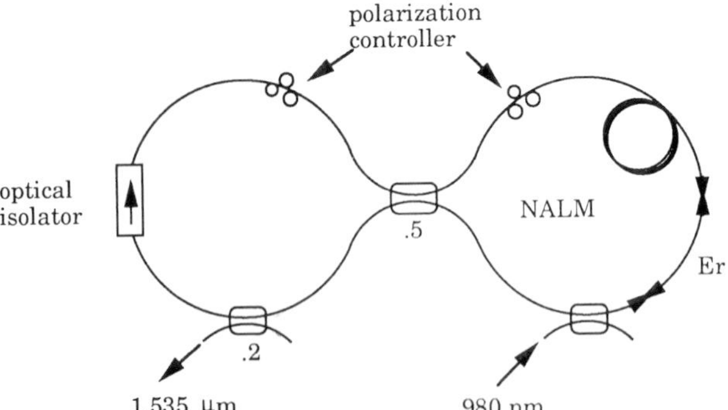

Figure 4.1. The figure eight laser consists of a unidirectional ring fiber laser (left-hand loop) with a nonlinear amplifying loop mirror as the gain element and the pulse shortening mechanism. Polarization controllers are included in the loops to select the polarization eigen states and adjust the transmission bias of the NALM.

transmission. Figure eight lasers have been constructed with both NALM and NOLM modulators (Wu *et al.*, 1993). Although most of the work has concentrated on the 1.5 μm transition in Er:fiber, an F8L operating at 1.3 μm has been constructed using Pr:fiber (Sugawa *et al.*, 1993).

4.2.1 *Theory of modelocking*

The design of a nonlinear optical loop mirror is shown schematically in Figure 4.2. The normal fiber loop mirror would have a 50% splitting fiber coupler with the two output ports spliced together. This is the configuration for a Sagnac interferometer. Light entering one of the fibers is split into two equal intensity counterpropagating beams. If we ignore polarization for the moment, the light will return to the splitter having traveled precisely the same path length and interfere such that the light retraces its path in the input fiber. It is for this reason that the Sagnac interferometer is often called a loop mirror or a fiber loop mirror (this is of course the same configuration as the antiresonant ring used in bulk components). In this balanced configuration the fiber loop will act as a perfect mirror for any input intensity. If, as is normally the case for any real fiber coupler, the splitting ratio is not precisely 50%, then the interference will not be complete and some of the light will 'leak' through into the other fiber. An equation for the reflectivity in this case can easily be derived,

$$R = 2\alpha(1 - \alpha)[1 + \cos(1 + 2\alpha)\phi] \tag{4.2}$$

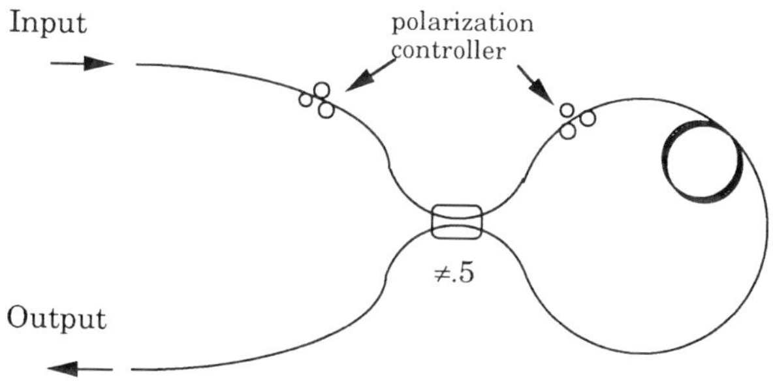

Figure 4.2. The nonlinear optical loop mirror consists of a coupler which is not 50% with its output ports fused together into a loop. Unequal intensities in the counterpropagating directions lead to nonlinear transmission.

where

$$\phi = \frac{2\pi n_2 P_i L}{\lambda A_{eff}} \tag{4.3}$$

and P_i is the input power, L is the loop length, λ is the wavelength, n_2 is the nonlinear index of the fiber, α is the power splitting ratio and A_{eff} is the effective fiber core area. Under these conditions the nonlinear index of the fiber can have an effect on the transmission characteristics of the loop mirror due to the different intensities of the counterpropagating beams.

Before discussing the nonlinear characteristics of the loop mirror it is necessary to discuss one other important linear effect. The polarization can play a significant role in the loop mirror. As discussed by Mortimore (1988), if a loop mirror is constructed from standard fiber (non-polarization maintaining) an adjustable fiber polarization controller must be included in the loop to attain the ideal characteristics mentioned above. The mechanism requiring this adjustment can be easily understood by examining Figure 4.3. The fiber in the loop has a certain amount of birefringence due to internal stress from fabrication and external stress from the position of the fiber. This birefringence can be modeled as a lumped bulk birefringent element with a fast and a slow axis. For this illustration we will only consider linear birefringence, but the explanation is equally applicable to the more generalized case of circular and elliptic birefringence as well (Burns and Kersey, 1992).

Consider the lumped element to be decomposed into three linear birefringent elements as pictured in the figure. Elements 1 and 3 have their

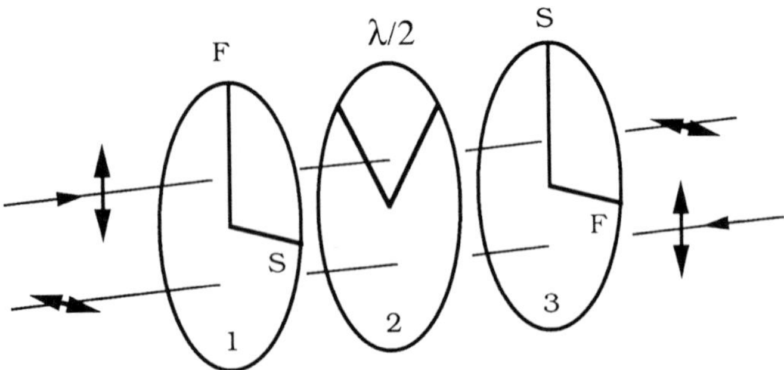

Figure 4.3. Three birefringent elements as arranged here demonstrate how light traveling in opposite directions through the same reciprocal optical elements can see different path lengths. For further explanation see the text.

optic axes oriented vertical and horizontal with the fast (F) and slow (S) axes reversed. Element 2 is a half wave plate with its axes inclined at 45°. All other components of the loop mirror are assumed to be polarization maintaining. As the beams counterpropagate around the loop they will arrive at the birefringent stack with the same polarization (in this case vertical) but traveling in opposite directions. Tracing the path of the light entering from the left, the light will propagate along the fast axis of the first element, be rotated 90° by the half wave plate, propagate along the fast axis of the third element and exit the structure in the horizontal polarization state. In the same manner, light entering from the right propagates through the three elements and exits in the horizontal polarization state, but it has propagated along the slow axes of elements 1 and 2. It is obvious from this example that although the two beams will be in a matched polarization state when they recombine at the coupler, one has traveled a different path length from the other, even though all of the elements are reciprocal, having only standard linear birefringence. The amount of differential phase delay in this example is dictated by the amount of birefringence in elements 1 and 2 which were not specified. In the actual loop mirror, a change of the polarization controller, in addition to changing the output state of the light, will induce a phase bias in the loop. This bias is not generally zero when the laser is operating and plays an important role in determining the self-starting of the laser. A careful study of the evolution of polarization in a standard fiber F8L is discussed in more detail in Stentz and Boyd (1993).

As mentioned in the previous section, since the index of refraction of a fiber is given by $n = n_0 + n_2 I$, higher intensity light propagating in one direction around the loop mirror will see a slightly higher refractive index than the low intensity counterpropagating light. This effective index difference will result in a differential phase shift between the counterpropagating beams, resulting in a further shifting of the output to the other fiber. The dependence of the transmission on intensity in Equation (4.2) is valid for continuous wave input. It is obvious that if a pulse of light is incident on the mirror which has a peak power close to the transmission maximum, then the wings of the pulse will be mostly reflected, while the peak of the pulse will be mostly transmitted. The transmitted pulse will therefore be shorter than the input pulse with the result that the loop mirror acts as a fast saturable absorber (Smith *et al.*, 1990a).

This simple picture of the nonlinear optical loop mirror (NOLM) is valid when dispersion can be ignored. As the pulse width decreases, however, the bandwidth increases to a point at which spreading due to dispersion in the loop will overcome the shortening due to the nonlinear switching, and the transmitted pulse will be longer than the input. In the case of a fiber loop with significant anomalous dispersion, one might think that the way to keep the pulses from spreading would be to increase the power until they become stable solitons and allow these pulses to enter the loop mirror. As has been pointed out by Doran and Wood (1988), the response of a loop mirror to a soliton contrasts markedly with that of a pulse where dispersion can be ignored. A soliton accumulates phase according to

$$\phi = 2\left(A - \frac{1}{2}\right)^2 z \tag{4.4}$$

where the soliton is given by $A\,\mathrm{sech}(t)$ and z is the distance of propagation in the fiber. This phase accumulation occurs equally in the peak and the wings of the pulse, which indicates that the pulses will not follow the simple picture of pulse shortening that we indicated earlier in this section, but can actually be switched as a whole through the loop mirror. It is for this reason that the NOLM has been refered to in some publications as a 'soliton filter' (Doran and Wood, 1988). This is of course a simplified picture, and it is necessary to resort to numerical analysis in order to map out the response of the loop mirror to soliton input.

Propagation in the nonlinear loop mirror was analyzed using the split step Fourier method of numerically solving the nonlinear Schrödinger equation (Agrawal, 1989). The results of the calculations are presented in normalized soliton units. The loop is measured in units of

$$z_0 = \frac{.322\pi^2 c\tau^2}{\lambda |D|} \tag{4.5}$$

(the characteristic dispersive or nonlinear scale length, where c is the speed of light in vacuum, τ is the pulse width, and D is the fiber dispersion) and the input energy is in units of the fundamental soliton energy

$$E_0 = 1.135\tau P_1, \tag{4.6}$$

where P_1 is the soliton peak power given by

$$P_1 = .776\lambda^3 \frac{|D| A_{\mathrm{eff}}}{\pi c n_2 \tau^2}, \tag{4.7}$$

(where A_{eff} is the effective area of the propagating mode).

Figure 4.4 displays the results of calculations of the transmission of the loop mirror as a function of input power for a variety of loop lengths. It

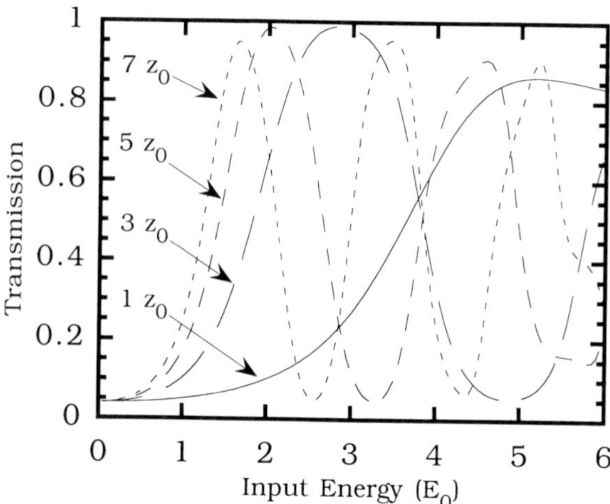

Figure 4.4. Transmission of the NOLM for different loop lengths measured in units of the soliton period (z_0) versus input energy measured in units of the fundamental soliton energy (E_1).

is clearly seen that the first peak of the transmission curve does not reach 100% for every loop length. If we plot the transmission value at the first maximum (see Figure 4.5), it is clear that there is a narrow range of loop lengths which will give complete transmission. Figure 4.5 also plots the output pulse width of the mirror relative to the input at the first

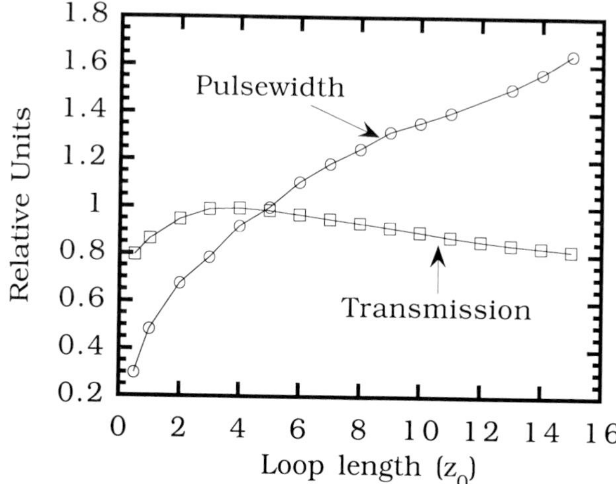

Figure 4.5. The relative pulse width and transmission at the first local maximum of transmission for a NOLM as a function of its loop length.

maximum of transmission, and we see that, depending on the loop length, the pulse can exit the loop mirror either longer or shorter than it entered. (Duling *et al.*, 1994).

This result is in sharp contrast to the notion of the loop mirror as a 'soliton filter'. The soliton is only able to be transmitted intact through the loop mirror for loop lengths in the range of 4–5z_0 and at an input energy level corresponding to approximately $2E_0$, which corresponds to the case where the counterpropagating pulses are as close to a fundamental soliton as possible. It was also found from the simulation that only if the input energy was below or close to the first transmission maximum did the pulse width maintain its shape.

To model the operation of the F8L using a NALM it has been necessary to resort to complete numerical analysis of the cavity to predict the actual operating parameters. The difficulty which this presents is that in order to take into account the reshaping of the soliton rather small time steps must be taken, while the slow saturation of the Er amplifier requires a large number of round trips. These criteria taken together necessitate a large amount of computer time for each simulation. Therefore most of the numerical work to date has included one approximation or another to reduce the computational burden.

The first numerical analysis to be carried out on the F8L was conducted by Bulushev *et al.* (1990). The laser analyzed was slightly different from that shown in Figure 4.1. The gain medium was placed in the external loop and the loop mirror was constructed from an unbalanced coupler. To provide an additional degree of freedom, a variable loss was inserted close to one end of the loop (this variable loss is equivalent to a non-saturable gain). By modeling the saturable gain medium and nonlinear propagation in the fiber, the authors were able to predict the regions of stable modelocking, forced self-starting and chaotic self-pulsing. These regions were plotted as a function of the central coupler splitting ratio and the variable loss in the loop mirror. The analysis did not include dispersion in the loop mirror and was therefore not applicable to soliton formation, but does indicate that the general laser concept is not fundamentally limited to the soliton dispersion regime.

The soliton effects present in the figure eight laser have been examined numerically in two publications. Tzelepis *et al.* (1993) examined the self-starting and pulse characteristics as a function of the central coupler splitting ratio, cavity length and dispersion. The central coupler was allowed to vary from .5 to provide initial lasing to seed the self-starting process. In an F8L that has a single polarization (PM fiber) there is normally no method for inserting the linear phase bias required by the loop to provide for this initial lasing other than relaxing the coupler value. In cavities of up to 100 m in length the authors were able to calculate that the laser should produce pulses below one picosecond. Their data also reproduced the spectral sidebands and corroborated the square-root dependence of the pulse width on the total cavity dispersion (see Section 4.3).

The approach of Singh (1993) was to analyze the components of the laser developing code for each process and then combine those elements to model the intended laser. In this manner he modeled not only the figure eight laser, but also the harmonically modelocked ring laser (Harvey and Mollenauer, 1993). In the modeling of the F8L, Singh ignored the question of self-starting, prefering to insert an active modulator into the laser to produce harmonic modelocking as in the ring laser.

4.2.2 *Experimental results*

The F8L has a rich variety of phenomena that it displays, due to the large number of interrelated processes present in the laser. It is the understanding and control of these operating regimes which has been the subject of recent research. The basic figure eight laser pictured in Figure 4.1 includes a NALM where the output is connected to the input by a feedback loop consisting of an output coupler and an optical isolator (Duling, 1991a, 1991b; Richardson, 1991a). Output couplings from .1 to .5 have been used, and both polarization selective and polarization independent isolators have been used. The output power is generally between 1 and 20 mW, with pulse widths ranging from 91 fs to 10 ps (Dennis and Duling, 1993c; Nakazawa *et al.* 1991, 1993a). With CW pumping to the amplifier, the laser is capable of operating with a CW output, spontaneous modelocking, self-sustaining modelocking, and a situation where the pulses consist of noise bursts of bandwidths up to 30 nm, with the repetition rate of the fundamental cavity (Richardson *et al.*, 1991c). In the case of spontaneous modelocking and self-sustaining modelocking, the laser will often operate with more than one pulse in the cavity at a time. Due to the passive nature of the artificial fast saturable absorber (AFSA), there is no process in the laser cavity to determine the relative location of the pulses. Therefore the pulses tend to move relative to each other or to stabilize to a spacing specified by minor imperfections in the cavity. The control of these pulses is treated later in this section.

The different modes of operation are found by adjustment of the polarization controllers in the cavity.

Although much of the work has been accomplished using Ti:sapphire laser pumping, due to the high pump powers available, there is also a large body of results obtained with diode pumping only (Nakazawa *et al.*, 1991, 1993a).

One of the striking features of the F8L and in fact all fiber soliton lasers is the uniformity of the pulse energy being produced for a specific cavity configuration. This quantization of the pulse energy is to be expected if there is either an amplitude limiting mechanism (nonlinear switching mechanisms increase in transmission to a point and then actually decrease, producing an amplitude limitation on the pulses) or a bandwidth limiting mechanism (physical filters or nonlinear dynamic mechanisms as discussed in Section 4.3 can produce bandwidth limiting) in the laser. The soliton condition in Equation (4.5) and

$$\tau \Delta \nu = .3148 \tag{4.8}$$

provide that either a bandwidth limit or an amplitude limit will determine a unique energy for the pulse.

This energy quantization has a significant consequence for the laser. In a normal modelocked laser the pulse energy will increase until the average power inside the laser cavity (pulse width times pulse repetition rate) is sufficient to saturate the gain sufficiently to equal the loss. In the Er:fiber laser system, since the pulse energy is quantized, the only way to increase the internal average power is to increase the number of pulses in the cavity. An advantageous quality of the Er:fiber amplifier is that the saturation power of the amplifier is a function of the pump power. The consequence is that the number of pulses in the cavity can be controlled by the level of pumping. It should be mentioned that this quantization is not absolute, in that the pulse width dependence of the pulse energy determines the quantization, and the pulse width can be varied within any given cavity. A complete study of the energy quantization in these lasers was carried out by Grudinin *et al.* (1992).

A consequence of the AFSA modelocking of the laser and the multiple pulses in the cavity necessary to saturate the gain is that there is no mechanism in the laser which dictates the position of the pulses relative to each other. The consequence is that the multiple pulses per cavity round trip are free to wander around in the cavity due to small variations in their amplitude or wavelength, or to settle into positions dictated by small subreflections in the cavity, pulse to pulse interactions, or in the frequency domain by the characteristics of the Lyot filter formed by the cavity polarizer and the natural birefringence of the fiber (Davey *et al.*, 1991b).

There are certain applications for which repetition rates higher than the cavity round-trip time would be an advantage. Since the fundamental cavity repetition rate cannot easily be higher than 50 MHz due to the required length of the Er amplifier itself, many of the applications that have been driven by dye or solid state lasers already operate at repetition rates higher than those available by fundamental modelocking of the fiber laser. Harmonic modelocking of the fiber laser requires some mechanism to specify the pulse repetition rate.

The first attempt at stabilizing the repetition rate was demonstrated by Yoshida *et al.* (1992), where an extra coupler (oriented similarly to the output coupler) was added to the feedback loop of the F8L and the ends connected to form a subring cavity. This subcavity then acted to transfer energy from one pulse into succeeding pulses, thus establishing prefered positions for the multiple pulses. Since the main and the subcavities are coupled, there is an interferometric criterion on the length relationship between the two cavities. In addition, the subcavity must be a rational fraction of the main cavity length. With proper stabilization, harmonic modelocking to 125 MHz (38 pulses in the cavity) was demonstrated. In order to remove the interferometric criterion it is possible to use cross-phase modulation in the loop mirror to determine the prefered pulse positions. Dennis and Duling (1992) describe a technique which uses the energy reflected from the loop mirror in the pulse forming process to bias the loop mirror at a particular point in time. An illustration of this

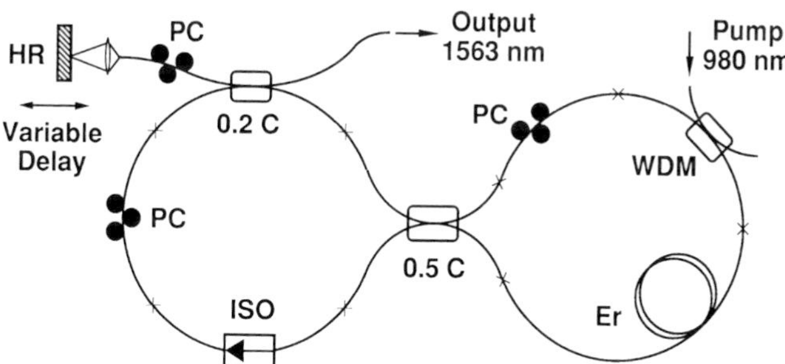

Figure 4.6. Feedback to the unused port of an F8L output coupler leads to control of the pulse spacing in the cavity. C, coupler; PC, polarization controller; WDM, wavelength division multiplexer; ISO, optical isolator; HR, high reflector.

arrangement is shown in Figure 4.6. In this arrangement the output coupler of the laser is placed so that a portion of the light reflected from the loop mirror will exit through the normally unused port of the output coupler. By reflecting this light back into the laser it is possible to establish a prefered position for subsequent pulses in the cavity.

With this method of feedback it was shown that the five pulses in the cavity could be controlled to occur equidistant from each other, producing a 142 MHz repetition rate pulse stream, or by adjusting the feedback delay it was also possible to reduce the spacing between the pulses to produce a short 121 GHz burst of pulses at the fundamental repetition rate of 28 MHz. In the case of the 121 GHz burst, the measurement must be made with an autocorrelator to resolve the separation between the pulses. Burst operation may be useful in coherent excitation of chemical processes or possibly packet production for high speed communications. The advantage of this system over that of the subring cavity is that, since the interaction occurs through the cross phase modulation, synchronization between the echoed pulse and the circulating cavity pulse must be done on the scale of the pulse width rather than interferometric addition.

Passive control of the repetition rate has been reported recently and attributed to the long-range electrostrictive interaction between soliton pulses (Dianov *et al.*, 1992) by Grudinin *et al.* (1993). In this manner pulse trains with repetition rates as high as 914 MHz have been generated that have a stable pulse separation with no drop-outs.

Active modulation of the F8L has been accomplished using a lithium niobate modulator in the external cavity, and harmonically modelocked pulse trains with repetition rates as high as 800 MHz have been generated (Krushchev *et al.*, 1993).

For some applications, pulses shorter than can be obtained directly from a fiber laser are desirable. This is where there is great advantage in having a laser which is already propagating in a fiber. It is a fairly straightforward procedure to implement the technique of fiber pulse compression (Shank *et al.*, 1982) to the pulses in the fiber laser. The technique can be employed in three different manners: internal to the fiber laser, or external to the laser with or without amplification.

External compression has been demonstrated by taking the output of the fiber laser and amplifying it in an EDFA and then propagating the pulse through a length of fiber. Since the pulse has more energy than is appropriate for its width (following the soliton critera of Equations (4.5) and (4.6)) it will reshape itself in propagation down a length of fiber.

Pulses as short as 30 fs have been obtained using this technique (Richardson *et al.*, 1992). An EDFA without the further propagation fiber was utilized by Fontana *et al.* (1992) to produce 75 fs pulses from an actively modelocked fiber laser which incorporated polarization rotation.

The output of the laser can be compressed without amplification if a dispersion decreasing fiber is used at the output of the laser. Compression of this type has been used in conjunction with an F8L to compress 3.5 ps pulses from an F8L to as short as 115 fs (Chernikov *et al.*, 1993b).

In some instances pulses larger than the appropriate soliton size are directly obtained from the output of the laser, or the output fiber can have a significantly lower dispersion than the average dispersion of the cavity. In these instances the pulse emitted by the laser will compress without amplification outside the cavity. It has been shown that using the first of these techniques 80 fs pulses could be obtained from transform-limited 1.5 ps pulses coming out of the laser. In this case the pulse had a substantial wing typical of soliton compression, but which is probably worse in this case due to the factor of nearly 20 compression.

An alternative method of compression which can be implemented is to accomplish the pulse-width reduction inside the fiber laser. An example of this type of compression was presented by Nakazawa *et al.* (1993a) and produced 98 fs pulses directly from the F8L. In this technique an amplifier and a section of dispersion shifted fiber were placed just prior to the output coupler inside the laser cavity. The increased energy of the pulse due to the amplifier caused soliton compression in the dispersion shifted fiber. The pulses shortened below the average pulse width for the laser. This result highlights a general consequence of the changing conditions that the pulse experiences in the cavity. For fiber lasers there is often a large variation in the pulse width as a function of position in the cavity. This change can be used to advantage as it was here, where the pulse was extracted at a point where it was near a minimum for the cavity, and in the case of the alternating dispersion laser described at the end of Chapter 1. The variation of energy and pulse width in the cavity of a fiber laser will also have detrimental effects, as described in the next section.

4.2.3 *Solitonic sidebands*

In Chapter 2, Section 2.2.4, a theoretical discussion is presented of the formation of spectral resonances associated with the periodic perturbations to the soliton caused by the cavity, and by third-order dispersion (TOD) in the cavity. These sidebands, observed in nearly all fiber lasers that include soliton-like pulse shaping in the cavity, have a profound effect on the operation of the laser and its ultimate performance.

Sharp sidebands in the spectra of fiber lasers incorporating nonlinear compensation for dispersion were noted as early as 1990 (Fermann *et al.*, 1990b; Richardson *et al.*, 1991c). Initially these sidebands were attributed to the effects of modulation instability (Richardson *et al.*, 1991b; Agrawal, 1992). The proper identification as a byproduct of the soliton propagation followed the work by Pandit and Kelly in 1992 (Pandit *et al.*, 1992; Kelly, 1992). A typical spectrum showing these spectral sidebands is shown in Figure 4.7. Both a linear plot and a log-scale plot of the data are portrayed. It is clear that the sidebands extend over the full extent of optical energy in the pulse, including that which is outside the fit for the appropriate hyperbolic secant shape of the soliton. An heuristic explanation of the source of these features is that, as the pulse reshapes due to discrete amplitude or dispersion shifts in the fiber, the non-soliton portion of the pulse will begin to propagate as a linear 'dispersive' wave. At the moment of generation the phase of the soliton and dispersive wave are the same, but with propagation in the cavity a phase shift will accumulate (this phase accumulation will be a function of frequency due to dispersion). On the next round trip the soliton will again produce a dispersive wave due to the perturbation. This second dispersive wave will now interfere with the dispersive wave from the previous round

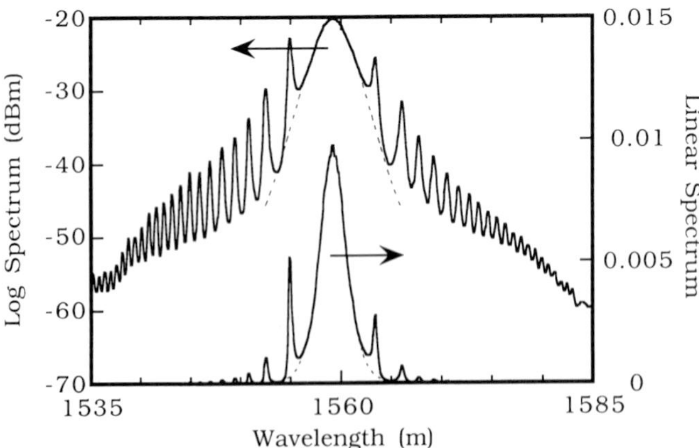

Figure 4.7. The optical pulse spectrum from a fiber laser producing short solitons with the typical resonance sidebands. Both log and linear scales are presented.

trip. Due to the wavelength dependent nature of the phase accumulation, there will be particular frequencies where constructive interference occurs (every 2π of accumulated phase shift).

Gordon (1992) has shown that a sinusoidal perturbation couples energy from the soliton to the dispersive wave at the two frequencies corresponding to $\pm 2\pi$ of accumulated phase shift. Discrete perturbations then correspond to a broad spectrum of Fourier components, each of which gives rise to a sideband. As the number of perturbations increases, a complicated picture involving interference of the different Fourier components emerges. Although early formulations of this problem included only second-order (group velocity) dispersion, as the pulse widths in the laser become shorter it is necessary to include the third-order dispersion as well. For lasers with a low second-order dispersion, the third-order term leads to an additional peak in the spectrum, as identified by Kelly (Brabec and Kelly, 1993).

A number of methods for measuring the dispersion of a laser *in situ* have been developed. Knox (1992) has shown that for a tunable mode-locked laser the dispersion of the cavity can be determined by measuring the repetition rate as a function of the central lasing frequency. Takada *et al.* (1992) have used the fluorescence of the active medium to do broad-band interferometry and determine the cavity dispersion in a method analogous to Fourier transform spectroscopy. In a fiber laser operating in the soliton regime it is possible to derive the dispersion of the cavity while the laser is operating, without the necessity of tuning the laser by examining the position of the resonant sidebands. This technique was first described by Dennis and Duling (1993a), and later expanded to include third-order dispersion (Dennis and Duling, 1993c).

In the presence of third-order dispersion the propagation constant for a linearly propagating wave can be expressed, using a Taylor expansion as,

$$\beta_{\mathrm{d}}(\omega) = \beta_0 + \beta_1 \Delta\omega + \frac{1}{2}\beta_2 \Delta\omega^2 + \frac{1}{6}\beta_3 \Delta\omega + \dots \tag{4.9}$$

where β_2 and β_3 are the group velocity and third-order dispersion per unit length of the fiber, and $\Delta\omega$ is the frequency offset from the soliton central angular frequency ω_0. Note that soliton propagation requires $\beta_2 < 0$ (anomalous dispersion). We will ignore the effect of third-order dispersion on the soliton propagation constant, taking

$$\beta_{\mathrm{s}}(\omega) = \beta_0 + \beta_1 \Delta\omega - \frac{1}{2}\beta_2 \tau_0^{-2} \tag{4.10}$$

for a soliton of length τ_0. The resonance condition requires that

$$2\pi N = \varphi_s - \varphi_d = L(\beta_s - \beta_d), \tag{4.11}$$

which yields

$$N = -\frac{1}{4\pi} L\beta_2\left(\Delta\omega_N^2 + \tau_0^{-2}\right) - \frac{1}{12\pi} L\beta_3 \Delta\omega_N^3. \tag{4.12}$$

Note that $L\beta_2$ and $L\beta_3$ are the total quadratic and cubic dispersions, respectively, for the cavity. If we note that

$$\beta_2 = \frac{-D\lambda^2}{2\pi c} \tag{4.13}$$

where D is the fiber dispersion parameter as usually measured experimentally, and the experimentally measured pulse width at half-maximum of the pulse is

$$\tau = \tau_0 2\ln\left(1 + \sqrt{2}\right). \tag{4.14}$$

Substituting (4.13) and (4.14) into (4.12) and solving for the wavelength offset yields

$$\Delta\lambda_N = \pm N\lambda_0\sqrt{\frac{2N}{cDL} - .0787\frac{\lambda_0^2}{(c\tau)^2}} \tag{4.15}$$

which is equivalent to the expressions given in the references (Kelly, 1992; Smith, N. J. *et al.*, 1992; Noske *et al.*, 1992).

From the perturbation theory analysis of Gordon (1992), the rate of energy loss from the soliton into the dispersive wave is proportional to the amplitude of the soliton spectrum at the location of the sideband. Since the laser will operate in the lowest loss state, the increasing loss with increasing bandwidth provides an effective bandwidth limit to the cavity. As a result of this bandwidth limiting there is a lower limit on the pulse width (Dennis and Duling, 1993b). Mollenauer, Gordon and Islam (1986) have shown numerically that periodic amplification of solitons leads to large fluctuations in the pulse energy and shape when the amplification period is approximately $8z_0$. Noting that the soliton phase evolves by $\pi/4$ in propagating a distance z_0, the $8z_0$ instability is equivalent to exact resonance between the central wavelength of the soliton and coupling to the dispersive wave with $N = 1$.

To test the validity of this theory a number of F8L cavities were constructed which consisted of a base cavity of all necessary components and then varying lengths of either standard dispersion single-mode fiber (17 ps/nm-km), low bend loss fiber (5 ps/nm-km), or dispersion shifted fiber (<.5 ps/nm-km). The base cavity could produce pulses varying from 300 fs to 500 fs dependent on the setting of the polarization controllers. With the shortest pulses, sidebands were observable out to $N = \pm 14$ with the low order sidebands similar in intensity to the main pulse spectrum, while the sidebands were essentially quenched for the longest pulses. For each cavity recordings of the autocorrelation and spectrum were obtained to determine the pulse length and the location of the soliton resonance sidebands. For most configurations, recordings were also taken for the longest pulse width that showed at least residual sidebands (typically about 30 dB down from the central peak), and for a pulse width roughly halfway between the extremes.

Plotting the sideband order versus the relative frequency location of the sidebands, we can see the effect of the addition of fiber with positive dispersion. In Figure 4.8 the curves are shown corresponding to the reduction of the second-order component of the cavity dispersion. The general parabolic shape of the curves is due to the dominant second-order dispersion, and as the second-order is reduced the now dominant third-order component is clearly visible. These curves have been fit by the function

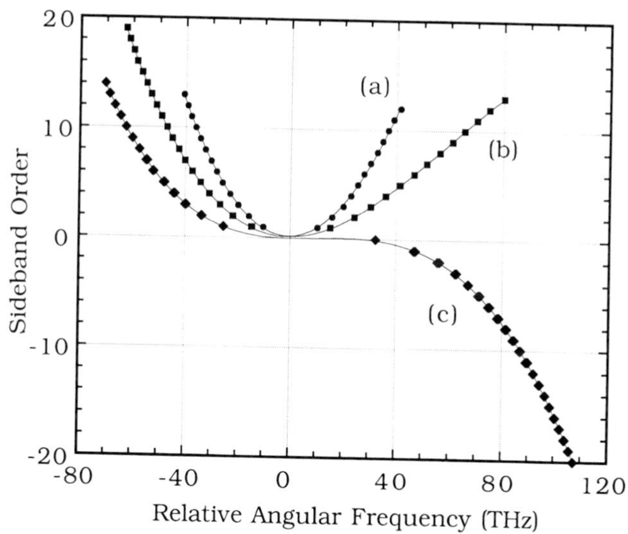

Figure 4.8. The sideband order (numbering from the peak of the spectrum) plotted against the frequency offset from the peak. The curves represent the same cavity with increasing amounts of dispersion compensating fiber inserted. The total cavity dispersions are (a) $-.093$ ps^2, (b) $-.045$ ps^2, (c) $-.015$ ps^2.

$$|N| = A(\Delta\omega_N^2 + B\Delta\omega_N^3 + C), \tag{4.16}$$

where the quadratic term due to the group velocity dispersion is

$$L\beta_2 = -4\pi A, \tag{4.17}$$

the third-order term is

$$L\beta_3 = -12\pi AB, \tag{4.18}$$

and the constant term relates to the nonlinear phase velocity of the soliton and will be discussed later,

$$\tau_0 = 1/\sqrt{C}. \tag{4.19}$$

It is clear from these plots that the addition of compensating fiber decreases the second-order dispersion significantly. With knowledge of the length of fiber added and the change in average dispersion of the cavity a measurement of the dispersion of the inserted fiber can be made. This is in fact the method by which all the fibers in the cavity were measured (Dennis and Duling, 1993c).

As a check of this measurement of dispersion, the technique of measuring the longitudinal mode beat-note frequency while tuning the operating wavelength (Knox, 1992) was used on a long (43 m) F8L, tunable through discrete bands over the Er gain spectrum by adjusting the polarization controller in the isolator loop. (The cavity birefringence and polarizing isolator form a Lyot filter.) The cavity round-trip time is plotted in Figure 4.9 with a quadratic fit to the data. The derivative of this fit yields the dispersion; values obtained from the sideband spectra at two wavelengths are also shown and are in reasonable agreement with the tuning data.

The vertical offsets of the curves in Figure 4.8 are expected to be proportional to $1/\tau_0^2$ from (4.19). To check that the offset was determined by the pulse width, we fit the curves of N versus $\Delta\omega_N$ for $-5 \leq N \leq 5$. Figure 4.10 plots the offset parameter C, i.e. the offset with the fiber dispersion factored out to facilitate comparisons across different cavities, against the measured pulse width for a wide variety of pulse widths and cavity configurations. The error bars indicate only the uncertainty in the values calculated from the fits to Equation (4.16). The solid line plots the theoretically expected offset parameter value from Equation (4.19). The scatter in this data may be due to the fact that the offset is a function of the average soliton value in the cavity, while the measurement is a sample of the pulse at the output coupler. Note that the offset arises physically from the difference between the

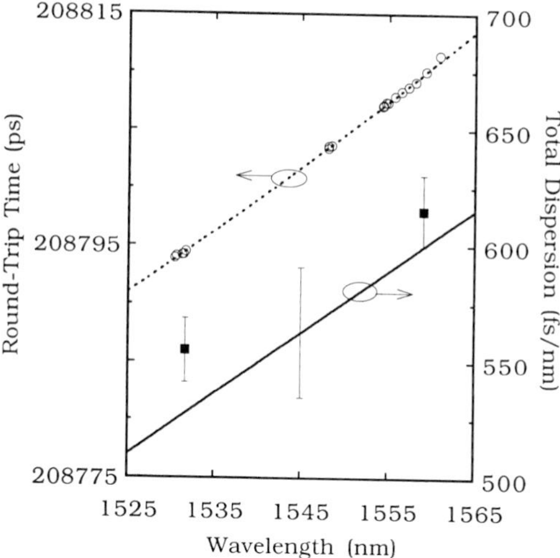

Figure 4.9. The open circles represent the round-trip time in the fiber laser cavity as the lasing wavelength is tuned. The filled squares are the dispersion value obtained from the sidebands for two of the wavelengths. The error bars are derived from the quality of the fit to the sideband spectra.

phase velocities of the soliton and of a linear wave at the same center frequency induced by the Kerr nonlinearity:

$$\Delta V_{\mathrm{p}} \equiv V_{\mathrm{p-sol}} - V_{\mathrm{p-lin}}(\omega) = -\frac{\beta_2}{2\tau_0^2} \tag{4.20}$$

Thus Figure 4.10 experimentally demonstrates the pulse-length dependence of the soliton phase velocity. The sign of the offset parameter also proves that the phase velocity of the soliton component is greater than that of the linear wave (if it were less, the fits in Figure 4.8 would drop below the origin at the central frequency.)

From the theory of Gordon (1992) it was determined that the effective loss from the soliton due to coupling to the dispersive wave is a function of the pulse width. While the loss is in fact a transfer of energy from the soliton into the dispersive wave, this is converted into a cavity loss by the F8L configuration as the low intensity dispersive wave is selectively rejected by the nonlinear amplifying loop mirror (NALM). The loss from the soliton is proportional to its spectral intensity at the sideband wavelength, and so increases exponentially with decreasing pulse duration.

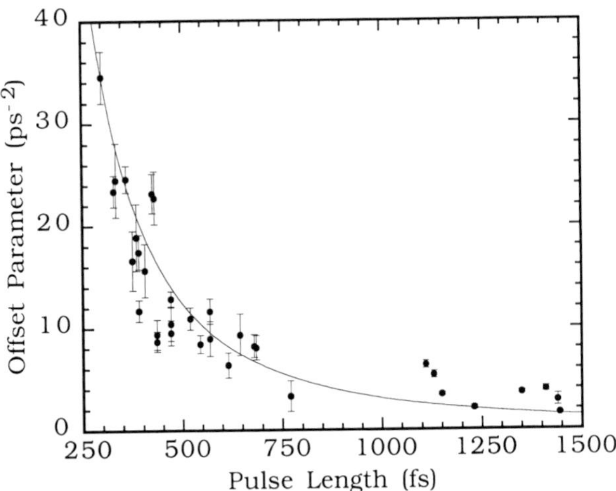

Figure 4.10. Dependence of the soliton phase velocity on pulse length. The data was derived from the fits to the sideband spectra with the error bars from those fits and the solid line is the predicted dependence with no free parameters.

The sideband power is particularly important for the potential application of modelocked fiber lasers as a source in soliton-based telecommunications systems. It has been shown that the timing of adjacent soliton pulses can be affected by interaction with much smaller satellite pulses and with dispersive wave components; over long distance propagation such an interaction can be expected to generate excess bit errors (Mitschke and Mollenauer, 1987; Smith and Mollenauer, 1989). While an ideal source would emit only the soliton component with no sidebands, all-fiber soliton lasers may be practical for telecommunications if the sidebands can be sufficiently suppressed.

The obvious relationship between the location of the first sideband and the shortest pulse that can be generated by a specific cavity can be seen in Figure 4.11. The spectra from three different F8L cavities are shown when producing the shortest pulse obtainable from that cavity. The constant ratio between the location of the first sideband and the pulse bandwidth is evident.

We have measured the fraction of the total output power in the dispersive wave component by recording the output spectrum as in Figure 4.11, fitting the central spectrum with that appropriate for a hyperbolic secant. Both the spectrum and the fit were then integrated over wavelength to obtain quantities proportional to the total power P_{tot} and the

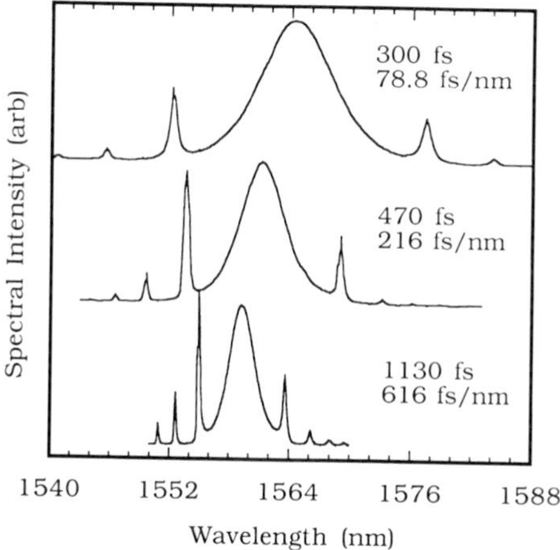

Figure 4.11. Pulse spectra from three different F8L cavities. The pulse bandwidth scales with the separation of the first two sidebands. The pulse duration and the total cavity dispersion are noted next to each plot.

power P_{sol} in the soliton component. We define the sideband fractional power as

$$F = \frac{P_{tot} - P_{sol}}{P_{sol}}. \tag{4.21}$$

Due to the complex nature of the various Fourier components of the coupling between the soliton and the dispersive wave it was not possible to compare different cavities, however, the sideband fractional power is observed to drop off exponentially with increasing pulse length, as illustrated in Figure 4.12, with the rate varying from cavity to cavity.

Values of F of 10–20% were typical with the shortest pulses, although values up to 35% were observed in two different configurations. Previously this fraction has not generally been taken into account when the pulse energy has been compared to the soliton energy for a pulse of the same length propagating in fiber with D equal to the average dispersion of the cavity (Grudinin *et al.*, 1992; Dennis and Duling, 1992; Dennis and Duling, 1993b). The reported pulse energies at minimum pulse width have typically been of the order of $1.1E_1$ to $1.4E_1$; our results imply that the dispersive wave may constitute much of the excess so that the primary pulse is very nearly equal to E_1.

Regardless of the cavity configuration, we find that the most important parameter in the operation of an F8L is the total dispersion. Not only

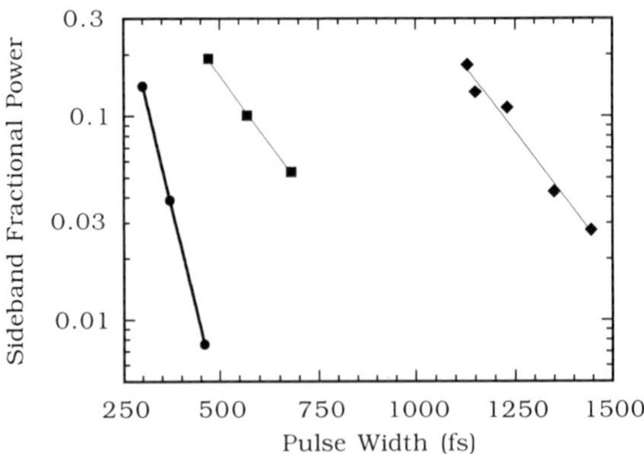

Figure 4.12. The fractional power in the sidebands is plotted as a function of pulse width for three different cavities. The exponential dependence of the loss to the sidebands is clearly seen.

does it determine (to first order) the location of the sidebands, but we find that the minimum pulse length obtained with a particular configuration is given approximately by

$$\tau_{\min} \approx \sqrt{L|\beta_{2-\text{ave}}|}. \tag{4.22}$$

Figure 4.13 illustrates this on a log-log plot, including data points from this work, from other all-fiber systems reported in the literature (where

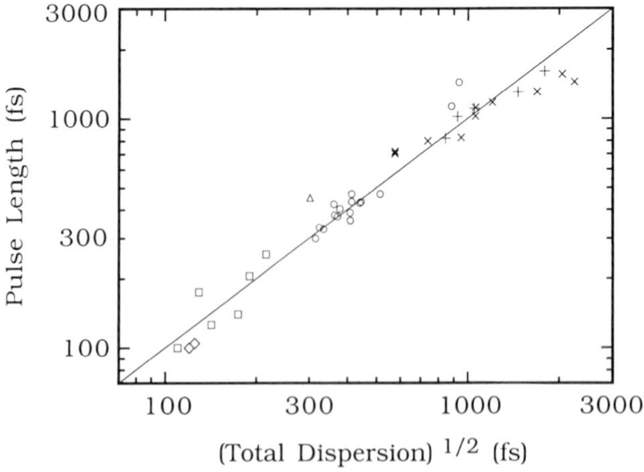

Figure 4.13. Dependence of minimum pulse length on the square root of the total cavity dispersion for a variety of cavity configurations. The data is collected from our own work, and from the literature.

total dispersion was not stated explicitly, it has been calculated from given or reasonable assumed fiber parameters), and from experiments we have conducted on modelocked systems, including dispersion compensating fiber within the cavity (Dennis and Duling, 1993c). The solid line has a slope of unity. The consistency of these results illustrates the role played by the total dispersion in limiting the pulse shortening via the nonlinear loss.

This square-root dependence of pulse width on total dispersion is in direct contradiction to the result expected from a theoretical analysis based on discrete dispersion, self-phase modulation and passive amplitude modulation, which predicts a linear dependence. We have previously attributed the minimum pulse width limitation to increased loss into the sidebands with decreasing pulse width (Dennis and Duling, 1993b). From Gordon (1992), the sideband loss is proportional to the soliton spectrum at the sideband wavelength. As the spectrum spreads with decreasing pulse width, we may thus expect that there will be a pulse width at which the sideband loss exceeds the enhanced transmission of the pulse through the NALM, frustrating further pulse shortening. From (4.15), the sideband wavelength offset is (to first order) inversely proportional to the square root of the total dispersion; as the pulse width is inversely proportional to the spectral width, the observed square-root dependence of the minimum pulse width on dispersion is obtained. We may reasonably expect that the excess loss required to defeat modelocking may vary from cavity to cavity, and as noted above the sideband power fraction at minimum pulse width varies substantially among the cavities we have studied. However, Figure 4.12 shows that the sideband loss increases very quickly with decreasing pulse width, so that the minimum pulse width as a function of the total dispersion can be expected to be approximately the same for the different cavities.

As a result of (4.22) and its confirmation in Figure 4.13, it is obvious that the route to shorter pulses is by reducing the total dispersion in the cavity. This can be accomplished in three ways: reduction in cavity length, replacement of standard fiber by dispersion shifted fiber in the cavity, and inclusion of fiber of the opposite sign of dispersion so that the average dispersion is reduced. By replacing all but the Er:fiber by dispersion shifted fiber and components the total dispersion in the cavity could be reduced to 70 fs/nm, yielding 160 fs pulses. In order to reduce the total dispersion further it is necessary to add fiber of the opposite sign. We have obtained pulse widths of <100 fs via this approach.

It is perhaps worthwhile at this point to mention that the availability of various dispersion fibers provides a wide flexibility in the construction and operation of fiber sources. We have seen that the periodic perturbation of the soliton results in an effective bandwidth limit in the cavity. The method of reducing dispersion in the cavity is an attempt to move the location of the sidebands out of the bandwidth of the pulse. An alternative approach would be to reduce the strength of the interaction by reducing the magnitude of the perturbation. Since it is reshaping of the soliton which causes the shedding of the dispersive wave, the dispersive wave generation can be eliminated if we create a situation where, although there are amplitude changes, the soliton does not tend to reshape. This can be accomplished by changing the dispersion in accordance with the amplitude changes such that the soliton condition (4.7) is always met. This would require that the dispersion be changed discretely at the output coupler, and that the dispersion in the amplifier be tailored to roughly follow the amplification. This tailoring can be accomplished in a piecewise manner as long as the segments of fiber remain shorter than the soliton period for the pulse.

4.3 Other modelocked fiber sources

4.3.1 *Fiber Raman lasers*

Some of the first subpicosecond fiber lasers were those based on the Raman gain generated by a modelocked laser injected into the fiber. Two good summaries of the work that was done in this field are contained in Kafka and Baer (1990) and Gouveia-Neto *et al.* (1988). The fundamental requirements of a laser of this type is to generate the Raman gain, compensate the group velocity dispersion between the pump and the lasing wavelengths and provide for the fiber cavity to be an integral multiple of the pump laser cavity. Fiber Raman lasers have been constructed that generate a variety of wavelengths. One of the key aspects of fiber Raman lasers is that they display time dispersion tuning. As the length of the cavity is changed, the wavelength of the laser will shift to provide the maximum spacio-temporal overlap of the laser pulse and the pump pulse. This effect can be avoided by including a tuning element in the cavity.

The simplest laser of this type utilizes a CW modelocked Nd:YAG laser operating at 1319 nm as the pump source (Kafka and Baer,

1987). By using a fiber with a zero dispersion point of just greater than 1360 nm, laser oscillation at 1410 nm can operate in the negative dispersion regime allowing soliton pulse shaping to enhance the active modelocking. It has also been suggested that the cross-phase modulation from the pump pulse contributes to the phase profile of the laser pulse which results in further pulse shortening. The cavity is shown in Figure 4.14. The laser consists of a specialized fiber coupler and an appropriate length of fiber. The coupler is chosen such that the pump light is coupled completely into the cavity and only a small fraction of the laser light is coupled out. The requirement for the fiber to be an integral number of pulses long can be accounted for by slight changes in the modelocking frequency of the pump laser. Changes made in this fashion are multiplied by the number of times the fiber cavity is longer than the pump cavity (often as many as 400). Lasers of this type have generated pulses as short as 300 fs.

The requirement of a high power pump laser precludes these lasers from being characterized as 'compact', but the power of diode pumped solid state lasers continues to rise, increasing the possibility that in the near future it will be possible to create a diode pumped fiber Raman laser.

4.3.2 *Nonlinear coupler modelocking*

One novel fiber laser which has yet to be demonstrated, but looks very promising, is that based on a cavity long nonlinear coupler (Winful and Walton, 1992; Walton and Winful, 1993). The general operation is that one core of a dual core fiber is doped to provide gain. The fiber length is chosen to provide small signal cross-coupling in one cavity

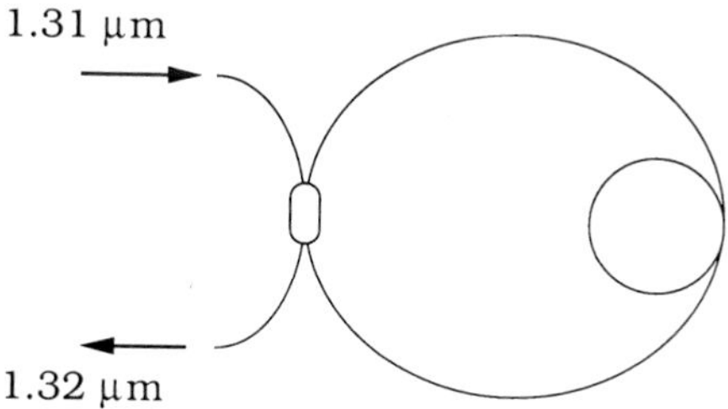

Figure 4.14. Cavity arrangement for a fiber Raman laser. A modelocked Nd:YAG laser operating at 1.31 μm synchronously supplies the gain for the lasing mode at 1.32 μm.

length. The ends of the doped core are mirrored to provide feedback to the cavity, with one end acting as an output coupler. The second core of the fiber provides increased loss for CW oscillation compared with pulsed operation, due to the reduction of the coupling coefficient with increasingly nonlinear propagation. In addition to providing loss for the CW component, the second core removes any dispersive wave component which might be shed by a soliton evolving in the main core. By removing this unwanted light from the cavity the laser gains stability and robustness.

4.3.3 *Semiconductor saturable absorbers*

As discussed in the section on diode pumped solid state lasers, work has been done on the modelocking with various types of semiconductor saturable absorbers. These passive modelockers use real carrier generation to produce saturable absorption in the cavity. Carrier-type modelocking (CTM) of fiber lasers, as first demonstrated by Zirngibl *et al.* (1991), are suitable for the generation of low repetition-rate pulses with widths between 300 fs and a few ps. CTM allows self-starting operation (with a single pulse in the cavity) in all-polarization maintaining fiber lasers (DeSouza *et al.*, 1993). CTM may be insensitive to polarization drifts in weakly-birefringent fiber (Loh *et al.*, 1993) and it allows stable operation with negligible self-phase modulation in the cavity. In CTM systems pulse formation is dominated by passive amplitude modulation arising from saturable absorption in a semiconductor near the band edge.

The saturation characteristics are determined by two processes. The first, carrier recombination, has a time constant of 10 ps to 30 ns depending on the semiconductor (Smith *et al.*, 1985), and the second, exciton screening (Haus and Silberberg, 1985) and carrier thermalization (Islam *et al.*, 1989), a time constant of ~300 fs. For pulses shorter than these time constants and for a cavity round-trip time very much longer than the relaxation times, the absorption saturates as a function of pulse energy in the form

$$\alpha = \sum \frac{\alpha_i}{1 + \frac{E_{\mathrm{p}}}{E_{si}}} + \alpha_{\mathrm{ns}} \tag{4.23}$$

where α_i is the saturable absorption due to the absorption mechanism, E_{p} is the pulse energy density and $E_{si} = h\nu/\sigma_{si}$ is the saturation energy density of the respective saturation mechanism with the absorption cross section σ_{si}, and α_{ns} is the non-bleachable absorption. Typical values for the saturation energy densities due to long-lived carriers are of the order of 1 mJ/cm^2, and for the two fast mechanisms saturation energy densities smaller by a factor of three are typically assumed. These values are proportional to the total linear absorption and increase linearly with the ratio of the absorption at the band edge to the absorption at the band tail, where modelocking is typically obtained. The long time constant governs the start-up dynamics and the short constant determines the shortest possible pulse widths (Islam *et al.*, 1989).

To date the best performance of a CTM fiber laser (Ober *et al.*, 1993) has been obtained by using a multiple quantum well (MQW) saturable absorber as part of an antiresonant nonlinear Fabry–Perot mirror (Keller *et al.*, 1992). In this work, one reflecting side was made from a totally reflecting stacked mirror structure and the other reflecting side was simply the front-surface reflection ($\sim$33%) from the MQW. The antiresonant structure reduces thermal loading of the MQW and increases its effective saturation energy. The self-starting modelocking threshold was obtained with 15 mW of absorbed pump power at an intracavity pulse energy of 180 pJ, allowing for diode pumped operation. The corresponding energy density inside the MQW was about 600 $\mu J/cm^2$. Higher pulse energies could be obtained by reducing the focusing onto the MQW.

The pulse energy on the MQW should be comparable with its saturation energy to ensure an appreciable amount of amplitude modulation. In this type of laser, passive modelocking is obtained with an intracavity nonlinear phase delay of less than $\pi/6$, which demonstrates that passive frequency modulation plays a minor role in the pulse formation process. As a consequence the pulse widths are far less sensitive to dispersion than in Kerr-type modelocked systems, and the pulse spectrum can be totally free of resonant sidebands.

High repetition rate operation in CTM is also possible. The two limitations are the high pulse energies required to sufficiently saturate the absorber and the rapid fall-off of the nonlinear response of the absorber at high repetition rates (Keller *et al.*, 1992). Since CTM fiber systems are less sensitive to dispersion, no dispersion compensation is required for short lengths of Nd fibers, and Nd oscillators could in principle be operated at a fundamental repetition rate of a few GHz. A Nd fiber laser could then be used as a master clock oscillator for the synchronization of an Er/Yb fiber laser (Fermann *et al.*, 1988). The incorporation of subcavities also provides an opportunity for high repetition rate pulse generation, as shown by Keller *et al.* (1992), who demonstrated the production of pulses at repetition rates up to 1 GHz in a bulk CTM Nd:YAG system. Here optical phase fluctuations are self-compensating by corresponding fluctuations of the optical carrier frequency. Recently harmonic partitioning of a CTM fiber laser was also demonstrated as a viable technique for moderately high repetition rate pulse generation (Fermann *et al.*, 1993). In this work the saturable absorber is located asymmetrically within a Fabry–Perot-type cavity, where the round-trip distance from the left cavity mirror to the saturable absorber is chosen to be an integer (n) multiple of the distance from the saturable absorber to the right cavity mirror. The saturable absorber is then optimally bleached when exactly $n + 1$ pulses oscillate in the cavity simultaneously, and stable higher-harmonic modelocking may be obtained. This technique is also insensitive to optical phase-fluctuations.

4.3.4 *Er:fiber saturable absorber*

Although producing pulses of much longer duration than has been discussed in the earlier sections of this book, an interesting mechanism for modelocking has been investigated by Nakazawa *et al.* (1993b). In this case the Er:fiber itself is used as the saturable absorber after being cooled to 4.2 K. The laser self-Q-switches and modelocks producing 20–30 ns pulses in 3.3 μs Q-switch trains, with a repetition period of 77 μs dictated by the gain recovery time of the laser. A ring laser configuration was used. The pulses that result are longer in duration than the dephasing time (T_2) so that the modelocking is due to saturable absorption rather than a coherent pulse formation. It is possible that a shorter section of Er:fiber might result in self-induced-transparency modelocking and shorter pulses.

4.3.5 *Linear external cavity*

In a method similar to the APM method of modelocking, Wigley *et al.* (1990) used a linear external cavity to provide an interferometric addition of a pulse which had experienced a nonlinear phase shift (the one in the fiber laser cavity) with one that had propagated linearly (the one in the external cavity). With this method Wigley was able to produce 200 ps pulses in a Nd:fiber laser. The laser modelocked while the end mirror of the linear cavity was translated.

4.3.6 *Synchronous pumping of the fiber laser*

In contrast to synchronously pumped dye lasers, where the pulsed nature of the pump leads to a modulation of the gain in the laser, the gain lifetime is so long in the fiber laser that there is effectively no gain modulation. On the other hand there can be transient phase modulation (or cross-phase modulation) between the pump and the light traveling in the cavity. This mechanism has been used to produce pulses as short as 2.4 ps at repetition rates on the order of 100 MHz (Avramopoulos *et al.*, 1990; Stock *et al.*, 1993; Noske *et al.*, 1993). The modulation window is limited by the pulse walk-off between the pump and the pulse in the cavity. In the work by Stock, a Nd:YAG laser was used to pump a Yb:Er:fiber. The modelocked laser had a fiber length of 30 m and the walk-off length was 38 m. As in most fiber lasers, in addition to the cross-phase modulation, soliton formation also contributes to the short pulse generation.

4.3.7 *Optical sources based on two frequency beat notes*

One method for producing high repetition rate trains of solitons which is not strictly a laser source is the work which has been accomplished using fibers which have been constructed to have a monotonically decreasing dispersion as a function of length (Chernikov *et al.*, 1992). These fibers range in length from 100 m to a few kilometers, and their dispersions vary over as much as an order of magnitude. The effect on a soliton of decreasing dispersion is equivalent to a distributed amplification. In this method of pulse production two single mode lasers which are locked with a finite frequency difference are combined in an optical fiber. The two lasers beat against each other, producing a sinusoidal output at their frequency offset. After amplification the sinusoidal signal is injected into the high dispersion end of the dispersion decreasing fiber (DDF). As the light propagates down the fiber the sinusoidal train evolves into a train of solitons. The repetition rate was varied from 80 to 130 GHz and pulse durations from 1.5 to 3 ps were generated.

By using significantly higher launch powers it is possible to have the effect of the DDF combined with the nonlinear effect of Raman self-scattering (RSS) to enhance the compression and generate pulses at repetition rates as high as 114 GHz and durations as short as 250 fs (Chernikov *et al.*, 1993a). Due to the RSS the soliton train translates in frequency and time away from the background pulses, resulting in an output which contains a clean train of solitons and a second train of longer pulses at a slightly shorter wavelength.

Another method of producing short pulse trains, which does not use dispersion decreasing fiber but does start with the beat between two single mode sources, uses the nonlinear transmission of a NOLM to shorten the beat note into a train of pulses. The NOLM not only imposes an amplitude modulation on the beat signal, but adds a phase modulation which can be compensated after the loop mirror. The result of this process is to produce a train of pulses which can be as short as 4.3 ps at a repetition rate of 32 GHz (Chernikov and Taylor, 1993).

4.4 Summary

We have described a number of mechanisms that have been used to modelock fiber lasers. By utilizing active, passive and nonlinear pulse propagation and combinations of these effects, pulses from 90 ps to 90 fs have been produced. For any given application the optimal modelocking solution may dictate a specific combination of these mechanisms.

We examined pure active modelocking and described the necessity of matching the cavity length to the drive frequency. The pulse shape will be Gaussian and the pulse duration will be governed by Kuizenga–Siegman modelocking theory. Passive modulation can get an assist from the chirping in the process which when compensated produces pulses approximately three times shorter than amplitude modulation.

Passive modelocking by use of a nonlinear optical loop mirror provides significantly shorter pulses while allowing all-polarization maintaining cavities to be built. The nonlinear transmission function provides pulse shortening on transmission.The consequence of the artificial fast saturable absorber in the cavity leads to the possibility of multiple randomly spaced pulses. The addition of subcavities or external feedback can control the pulse spacing to some extent.

The minimum pulse that can be obtained in a fiber laser operating in the soliton forming dispersion regime is dictated by the total dispersion in the cavity and not the average cavity dispersion. This is true for cavities where the pulses start from the noise, or where the active modelocking mechanism is weak. This limitation from total dispersion is due to the periodic perturbation of the soliton as it traverses the cavity. The soliton sheds a dispersive wave on each roundtrip. The bandwidth of the pulse will broaden until the loss to the dispersive wave balances the gain due to self-amplitude modulation. The interference between the dispersive waves generated on successive round trips leads to sharp resonances in the laser spectrum.

We have presented a number of other laser systems including synchronous pumping of Raman gain, nonlinear coupler modelocking, semiconductor saturable absorbers, erbium-doped fiber saturable absorbers, phase modulation through synchronous pumping of the erbium gain fiber. In addition the compact source utilizing two narrow band sources to produce an optical beat note which is transformed into a pulse train by nonlinear transmission was also described. The wide range of modelocking techniques used in fiber lasers has led to a rapidly growing field and a large number of useful sources. As these techniques of modelocking are developed and as new methods are implemented in fiber, the fiber laser will become as common as the helium neon laser and ultrashort pulse lasers will become economically available to a wider range of researchers, leading to further advancement of the field.

References

Agrawal, G. P. (1989) *Nonlinear Fiber Optics* (Academic Press: New York).

Agrawal, G. P. (1992) *Phot. Tech. Lett.*, **4**, 562–4.

Alcock, I. P., Tropper, A. C., Ferguson, A. I. and Hanna, D. C. (1987) *IEE Proceedings*, **134**, 183–6.

Avramopoulos, H., Houh, H., Whitaker, N. A., Gabriel, M. C. and Morse, T. (1990) *IEEE Tech. Dig. Series*, **13**, PdP8.

Brabec, T. and Kelly, S. M. J. (1993) *Opt. Lett.*, **18**, 2002–4.

Bulushev, A. G., Dianov, E. M. and Okhotnikov, O. G. (1990) *Opt. Lett.*, **15**, 968–70.

Burns, D. and Sibbett, W. (1990) *Electron. Lett.*, **26**, 505–6.

Burns, W. K. and Kersey, A. D. (1992) *J. Lightwave Tech.*, **10**, 992–9.

Chernikov, S. V. and Taylor, J. R. (1993) *Electron. Lett.*, **29**, 658–9.

Chernikov, S. V., Taylor, J. R., Mamyshev, P. V. and Dianov, E. M. (1992) *Electron. Lett.*, **28**, 931–2.

Chernikov, S. V., Dianov, E. M., Richardson, D. J., Laming, R. I. and Payne, D. N. (1993a) *Appl. Phys. Lett.*, **63**, 293–5.

Chernikov, S. V., Dianov, E. M., Richardson, D. J. and Payne, D. N. (1993b) *Opt. Lett.*, **18**, 476–8.

Cochlain, C. R., Mears, R. J. and Sherlock, G. (1993) *IEEE Photonics Technol. Lett.*, **5**, 25–7.

Davey, R. P., Fleming, R. P. E., Smith, K., Kashyap, R. and Armitage, J. R. (1991a) *Electron. Lett.*, **27**, 2087–8.

Davey, R. P., Langford, N. and Ferguson, A. I. (1991b) *Electron. Lett.*, **27**, 1257–9.
Davey, R. P., Langford, N. and Ferguson, A. I. (1991c) *Electron. Lett.*, **27**, 726–8.
Davey, R. P., Smith, K. and McGuire, A. (1992) *Electron. Lett.*, **28**, 482–4.
Dennis, M. L. and Duling III, I. N. (1992) *Electron. Lett.*, **28**, 1894–6.
Dennis, M. L. and Duling III, I. N. (1993a) *Electron. Lett.*, **29**, 409–11.
Dennis, M. L. and Duling III, I. N. (1993b) *Appl. Phys. Lett.*, **62**, 2911–3.
Dennis, M. L. and Duling III, I. N. (1993c) *IEEE J. Quantum Electron.*, **30**, 1469–77.
DeSouza, E. A., Islam, M. N., Soccolich, C. E., Pleibel, W., Stolen, R. H., Simpsona, J. R. and Giovanni, D. J. (1993) *Electron. Lett.*, **29**, 447.
Deutsch, B. and Pfeiffer, Th. (1992) *Electron. Lett.*, **28**, 303–5.
Dianov, E. M., Luchnikov, A. V., Pilipetskii, A. N. and Prokhorov, A. M. (1992) *Applied physics. B, Photophysics and laser chemistry*, **B54**, 175–80.
Doran, N. J. and Wood, D. (1988) *Opt. Lett.*, **13**, 56–8.
Duling, I. N., Goldberg, L. and Weller, J. F. (1988) *Electron. Lett.*, **24**, 1333–4.
Duling III, I. N. (1991a) *Electron. Lett.*, **27**, 544–5.
Duling III, I. N. (1991b) *Opt. Lett.*, **16**, 539–41.
Duling III, I. N., Chen, C. J., Wai, A. K. and Menyuk, C. R. (1994) *IEEE J. Quantum Electron.*, **30**, 194–9.
Dzhibladze, M. I., Esiashvili, Z. G., Teplitskii, E. Sh., Isaev, S. K. and Sagaradze, V. R. (1983) *Kvantovaya Elektron.* (Moscow), **10**, 432–4
Fermann, M. E., Hanna, D. C., Shepherd, D. P., Suni, P. J. and Townsend, J. E. (1988) *Electron. Lett.*, **24**, 894.
Fermann, M. E., Haberl, F., Hofer, M. and Hochreiter, H. (1990a) *Opt. Lett.*, **15**, 752–4.
Fermann, M. E., Hofer, M., Haberl, F. and Schmidt, A. J. (1990b) *Proceedings of the 16th European Conference on Optical Communications*, 1053–6.
Fermann, M. E., Stock, M. L., Yang, L.-M., Andrejco, M. J. and Harter, D. (1993) *Opt. Soc. Am. Topical Meeting on Nonlinear Guided Wave Phenomena, WA4.*
Fermann, M. E., Yang, L.-M., Stock, M. L. and Andrejco, M. J. (1994) *Opt. Lett.*, **19**, 43.
Finlayson, N., Nayar, B. K. and Doran, N. J. (1992) *Opt. Lett.*, **17**, 112–14.
Fontana, F., Grasso, G., Manfredini, N., Romagnoli, M. and Daino, B. (1992) *Electron. Lett.*, **28**, 1291–3.
Frisken, S. J., Telford, C. A., Betts, R. A. and Atherton, P. S. (1991) *Electron. Lett.*, **27**, 887–8.
Geister, G. and Ulrich, R. (1988) *Opt. Commun.*, **68**, 187–9.
Gordon, J. P. (1992) *J. Opt. Soc. Am. B*, **9**, 91–7.
Gouveia-Neto, A. S., Gomes, A. S. L. and Taylor, J. R. (1988) *IEEE J. Quantum Electron.*, **24**, 332.
Grudinin, A. B., Richardson, D. J. and Payne, D. N. (1993) *Electron. Lett.*, **28**, 67–8.
Hanna, D. C., Kazer, A., Phillips, M. W., Shepherd, D. P. and Suni, P. J. (1989) *Electron. Lett.*, **25**, 995–6.

Harvey, G. T. and Mollenauer, L. F. (1993) *Opt. Lett.*, **18**, 107–9.

Haus, H. A. and Silberberg, Y. (1985) *J. Opt. Soc. Am. B*, **2**, 1237.

Islam, M. N., Sunderman, E. R., Soccolich, C. E., Bar-Joseph, I., Sauer, N., Chang, T. Y. and Miller, B. I. (1989) *IEEE J. Quant. Electron.*, **25**, 2454.

Kafka, J. D. and Baer, T. (1987) *Opt. Lett.*, **12**, 181–3.

Kafka, J. D. and Baer, T. (1990) *SPIE 1373 Fiber Laser Sources and Amplifiers II*, 140–9.

Kafka, J. D., Baer, T. and Hall, D. W. (1989) *Opt. Lett.*, **14**, 1269–71.

Keller, U., Miller, D. A. B., Boyd, G. D., Chiu, T. H., Ferguson, J. F. and Asom, M. T. (1992a) *Opt. Lett.*, **17**, 505.

Keller, U., Nelson, L. E. and Chiu, T. H. (1992b) *OSA Proceedings on Advanced Solid State Lasers*, **13**, 94.

Kelly, S. M. J. (1992) *Electron. Lett.*, **28**, 806–7.

Kelly, S. M. J., Smith, K., Blow, K. J. and Doran, N. J. (1991) *Opt. Lett.*, **16**, 1337–9.

Knox, W. H. (1992) *Opt. Lett.*, **17**, 514–6.

Krausz, F., Fermann, M., Brabec, T., Curley, P. F., Hofer, M., Ober, M. H., Spielmann, C., Wintner, E. and Schmidt, A. J. (1992) *IEEE J. Quant. Electron.*, **28**, 2097–2122.

Krushchev, I. Y., Richardson, D. J. and Dianov, E. M. (1993) *Proc. European Conference on Optical Communications*, **2**, 33–5.

Loh, W. H., Atkinson, D., Morkel, P. R., Hopkinson, M., Rivers, A. and Payne, D. N. (1993) *IEEE Photonics Tech. Lett.*, **5**, 35.

Mitschke, F. M. and Mollenauer, L. F. (1987) *Opt. Lett.*, **12**, 355–7.

Mollenauer, L. F., Gordon, J. P. and Islam, M. N. (1986) *IEEE J. Quant. Electron.*, **22**, 157–73.

Mortimore, D. B. (1988) *J. Lightwave Technol.*, **6**, 1217–24.

Nakazawa, M., Yoshida, E. and Kimura, Y. (1991) *Appl. Phys. Lett.*, **59**, 2073–5.

Nakazawa, M., Yoshida, E. and Kimura, Y. (1993a) *Electron. Lett.*, **29**, 63–5.

Nakazawa, M., Suzuki, K., Kubota, H. and Kimura, Y. (1993b) *Opt. Lett.*, **18**, 613–15.

Noske, D. U., Pandit, N. and Taylor, J. R. (1992) *Opt. Lett.*, **17**, 1515–17.

Noske, D. U., Boskovic, A., Guy, M. J. and Taylor, J. R. (1993) *Electron. Lett.*, **29**, 1863–4.

Ober, M. H., Hofer, M., Chiu, T. H., and Keller, U. (1993) *Opt. Lett.*, **18**, 1532.

Pandit, N., Noske, D. U., Kelly, S. M. J. and Taylor, J. R. (1992) *Electron. Lett.*, **28**, 455–7.

Phillips, M. W., Ferguson, A. I. and Hanna, D. C. (1989a) *Opt. Lett.*, **14**, 219–21.

Phillips, M. W., Ferguson, A. I., Kino, G. S. and Patterson, D. B. (1989b) *Opt. Lett.*, **14**, 680–2.

Richardson, D. J., Laming, R. I., Payne, D. N., Matsas, V. and Phillips, M. W. (1991a) *Electron. Lett.*, **27**, 542–4.

Richardson, D. J., Laming, R. I., Payne, D. N., Phillips, M. W. and Matsas, V. J. (1991b) *Electron. Lett.*, **27**, 730–2.

Richardson, D. J., Laming, R. I., Payne, D. N., Matsas, V. J. and Phillips, M. W. (1991c) *Electron. Lett.*, **27**, 1451–3.
Richardson, D. J., Afanasjev, V. V., Grudinin, A. B. and Payne, D. N. (1992a) *Opt. Lett.*, **17**, 1596.
Richardson, D. J., Grudinin, A. B. and Payne, D. N. (1992b) *Electron. Lett.*, **28**, 778–9.
Schlager, J. B., Kawanishi, S. and Saruwatari M. (1991) *Electron. Lett.*, **27**, 2072–3.
Shan, X., Cleland, D. and Ellis, A. (1992) *Electron. Lett.*, **27**, 182–4.
Shank, C. V., Fork, R. L., Yen, R., Stolen, R. H. and Tomlinson, W. J. (1982) *Appl. Phys. Lett.*, **40**, 761–3.
Siegman, A. E. and Kuizenga, D. J. (1974) *Optoelectron.*, **6**, 43–66.
Singh, K. (1993) Doctoral Thesis, Rensselaer Polytechnic Institute, Troy NY.
Smith, K. and Mollenauer, L. F. (1989) *Opt. Lett.*, **14**, 1284–6.
Smith, K., Doran, N. J. and Wigley, P. G. J. (1990a) *Opt. Lett.*, **15**, 1294–6.
Smith, K., Armitage, J. R., Wyatt, R., Doran, N. J. and Kelly, S. M. J. (1990b) *Electron. Lett.*, **26**, 1149–50.
Smith, K., Greer, E. J., Wyatt, R., Wheatley, P., Doran, N. J. and Lawrence, M. (1991) *Electron. Lett.*, **27**, 244–5.
Smith, N. J., Blow, K. J. and Andonovic, I. (1992) *J. Lightwave Tech.*, **10**, 1329–33.
Smith, P. W., Silberberg, Y. and Miller, D. A. B. (1985) *J. Opt. Soc. Am. B*, **2**, 1228.
Stentz, A. J. and Boyd, R. W. (1993) *Conference on Lasers and Electro-Optics*, CFJ4.
Stock, M. L., Yang, L.-M., Andrejco, M. J. and Fermann, M. E. (1993) *Opt. Lett.*, **18**, 1529–31.
Sugawa, T., Yoshida, E., Miyajima, Y. and Nakazawa, M. (1993) *Electron. Lett.*, **29**, 902–3.
Takada, A. and Miyazawa, H. (1990) *Electron. Lett.*, **26**, 216–17.
Takada, K., Kitagawa, T., Hattori, K., Yamada, M. and Horiguchi, M. (1992) *Electron. Lett.*, **28**, 1889.
Tzelepis, V., Markatos, S., Kalpogiannis, S., Sphicopoulos, Th. and Caroubalos, C. (1993) *J. Light. Tech.*, **11**, 1729.
Walton, D. T. and Winful, H. G. (1993) *Opt. Lett.*, **18**, 720–2.
Wigley, P. G. J., French, P. M. W. and Taylor, J. R. (1990) *Electron. Lett.*, **26**, 1238–40.
Winful, H. G. and Walton, D. T. (1992) *Opt. Lett.*, **17**, 1688–90.
Wu, S., Strait, J., Fork, R. L. and Morse, T. F. (1993) *Opt. Lett.*, **18**, 1444.
Yoshida, E., Kimura, Y. and Nakazawa, M. (1992) *Appl. Phys. Lett.*, **60**, 932–4.
Zirngibl, M., Stulz, L. W., Stone, J., Hugi, J., DiGiovanni, D. and Hansen, P. B. (1991) *Electron. Lett.*, **27**, 1734–5.

5

Nonlinear polarization evolution in passively modelocked fiber lasers

MARTIN E. FERMANN

5.1 Introduction

Single-mode fibers typically contain two eigenmodes with the same light intensity distribution, but with orthogonal polarization states. In perfectly isotropic fibers these modes are degenerate and have the same propagation constants. Hence isotropic fibers are polarization maintaining, i.e. the output polarization state is the same as the input polarization state. However, this statement is not generally valid in the presence of nonlinearities. Though an isotropic fiber is still polarization maintaining for linearly and circularly polarized light independent of the light intensity, an elliptical polarization state will rotate as a function of power.

The phenomenon of ellipse rotation was already well known in the 1960s and was commonly used to measure the refractive index of isotropic materials with Kerr-type nonlinearities (Maker and Terhune, 1965). By incorporating suitable polarization optics, ellipse rotation can also be used as an ultrafast saturable absorber for passive modelocking, as first realized by Dahlström (1972). However in this work stable CW modelocking was not obtained due to the low available power levels. The first application of nonlinear polarization evolution in optical fibers dates back to Stolen *et al.* (1982), who used a highly-birefringent fiber for shortening and cleaning the pulses from a separate external ultra-short pulse source. Nonlinear polarization evolution for the general case of low birefringence fibers was first analyzed by Winful (1985), who also predicted the phenomenon of polarization instability when high intensity light is launched close to the fast axis of the fiber. Nonlinear polarization evolution in elliptically birefringent fibers was analyzed for the first time by Menyuk (1989), who also predicted the effect of soliton trapping caused by nonlinear polarization evolution in the presence of negative fiber dispersion (Menyuk, 1987, 1988; Menyuk and Wai, 1992). Soliton trapping was demonstrated experimentally for the first time by Islam *et al.* (1989) in single-pass soliton transmission experiments.

With the advent of wide-bandwidth solid state lasers in the late 1980s, the intracavity power levels could be sufficiently raised to allow the observation of pulse shortening by nonlinear polarization evolution in CW modelocked laser oscillators (Squier *et al.*, 1990; Carruthers and Duling, 1990). The first passively modelocked fiber laser based on nonlinear polarization evolution was described by Hofer *et al.* (1991), where 70 fs pulses were directly generated from a neodymium fiber laser.

Recent work (Fermann *et al.*, 1994) has shown that modelocked fiber lasers based on nonlinear polarization evolution are particularly useful since they can be constructed to be environmentally stable. With the prospect of stable compact femtosecond fiber laser sources becoming practical tools for the laboratory, nonlinear polarization evolution is bound to receive continued attention in the future and to be established as one of the most important modelocking techniques for fiber lasers.

5.2 Linear polarization evolution in fiber lasers

5.2.1 *Fiber properties*

Perfectly isotropic fibers do not exist in practice and therefore, in general, the fibers are birefringent, where the polarization eigenmodes can also vary along the fiber length. The polarization eigenmodes are either linearly, elliptically or circularly polarized, corresponding to linear, elliptical and circular birefringence in the fiber (Ulrich and Simon, 1979). Fibers with dominant linear birefringence (Payne *et al.*, 1982) and elliptical birefringence (Laming *et al.*, 1989) have been manufactured, but to date the fabrication of circularly birefringent fibers, with nonlinear propagation characteristics similar to isotropic fibers, remains elusive.

The relative strengths of the birefringence-causing mechanisms may be described by the induced mode beat length L_B, i.e. the length period at which the polarization state is reproduced. L_B is given by

$$L_B = \frac{2\pi}{\delta\beta} \tag{5.1}$$

where $\delta\beta$ is the difference in the propagation constant of the two polarization eigenmodes. For linearly birefringent fiber $\delta\beta = 2\pi(n_x - n_y)/\lambda$, where $n_{x,y}$ are the refractive indices along the x and y axes at the wavelength λ. Typical beat lengths can vary from 1 mm in high birefringence fibers up to 100 m in low birefringence fibers. The various intrinsic and extrinsic effects causing birefringence in fibers were reviewed in detail by Rashleigh (1983). The most important intrinsic mechanisms are linear stress and form birefringence. The former is due to different thermal expansion coefficients of the core, cladding and substrate structure, which causes stress when the fiber is cooled in the fiber drawing machine; the latter arises from a deviation of the fiber core from concentricity. The linear birefringence due to a small core ellipticity, $(d_1/d_2 - 1) << 1$, can be approximated by

$$\delta\beta \approx 0.2k_0\left(\frac{d_1}{d_2} - 1\right)\Delta n^2, \tag{5.2}$$

where d_1, d_2 are the major and minor lengths of the core ellipse, $k_0 = 2\pi/\lambda$ and $\Delta n = n_{\text{co}} - n_{\text{cl}}$ is the index difference between the core and the cladding. Note that for values of $2 < (d_1/d_2 - 1) < 6$, form birefringence near the higher mode cut-off becomes insensitive to the core ellipticity and approaches a limit of $\delta\beta \approx 0.25k_0(\Delta n)^2$. For a value of $\Delta n = 0.015$, which is typical for an erbium amplifier, this corresponds to a beat length of 3 cm at a wavelength of 1.55 μm.

Circular birefringence from 'frozen-in' fiber twists is normally negligible unless the fiber is spun in the drawing machine (Payne *et al.*, 1982), but even in this case the intrinsic linear birefringence of the fiber still persists and adds to the circular birefringence to result in elliptically polarized eigenmodes. The most important extrinsic birefringence mechanisms arise from fiber bends, fiber twists and externally applied pressure. However, in high birefringence fibers the internal birefringence mechanisms dominate over the external ones and it is adequate to describe linear polarization evolution only in terms of internally defined polarization eigenmodes.

Equation (5.2) allows us to calculate the expected beat length due to form birefringence of a typical erbium amplifier fiber. Assuming a fiber with a core concentricity deviation of 1 % and a Δn of 0.015, we obtain a beat length of 3 m at a wavelength of 1.55 μm. Beat lengths of less than 10 cm are not unusual in near-circular doped fibers (Hofer *et al.* 1990), which demonstrates that large internal stresses are commonplace in doped fibers. Typically a beat length of 10 cm at a wavelength of 1.55 μm is sufficient to limit random cross-coupling between the polarization eigenmodes to less than 1% in a 1 m length of 'weakly coiled' fiber

(Fermann *et al.*, 1993b). Since form birefringence is temperature insensitive, form birefringent fibers are generally preferable to stress birefringent fibers.

Clearly, reproducible polarization control requires a high birefringence fiber and an example of such a (linearly birefringent) fiber with well-defined polarization axes is shown in Figure 5.1. The fiber has a rectangular shape obtained by squashing the fiber preform and drawing the fiber at a low temperature (Stolen *et al.*, 1984). By matching the cladding to the substrate composition, the fiber can be dominantly form birefringent with a beat length of 5 to 10 cm at 1.55 μm (Fermann *et al.*, 1993b).

5.2.2 *Mathematical description of linear polarization evolution*

The field propagating in the fiber laser cavity may be conveniently described in terms of Jones vectors

$$A = \begin{pmatrix} A_x \\ A_y \end{pmatrix}, \tag{5.3}$$

where A_x and A_y correspond to the field along the x and y axes (Figure 5.1) and where we normalize A such that $|A|^2$ is the total light power. Forward propagation through the fiber is then represented by a Jones matrix of the form

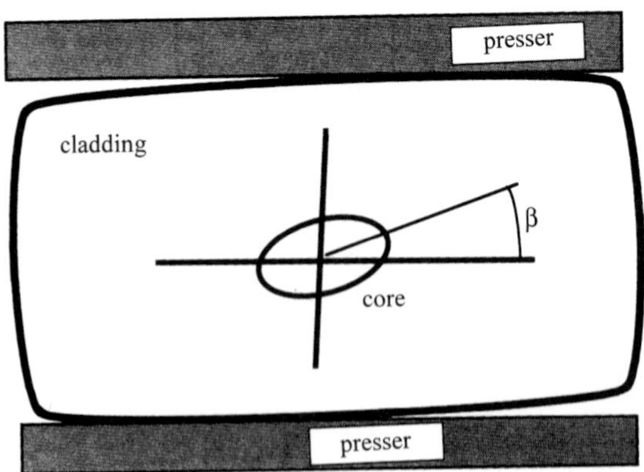

Figure 5.1 Cross-section of a weakly form birefringent fiber. Also shown is a fiber presser that adjusts the phase between the eigenmodes of the fiber. Typically the axes of the fiber core are misaligned by a small angle with respect to the fiber presser.

$$M_{\mathrm{f}} = \begin{pmatrix} a & b \\ -b^* & a^* \end{pmatrix}. \tag{5.4}$$

Clearly, in this basis the Jones matrix of a linearly birefringent fiber is diagonal, where $b = 0$ and $a = \exp(i\delta\beta z/2)$. The Jones matrix M_{b} for a backward pass through the fiber is given by the transpose of M_{f} with a transformation T to a coordinate system with a reversed sign of the y-axis (Mortimer, 1988; Haus *et al.*, 1994). The coordinate transformation is represented as

$$T = \begin{pmatrix} 1 & 0 \\ 0 & -1 \end{pmatrix}. \tag{5.5}$$

The Jones matrix for a backward pass is thus obtained as $M_{\mathrm{b}} = T^{\mathrm{t}} M_{\mathrm{f}} T$. Note that a mirror also causes a reversal in the sign of the y-axis and therefore the Jones matrix for a mirror M_{m} is equal to T, i.e. $M_{\mathrm{m}} = T$. In the case of a Fabry–Perot cavity, the light propagates forward through the fiber and after reflection at a mirror propagates backward to the origin; the corresponding Jones matrix is thence obtained as

$$M_{\mathrm{d}} = M_{\mathrm{b}} T M_{\mathrm{f}}. \tag{5.6}$$

In a balanced nonlinear Sagnac fiber ring interferometer (Fermann *et al.*, 1990), operated in reflection as shown in Figure 5.2, one arm is propagated in the forward direction, whereas the other arm is propagated in the backward direction; the resulting Jones matrix has the form

$$M_{\mathrm{sr}} = \frac{1}{2}(M_{\mathrm{f}} + M_{\mathrm{b}}) \tag{5.7}$$

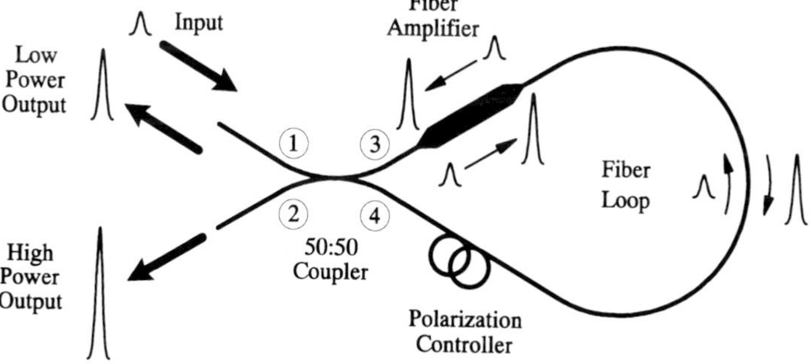

Figure 5.2. Operation principle of a balanced amplifying nonlinear Sagnac fiber ring interferometer. The circled numbers identify the four ports of the fiber coupler used to construct the loop mirror.

and we obtain a sum rather than a product of Jones matrices. In a Sagnac ring operated in transmission, M_b is replaced by $-M_b$ in Equation (5.7). A final useful element is a Faraday mirror, which rotates the polarization axes by 90° in reflection. The corresponding Jones matrix is given by

$$F = \begin{pmatrix} 0 & \pm 1 \\ \pm 1 & 0 \end{pmatrix} \tag{5.8}$$

where the ±sign corresponds to rotation to the left or right. It is then easily shown that replacing the mirror by a Faraday mirror, the polarization after a double-pass through the fiber is rotated by 90° with respect to the input, irrespective of any arbitrary polarization changes in the fiber (Duling and Esman, 1992).

An arbitrary polarization transformation may be generally represented by a waveplate with a phase delay φ and a rotation by an angle α. An arbitrary polarization transformation in the fiber may be accomplished by using three rotatable fiber loops (Lefevre, 1980; Koehler and Bowers, 1985) or three piezoelectric cylinders pressing onto the side of the fiber (Shimizu *et al.*, 1991).

5.3 Nonlinear polarization evolution in fiber lasers

5.3.1 *Polarization evolution without dispersion*

Nonlinear pulse propagation in a linearly birefringent fiber may be described by two coupled nonlinear Schrödinger equations. Initially, we are interested in the fiber parameters present in the early stages of the pulse buildup and can therefore neglect dispersion and higher-order nonlinear effects. We thus obtain the following coupled nonlinear equations for the amplitude evolution in the fiber (Winful, 1985; Agrawal, 1989)

$$\begin{aligned} \frac{\partial A_x}{\partial z} &= i\gamma\left[\left(|A_x|^2+\frac{2}{3}|A_y|^2\right)A_x+\frac{1}{3}A_y^2A_x^*e^{+i2\delta\beta z}\right] \\ \frac{\partial A_y}{\partial z} &= i\gamma\left[\left(|A_y|^2+\frac{2}{3}|A_x|^2\right)A_y+\frac{1}{3}A_x^2A_y^*e^{-i2\delta\beta z}\right] \end{aligned} \tag{5.9}$$

where $\gamma = k_0n_2/A$ is the nonlinearity parameter of the fiber, $n_2 = 3.2\times10^{-20}$ m/W is the nonlinear refractive index in (silica) glass and A is the area within the $1/e$ intensity distribution of the light in the fiber core. These equations are invariant with respect to a sign change of y and are invariant in reflection.

Pure nonlinear ellipse rotation is obtained when $\delta\beta = 0$ and in this case the above equations are directly integrable in terms of circularly polarized eigenmodes $C_{\pm} = (x \pm iy)/2$. The nonlinear phase delay for $C_{\pm}$ is then obtained as (Maker and Terhune, 1965)

$$\Phi_{\pm} = \frac{2}{3}\gamma\left(|C_{\pm}|^2 + 2|C_{\mp}|^2\right)L, \tag{5.10}$$

where L is the fiber length. The nonlinear phase delay may simply be represented in a diagonal Jones matrix $N_{c\pm}$ using $C_{\pm}$ as the basis. In reflection C_+ changes to C_-, whereas the Jones matrix for the nonlinear phase delay stays the same, and therefore the nonlinear phase delay between C_+ and C_- accumulates, i.e. a polarization ellipse continues to rotate away from its original orientation, and hence we may regard nonlinear ellipse rotation as a nonlinear non-reciprocal process.

Nonlinear ellipse rotation is strongly affected by even the smallest amount of linear birefringence and is negligible when the polarization beat length is comparable with or shorter than the fiber laser length (Hofer *et al.*, 1991). As the coherence term in Equation (5.9) averages to zero, i.e. in the rotating wave approximation (RWA), Equation (5.9) is directly integrable in terms of linearly polarized eigenmodes, and we obtain the nonlinear phase delay along the x and y axes as (Stolen *et al.*, 1982)

$$\Phi_{x,y} = \gamma\left(|A_{x,y}|^2 + \frac{2}{3}|A_{y,x}|^2\right)L, \tag{5.11}$$

which again allows a representation of $\Phi_{x,y}$ in a diagonal Jones matrix $N_{x,y}$.

5.3.2 *Polarization evolution in the presence of dispersion*

In the presence of dispersion, nonlinear polarization evolution in passively modelocked fiber lasers is generally affected by soliton formation. Particularly when the fiber laser has negative overall dispersion evenly distributed along the cavity, the pulses oscillating in the fiber lasers will approximate solitons. The peak pulse power is thus given by the soliton power $P_s = 3.13\beta_2/\gamma\Delta\tau^2$ and the pulse width is given by the soliton pulse width τ. In this $\Delta\tau = 1.763\tau$ is the FWHM pulse width and $\beta_2 = \mathrm{d}^2\beta/\mathrm{d}\omega^2$ is the group-velocity dispersion of the fiber at the carrier frequency ω.

The nonlinearity of the resonator may be characterized by the number of soliton periods z_s in the cavity, where $z_s = (\pi/2)\tau^2/\beta_2$. Resonators with evenly distributed dispersion typically oscillate with 1 to 2 soliton periods in the cavity (Taverner *et al.*, 1993). The reasons for a lower limit are not well understood, but it is believed that intracavity scattering centers play a major part, since they favor CW over modelocked operation (Haus and Ippen, 1991). The upper limit is set by the shedding of energy into a dispersive wave component which interacts with the main soliton pulse and removes energy from it until it becomes unstable (Gordon, 1992; Kelly, 1992; Taverner *et al.*, 1993).

In a birefringent fiber the pulses are also affected by group-velocity walk-off between its polarization axes. If the corresponding terms are included into Equation (5.9) the result is (Menyuk, 1987, 1988; Menyuk and Wai, 1992)

$$\begin{aligned}\frac{\partial A_x}{\partial z}+\frac{\Delta\beta}{2}\frac{\partial A_x}{\partial t}+\frac{1}{2}\beta_2\frac{\partial^2 A_x}{\partial t^2}&=i\gamma\left(|A_x|^2+\frac{2}{3}|A_y|^2\right)A_x+\frac{g}{2}A_x\\ \frac{\partial A_y}{\partial z}-\frac{\Delta\beta}{2}\frac{\partial A_y}{\partial t}+\frac{1}{2}\beta_2\frac{\partial^2 A_y}{\partial t^2}&=i\gamma\left(|A_y|^2+\frac{2}{3}|A_x|^2\right)A_y+\frac{g}{2}A_y,\end{aligned} \tag{5.12}$$

where $\Delta\beta=\delta\beta/\omega$ is the difference between the inverse group velocities along the x and y axes. We have also included an amplitude gain parameter per unit length g and we have neglected a bandwidth limitation from the gain medium. The birefringence is assumed to be linear and large enough that the coherence term in Equation (5.9) averages to zero.

For pulses propagating along either axis, Equation (5.12) has the well-known soliton solutions. On the other hand a pulse launched at 45° with respect to the fiber axes will break apart due to group-velocity walk-off between the fiber axes and will not interfere at a polarizer after the walk-off distance $L_{\mathrm{wo}}\approx 2.3\Delta\tau/\Delta\beta$.

For a sufficiently small fiber birefringence, however, cross-phase modulation between the pulses can just compensate for the pulse walk-off and the pulses can stay bound together indefinitely, a process commonly known as soliton trapping (Menyuk, 1987, 1988; Islam *et al.*, 1989). In soliton trapping the pulse propagating along the fast axis stays slightly ahead of the pulse propagating along the slow axis, so that the front pulse can be down-chirped by the leading edge of the back pulse, slowing it down to their average group-velocity. The reverse effect up-chirps the back pulse, speeding it up. As a result the spectrum of the two pulses is split, where the fast-axis pulse spectrum gets down-shifted and the slow-axis pulse spectrum gets up-shifted.

For a pulse launched along 45° with respect to the birefringence axes, it is found that the soliton power is increased by a factor of (6/5). The effective nonlinearity is weaker at 45° and a higher pulse power is required to compensate for the pulse spreading due to dispersion (Menyuk, 1988). The effect of the group velocity walk-off may be characterized by the parameter (Menyuk, 1988)

$$\Gamma = \frac{\Delta\beta}{2\beta_2}. \tag{5.13}$$

In the absence of gain and for $\Gamma = 0.05$ the threshold pulse power P_{th} required to compensate for the group-velocity walk-off in the fiber is just the soliton power. In real values this corresponds to a 300 fs pulse, a polarization beat length of 10 cm and a fiber dispersion of $\beta_2 = -10\,000\ \mathrm{fs}^2/\mathrm{m}$. For shorter beat lengths the threshold power has to exceed the fundamental soliton power and a more and more oscillatory evolution of the pulse shape with time is observed (Menyuk, 1988). We can therefore expect that in a fiber resonator stable pulse formation is only possible when the pulse power required to compensate for the group-velocity walk-off is less than the soliton power, i.e.

$$P_{\mathrm{th}} < P_{\mathrm{s}}. \tag{5.14}$$

Assuming that the soliton pulses in a fiber resonator oscillate with about one soliton period per cavity, we can calculate that, for the parameters above, a maximum fiber length of 4.5 m is allowed before group-velocity walk-off along the fiber axes inhibits stable pulse formation. For a beat length of 10 cm this corresponds to a pulse walk-off of 230 fs at a wavelength of 1.55 μm. In practice passive modelocking is obtained with unequal excitation of the fiber axes and it was found in recent experiments (Fermann *et al.*, 1993b) that the stability of passively modelocked lasers based on nonlinear polarization evolution is only ensured when $P_{\mathrm{th}} << P_{\mathrm{s}}$, so Equation (5.14) serves only as an upper limit for the allowable group-velocity walk-off in the cavity. Of course, no limit should exist if the group-velocity walk-off along the fiber axes is compensated within the fiber laser cavity.

5.4 Fiber laser cavities

A variety of cavity configurations allow the exploitation of nonlinear polarization evolution (or polarization switching) for passive modelocking. The first to be demonstrated was a Fabry–Perot-type cavity as shown schematically in Figure 5.3 (Hofer *et al.*, 1991, 1992). In this a polarizer is incorporated in front of the left cavity mirror to convert the nonlinear phase modulation into an amplitude modulation. Assuming a linearly birefringent fiber with a polarizer at an angle α with respect to the fiber x-axis, the output from the polarizer after one round trip through the fiber is calculated as

$$\begin{pmatrix} A'_x \\ A'_y \end{pmatrix} = TP_\alpha N^2_{x,y} M^2_{\mathrm{f}} \begin{pmatrix} \cos\alpha \\ \sin\alpha \end{pmatrix}, \tag{5.15}$$

where P_α is the Jones matrix for a polarizer at an angle α with respect to the x-axis. The output polarization state will be elliptical, where the angle β of the major axis of the polarization ellipse to the x-axis is given by (Hofer *et al.*, 1992; Davey *et al.*, 1993)

$$\tan 2\beta = \tan 2\alpha \cos\left[2(\varphi + \Delta\Phi_{x,y})\right], \tag{5.16}$$

where $\varphi = \delta\beta L$ and $\Delta\Phi_{x,y}$ is the difference between the average nonlinear phase delays along the x and y-axes. The linear reflectivity of the cavity may be obtained as

$$R = R_0 + \frac{1}{2}\sin^2 2\alpha \cos 2\varphi, \tag{5.17}$$

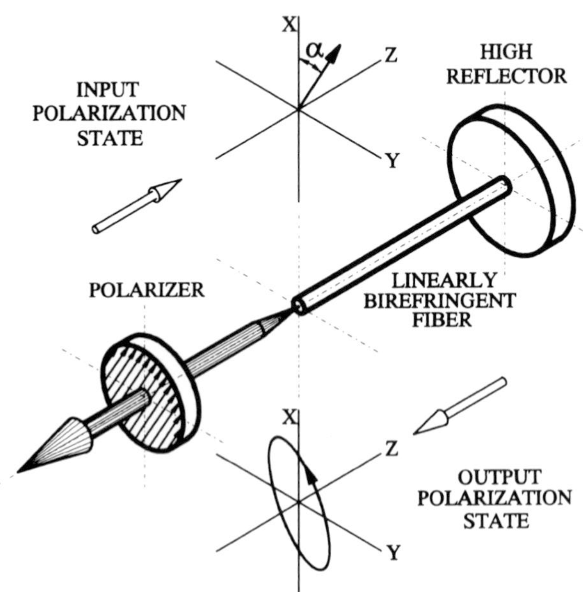

Figure 5.3. Operation principle of nonlinear polarization evolution in a passively modelocked Fabry–Perot cavity.

where $R_0 = \cos^4\alpha + \sin^4\alpha$. From Equation (5.15) the corresponding nonlinear reflectivity of the cavity may be expanded to first order in the pulse power P as $R_{nl} = R + \kappa P$, where

$$\kappa = -\frac{\gamma L}{6}\sin 4\alpha \sin 2\alpha \sin 2\varphi. \tag{5.18}$$

Note that the maximum value for κ is obtained when $\varphi = -\pi/4 (\mathrm{mod}\, n2\pi)$ and when $\alpha = 27.3°$, where α and φ can be selected by adjusting the polarization state of the fiber with appropriate polarization controllers.

For a ring cavity a similar equation is obtained. R and κ are independent of the sign of α and, therefore, the nonlinear reflectivity function has the same value for both clockwise and counter-clockwise propagating light. Hence a linearly birefringent fiber ring laser is reciprocal even in nonlinear operation, and in principle bi-directional operation of a fiber laser passively modelocked with nonlinear poarization evolution should be possible. This is quite distinct from a figure eight laser (F8L), which, though it is reciprocal, can only be operated in one direction (Duling, 1991). Note, however, that the reciprocity of a nonlinear linearly-birefringent fiber laser is lost in the presence of fiber sections with 'zero' linear birefringence, since nonreciprocal ellipse rotation can then set in, breaking the symmetry of the cavity. It is therefore expected that bi-directional operation of a passively modelocked linearly birefringent fiber ring laser will depend very critically on the polarization control.

The polarization control may be greatly simplified when resorting to high birefringence fiber. In this case the polarization axes are well defined and only two polarization controllers are required to set the intracavity polarization state to obtain passive modelocking. As explained in Section 5.3.2 a difficulty arises, since the large linear phase delay between the x and y axes greatly reduces the stable operation range of the modelocked laser (Fermann *et al.*, 1993b, 1993c). The linear phase delay may be compensated, in principle, by splicing together two sections of identical fiber with reversed polarization axes. However, the linear phase delay along the fiber axes is then still temperature sensitive. Equally, due to unavoidable misalignment of the polarization axes, the linear polarization evolution in the fiber will be wavelength dependent. For a ring cavity with a single polarizer at an angle α with respect to the polarization axes and two identical fiber sections of length L, we may show that the linear transmission in the cavity will be modulated with a modulation depth of

$$t_{\mathrm{mis}} = 4\varsigma \sin 2\alpha \cos \delta\beta L, \tag{5.19}$$

where ς is the angle of misalignment of the polarization axes. The corresponding FWHM bandwidth of the transmission function is then $\Delta B = L_B \lambda / L$. Under optimum conditions ς can be as small as a tenth of a degree, giving a maximum t_{mis} of the order of 0.7%. For high birefringence fiber lengths of 1 m and a beat length of 1 cm at a wavelength of 1.55 μm the corresponding modulation bandwidth is 15.5 nm, which shows clearly that high birefringence fiber cavities should use short fiber lengths.

The linear phase delay along the fiber axes and its temperature sensitivity may be eliminated by the incorporation of two Faraday rotators into the cavity, as shown in Figure 5.4 (Fermann *et al.*, 1993c). The first Faraday rotator is chosen to act as a Faraday mirror, rotating the polarization state by 90° in reflection. The Faraday mirror also eliminates spatial hole-burning in the cavity, which is believed to lower the startup threshold of passively modelocked fiber lasers (Krausz and Brabec, 1993). Equally, the Faraday mirror suppresses spurious back-reflections from the intracavity fiber ends, which is essential for the elimination of a CW lasing background (Tamura *et al.*, 1993). The second Faraday rotator compensates the polarization rotation of the first rotator. Under linear operation the polarization state of the output is always equal to the polarization state of the input, since any linear phase delay along the polarization axes is compensated. On the other hand, the nonlinear phase delay in forward and backward transmission through the fiber remains uncompensated and accumulates. By the incorporation of an additional waveplate between the polarizer and the second Faraday rotator an appropriate linear phase delay along the two polarization axes can be incorporated, and temperature insensitive operation of the nonlinear polarization switching mechanism can be obtained.

The details of the polarization evolution in the cavity may be understood as follows. For simplicity we initially assume the presence of two

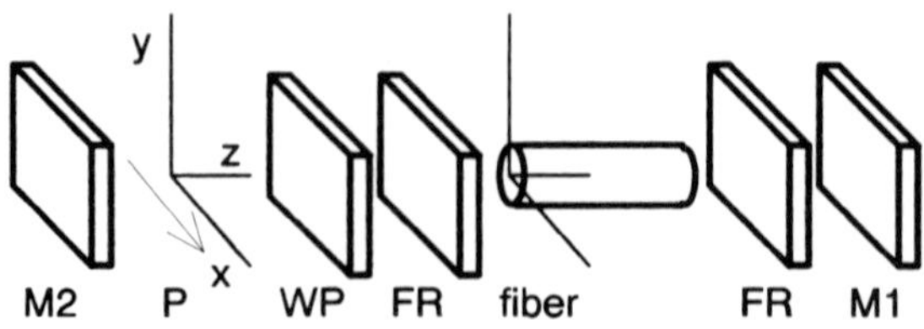

Figure 5.4. Principal elements to obtain environmentally stable nonlinear polarization evolution in a passively modelocked Fabry–Perot fiber laser cavity. M1, M2 are cavity mirrors, P is a polarizer, WP is a waveplate, FR are 45° Faraday rotators.

waveplates. We start from cavity mirror two and discuss one round trip. Without loss of generality we assume that the polarizer is aligned along the major axis of the highly-birefringent fiber. The first waveplate has its axes rotated by a fixed angle $\xi = 45°$ with respect to the major axis of the fiber. The first waveplate induces a single-pass linear phase delay φ along its axes. The amount of the phase delay φ is controlled by a rotation of the first waveplate by an adjustable angle in the plane defined by the propagation direction and the major axis of the fiber. The second waveplate is assumed to be simply a $\lambda/2$ plate which rotates the polarization state by an angle α. The single-pass nonlinear phase delay accumulated between the polarization eigenmodes of the fiber is then given by

$$\Delta\Phi_{x,y} = \frac{\gamma PL}{3} \sin 2\alpha \cos \varphi. \tag{5.20}$$

Using the formalism described in Section 5.2.2, the nonlinear reflectivity of the cavity is calculated as

$$\begin{aligned} R(P) = &\cos^2\varphi \cos^2\Delta\Phi + \sin^2\Delta\Phi\left(\sin^2 2\alpha + \cos^2 2\alpha \sin^2\varphi\right) \\ &- \frac{1}{2}\sin 2\Delta\Phi \cos 2\alpha \sin 2\varphi, \end{aligned} \tag{5.21}$$

where we have omitted the subscript x, y. Note that Equations (5.20) and (5.21) fully describe all possible nonlinear responses of the cavity. This may be understood with reference to the Poincaré sphere (Rashleigh, 1983). Here the first waveplate simply sets the latitude and the second waveplate sets the longitude of the polarization transformation, and therefore all possible points on the Poincaré sphere may so be reached. The nonlinear reflectivity of the cavity is uniquely defined only by this polarization transformation, independent of the details of the polarization controllers. When the $\lambda/2$ plate is omitted the rotation by the angle α is obtained by rotating the fiber or by varying ξ. The equivalence of these techniques may be shown by calculating the Jones matrix for a rotated and tilted waveplate from its index ellipsoid.

5.5 Experiments

5.5.1 *Dispersion compensated fiber lasers*

The first passively modelocked fiber lasers based on nonlinear polarization evolution were based on dispersion compensated cavities

(Hofer *et al.*, 1991, 1992). The experimental setup is shown in Figure 5.5. Here a 20 cm length of neodymium doped weakly linearly birefringent fiber was used in a Fabry–Perot-type cavity. Intracavity dispersion compensation was obtained with a quadruple pass through four bulk SF 10 prisms. Three piezoelectric cylinders were used for polarization control. Due to small residual intracavity reflections and spatial hole-burning (Krausz and Brabec, 1993) in the cavity, this fiber laser was not self-starting. Pulse startup was initiated with an acousto-optic modulator (Hofer *et al.*, 1991, 1992) or a moving mirror (Fermann *et al.*, 1993a; Ober *et al.*, 1993). The optimum settings of the polarization controllers were determined from trial and error, and with optimal polarization adjustment, passive modelocking could be sustained for long periods of time with only occasional cavity alignment.

Since the fiber was linearly birefringent, its nonlinear response is defined by Equations (5.15–5.17). To compare the theory with experiment, polarization measurements were made for a range of fiber positions (Hofer *et al.*, 1992). In this, the polarization state of the modelocked pulses was measured for pulses emerging from fiber end 1 (output 1). Subsequently, the linearly polarized eigenmodes for that particular fiber position were found, and the angle α and the linear phase delay

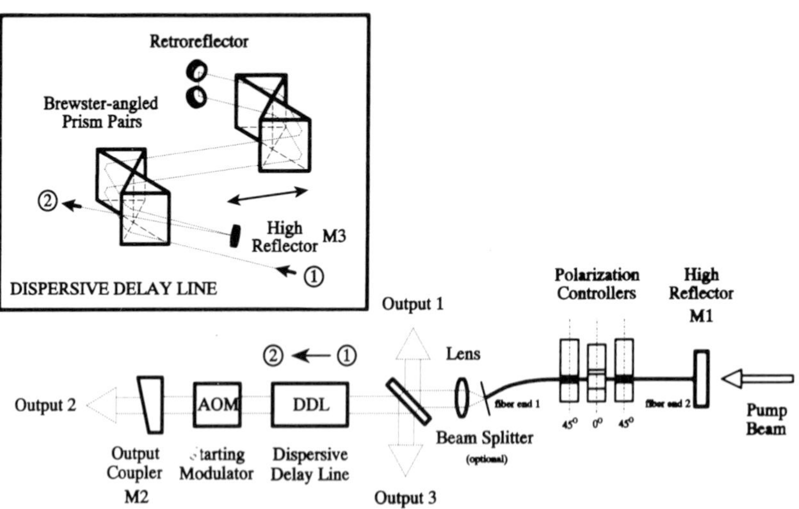

Figure 5.5. Dispersion compensated neodymium fiber laser cavity passively modelocked with nonlinear polarization evolution. The insert shows the details of the employed dispersive delay line. In a round trip each prism is passed four times by the laser pulse.

$\pm n\pi \pm \varphi$ ($n = 0, 1, 2...$) were calculated. A typical result is shown in Figure 5.6. Here optimum modelocking was obtained for $\alpha \approx 38°$ and $\varphi \approx -0.4\pi$ (mod $n2\pi$). The polarization ellipse rotated by 17° between the CW and modelocked state. From Equation (5.16) (Davey *et al.*, 1993) the average differential round-trip nonlinear phase delay between the two polarization axes is then calculated as $2\Delta\Phi_{x,y} = 0.23\pi$. Using the value for α of 38° a corresponding total nonlinear phase delay of 2.8π is calculated to exist in the cavity. From separate measurements of the pulse energy and width in the cavity, an average nonlinear phase delay of $\sim 2.5\pi$ was estimated, which is in good agreement with the above estimate.

Passive modelocking could be obtained for an input angle α ranging from 20° to 45° and for phase delays φ ranging from 0.3π to 0.4π. Calculated nonlinear reflectivity curves for values of $\alpha = 38°$ and with $\varphi = -0.40$, -3.40π are given in Figure 5.7(a). In this, the curve with $\varphi = -3.40\pi$ is not distinguishable from the rotating wave approximation. Figure 5.7(a) shows that the change in reflectivity between modelocked and CW operation may reach nearly 90%. Nonlinear reflectivity curves for a value of $\alpha = 45°$ with $\varphi = -0.38\pi$, -3.38π are given in Figure 5.7(b). Here the RWA does not approximate the nonlinear reflectivity curve well, since the RWA produces a reflectivity independent of laser

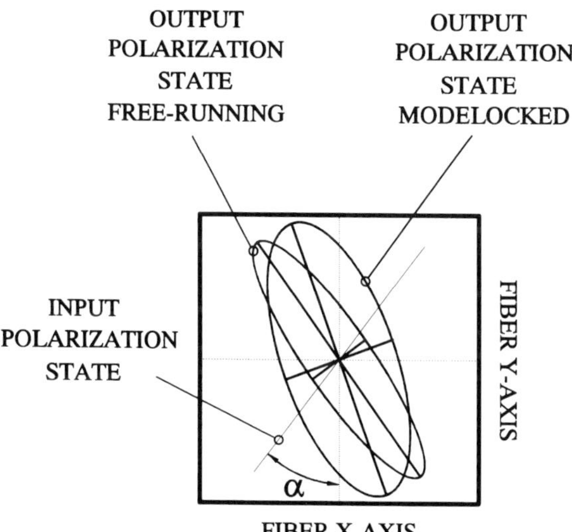

Figure 5.6. Input polarization state and output polarization state for CW and modelocked neodymium fiber laser operation measured at fiber end 1 (see Figure 5.5).

(a)

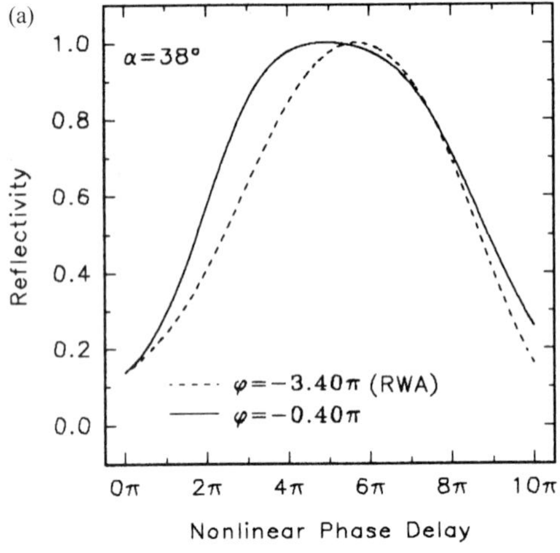

(b)

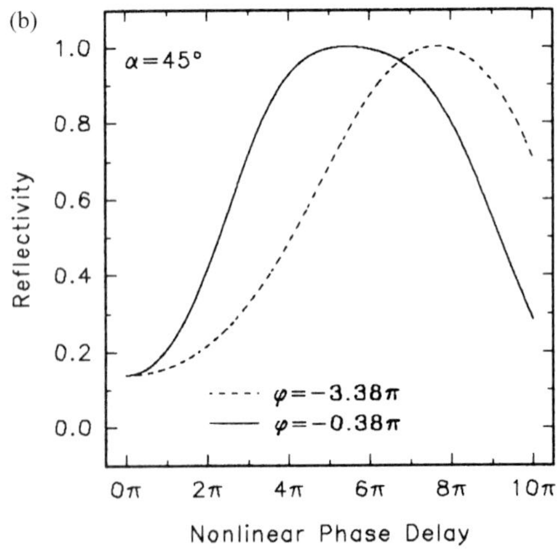

Figure 5.7. Calculated nonlinear reflectivity as a function of intracavity nonlinear phase delay in a neodymium fiber laser for a value of (a) $\alpha = 38°$ with $\varphi = -0.4\pi, -3.4\pi$ (with respect to the slow axis of the fiber) and (b) $\alpha = 45°$ with $\varphi = -0.38\pi, -3.38\pi$.

power. For near-equal excitation of the fiber axes the nonlinear reflectivity curves vary only slightly with a variation in α and φ, which allows a particularly stable operation of the modelocked laser. In addition, an early saturation of the nonlinear reflectivity may be avoided by increasing the absolute value of the linear phase delay φ. In Figure 5.7(b) a nonlinear phase delay of 8π is required to saturate the nonlinear reflectivity for a value of $\varphi = -3.38\pi$, and thus high intracavity powers can be supported by the modelocked laser. As may be seen in Figure 5.7(b) the slope of the nonlinear reflectivity curve then reduces to zero for small intensities. Recent experiments, however, confirm that passive modelocking may be sustained with nonlinear reflectivity curves with zero slope at zero power (Taverner *et al.*, 1993).

With a moving mirror employed for pulse start-up, the cavity allowed the generation of near-bandwidth-limited pulses as short as 53 fs (Ober *et al.*, 1993). The corresponding autocorrelation traces are shown in Figure 5.8. Pulses as short as 34 fs (Figure 5.9) could be generated when employing an acousto-optic modulator for startup (Ober *et al.*, 1993). The time–bandwidth product of these pulses was 0.60 (assuming a sech^2-shape), which demonstrates that even shorter pulses should be

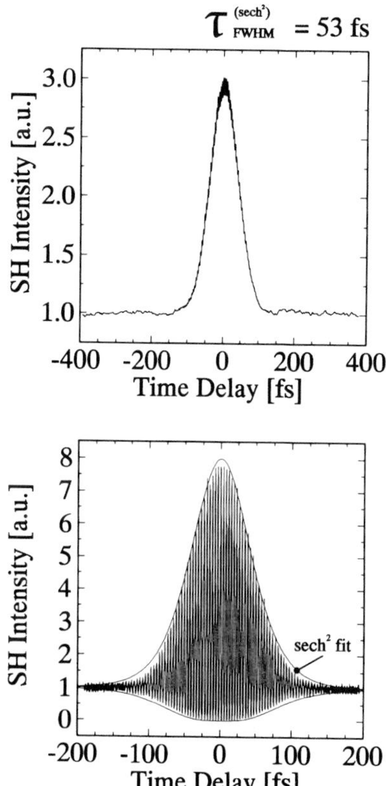

Figure 5.8. Autocorrelation of a typical pulse generated with a neodymium fiber laser started with a moving mirror. The FWHM pulse width is 53 fs assuming a sech^2 shape.

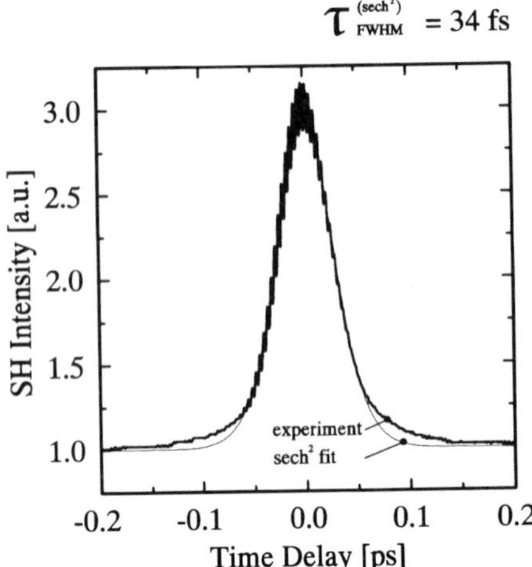

Figure 5.9. Autocorrelation of the shortest pulses generated from a (neodymium) fiber laser to date. The FWHM pulse width is 34 fs assuming a sech^2 shape.

expected when reducing the amount of third-order dispersion in the cavity by an appropriate selection of a prism material for dispersion compensation.

The output pulse energy of this type of cavity can be as high as 1 nJ for a 50% output coupler. Assuming a pulse width of 50 fs, the corresponding intracavity nonlinear phase delay would thus be as high as 25π, which would clearly saturate the nonlinear amplitude switching mechanism in the cavity and make the laser unstable. However, dispersion compensation allows a large variation of the pulse width inside the cavity. Short pulses are only observed at the dispersive end of the cavity, whereas inside the fiber the pulses are much longer and strongly chirped. A measurement (Hofer *et al.*, 1992) of the observed pulse width variations inside the cavity is shown in Figure 5.10. Intracavity compression factors in excess of 20 have been observed (Ober and Hofer, 1993). Therefore, as in chirped pulse amplification (Strickland and Mourou, 1985), strongly chirped oscillator pulses can greatly reduce the nonlinearity of the system and allow the generation of pulses with very high energies. Recently, an all-fiber equivalent of such a system was also demonstrated, where positive and negative dispersion fibers were used in the cavity to raise the pulse energy by more than one order of magnitude compared to a fiber laser based only on negative dispersion fiber (Tamura *et al.*, 1993).

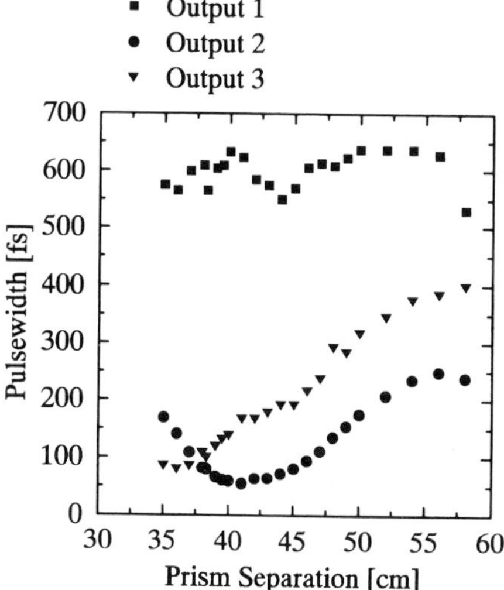

Figure 5.10. Intracavity pulse width variations in a dispersion compensated neodymium fiber laser (as in Figure 5.5). The pulse widths were measured at output ports 1, 2 and 3 as a function of prism separation.

5.5.2 *Lasers based on fibers with negative dispersion*

The operation range of fiber lasers with only negative dispersion fiber inside the cavity is generally limited, since the system requires that the intracavity pulse power corresponds closely to the fundamental soliton power. Equally, stable oscillation of the system requires that the length of the cavity does not exceed more than about two soliton periods (Fermann *et al.* 1993a; Taverner *et al.* 1993; Tamura *et al.*, 1993). Since one soliton period corresponds to a nonlinear phase delay of $\pi/2$, the nonlinear phase delay is not expected to exceed π inside the cavity, which is about three times lower than that possible in a dispersion-compensated fiber cavity. Therefore a much less dramatic nonlinear polarization switching mechanism is expected compared with a dispersion-compensated cavity. Recent experiments in weakly birefringent negative dispersion fiber confirmed this expectation, where the polarization ellipse was observed to rotate by only 1–5° (averaged over the continuous laser power) between the CW and the modelocked state (Davey *et al.*, 1993). Note, however, that a study by Nakazawa *et al.* (1993) suggests that in cavities with near-zero birefringence negative-dispersion fiber a peak nonlinear ellipse rotation up to 180° is possible. Equations (5.10) and

(5.11) reveal that the differential nonlinear phase delay for a given differential excitation of the polarization eigenmodes is about twice as large in near-zero birefringence fibers as in high-birefringence fibers. In practice a better evaluation of the nonlinear switching mechanism in various cavity designs is the resulting nonlinear reflectivity (Equation 5.18). So far measurements of the nonlinear reflectivity for cavities with near-zero birefringence fibers are not available; nevertheless the results by Nakazawa *et al.* (1993) indicate that κ should be close to its maximum value for this case. However, whenever near-zero birefringence fiber is used, the operation of the cavity will become susceptible to temperature and pressure variations; in particular the operation characteristics are expected to change dramatically as soon as the fiber is coiled into a small loop. The ability to coil fiber lasers is a necessary requirement for their practical applications.

The small nonlinear phase delay possible in negative dispersion fibers is clearly advantageous when constructing very high repetition rate fiber lasers, since less energy per pulse is required and more pulses can be generated with a given pump power. Since fiber lasers have to have a relatively long minimum length (about 1 m is typical for erbium doped fibers), high repetition rates have to be obtained by resorting to subcavities. Subcavities are typically phase-sensitive (Yoshida *et al.*, 1992). However, recently a system was demonstrated where the rejected part from the nonlinear switch in the F8L was fed back to the main cavity to produce phase-insensitive high harmonic modelocking (Dennis and Duling, 1992). It can be envisioned that the light rejected by the polarizer in Figure 5.5 can also be fed back to generate pulses at higher repetition rates.

5.5.3 *Polarization maintaining passively modelocked fiber lasers*

Passive modelocking by nonlinear polarization evolution is possible in short lengths of high birefringence fiber as long as the group-velocity walk-off along the polarization axes is short compared to the pulse width (see Section 5.3.2). An example of a possible cavity design is shown in Figure 5.11 (Fermann *et al.*, 1993b). Here a fiber with a rectangular shape was used and the cross-section was shown in Figure 5.1. The round-trip fiber length was 1.8 m and the polarization beat length was 10 cm. The fiber was doped with 5×10^{18} erbium ions/cm^3; the numerical aperture was 0.19 and the core area was 28 μm^2. From the location of the sidebands in the modelocked spectrum (Kelly, 1992) the

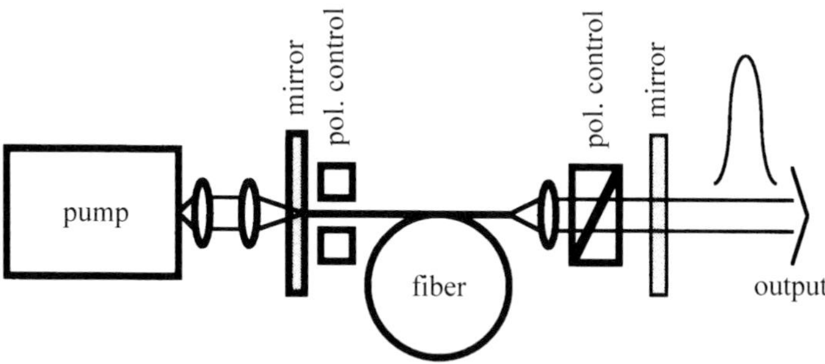

Figure 5.11. Cavity design for a passively modelocked polarization maintaining erbium fiber with two adjustments for reproducible polarization control.

dispersion of the fiber was estimated as $\beta_2 = -17000\ \text{fs}^2/\text{m}$. The polarization state in the cavity was set by a rotating polarizer and by a piezoelectric cylinder that was pressing onto the side of the fiber.

The nonlinear reflectivity of the cavity is adequately described by Equations (5.17) and (5.18). Thus the polarizer was used to set α, i.e. to differentially excite the polarization axes of the fiber, and the piezoelectric element was employed to set the linear phase delay φ between them. Typically passive modelocking was obtained with $\alpha \approx 21°$ and $\varphi \approx -0.33\pi$, which gives a linear reflectivity of $R = 66\%$. Modelocking was initiated with a moving mirror, i.e. typically a cavity mirror mounted on a vibrating piezoelectric element. Once modelocking was initiated, it was self-sustaining, and, by adjusting the pump power level, pulses at the fundamental cavity frequency without a CW component could be obtained. Typically bandwidth-limited pulses with a width between 200 and 400 fs could be generated.

Since the fiber was form birefringent, the stability of the system was quite good and only one of the polarization controllers required occasional readjustment to account for a slow temperature drift of the phase delay φ along the fiber axes.

When the fiber length was increased, the stability of the generated pulses decreased and pulse formation without a CW component became increasingly difficult. For a round-trip fiber length longer than 4.5 m self-sustaining pulse generation became impossible. However, as the fiber length increased modelocking was obtained with ever smaller values of α, i.e. with more and more linearly polarized pulses. By comparing the present fiber parameters with Section 5.3.2 we find that soliton trapping is expected to play a role in determining the stability of the oscillator. As the polarization walk-off increases, the amount of energy in the weaker polarization direction trapped by the stronger polarization direction must therefore decrease.

5.5.4 *Environmentally stable passively modelocked fiber lasers*

The ultimate aim for any ultrashort pulse source must be that it can work without any continuous adjustments. It is only then that a particular laser can find widespread use. Clearly the presence of two interfering eigenmodes in Kerr-type modelocked fiber lasers makes them sensitive to temperature and pressure variations. Ideally, the operation of a modelocked fiber laser should be guaranteed under these environmental changes without any modifications to the cavity.

In Section 5.5.3 we have seen that the use of polarization maintaining fiber can fix the polarization axes of a fiber laser, and thus the position of the interfering arms of the nonlinear fiber interferometer is fixed. For a given differential excitation of the polarization axes, the nonlinear phase delay along the axes thus becomes insensitive to pressure and temperature variations. What remains to be done is to fix the linear phase delay between the polarization axes. As seen in Section 5.4, the linear phase delay can be set permanently by the inclusion of two Faraday rotators and a waveplate. Here the Faraday rotators compensate any linear phase drift inside the cavity and leave the nonlinear pulse shaping unaffected. The waveplate simply allows us to fix the linear phase delay between the polarization axes.

Note also that an environmentally insensitive phase delay between the polarization axes of the fiber can be incorporated in the presence of fiber components with low birefringence as long as their length is short compared with the length of the highly birefringent fiber components. With this requirement nonlinear pulse shaping in the weakly birefringent fiber is negligible and dominated by the highly birefringent fiber. However, to ensure that the birefringence axes of the highly birefringent fiber stay aligned with the intracavity polarizer, the highly birefringent fiber section should include the intracavity fiber end, i.e. the weakly birefringent fiber is only allowed in front of the Faraday mirror.

An exemplary cavity design in shown in Figure 5.12. In a first demonstration of the operation principle a single-pass length of 2.6 m of high birefringence fiber and a single-pass length of 0.6 m of standard communications type low birefringence fiber were used. The high birefringence

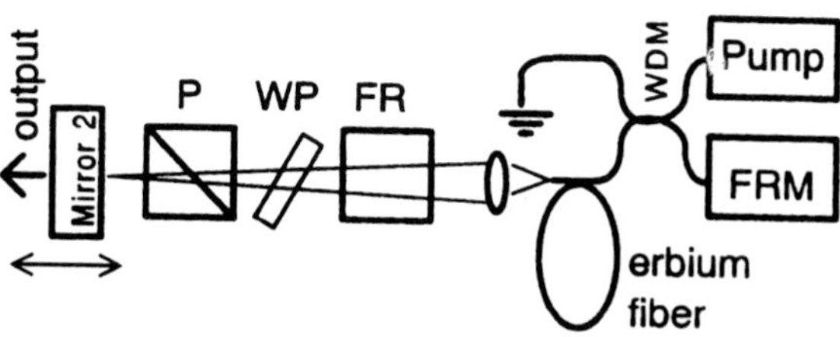

Figure 5.12. Setup of an environmentally stable passively modelocked erbium fiber laser.

fiber was doped with erbium and had the same parameters as described in Section 5.5.3. The core area of the low birefringence fiber was about twice as large as that of the high birefringence fiber, which reduced its effective nonlinear length by a further factor of two. The low birefringence fiber was dopant free. The fiber laser was pumped with 980 nm light via a wavelength division multiplexing coupler. An 80% reflecting mirror was used as the output coupler.

Stable modelocking was obtained with the values $\varphi \approx 132^\circ$ and $\alpha \approx 10^\circ$ (see Equations 5.20, 5.21). Once the waveplate and the polarizer were set, they did not require any additional adjustment and remained permanently fixed. The laser was completely insensitive to perturbations of the low birefringence fiber and allowed perturbations of the high birefringence fiber, as long as the perturbation period was large compared to its beat length. In practice this meant that the fiber could be moved around on an optical table or it could be lifted without a loss of modelocking.

One remaining difficulty of the cavity was that modelocking was not self-starting due to the presence of spurious reflections. Therefore a moving mirror (see Section 5.5.3) had to be employed to start up the pulses. The modelocking threshold was some 70% higher than the pump power level of 60 mW at which clean CW-free single pulses were obtained in the cavity. The main source of the reflections came from the intracavity lens and the intracavity fiber end, which was not AR-coated. Therefore a lower modelocking threshold could be expected when optimizing these cavity elements.

Once the pump power level was set to its optimum value it could be changed by $\pm 7\%$ without the onset of a visible CW background or other pulse instabilities. An example of the obtained pulse spectra at the edges of the stability range is shown in Figure 5.13. Note that as the pump power is increased the pulses get shorter and their spectral width broadens, leading to an increased number of soliton periods per cavity length,

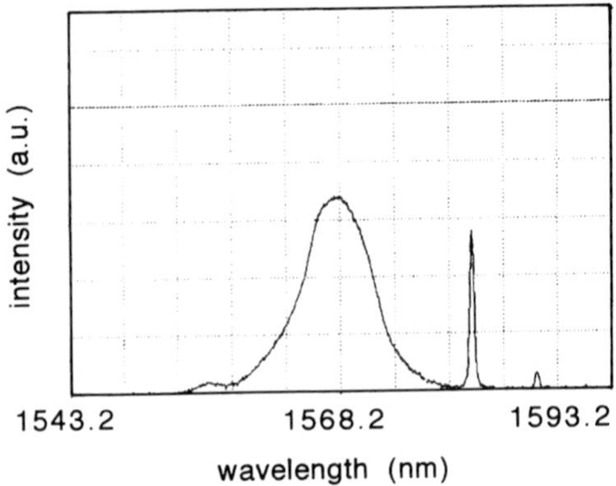

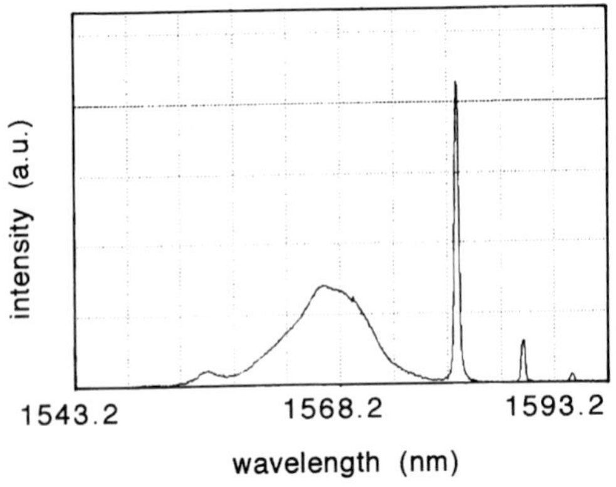

Figure 5.13. Typical pulse spectra obtained at either end of the stability range of an environmentally stable erbium fiber laser. The corresponding FWHM spectral widths are 5.8 and 6.8 nm.

with a corresponding increased shedding of energy into a dispersive wave (Kelly, 1992) as indicated by the increased height of the spectral resonances. A typical autocorrelation trace is shown in Figure 5.14. The generated pulses had FWHM width of 360 fs with a time–bandwidth product of ~ 0.30 (assuming a sech^2 shape) and were free of pedestals. The repetition rate of the pulses was 27 MHz and the average pulse energy measured after the output coupler was 10 pJ. Note that a pulse energy of 60 pJ could be extracted when using the light rejected by the polarizer. With an estimated intracavity loss of 80%, these values translate into an average intracavity pulse energy of 55 pJ. The corresponding round-trip total nonlinear phase delay is about 1.1π, which is comparable with results

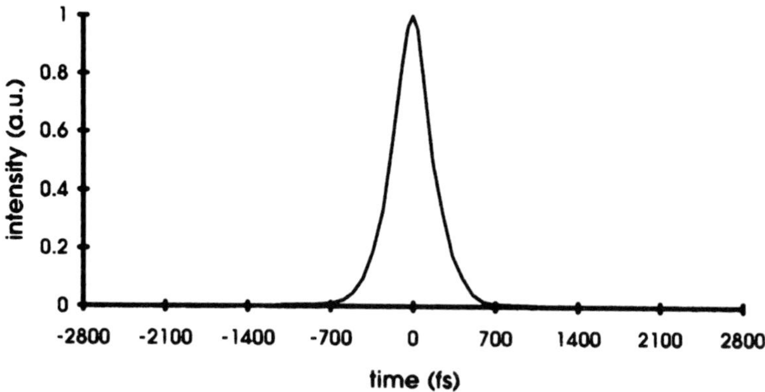

Figure 5.14. Typical autocorrelation of the pulses generated with an environmentally stable erbium fiber laser. The FWHM pulse width is 360 fsec assuming a sech^2 shape.

obtained in standard non-polarization-maintaining Kerr-type mode-locked fiber lasers (Fermann *et al.*, 1993a). Note that off-axis excitation of the fiber eigenmodes (Menyuk, 1987) and group-velocity walk-off between them is expected to reduce the effective nonlinearity of the fiber, and therefore the real value of the total nonlinear phase delay is probably smaller.

From the measurements of φ and α we obtain the single pass differential nonlinear phase delay between the two polarization eigenmodes of the fiber as $\Delta\Phi \approx 7.5°$. However, due to the presence of soliton shaping in the cavity, the differential nonlinear phase delay between the polarization eigenmodes can be reduced up to a factor of two. Therefore a more realistic estimate for $\Delta\Phi$ is $4° < \Delta\Phi_{\mathrm{eff}} < 7.5°$. From Equation (5.21) we can calculate the expected change in the reflectivity of the output coupler between the CW and modelocked state of the cavity. We obtain the CW reflectivity $R(0) = 45\%$. Using the value for $\Delta\Phi_{\mathrm{eff}}$ we obtain $51\% < R(P) < 57\%$. In practice we estimate the maximum change in the reflectivity amounted to not more than 5%. The small discrepancy can arise because the differential nonlinear phase delay $\Delta\Phi_{\mathrm{eff}}$ is smaller than estimated.

Recently, a high birefringence fiber with a group-velocity dispersion of only $-13000\ \mathrm{fs}^2/\mathrm{m}$ was incorporated into the cavity. With a high-birefringence fiber length of 1.4 m and a low-birefringence fiber length of 35 cm, pulses as short as 125 fs (assuming a sech^2 shape) were generated. The autocorrelation trace is shown in Figure 5.15. The time–bandwidth product was 0.26, indicating a small deviation from a sech^2 pulse shape. With

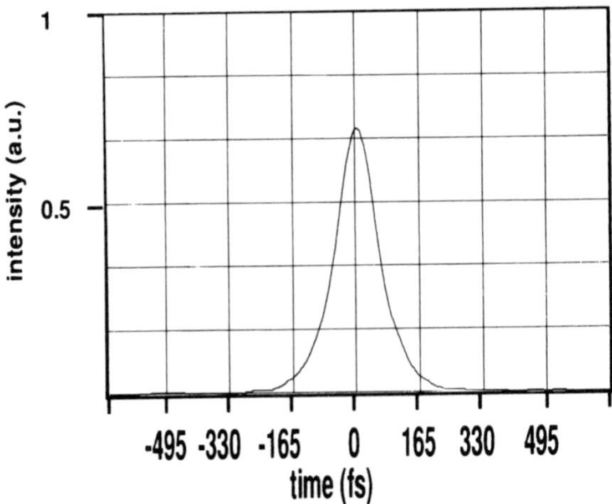

Figure 5.15. Autocorrelation of the shortest pulses generated with an environmentally stable erbium fiber laser. The FWHM pulse width is 125 fs assuming a sech^2 shape.

a 20% output coupler the output pulse energy was $\sim$ 10 pJ and $\sim$ 100 pJ pulses could be extracted by using the light rejected by the polarizer.

In another version of the laser source (Fermann, 1993) the moving mirror could be replaced by a semiconductor saturable absorber (Zirngibl *et al.*, 1991) to produce self-starting passive modelocking. Here pulses as short as 300 fs could be obtained, where measurements of the polarization evolution in the cavity verified that the pulse width was determined by nonlinear polarization evolution and the saturable absorber was only active in the early stages of the pulse buildup. The use of such a hybrid cavity is advantageous since, due to spatial variations of the semiconductor composition, it is typically very difficult to obtain femtosecond pedestal-free pulses with fiber lasers with semiconductor saturable absorbers as the only modelocking element (Loh *et al.*, 1993).

In addition, with further optimizations the cavity design should allow the generation of pulses shorter than 100 fs, i.e. we can expect that nonlinear polarization evolution will produce pulses about one order of magnitude shorter than is possible with semiconductor saturable absorbers.

5.6 Summary

With the demonstration of environmentally stable passively modelocked fiber lasers, the first step towards a wide distribution of these pulse sources has been made. Therefore the continuation of an intensive research effort in this area seems to be justified. As pointed out at various stages in this chapter, many details of the pulse evolution in passively modelocked fiber lasers are still not very well understood. The measurements presented here should enable the writing of detailed computer programs that can address some of the open questions. In particular, the influence of dispersion compensation, group-velocity walk-off, soliton trapping or even soliton shaping in the cavity still needs to be quantified. Further studies are also required to quantify the conditions for pulse buildup, especially in the presence of semiconductor saturable absorbers.

However, despite these limitations, practical diode-pumped femtosecond fiber pulse sources are a reality even now. With a short length of polarization maintaining erbium doped fiber and a few standard fiber components, a reliable 100 fs pulse source can easily be constructed. The conditions for pulse startup can be measured once and then fixed indefinitely. Pulse formation can be initiated either with a simple piezoelectrically-driven moving mirror, or, when available, with semiconductor saturable absorbers. Further engineering of the fiber laser will increase its reliability and stable operation range.

The total cost of such a fiber laser pulse source is currently dominated by the pump laser. Since the cost of the pump lasers is expected to fall steadily as fiber amplifiers become standard components in optical communications, the cost of ultrafast fiber lasers is set to come down significantly during the next couple of years. Ultrafast optical devices that currently appear exotic due to their high cost should therefore equally become more and more practical. As a result future research will increasingly be centered on applications of ultrafast fiber laser pulse sources.

Acknowledgments

I am indebted to M. H. Ober and M. Hofer for help in preparing this manuscript. I am also grateful to M. L. Stock and D. Harter for valuable comments. Finally I acknowledge financial support from the Alexander von Humboldt Stiftung.

References

Agrawal, G. P. (1989) *Nonlinear Fiber Optics*, Academic Press, Boston.
Carruthers, T. F. and Duling, I. N. (1990) *Opt. Lett.*, **15**, 804–6.

Dahlström, L. (1972) *Opt. Commun.*, **5**, 157–60.
Davey, R. P., Langford, N. and Ferguson, A. I. (1993) *Electron. Lett.*, **29**, 758–60.
Dennis, M. L. and Duling, I. N. (1992) *Electron. Lett.*, **28**, 1894–5.
Duling, I. N. (1991) *Opt. Lett.*, **16**, 539–41.
Duling, I. N. and Esman, R. D. (1992) *Conference on Lasers and Electro-Optics*, Vol. 12 of *1992 OSA Technical Digest Series*, paper CPDP28.
Fermann, M. E. (1993) unpublished data.
Fermann, M. E., Haberl, F., Hofer, M. and Hochreiter, H. (1990) *Opt. Lett.*, **15**, 752–4.
Fermann, M. E., Andrejco, M. J., Stock, M. L., Silberberg, Y. and Weiner, A. M. (1993a) *Appl. Phys. Lett.*, **62**, 910–12.
Fermann, M. E., Andrejco, M. J., Silberberg, Y. and Stock, M. L. (1993b) *Opt. Lett.*, **18**, 894–6.
Fermann, M. E., Yang, L. M., Stock, M. L. and Andrejco, M. J. (1994) *Opt. Lett.*, **19**, 43–5
Gordon, J. (1992) *J. Opt. Soc. Am.*, **B9**, 91–97.
Haus, H. A. and Ippen, E. P. (1991) *Opt. Lett.*, **16**, 1331–3.
Haus, H. A., Ippen, E. P. and Tamura, K. (1994) *IEEE J. Quantum Electron.*, **QE-30**, 200–8.
Hofer, M., Fermann, M. E., Haberl, F. and Townsend, J. E. (1990) *Opt. Lett.*, **15**, 1467–9.
Hofer, M., Fermann, M. E., Haberl, F., Ober, M. H. and Schmidt, A. J. (1991) *Opt. Lett.*, **16**, 502–4.
Hofer, M., Ober, M. H., Haberl, F. and Fermann, M. E. (1992) *IEEE J. Quantum Electron.*, **QE-28**, 720–8.
Islam, M. N., Poole, C. D. and Gordon, J. P. (1989) *Opt. Lett.*, **14**, 1011–14.
Kelly, S. M. J. (1992) *Electron. Lett.*, **28**, 806–7.
Koehler, B. G. and Bowers, J. E. (1985) *Appl. Opt.*, **24**, 349–53.
Krausz, F. and Brabec, T. (1993) *Opt. Lett.*, **18**, 888–90.
Laming, R. I., Payne, D. N. and Li, L. (1989) *J. Lightwave Technol.*, **LT-7**, 2084–91.
Lefevre, H. C. (1980) *Electron. Lett.*, **16**, 778–80.
Loh, W. H., Atkinson, D., Morkel, P. R., Hopkinson, M., Rivers, A., Seeds, A. J. and Payne, D. N. (1993) *Appl. Phys. Lett.*, **63**, 4–6.
Maker, P. D. and Terhune, R. W. (1965) *Phys. Rev. A*, **137**, 801–13.
Menyuk, C. R. (1987) *Opt. Lett.*, **12**, 614–6.
Menyuk, C. R. (1988) *J. Opt. Soc. Am. B*, **5**, 392–402.
Menyuk, C. R. (1989) *IEEE J. Quantum Electron.*, **QE-25**, 2674–82.
Menyuk, C. R. and Wai, P. K. A. (1992) in *Optical Solitons – Theory and Experiment*, ed. Taylor, J. R. (Cambridge University Press, Cambridge).
Mortimer, D. (1988) *J. Lightwave Technol.*, **LT-6**, 1217–24.
Nakazawa, M., Yoshida, E., Sugawa, T. and Kimura, Y. (1993) *Electron. Lett.*, **29**, 1327–9.
Ober, M. H. and Hofer, M. (1993) unpublished data.
Ober, M. H., Hofer, M. and Fermann, M. E. (1993) *Opt. Lett.*, **18**, 367–9.

Payne, D. N., Barlow, A. J. and Ramskov-Hansen, J. J. (1982) *IEEE J. Quantum Electron.*, **QE-18**, 477–88.

Rashleigh, S. C. (1983) *J. Lightwave Technol.*, **LT-1**, 312–31.

Shimizu, H., Yamazaki, S., Ono, T. and Emura, K. (1991) *J. Lightwave Technol.*, **LT-9**, 1217–24.

Squier, J., Salin, F., Rouer, C., Coe, S. and Mourou, G. (1990) *Conference on Lasers and Electro-Optics*, Vol. 7 of *1990 OSA Technical Digest Series*, paper CPDP9.

Stolen, R. H., Botineau, J., Ashkin, A. (1982) *Opt. Lett.*, **7**, 512–14.

Stolen, R. H., Pleibel, W. and Simpson, J. R. (1984) *J. Lightwave Technol.*, **LT-2**, 639–41.

Strickland, D. and Mourou, G. (1985) *Opt. Commun.*, **56**, 219–21.

Tamura, K., Ippen, E. P., Haus, H. A. and Nelson, N. E. (1993) *Opt. Lett.*, **18**, 1080–2.

Taverner, D., Richardson, D. J. and Payne, D. N. (1993) *Opt. Soc. Am. Topical Meeting on Nonlinear Guided Wave Phenomena, Cambridge*, paper WC3.

Ulrich, R. and Simon, A. (1979) *Appl. Opt.*, **18**, 2241–51.

Winful, H. G. (1985) *Appl. Phys. Lett.*, **47**, 213–15.

Yoshida, E., Kimura, Y. and Nakazawa, M. (1992) *Appl. Phys. Lett.*, **60**, 932–4.

Zirngibl, M., Stulz, L. W., Stone, J., Hugi, J., DiGiovanni, D. and Hansen, P. B. (1991) *Electron. Lett.*, **27**, 1734–5.

6

Ultrafast vertical cavity semiconductor lasers

WENBIN JIANG AND JOHN BOWERS

6.1 Introduction

Modelocking was at first demonstrated in the mid-1960s using He-Ne lasers [1], ruby lasers [2], and Nd^+ glass lasers [3]. The pulse width was initially quite long, but by the end of the 1970s subpicosecond pulses were generated using passive modelocked dye lasers [4–6]. The colliding pulse modelocking (CPM) of dye lasers in early 1980s [7] made sub-100 femtosecond pulses routinely available, and additive pulse modelocking (APM) of color center lasers pushed the wavelength of the sub-100 femtosecond pulses from the visible to the infrared [8, 9]. Kerr lens modelocking of Ti: sapphire lasers was discovered in 1990 [10–11] and produced the shortest laser pulses [12] directly from the laser cavity without any additional external cavity pulse compression. The advancement of short pulse generation technologies has provided very useful tools for researchers in physics, chemistry, biology, and optical communications. All of those ultrashort laser pulse systems are, however, large and expensive. A compact pulse source such as a modelocked semiconductor laser or fiber laser with comparable specifications to those big laser systems will be attractive because of its simplicity and cost effectiveness. Fiber lasers have limited cavity frequencies, typically below 100 MHz, and this limits their usefulness for many applications such as high bit-rate soliton transmission systems. Our goal with semiconductor modelocked lasers is to integrate the many fiber components into a compact, monolithic source.

6.1.1 *Modelocked in-plane semiconductor lasers*

Semiconductor lasers were demonstrated in 1962 by four groups almost simultaneously [13–16] only two years after the demonstration of the first laser [17]. Efforts have since been made in generating short optical pulses with semiconductor laser modelocking. Limited success was reported with active modelocking by Harris [18] and passive modelocking by Morozov *et al.* [19], who observed deep modulations at the frequency of the cavity round trip with GaAs p-n junction structure lasers in external cavities. Semiconductor laser modelocking, however, had not gained very much success until the late 1970s.

The first successful operation of semiconductor laser modelocking was reported by Ho *et al.* [20], who actively modelocked a GaAlAs double heterojunction (DH) laser in an external cavity at a wavelength of 810 nm. Similar results were reported thereafter with either different diode lasers or different current bias schemes [21–23]. The pulse duration from an actively modelocked semiconductor laser was typically tens of picoseconds. Passive modelocking with saturable absorbers was necessary in order to generate pulses of several picoseconds [24] or even subpicoseconds [25].

The pulses generated with the above techniques usually had substructure due to the reflection from the diode facet facing the external cavity. One approach to eliminate such an effect was to include a bandwidth limiting etalon in the laser cavity and/or to antireflectively (AR) coat the facet [26–29]. An electro-optic tuner (EOT) could be used in the cavity instead of the etalon for the bandwidth control as well as wavelength tuning, as reported by Olsson and Tang [30] in either a linear cavity or a ring cavity configuration.

A poor AR coating induces multiple pulses [31]. The simulation by Morton *et al.* [32] indicated that multiple pulses would be induced even with an AR coating of only 0.1% reflectivity. The imperfect AR coating was also the origin of the instability for an actively modelocked semiconductor laser with strong pulse current injection [33]. Such a stringent requirement to AR coatings makes it imperative to look for other approaches to overcome the defects of AR coatings. Holbrook *et al.* [34, 35] actively modelocked an angled stripe laser in an external cavity. Bandwidth-limited pulses of 16 ps duration at a peak power of 1 W and a time–bandwidth product $\Delta\upsilon\Delta t$ of 0.36 were obtained. Chen *et al.* [36] and Chang and Vukusic [37] reported Brewster-angled semiconductor diode lasers to be actively modelocked in external cavities. Mar *et al.* [38] eliminated multiple pulsing by using long laser chips to delay the reflected pulses into the off-region of the modulation. Helkey *et al.* [39] demonstrated a curved waveguide diode laser to reduce the reflection from an AR coated facet and modelocked the laser in an external cavity.

Derickson *et al.* [40] suppressed the multiple pulses caused by the imperfect AR coating using the intra-waveguide saturable absorber for a two-section diode laser in an external cavity. A saturated traveling-wave laser amplifier (TWA) could also be used to remove unwanted secondary pulses while amplifying and limiting the amplitudes of pulses from an actively modelocked semiconductor laser [41].

Some applications require that the modelocked pulse sources be tunable in wavelength. Although an EOT or a narrow band filter could be used in the cavity for the wavelength tuning [30, 42], the most widely used technique is to replace the external cavity mirror by a grating [36, 37, 43–51]. The wavelength tunability is typically around 20 nm. A wider tuning range from 1.26–1.32 μm with a pulse width shorter than 6 ps was achieved with the combination of a modelocked laser terminated by a reflective grating in the external cavity and a grating-pair compressor [52].

Modelocked semiconductor lasers with higher repetition rate are useful in high bit-rate optical communications and high speed optical computing. To increase the repetition rate, one approach is to modelock the laser at the harmonic frequency of the cavity fundamental mode [28, 53–60]. A repetition rate of 20 GHz (tenth harmonic of the cavity fundamental mode) at a wavelength of 1.55 μm with a pulse width of 5 ps was generated in this way [54]. A compact external resonator using a Selfoc rod lens [61–63] or an optical-fiber [53–55, 64–66] is another approach to generate GHz repetition rate pulses. A more feasible approach to generate high repetition rate pulses is to modelock a monolithic diode laser. The monolithic diode laser can be actively modelocked, passively modelocked, or hybridly modelocked, owing to an earlier idea of a segmented-contact device with independent current control to effect a localized absorbing region in the cavity [67]. Lau *et al.* [68] reported the passive modelocking at 18 GHz by operating the single contact monolithic buried DH GaAlAs laser close to the catastrophic damage limit. Hybrid modelocking was also demonstrated when an RF signal was applied to the laser. Higher repetition rate results were reported with active modelocking, passive modelocking or hybrid modelocking [69–83]. The highest repetition rate of 350 GHz was obtained with a monolithic colliding pulse modelocked (CPM) InGaAsP multiple quantum well (MQW) laser structure [79] (see Chapter 9).

Efforts have also been made in generating the shortest possible pulses from modelocked semiconductor lasers. The typical pulse widths from actively or passively modelocked semiconductor lasers are a few picoseconds, although there are also reports of subpicosecond pulse generation from either passive modelocking [25, 31] or active modelocking [57, 58]. A drawback of passive modelocking is the large timing jitter between pulses [84]. Electrical feedback stabilization of a passively modelocked semiconductor laser was demonstrated to reduce the jitter from more than 30 ps to 4 ps [85]. Hybrid modelocking [86, 87] is also a good compromise that generates pulses of comparable pulse width to passive modelocking with small timing jitter between pulses [84]. CPM [7] is another important technique to generate subpicosecond pulses from semiconductor lasers (treated explicitly in Chapter 9).

The first semiconductor CPM laser was reported by van der Ziel *et al.* [88] in an external cavity with a saturable absorber at one facet of the device introduced by proton bombardment (as described in Chapter 7). Bursts of pulses of 0.56 ps FWHM at a repetition rates of 625 MHz were obtained. Vasil'ev *et al.* [89] reported a linear cavity CPM laser with a three-segment device in which the saturable absorber was located between the two gain sections. Pulses of 0.8 ps at a repetition rate of 710 MHz were generated. Wu and Chen *et al.* [78, 79] monolithically incorporated the CPM technique in quantum well lasers. Pulses of 0.64 ps at a repetition rate of 350 GHz were generated.

The fundamental limit to the pulse width from a modelocked semiconductor laser is the gain bandwidth, which is typically around 50 fs. The actual pulse widths from all the modelocked semiconductor lasers are substantially longer. Carrier variation induced pulse chirping [47, 90] is the main cause of the long pulses. Various chirp compensation techniques [50, 52, 91–94] can be used to compress the pulses. Subpicosecond pulses can routinely be generated by pulse compression outside of the laser cavity.

Due to the small gain saturation energy, the output power from a modelocked semiconductor laser is small. The peak power is limited to a few watts in the best cases [34, 35, 45, 94]. Applications such as soliton optical communications, optical switching, optical clock distribution, etc, require high power pulses. Amplification using traveling wave amplifiers (TWA) [41, 95–97] is an attractive way of boosting the pulse energy. A peak power of 165 W at the pulse width of 200 fs was obtained in this way [98]. High CW output power can be generated from phase locked diode array lasers. Modelocking of those array lasers looks promising in generating high energy pulses [81, 99–103].

Compared with other modelocked solid state lasers and modelocked dye lasers, semiconductor lasers are still low in output power and long in pulse width. They have, however, the advantage of being electrically pumped. Modelocked semiconductor lasers are compact, easy to use and cost-effective. They can be used in applications that do not require high powers. For example, those lasers can be used in fiber optical communications for data transmission [104]. They can be used for electro-optic sampling [105], optical computing [106], gyroscopes [107], etc. Modern epitaxial techniques can tailor the materials for wavelengths ranging from the visible to the infrared, which gives modelocked semiconductor lasers an extra advantage to be applied where other types of lasers do not work.

6.1.2 *Vertical cavity surface emitting lasers*

In the late 1970s, Iga *et al.* [108] proposed having the semiconductor laser oscillate perpendicular to the plane of the device surface to overcome the difficulties facing cleaved in-plane semiconductor lasers. Vertical cavity surface emitting lasers (VCSELs) offer many advantages over in-plane lasers. First, the monolithic fabrication process and wafer scale probe test will substantially reduce the manufacturing cost since only good devices will be kept for further packaging [109]. Second, a densely packed two-dimensional laser array can be fabricated since the device occupies no larger area than commonly used electronic devices [110]. This is very important for applications in opto-electronic integrated circuits. Third, the microcavity length allows inherently single longitudinal cavity mode operation due to its large mode spacing. Temperature-insensitive devices can be fabricated with an offset between the wavelength of the cavity mode and the active gain peak [111]. Finally, the device can be designed with a low numerical aperture, circular output beam to match the optical mode of an optical fiber, thereby permitting efficient coupling without additional optics [112].

A conventional in-plane semiconductor laser utilizes its cleave edges as the laser cavity reflectors since the length of the active layer is usually several hundred micrometers, which provides enough gain to overcome the cavity reflector loss. A VCSEL, however, needs both of its surfaces to be highly reflective to decrease the cavity mirror loss as its active layer is usually less than 1 μm thick. The first VCSEL was demonstrated with GaInAsP/InP in 1979, which was operating pulsed at 77 K with annealed Au at both sides as reflectors [108]. A room temperature pulsed operating VCSEL was demonstrated with a GaAs active region in 1984 [113]. Room temperature CW operating GaAs VCSELs succeeded by improving both the mirror reflectivity and current confinement [114].

At present, an output power of over 100 mW has been obtained from an InGaAs/GaAs VCSEL with GaAs/AlAs monolithic diffractive Bragg reflectors (DBR) mounted on a diamond heat sink [115]. Lasing threshold of sub-mA or high wall-plug efficiency have also been reported with similar VCSELs [116, 117]. Room temperature CW InGaAsP/InP VCSELs have met some difficulties primarily due to a low index difference between GaInAsP and InP, which causes difficulty in preparing highly reflective monolithic DBRs [118]. Nevertheless, a CW InGaAsP VCSEL was recently reported operating at 15°C with dielectric mirrors [119]. Record low threshold current of 9 mA at room temperature has been demonstrated in InGaAsP VCSELs using GaAs/AlAs DBR mirrors [120].

Two dimensional (2-D) arrayed VCSELs can find important applications in stacked planar optics, such as the simultaneous alignment of a tremendous number of optical components used in parallel multiplexing lightwave systems and parallel optical logic systems, etc. High power lasers can also be made with phase locked 2-D arrayed VCSELs. Some 2-D arrayed devices have been demonstrated [110, 121] and efforts have been made in coherently coupling these arrayed lasers [122].

6.1.3 *Surface emitting laser modelocking*

Many applications require subpicosecond pulses with high peak powers from modelocked semiconductor lasers. The peak power of a modelocked in-plane semiconductor laser is, however, less than one percent that of the dye lasers or other solid state lasers, and so is the pulse energy. Is there any way to generate high peak power laser pulses from semiconductor lasers? External cavity amplification and phase-locked array laser modelocking are two options as discussed above. In this chapter, a simple approach to generate high power and ultrashort pulses is presented, based on an idea of using a vertical cavity to increase the gain saturation energy. The gain saturation energy of an active medium is expressed by

$$\varepsilon_{\mathrm{s}} = \frac{h\nu\,\sigma}{\Gamma \mathrm{d}g/\mathrm{d}N} \tag{6.1}$$

where $h\nu$ is the laser photon energy, σ is the gain cross-sectional area, Γ is the transverse mode confinement factor, and $\mathrm{d}g/\mathrm{d}N$ is the differential gain.

For a semiconductor laser with $\sigma = 0.1 \times 1\,\mu\text{m}^2, \Gamma = 0.3, \lambda = 0.88\,\mu\text{m}$, and $\text{d}g/\text{d}N = 3 \times 10^{-16}\text{cm}^2$, the gain saturation energy is $\varepsilon_s = 2.5$ pJ, a typical value for the intracavity pulse energy in modelocked semiconductor lasers. In order to generate high energy pulses from a modelocked semiconductor laser, the gain saturation energy ε_s has to be increased. Both the differential gain $\text{d}g/\text{d}N$ and the lasing frequency ν in the expression (6.1) are intrinsic to a material and hard to change, but the cross-sectional area σ is an easily variable parameter. For a VCSEL with $\sigma = 10 \times 10\,\mu\text{m}^2$, for example, the gain saturation energy ε_s will be 750 pJ if Γ is taken as unity. The intracavity laser pulse energy from a VCSEL can therefore be at least two orders of magnitude larger than that from an in-plane semiconductor laser. The area of a VCSEL could be $50 \times 50\,\mu\text{m}^2$ or even larger. To this end, it becomes clear that VCSEL modelocking will be a good approach to generate ultrashort pulses with high pulse energy. The better beam quality from such a modelocked VCSEL will also be more useful in most applications.

6.1.4 *Gain switched surface emitting laser*

Gain switching is a parallel technique to modelocking in generating short pulses from a semiconductor laser, as shown by the diagram

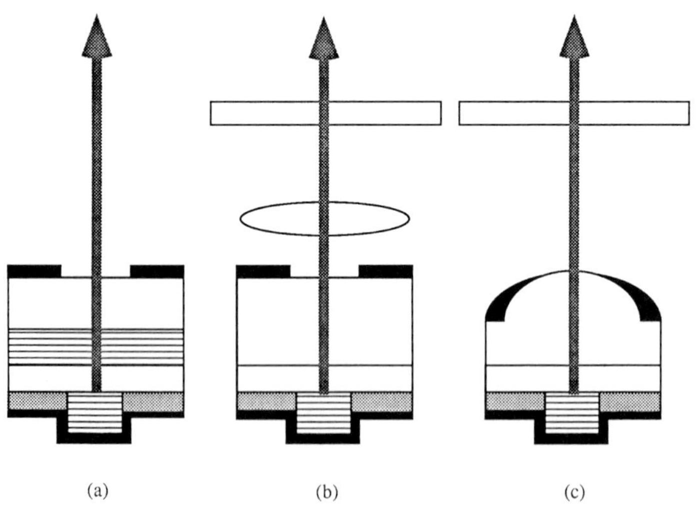

Figure 6.1. (a) Gain switched VCSEL diagram, (b) Modelocked VCSEL diagram with an intracavity lens, and (c) Modelocked VCSEL diagram with a monolithic lens. All the substrates are assumed to be transparent to the VCSEL wavelength.

in Figure 6.1 along with the modelocking cavity configurations. VCSEL gain switching can be used alternatively to generate short optical pulses with high peak powers. An early version of a gain switched VCSEL was an optically pumped film laser made of a single-crystal GaAs film of 1–2 μm thick sandwiched between highly reflective dielectric mirrors [123, 124]. When pumped by a cavity dumped passively modelocked dye laser at 615 nm with a pulse width of 1 ps and a peak power of 2 kW, pulses of 1 ps pulse width with a peak power of 1–10 W were generated. The active area had a spot size of 3 to 5 μm. The film laser was tunable in wavelength from 890 nm to 770 nm by using a wedged cavity [125]. Thin-film $In_{1-x}Ga_xAs_yP_{1-y}$ lasers of different compositions, including InP and $In_{0.53}Ga_{0.47}As$, were also made to lase between 0.77 and 1.65 μm [126, 127]. Each film laser sample could operate over an energy range exceeding 200 meV because of the intense photoexcitation. Pulses from film lasers were chirped and could be compressed in a dispersive fiber [128, 129].

Although those gain switched film lasers worked well at room temperature, they were pumped by pulses at a low repetition rate of dozens of kHz to prevent device heating. With the improvement of semiconductor epitaxial techniques, better heat dissipation has been achieved by replacing dielectric mirrors with monolithic semiconductor DBR mirrors. A gain-switched periodic-gain VCSEL with the monolithic DBR mirrors was able to generate 4 ps pulses at room temperature when optically pumped by pulses from a modelocked dye laser at a repetition rate of 80 MHz [130, 131]. A separate experiment with a high-Q cavity ($R_1R_2 \sim$ 99.6%) generated pulses of 13.4 ps at 82 MHz [132].

The electrically gain-switched VCSEL was at first reported with a bulk GaAs active layer [133]. The laser structure was composed of a p-type multilayer mirror of 16 pairs, an n-type multilayer mirror of 30 pairs and a two-wavelength thick bulk active layer. The multilayers were made of $Al_{0.1}Ga_{0.9}As$/AlAs with graded interfaces. The current was confined to a $16 \times 16\,\mu m^2$ region by ion implantation. A 10 μm window was made by lift-off of the Cr–Au contact. When the device was driven by a current pulse train with a pulse width of 200 ps at a repetition rate of 100 MHz, it emitted optical pulses with a pulse width of 28.5 ps at a peak power over 75 mW. A relaxation oscillation of 39 GHz was recorded from the gain switching experiment [134], many times larger than the small signal bandwidths of 8 GHz in an index guided VCSEL structure [135, 136] and 5.4 GHz in a gain guided VCSEL structure [137] reported earlier. An optical pulse width of 13.5 ps was reported by the same group later [138].

An independent gain switching experiment used an active layer consisting of four 10 nm thick GaAs quantum wells sandwiched between two AlGaAs/AlAs DBR mirrors [139]. Top-surfacing emitting, 15 μm diameter gain-guided lasers were fabricated using proton implantation for lateral current confinement. The VCSEL was packaged and bonded inside a coaxial high frequency (SMA) connector to avoid parasitics. The VCSEL was gain-switched using large amplitude sinusoidal modulation on top of a DC bias close to threshold. With an input RF power of 27 dBm at a repetition rate of 2 GHz, the laser emitted optical pulses with a pulse width of 24 ps and a wavelength of 820 nm. The time–bandwidth product of the pulses was 0.5, indicating little pulse chirping.

A gain switched VCSEL with an active region of three 8 nm $In_{0.2}Ga_{0.8}As$ quantum wells sandwiched between a 16-period AlAs/GaAs top DBR mirror and a 18.5-period bottom DBR mirror was also reported [140]. The laser was driven by 100 ps electrical pulses at a repetition rate of 80 MHz. 17 ps optical pulses were generated from the VCSEL, and relaxation oscillation frequencies of over 80 GHz were observed [141].

6.2 Optically pumped modelocked vertical cavity lasers

Modelocking is generally classified into three categories – active modelocking, passive modelocking and hybrid modelocking. Active modelocking with RF current modulation or synchronous electrical pulse pumping and passive modelocking with saturable absorbers are the commonly used methods to generate ultrashort pulses from in-plane semiconductor lasers. To demonstrate the modelocking of VCSELs, optical pumping similar to synchronous pumping of a dye laser or a color center laser was initially used.

Synchronous optical pumping has the advantage of higher pump efficiency at reduced heat generation, as well as the simple laser cavity configuration. The laser wavelength from a synchronously modelocked laser is generally tunable. Two synchronized pulse trains at different wavelengths (one pump, one laser) are good sources for certain pump-probe measurements. Synchronous optical pumping shows effectiveness in modelocking a VCSEL to generate ultrashort optical pulses with high peak power or high pulse energy. Actually almost any direct bandgap semiconductor materials can be used for this purpose, thus ultrashort pulses with a wavelength ranging from visible to the far infrared can in principle be obtained simply by using different bandgap semiconductor materials. Synchronous modelocking of a semiconductor laser was earlier reported with a bulk GaAs crystal of $4 \times 5 \times 10$ mm as the gain in an external cavity using two photon absorption pumping [142, 143]. Tunable picosecond pulses covering a wavelength range of 840–885 nm were generated over a temperature range from 97 to 260 K. Peak power of 1 MW at 97 K was achieved due to the large active volume involved. Synchronous modelocking of platelet lasers, the prototype of VCSEL modelocking, was reported with CdS, CdSe, CdSSe, InGaAsP, and ZnCdS semiconductor materials as gain mediums when pumped by either a modelocked Ar^+ laser or a modelocked YAG laser [144–147]. Those semiconductor platelets of generally a few micrometers thickness were kept at liquid nitrogen temperature during the laser operation. The output pulse widths from those modelocked lasers were several picoseconds, and the highest peak power was 50 W from the InGaAsP platelet laser. The modelocked platelet lasers were tunable in wavelength, ranging from 490 nm to 2 μm [146, 148].

6.2.1 *Modelocked GaAs vertical cavity lasers*

The first modelocked VCSEL operating at room temperature used a MQW GaAs/$Ga_{0.7}Al_{0.3}As$ structure with 120 quantum wells as the active medium [149]. The well-width was 15 nm and the barrier-width was 7 nm. The integrated semiconductor DBR mirror of 22.5 periods of $Ga_{0.8}Al_{0.2}As$/AlAs was grown below the MQWs on the semi-insulating GaAs substrate. On top of the epitaxial layer was a layer of SiN_xO_y AR coating.

The laser cavity studied was Z-shaped as shown in Figure 6.2. An epitaxially grown DBR mirror was used as one of the cavity end mirrors. M1 was a dichroic mirror that was highly reflective at the lasing wavelength and highly transmissive to the pump. M2 was a total reflector. M3 was an output coupler with a reflectivity of 97%. The focal length of the AR coated lens was 38.1 mm. A Brewster angle plate was used in the cavity to control the laser polarization. The sample was mounted on a copper heat sink that was maintained at room temperature during the laser operation. The pump was a synchronously modelocked dye laser.

To test the MQW sample, a CW Ti:sapphire laser at the wavelength of 800 nm was used to pump the external cavity surface-emitting laser. Room temperature CW operation at a pump power less than 80 mW

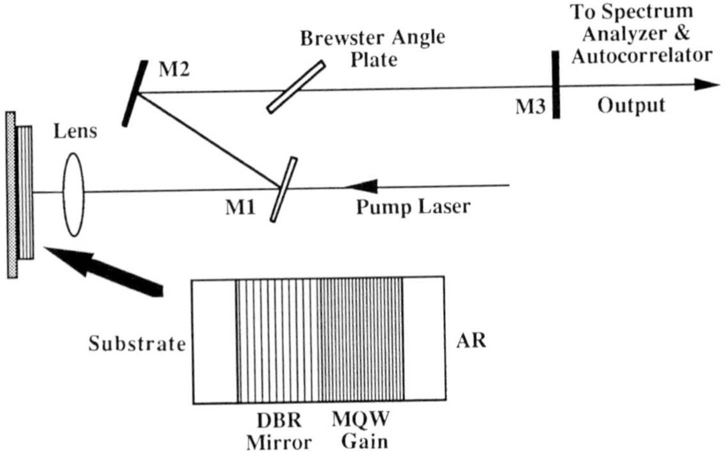

Figure 6.2. Cavity configuration of a GaAs modelocked VCSEL.

was obtained with a 25.4 mm focal length lens and an output coupler of 99% reflectivity. It was, however, not stable due to the device heating. The stability was improved by chopping the pump laser at 1:1 duty ratio, while maintaining the heat sink at 0°C. Under these conditions, the threshold pump power was 90 mW with an output coupler of 97% (See curve (a) in Figure 6.3). The maximum output power was 5.6 mW at the pump power of 140 mW. The decrease of the output power at high pump powers was caused by heating. The material gain also decreased with the increase of the temperature.

An independent work using a bulk GaAs active layer of 2 μm at 77 K and pumped by an all-line infrared (752 and 799 nm) krypton-ion laser emitted CW output power of 700 mW at a wavelength of 830–840 nm [150]. The pump power was 1.8 W. The absolute quantum efficiency of conversion from pump photons to lasing photons was 44% and the differential quantum efficiency was 58%.

To modelock the external cavity surface-emitting laser, a synchronously modelocked Coherent 700-series dye laser was used as the pump. The dye laser generated 5–10 ps pulses at a wavelength of 800 nm and a repetition rate of 80 MHz with the maximum average power of 190 mW. Curve (b) in Figure 6.3 shows the average output power of the modelocked VCSEL varying with the pump power at room temperature. The focal length of the cavity lens was 38.1 mm and the reflectivity of the output coupler was 97%. The threshold pump power was 34 mW. The maximum output power under synchronous

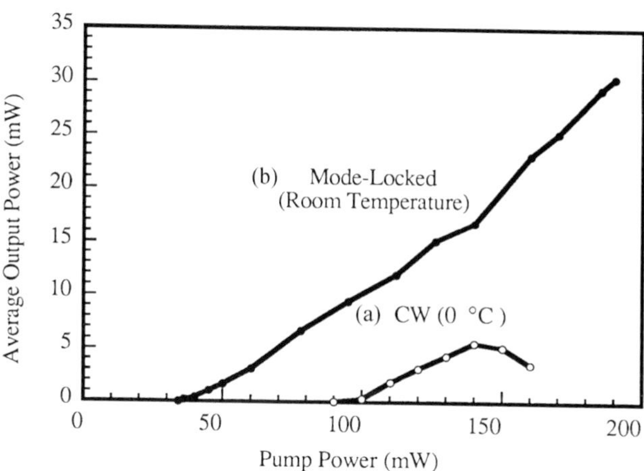

Figure 6.3. Average output power from an optically pumped external-cavity GaAs VCSEL as a function of input power.

pumping was 30.5 mW at a pump power of 190 mW. It was limited by the available pump power. The maximum external differential quantum efficiency was 27%. No output power saturation was observed within the available pump power.

Autocorrelation measurements were conducted to determine the laser pulse width indirectly. A typical autocorrelation trace of the pulse from the modelocked GaAs VCSEL is shown in Figure 6.4. The average power of this pulse was 14 mW. The pulse width was 14 ps if a Gaussian pulse shape was assumed. The coherent spike of the autocorrelation trace indicated that the spectral width of the pulse was much wider than the transform-limited pulse spectrum, which was confirmed by the measured power spectrum as shown in the inset of Figure 6.4 with a spectral width of 6.6 nm. The time–bandwidth product of the pulse was 35.8, two orders of magnitude larger than the transform-limit. The laser pulse width and the output power were dependent on the cavity length matching between the pump and the laser [151], as shown in Figure 6.5. The minimum pulse width was obtained when the VCSEL cavity was slightly longer than the length at which the maximum output power was obtained. The VCSEL operation was unstable when the VCSEL cavity was shorter than the pump cavity.

The large time–bandwidth product of the pulses implies a large pulse chirping. Group velocity dispersion (GVD) can be introduced outside of the laser cavity to compensate for the pulse chirping [152, 153], and thereby compress the laser pulses. A parallel grating-pair that gave

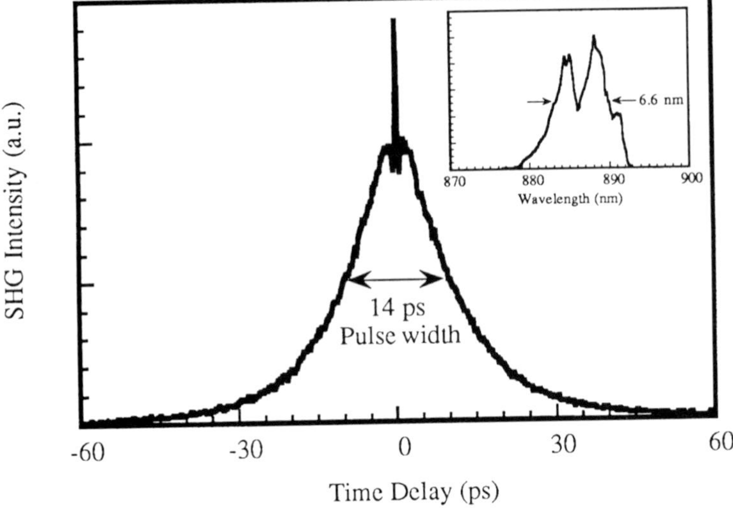

Figure 6.4. A typical intensity autocorrelation trace of pulses from a modelocked GaAs VCSEL. The pulse width at FWHM is 14 ps if a Gaussian pulse shape is assumed. The corresponding power spectrum is shown in the inset.

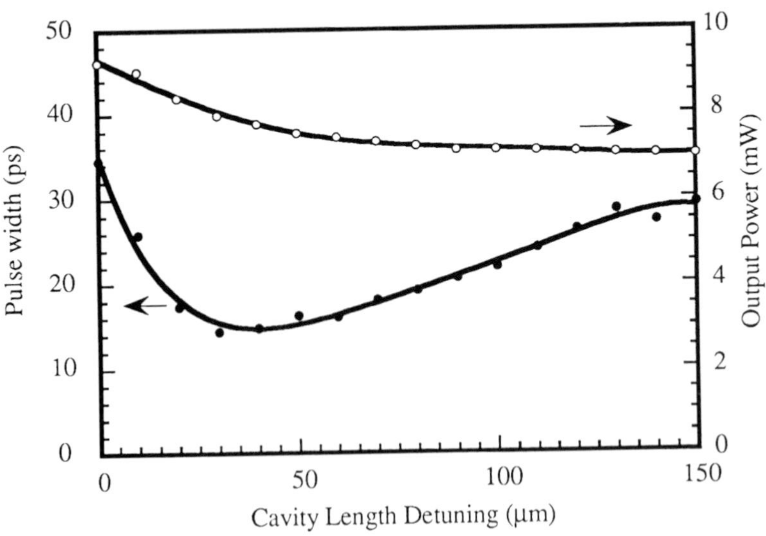

Figure 6.5. Pulse width and the corresponding average output power from a modelocked GaAs VCSEL varying with the laser cavity length tuning.

only negative GVD [154] was unable to compress the chirped pulses from the modelocked VCSEL. A grating-pair with a telescope configuration [155] can generate either positive GVD or negative GVD, depending on the distances between the two gratings and the two lenses. Such an arrangement was used for the pulse compression instead of the parallel grating-pair. Spectral windowing was used to modify the spectrum of the compressed pulses. When the compressor was in the positive GVD region, the laser pulses were shortened, indicating the down-chirp characteristic of the pulses. Figure 6.6 shows the autocorrelation of the pulses under optimum compression by the solid curve. The fitting to the autocorrelation by a single side exponential pulse shape, $P(t) = P_1 \exp(-t/\tau_1) + P_2 \exp[-t/\tau_2]$, is shown by the dashed line, where $\tau_1 = 0.35$ ps and $\tau_2 = 2.4$ ps. The full pulse width at half-maximum (FWHM) was 324 fs. The average power was 4 mW, thus $P_1 = 50$ W and $P_2 = 14$ W. The peak power of the pulses was 64 W. The corresponding spectrum of the compressed pulse is shown in the inset of Figure 6.6. The spectral width was 2.5 nm, so the time–bandwidth product was 0.31. (Compare these figures with those for solid state lasers, Chapter 3, Table 3.1.)

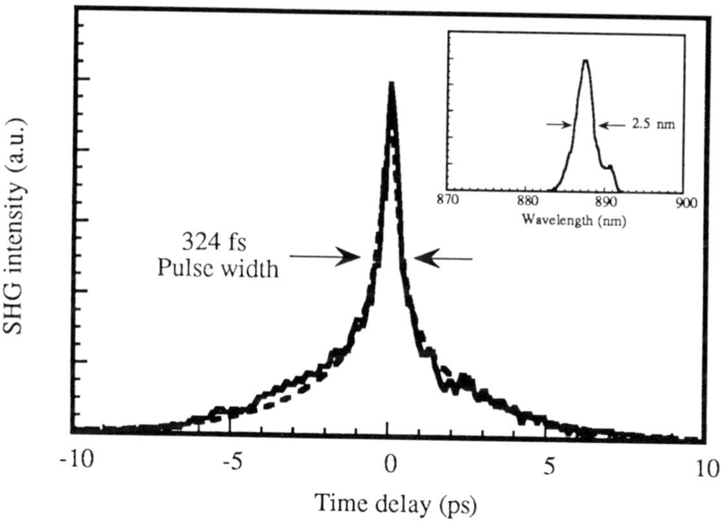

Figure 6.6. The intensity autocorrelation trace of pulses in the optimum compression (solid line) and the fitting to the autocorrelation by a single side exponential pulse shape (dashed line). The corresponding power spectrum of the compressed pulses is shown in the inset.

6.2.2 *Femtosecond periodic gain vertical cavity lasers*

Although subpicosecond pulses have been obtained by external cavity pulse compression, an equally interesting focus is to explore the generation of the shortest possible pulses **directly** from VCSELs.

An important factor that prevents subpicosecond pulse generation directly from the modelocked VCSELs is significant laser pulse chirping. Intracavity chirp compensation with a prism-pair is usually an efficient way for subpicosecond pulse generation directly from a laser oscillator, but does not work for such a laser with large pulse chirping. An intracavity filter has thus been used to limit the extra bandwidth created by the pump induced cross-phase modulation (XPM) and the gain saturation induced self-phase modulation (SPM) [156, 157]. This method is effective in controlling the time–bandwidth product of the laser pulses. The output laser pulse width usually remains unchanged or is sometimes shorter with the filter.

The experimental configuration is the same as that shown in Figure 6.2 with a GaAs/AlGaAs MQW active medium at room temperature. To control the pulse chirping, a Brewster-angle placed birefringent filter was used in the laser cavity instead of the Brewster-angle glass plate. The filter thickness was 0.5 mm. The output pulse from the modelocked VCSEL was asymmetrically shaped at a pulse width of 10 ps with a relatively long tail, as measured by a synchronous-scan streak camera. The corresponding laser pulse spectral width was 0.9 nm. The time–bandwidth product of the pulses was 3.6, one tenth of that shown in Figure 6.4.

For synchronous modelocking, the leading edge of the laser pulse is shaped by the rising edge of the pump pulse, and the trailing edge is shaped by the gain saturation. The long pulse tail indicates poor gain saturation. On the other hand, the intracavity laser pulse energy was 4.1 nJ, three times larger than the gain saturation energy E_{sat} of 1.2 nJ [151]. The gain should have been deeply saturated by the laser pulse.

To have a further insight into this contradiction, a modelocked Ti:sapphire laser producing 100–200 fs pulses was used as the pump to replace the dye laser. The VCSEL cavity was converted to 89 MHz to match the repetition rate of the Ti:sapphire laser. To decrease the intracavity pulse energy for the same amount of output power, thereby minimizing the gain-saturation-induced SPM, the reflectivity of the output coupler was reduced to 95%. The intensity autocorrelation trace of the output pulse from the modelocked VCSEL without the birefringent filter in the cavity showed a pulse width of 2.3 ps and a spectrum width of 2.9 nm. The time–bandwidth product was 2.3. The average output power was 13 mW at the pump power of 200 mW. It turns out that reducing the intracavity pulse energy has helped in reducing the pulse chirping, as shown by the smaller time–bandwidth product. The laser operation, though, was very noisy. The laser stability was considerably improved when a birefringent filter with a thickness of 0.5 mm was placed in the cavity. The shortest laser pulses directly from the laser oscillator were obtained as shown by the autocorrelation trace in Figure 6.7. Assuming a hyperbolic secant pulse shape, the laser pulse width was 0.92 ps. The average laser output power was 11 mW at the pump power of 200 mW. The side lobe 5.5 ps away from the main peak in the autocorrelation trace indicated the existence of secondary pulses. This effect could be suppressed at the price of broadened laser pulse width by detuning the laser cavity length, as well as tuning the filter. The effect could also be strengthened by the cavity detuning in the other direction. More sub-pulses would then be observed. The spectrum of the 0.92 ps pulses is shown in the inset of Figure 6.7. The spectral width was 2.4 nm. The time–bandwidth product of the pulses was 0.85, which was a factor of three improvement over that without the filter. The ripple in the spectrum

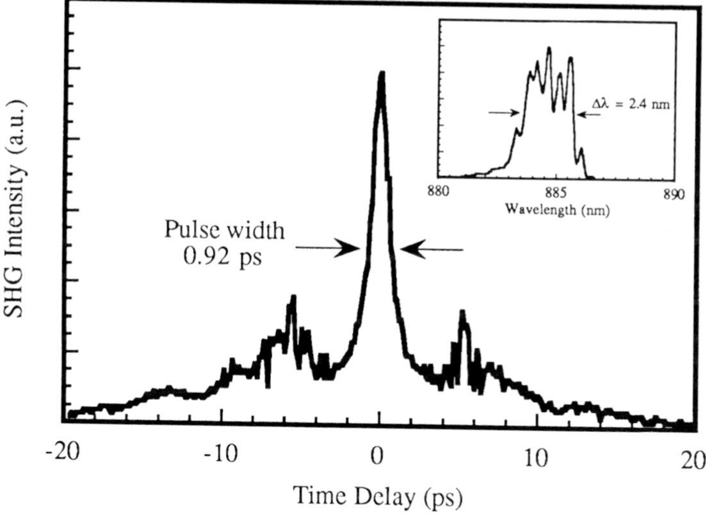

Figure 6.7. The intensity autocorrelation traces of pulses from a modelocked GaAs VCSEL with a birefringent filter of 0.5 mm in the cavity when pumped by a modelocked Ti:sapphire laser. The corresponding power spectrum is shown in the inset.

was the result of the interferences between the main pulses and the secondary pulses. The modulation mode spacing was 0.5 nm, corresponding to a time-spacing of 5.2 ps between the multiple pulses, which was in good agreement with the autocorrelation measurement of 5.5 ps.

The laser pulses were much longer than the pump pulses, which again indicated the poor tail shaping, or in other words, poor gain saturation. The intracavity pulse energy in this case was actually around 2.3 nJ, still larger than the gain saturation energy E_{sat}. What is the cause of this anomaly? There must be an anti-gain-saturation mechanism intrinsic to this laser. Note that the gain medium in this laser has been directly attached to one of the cavity mirrors. Spatial hole-burning has occurred during the laser operation because of a standing wave in the gain medium formed by the incident and the reflected laser beams (Figure 6.8). The carriers in the quantum wells outside of the strong field regions will transport into these regions through thermionic emission, diffusion and tunneling. The carrier transport will not stop until the carrier densities become equal everywhere. The duration needed to reach such an equilibrium is the dominant limitation in obtaining femtosecond pulses, if the pulse chirping has been carefully excluded.

The carrier transport time for a MQW sample is on the order of several picoseconds or longer [158]. The gain in the MQW will remain high for

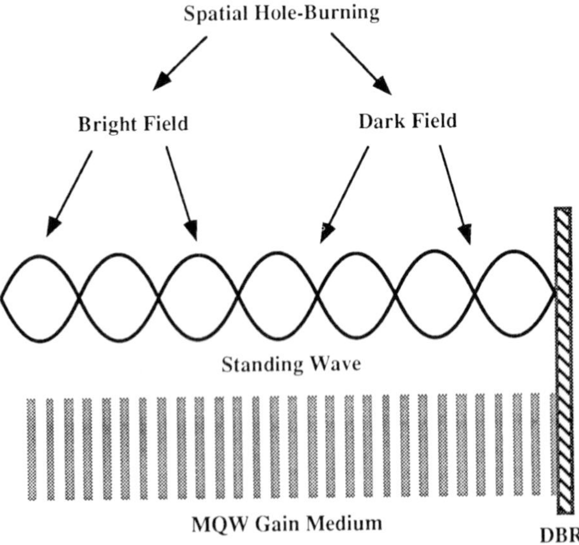

Figure 6.8. Spatial hole-burning caused by the standing wave pattern inside the MQW gain medium.

this picosecond time duration, within which the gain saturation is not complete even though the intracavity laser pulse energy is larger than the gain saturation energy. The laser pulses generated will either show a long tail extending over this duration, or be accompanied by subpulses if extra resonances are possible due to a Fabry–Perot effect. The pulse duration will also be limited on the order of picoseconds no matter how short the pump pulse is.

A solution to the carrier transport problem is to place the gain only at the peaks of the optical standing wave pattern. This periodic gain structure also has the advantage of lowering the lasing threshold [159]. The gain structure included a DBR mirror on top of the GaAs substrate. The active region was a periodic structure with $Ga_{0.8}Al_{0.2}As$/GaAs of 30 pairs. The barrier $Ga_{0.8}Al_{0.2}As$ was 63 nm thick and the well GaAs was 60 nm thick. Between the active layers and the DBR mirror was a 32 nm thick barrier $Ga_{0.8}Al_{0.2}As$ used for the mode matching between the standing waves in the DBR mirror and in the gain region. This structure was designed for a resonant wavelength of 883 nm. The epitaxial side of the MBE grown wafer was AR coated. The device was mounted on a heat sink in the laser cavity to replace the MQW sample used earlier. The output coupler of the cavity had a reflectivity of 97%. Birefringent filters were not necessary in the cavity since the periodic gain structure itself effectively served as a low-pass filter in the lasing wavelength region. During the laser operation, the shorter wavelength (higher frequency) part of the spectrum was attenuated by the filter effect of the periodic gain structure, while the longer wavelength (lower frequency) part of the spectrum was limited by the gain profile.

The VCSEL was pumped by a modelocked Ti:sapphire laser at the wavelength of 800 nm. The output power from the modelocked VCSEL was 6 mW at a pump power of 200 mW. The intensity autocorrelation trace of the modelocked pulses is shown in Figure 6.9. The full width at half-maximum (FWHM) of the autocorrelation trace was 300 fs. The corresponding pulse spectrum is shown in the inset of Figure 6.9 with a spectral width of 2.5 nm. If a pulse shape of hyperbolic secant were assumed, the laser pulse width would be 190 fs and the time–bandwidth product $\Delta\nu\Delta t$ would become 0.18, which is smaller than the transform limited value of 0.32. If a pulse shape of single side exponential were assumed, the laser pulse width would be 150 fs and the time–bandwidth product $\Delta\nu\Delta t$ would become 0.14, which is slightly larger than the transform limited value of 0.11. The actual pulse shape probably had a very rapid rising edge close to the Gaussian shape, but a slower falling tail

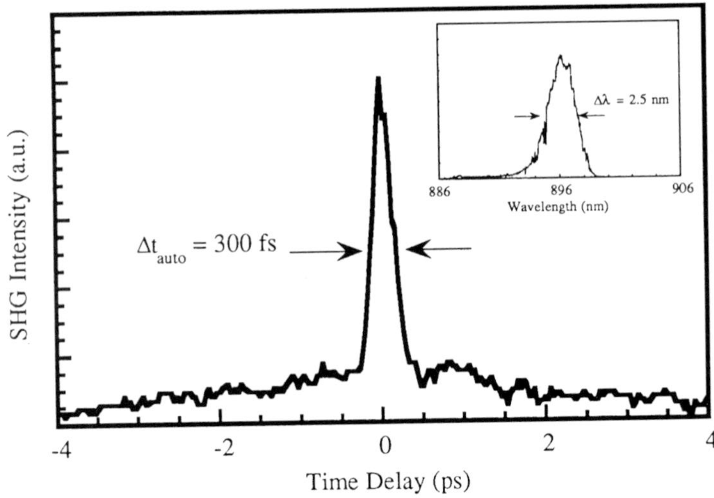

Figure 6.9. The intensity autocorrelation of pulses from a modelocked GaAs periodic gain-VCSEL when pumped by a modelocked Ti:sapphire laser. The autocorrelation width is 300 fs. The corresponding power spectrum of the pulses is shown in the inset.

following the exponential shape. The large pedestals were probably due to the residual longitudinal and transverse carrier transport effects and carrier heating.

6.2.3 *Modelocked InGaAs/InP vertical cavity lasers*

InGaAs/InP based laser systems at a wavelength of 1.5 μm have important applications in optical communications and soliton transmissions. A group at NTT (Nippon Telephone and Telegraph) used an InGaAs/InP MQW as the gain medium in a modelocked VCSEL [160]. The MQW was grown by molecular beam epitaxy, and consisted of 200 periods of InGaAs (15 nm) separated by InP (5 nm), which exhibited a room-temperature electron-heavy-hole (e-hh) absorption peak at 1.589 μm. The substrate side of the sample was AR coated to reduce reflection and the epitaxial side was gold coated, providing an end mirror for the laser cavity with a reflectivity of 96% and good thermal conductivity. Nonradiative Auger recombination losses were large at room temperature, so the sample was cooled to liquid-nitrogen temperatures in a cryostat with AR-coated windows. The VCSEL cavity was completed by a 10% output coupler, a 1-mm-thick quartz birefringent tuning element (with a bandwidth of 30 nm) to tune the lasing frequency, and an AR-coated lens (focal length of 38.1 mm) focused on the MQW sample. A dichroic steering mirror coupled optical pump light from the 1.32 μm Nd:YAG pump laser into the 1.5 m VCSEL cavity.

The VCSEL gave large output powers when operating CW. Figure 6.10(a) shows output powers as a function of pump power. The maximum output power was 190 mW at 1.3 W pump power, and the lasing threshold was at 40 mW pumping. These power levels are comparable with those of color-center lasers at this wavelength [161]. The absolute quantum efficiency of conversion from pump photons to lasing photons was as high as 18%, and the maximum differential quantum efficiency was 23%. Figure 6.10(b) shows the wavelength of the VCSEL, which shifts from 1.495 μm at threshold to 1.518 μm. Thermal loading, due to inadequate thermal contact of the MQW to the cryostat cold finger limited further increases of the output power. The VCSEL wavelength was tunable from 1.44 μm to 1.53 μm using the birefringent tuning element, as shown in Figure 6.11 at a pump power of 850 mW. The short wavelength lasing limit at 1.44 μm was caused by the rapidly deteriorating AR coating at this wavelength, and the long wavelength limit at 1.53 μm was due to the band edge.

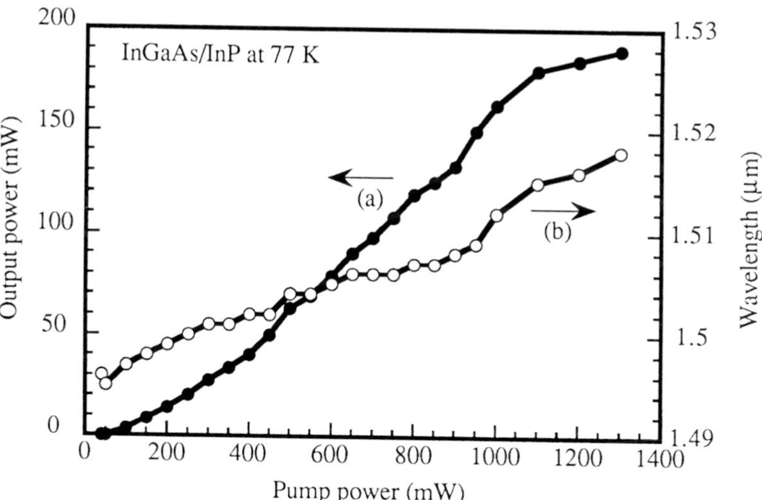

Figure 6.10. (a) Average output power from an InGaAs/InP VCSEL as a function of the pump power when the laser is operating CW. (b) The corresponding central wavelength varying with the pump.

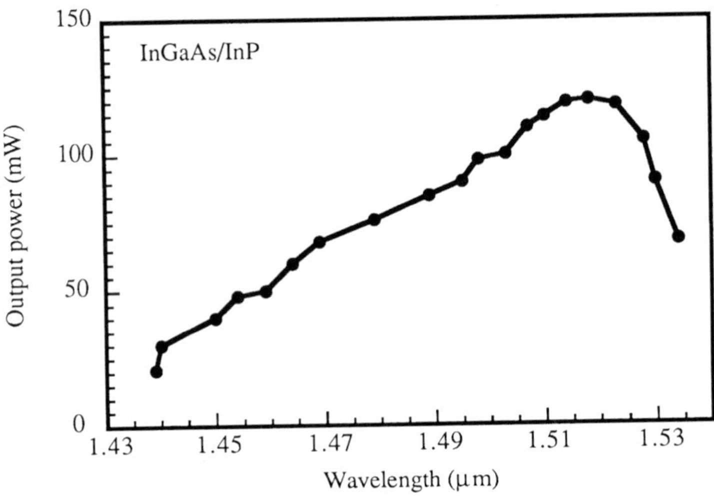

Figure 6.11. The wavelength tuning range of the InGaAs VCSEL at 77 K.

Synchronous pumping at 100 MHz with 110 ps pulses from the 1.32 μm Nd:YAG gave 100 ps pulses and average output powers nearly identical to the CW results, but with a reduced threshold pump power of 20 mW. The etalon effect due to the residual reflection of AR coating on the InP substrate to both the pump and laser prohibited shorter pulse generation. To solve the etalon problem, the InP substrate was slanted by polishing [162]. Synchronous pumping the VCSEL with this 5° slanted substrate active medium by the modelocked 1.32 μm YAG laser generated pulses of durations ranging from 5 to 30 ps with an average power of 260 mW. The time–bandwidth product of the pulses was 100 times larger than the transform limit. The subsequent pulse compression with a parallel grating-pair reduced the pulse width to 150 fs (Figure 6.12). Spectral filtering was helpful in eliminating the pedestals during the compression [163]. The compressed pulses had a peak power over 3 kW. Soliton compression inside of fiber [164] was then used to further shorten the pulse width. The second stage compression with a 400 m fiber reduced the pulse width from 150 fs to 21 fs [165].

Efforts were made to stabilize the laser operation and to generate transform limited pulses from the synchronously modelocked VCSEL. A coherent photon-seeding technique using an integrated output-coupling mirror was used to passively stabilize the laser [166]. A CW modelocked 1.06 μm YAG laser was used instead of the 1.32 μm YAG

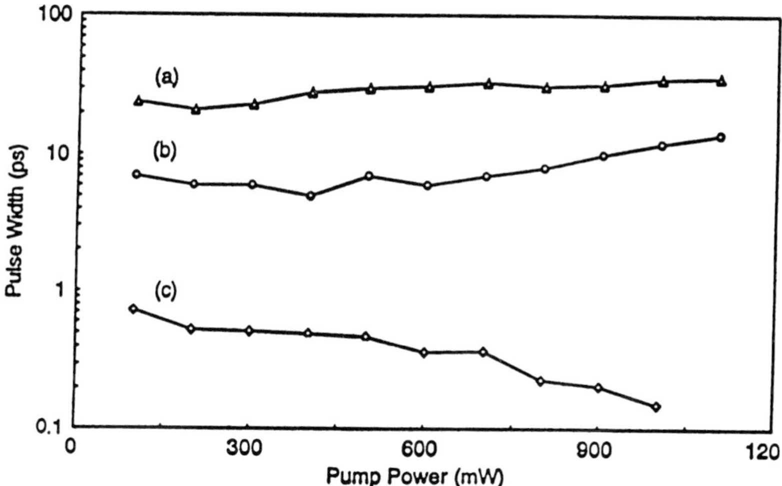

Figure 6.12. Pulse widths versus the pump power for the EXSEL. The data denoted by curves (a) and (c) are the pulse widths before and after compression, respectively, when the EXSEL cavity is adjusted to maximize the spectral widths. The data denoted by curve (b) are the pulse widths before compression when the cavity is adjusted to minimize the pulse widths. (Copied from *Opt. Lett.* **16**, 1394 (1991).

laser as the pump. The pulse width was reduced from 17 ps to 9.5 ps, and the time–bandwidth product was 0.55, a factor of 30 reduction. An APM technique combined with the synchronous pulse pumping was also used to create transform limited pulses [167]. A bulk $In_{0.53}Ga_{0.47}As$ layer 3 μm thick was used instead of the MQW as the active medium. The external auxiliary cavity consisted of a 1.3 m long fiber and an aluminum mirror with PZT control. Without the APM, the laser put out pulses at a width of 19 ps. The APM reduced the pulse width to 6 ps, and the time–bandwidth product was reduced to 0.3.

6.3 Analysis of laser pulse chirping in modelocked VCSELs

Modelocking of vertical cavity surface emitting lasers with either GaAs or InGaAs semiconductor active materials using synchronous optical pumping has generated pulses with a peak power over 100 W because of the large active cross-sectional area in the VCSELs. The pulses generated in the VCSELs are usually highly chirped and have a large time–bandwidth product. This large chirping makes it difficult to directly generate sub-picosecond pulses from the lasers.

For a modelocked VCSEL as shown in Figure 6.2, four primary mechanisms will affect the laser pulse chirp: (a) the SPM due to the gain saturation and the XPM due to the pulsed optical pumping, (b) the phase dispersion due to the finite gain shape, (c) the dispersion due to the DBR structure of both the semiconductor mirror and the dielectric mirrors, and (d) the material dispersion of the intracavity components. The dispersion caused by DBR mirrors has been well analyzed [168, 169] and is on the order of 10^{-28}s^2. It varies depending on the design and the layer uniformity. The material dispersion for the related semiconductor materials in the transparent region and other intracavity components like fused silica glass are also known [170, 171]. They are around $1 \times 10^{-28}\text{s}^2$ for 2 μm thick GaAs/AlGaAs [170] or for 5 mm thick fused silica [171]. This section will analyze the laser pulse chirping due to (a) and (b) after a single pass through the cavity.

6.3.1 *SPM and XPM of laser pulses*

When a laser pulse propagates through a semiconductor gain medium, its phase will experience a change, as well as its amplitude. The pulse will be chirped due to the gain saturation induced SPM and the pulsed optical pumping induced XPM. The wave propagation equations in the local time domain for the laser amplitude $A(z, \tau)$ and the pump power $P_p(z, \tau + \Delta\tau)$ are given by [151]

$$\frac{\partial A(z,\tau)}{\partial z} = \frac{1}{2}(1 - i\alpha)g(z,\tau)A(z,\tau) \tag{6.2}$$

$$\frac{\partial P_p(z,\tau + \Delta\tau)}{\partial z} = -\alpha_p(z,\tau)P_p(z,\tau + \Delta\tau) \tag{6.3}$$

where the slowly varying pulse amplitude envelope of the electric field $A(z, \tau)$ is normalized so that $|A(z,\tau)|^2$ stands for the instantaneous laser power, $g(z, \tau)$ is the amplifier power gain at the laser frequency $\omega_0, \alpha_p(z, \tau)$ is the absorption of the active material to the pump pulse at pump frequency ω_p, and α is the linewidth enhancement factor. $\Delta\tau$ determines the position of the pump pulse relative to the laser pulse. If $\Delta\tau > 0$, the pump pulse arrives earlier than the laser pulse, and if

$\Delta\tau < 0$, the laser pulse arrives earlier than the pump pulse. The relationship of the gain to the carrier density is modeled to be linear as given by

$$g(z,\tau) = a[N(z,\tau) - N_0] \tag{6.4}$$

where a is the differential gain coefficient and N_0 is the carrier density value at transparency. The transverse mode confinement factor Γ is taken as unity, which is reasonable for VCSELs. $\alpha_\mathrm{p}(z,\tau)$ is also assumed to be linearly proportional to the carrier density $N(z,\tau)$, as shown by

$$\alpha_\mathrm{p}(z,\tau) = -b\left[N(z,\tau) - N_\mathrm{p0}\right] \tag{6.5}$$

where b is the differential gain coefficient at the pumping frequency, ω_p, and N_p0 is the transparency carrier density value at ω_p.

The carrier density variation with time at position z in the gain medium is determined by the rate equation

$$\frac{\partial N(z,\tau)}{\partial\tau} = \frac{\alpha_\mathrm{p}(z,\tau)P_\mathrm{p}(z,\tau+\Delta\tau)}{\hbar\omega_\mathrm{p}\sigma} - \frac{g(z,\tau)}{\hbar\omega_0\sigma}|A(z,\tau)|^2 \tag{6.6}$$

where σ is the active cross-sectional area. Equation (6.6) can be rewritten as

$$\frac{\partial g(z,\tau)}{\partial\tau} = \frac{\alpha_\mathrm{p}(z,\tau)P_\mathrm{p}(z,\tau+\Delta\tau)}{\gamma E_\mathrm{psat}} - \frac{g(z,\tau)}{E_\mathrm{sat}}|A(z,\tau)|^2 \tag{6.7}$$

where

$$E_\mathrm{sat} = \frac{\hbar\omega_0\sigma}{a} \tag{6.8}$$

$$E_\mathrm{psat} = \frac{\hbar\omega_\mathrm{p}\sigma}{b} \tag{6.9}$$

are the saturation energies of the gain medium at laser frequency ω_0 and pump frequency ω_p, respectively, and

$$\gamma = b/a \tag{6.10}$$

is the ratio between the differential gain coefficients at ω_p and ω_0. Both the pump and the initial laser input pulses will be assumed to be Gaussian, $E_0/(\sqrt{\pi}\tau_0)\exp[-(\tau/\tau_0)^2]$, where E_0 is the pulse energy and $2\sqrt{\ln 2}\tau_0$ is the FWHM (full width at half-maximum) pulse width. The normalized complex electric field $A(z,\tau)$ is rewritten in terms of the real values $P(z,\tau)$, the laser power, and $\phi(z,\tau)$, the phase,

$$A(z,\tau) = \sqrt{P(z,\tau)}\exp[i\phi(z,\tau)] \tag{6.11}$$

so that Equation (6.2) can be analytically solved, as given by

$$P_{\text{out}}(\tau) = P_{\text{in}}(\tau)e^{G(\tau)} \tag{6.12}$$

$$\phi_{\text{out}}(\tau) = \phi_{\text{in}}(\tau) - \frac{1}{2}\alpha G(\tau) \tag{6.13}$$

where $P_{\text{in}}(\tau)$ and $P_{\text{out}}(\tau)$ stand for the power of the input and output laser pulses from the gain medium, $\phi_{\text{in}}(\tau)$ and $\phi_{\text{out}}(\tau)$ stand for the phase of the input and output laser pulses, and $G(\tau)$ is the total gain experienced by the laser after passing through the gain medium, as defined by

$$G(\tau) = \int_0^L g(z,\tau)\mathrm{d}z \tag{6.14}$$

Equation (6.3) can also be solved, as given by

$$P_{\text{p}}(L,\tau+\Delta\tau) = P_{\text{p}}(0,\tau+\Delta\tau)\exp\left[-\int_0^L \alpha_{\text{p}}(z,\tau)z\mathrm{d}z\right] \tag{6.15}$$

where $P_{\text{p}}(0,\tau+\Delta\tau)$ is the input pump power and $P_{\text{p}}(L,\tau+\Delta\tau)$ is the output pump power. By integrating Equation (6.7) over z from 0 to L,

$$\begin{aligned}\frac{\mathrm{d}G(\tau)}{\mathrm{d}\tau} = {} & \frac{P_{\text{p}}(\tau+\Delta\tau)}{\gamma E_{\text{psat}}}\{1-\exp[\gamma G(\tau) - b(N_{\text{p0}} - N_0)L]\} \\ & - \frac{P_{\text{in}}(\tau)}{E_{\text{sat}}}\{\exp[G(\tau)] - 1\}\end{aligned} \tag{6.16}$$

is obtained. By solving this differential equation, the output pulse information may be obtained using Equations (6.12) and (6.13).

The instantaneous frequency sweep $\Delta\omega$ of the laser pulse after passing through the gain medium is related to $G(\tau)$ by

$$\Delta\omega_{\text{out}}(\tau) = -\frac{\mathrm{d}\phi_{\text{out}}(\tau)}{\mathrm{d}\tau} = \Delta\omega_{\text{in}} + \frac{1}{2}\alpha\frac{\mathrm{d}G(\tau)}{\mathrm{d}\tau} \tag{6.17}$$

If the input laser pulse is non-chirped ($\Delta\omega_{\text{in}} = 0$), $\Delta\omega_{\text{out}}$ may be expressed by

$$\begin{aligned}\Delta\omega_{\text{out}}(\tau) = {} & \frac{\alpha P_{\text{p}}(\tau+\Delta\tau)}{2\gamma E_{\text{psat}}}\left\{1-\exp[\gamma G(\tau) - b(N_{\text{p0}} - N_0)L]\right\} \\ & - \frac{\alpha P_{\text{in}}(\tau)}{2E_{\text{sat}}}\left\{\exp[G(\tau)] - 1\right\}\end{aligned} \tag{6.18}$$

Examine Equation (6.18). The first term on the right-hand side is related to the pulsed optical pumping, which induces XPM to the pulses. The second term is related to the gain saturation, which induces SPM to the pulses. Before going through the detailed numerical evaluation, two extreme cases that have analytical solutions are discussed as follows.

(a) The first case is for no gain saturation, but including the pulsed optical pumping effect. In this case, $E_{\mathrm{in}} \ll E_{\mathrm{sat}}$, where $E_{\mathrm{in}} = \int_{-\infty}^{+\infty} P_{\mathrm{in}}(\tau)\mathrm{d}\tau$ is the input laser pulse energy. Equation (6.16) can be analytically solved so that the frequency sweep $\Delta\omega_{\mathrm{out}}$ of the output pulse will be given by

$$\Delta\omega_{\mathrm{out}}(\tau) = R(\tau + \Delta\tau)\frac{P_{\mathrm{p}}(\tau + \Delta\tau)}{\gamma E_{\mathrm{psat}}} \tag{6.19}$$

using Equation (6.18), where

$$R(\tau + \Delta\tau) = \frac{1}{2}\alpha\frac{\left(e^{bN_{\mathrm{p0}}L} - 1\right)\exp\left[-\frac{E_{\mathrm{p}}(\tau+\Delta\tau)}{E_{\mathrm{psat}}}\right]}{1 + \left(e^{bN_{\mathrm{p0}}L} - 1\right)\exp\left[-\frac{E_{\mathrm{p}}(\tau+\Delta\tau)}{E_{\mathrm{psat}}}\right]} \tag{6.20}$$

and $E_{\mathrm{p}}(\tau + \Delta\tau) = \int_{-\infty}^{\tau} P_{\mathrm{p}}(\tau + \Delta\tau)\mathrm{d}\tau$. The case can be physically pictured in this way. A very weak laser pulse passes through a gain medium pumped by an optical pulse (Figure 6.13(a)). The absorption to the pump causes the variation of carrier density, as shown in Figure 6.13(b). The variation of carrier density causes the variation of refractive index, as shown in Figure 6.13(c), thus inducing the phase modulation to the laser pulse, as shown in Figure 6.13(d), where the laser pulse is assumed to enter the gain medium simultaneously with the pump. This cross-phase modulation (XPM) is solely due to the pulsed pumping. How much the laser pulse phase is affected by the XPM depends on when the laser pulse is launched into the gain medium. In other words, the laser pulse can be positioned anywhere in Figure 6.13(d), but the phase variation of the laser pulse is as it is in the figure. For a synchronously pumped laser in stable operation, a laser pulse usually arrives in the gain region later than a pump pulse ($\Delta\tau > 0$) [172]. The laser pulse will thus become down-chirped because of the XPM.

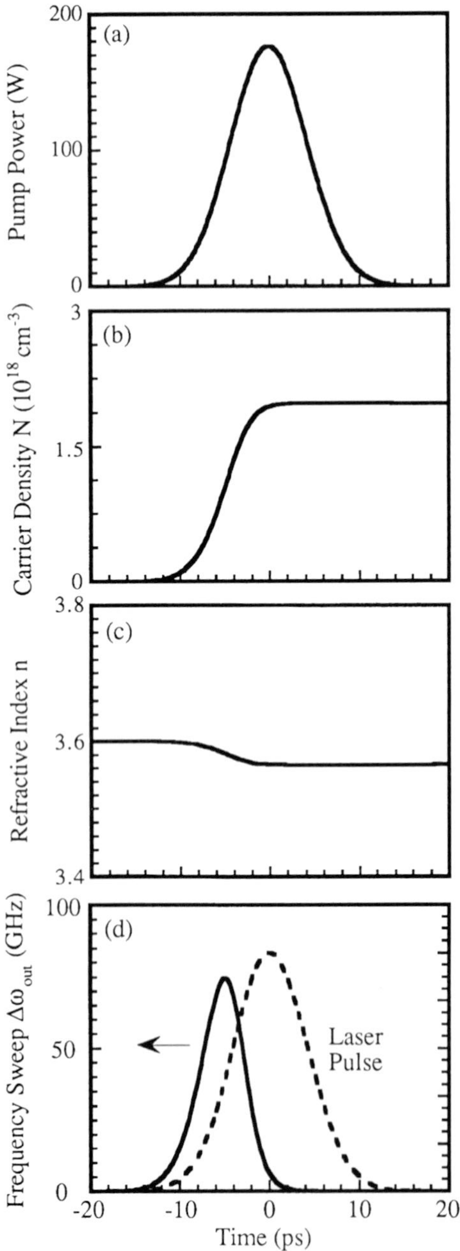

Figure 6.13. XPM induced laser pulse chirping due to pulsed optical pumping if $\Delta\tau = 0$. The pump power is 150 mW at a repetition rate of 80 MHz and its pulse width is 10 ps.

(b) The second case is for the absence of the pump, but with an initial gain G_0. By ignoring the first term in Equation (6.16) and assuming $G_0 \ll 1$,

$$G(\tau) = G_0 \exp\left[-\frac{E(\tau)}{E_{\text{sat}}}\right] \tag{6.21}$$

is obtained, where $E(\tau) = \int_{-\infty}^{\tau} P_{\text{in}}(\tau)\mathrm{d}\tau$. The frequency sweep of the output laser pulse $\Delta\omega_{\text{out}}$ is given by

$$\Delta\omega_{\text{out}}(\tau) = -\frac{1}{2}\alpha G_0 \exp\left[-\frac{E(\tau)}{E_{\text{sat}}}\right]\frac{P_{\text{in}}(\tau)}{E_{\text{sat}}} \tag{6.22}$$

This case is essentially similar to that discussed in Ref. [173]. Physically it can be pictured in this way. A high energy laser pulse (Figure 6.14(a)) passes through a pre-amplified gain medium and creates a carrier density variation, as shown in Figure 6.14(b). The refractive index of the gain medium will vary as shown in Figure 6.14(c). Such an index variation introduces an SPM to the laser pulse, as shown in Figure 6.14(d). The laser pulse thus becomes up-chirped because of the SPM.

Figure 6.15(a) shows the instantaneous frequency sweep $\Delta\omega_{\text{out}}$ of the laser pulse for various input pulse energies using the parameters given in Table 6.1 with $G_0 = a(N_{\text{p}0} - N_0)L$, the initial gain at which the gain medium is transparent to the pump pulse. When $E_{\text{in}} \ll E_{\text{sat}}$, $\Delta\omega_{\text{out}}$ shifts toward the negative following the laser pulse shape, so the chirp will be nonlinear over the central portion of the laser pulse. When $E_{\text{in}} > E_{\text{sat}}$, however, the saturation of the gain will alter the laser phase behavior so that the laser pulse becomes up-chirped over its central portion, as shown by the curve for $E_{\text{in}} = 5$ nJ.

A saturable absorber case worth mentioning is when $G_0 = -aN_0L$, as shown in Figure 6.15(b). The sign of the chirp acquired by the laser pulse from the saturable absorber is opposite to that acquired from the gain medium due to the gain saturation.

Concluding the discussions of (a) and (b), for a pump power of 150 mW and laser pulse energy of 5 nJ, the pulsed optical pumping introduces a down-chirp to the laser pulse, and the gain saturation introduces an up-chirp to the laser pulse. To evaluate the laser pulse chirp quantitatively for the actual case of the modelocked VCSEL, both effects should be lumped together. $G(\tau)$ is first obtained numerically from Equation (6.16) with both of the two right-hand side terms involved. The frequency sweep $\Delta\omega_{\text{out}}$ can then be calculated using Equation (6.18). Under certain

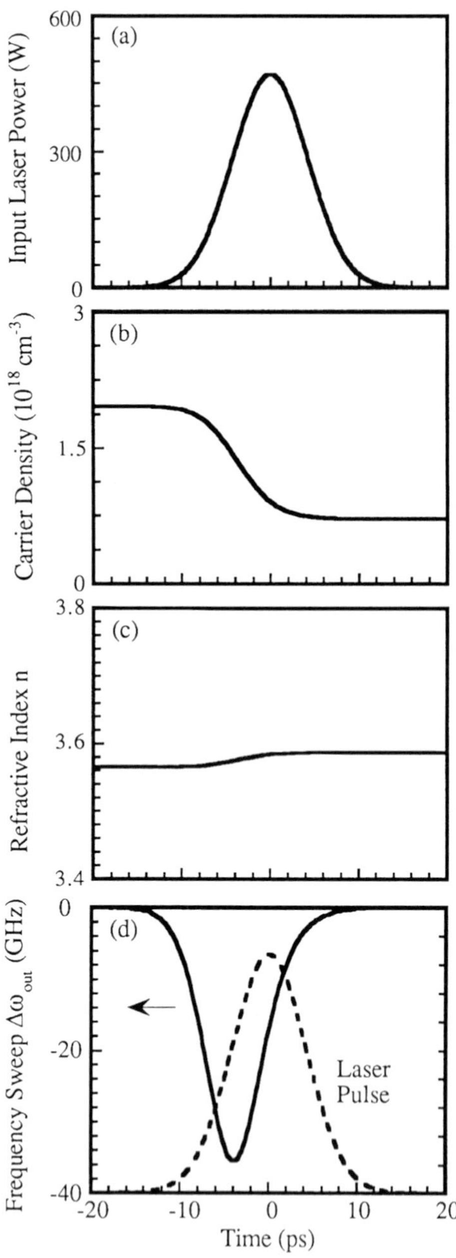

Figure 6.14. SPM induced laser pulse chirping due to gain saturation. The laser has a pulse energy of 5 nJ and a pulse width of 10 ps.

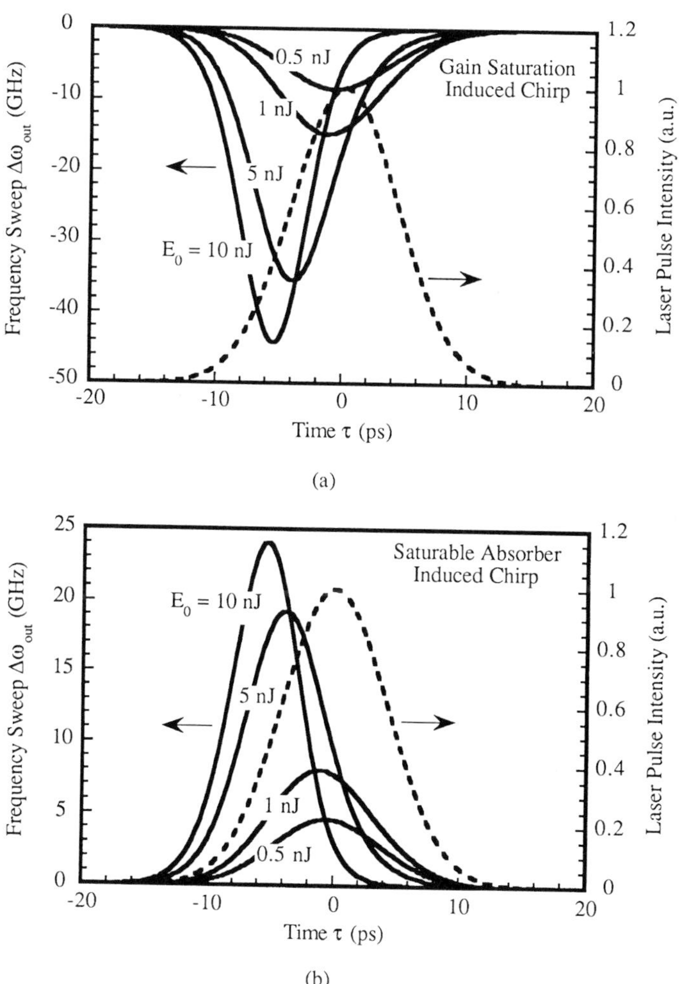

Figure 6.15. The instantaneous frequency sweep $\Delta\omega_{\text{out}}$ of a laser pulse with several different pulse energies after a single pass through a 120 GaAs/AlGaAs multiple quantum well (MQW) sample without the pump, but with an initial gain (a) $G_0 = a(N_{\text{p0}} - N_0)L$, and (b) $G_0 = -aN_0L$.

circumstances, either linear down-chirp or linear up-chirp can be obtained, which is important for laser pulse compression.

As an example, Figure 6.16 shows the instantaneous frequency sweep $\Delta\omega_{\text{out}}$ of a 15 ps pulse with a pulse energy of 5 nJ after passing through a sample with 120 GaAs/AlGaAs multiple quantum wells (MQWs). The pulse energy of 5 nJ corresponds to 12 mW average output power if the

Table 6.1. *Parameters used for modeling*

Pump wavelength	λ_p (nm)	800
VCSEL central wavelength	λ_0 (nm)	860
Effective length of gain medium	L (μm)	1.8
Diameter of active medium	d (μm)	19
Transparent carrier density at ω_0	N_0 (cm^{-3})	6.97×10^{17}
Transparent carrier density at ω_p	N_{p0} (cm^{-3})	1.98×10^{18}
Differential gain at ω_0	a (cm^2)	5.27×10^{-16}
Differential gain at ω_p	b (cm^2)	3.57×10^{-15}
Differential gain ratio	γ	6.77
Intraband relaxation time	T_2 (ps)	0.1
Carrier life time	τ_c (ns)	1
Linewidth enhancement factor	α	5
Saturation energy at ω_0	E_{sat} (nJ)	1.24
Saturation energy at ω_p	E_{psat} (nJ)	0.20
Net cavity loss	Λ	7%
Spontaneous emission energy	E_{sp} (fJ)	50
Spontaneous emission bandwidth	$\Delta\omega_{sp}$ (s^{-1})	50×10^{12}

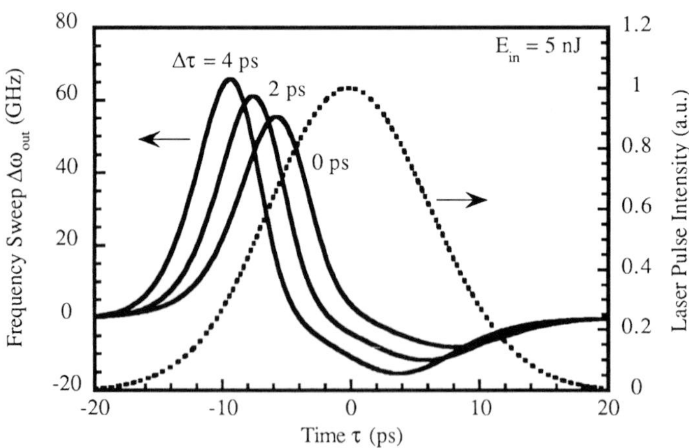

Figure 6.16. The instantaneous frequency sweep $\Delta\omega_{out}$ of a 15 ps pulse with a pulse energy of 5 nJ after a single pass through a 120 GaAs/AlGaAs MQW sample with $\Delta\tau$ as a varying parameter. The pump pulse width is 10 ps and the pump power is 150 mW at a repetition rate of 80 MHz.

coupling mirror has a 97% reflectivity in the modelocked VCSEL [149]. The typical pump power is 150 mW with 10 ps pulse width at 80 MHz repetition rate. The rest of the necessary parameters can be found in Table 6.1. When $\Delta\tau$ is small, the XPM induced pulse chirping due to the pulsed optical pumping dominates. When $\Delta\tau$ is large, the SPM induced pulse chirping due to the gain saturation dominates. An appropriate $\Delta\tau$ will create a linear down-chirp to the central portion of the laser pulses because of the balance between the XPM and the SPM.

The sign of the chirp on a single pass through the laser has a strong dependence on the laser pulse width if the pump pulse width is fixed. Figure 6.17 shows the chirp C and the instantaneous frequency sweep $\Delta\omega_{\text{out}}$ at the pulse peak varying with the laser pulse width. The laser pulse energy is 5 nJ and $\Delta\tau = 0$. The pump pulse width is 10 ps, and pump power is 150 mW. The laser pulse chirp at its peak is negative if the pulse width is longer than 8 ps and becomes positive if the pulse width is shorter. There is a similar tendency over other values of $\Delta\tau$. The chirp C is defined by

$$C = -\frac{\mathrm{d}^2\phi_{\text{out}}(\tau)}{\mathrm{d}\tau^2} \tag{6.23}$$

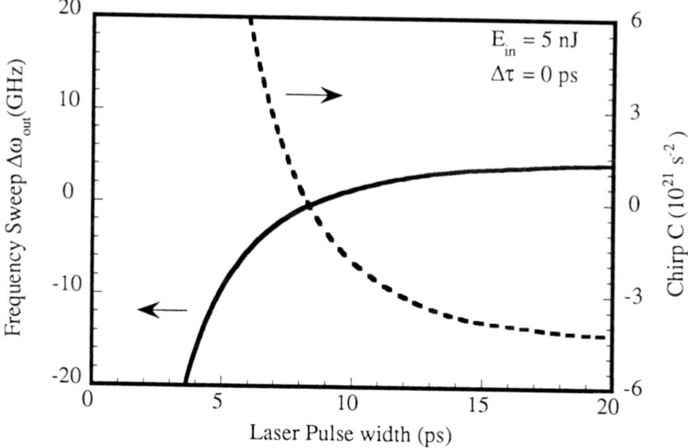

Figure 6.17. The instantaneous frequency sweep $\Delta\omega_{\text{out}}$ of a pulse and its chirp C at the pulse peak after the single pass through a 120 GaAs/AlGaAs MQW sample varying with the laser pulse width. The laser pulse energy is 5 nJ and $\Delta\tau = 0$. The pump pulse width is 10 ps and the pump power is 150 mW at a repetition rate of 80 MHz.

Similar chirp behavior relating to the laser pulse width can be found for the wider pump pulse width.

To gain a knowledge on how much linear group velocity dispersion (GVD) is needed to compensate for the laser pulse chirping because of the gain saturation and the pulsed pumping, the phase of the output pulse in the frequency domain is examined. The phase function $\Phi(\omega)$ in the frequency domain is defined by $\Phi(\omega) = \beta(\omega)L$, where $\beta(\omega)$ is the wave propagation constant. The first derivative of the pulse phase with regard to the frequency $\mathrm{d}\Phi(\omega_0)/\mathrm{d}\omega$ is defined as the group delay. The net effect of this group delay is to create a local time shift $\Delta\tau$ for the whole pulse in the time domain. It can be compensated by varying the laser cavity length. The second derivative of the pulse phase $\mathrm{d}^2\Phi(\omega_0)/\mathrm{d}\omega^2$ is related to the dispersion parameter β_2 by $\beta_2 = \mathrm{d}^2\Phi(\omega_0)/\mathrm{d}\omega^2/L$. Figure 6.18 shows the variation of $\mathrm{d}^2\Phi(\omega_0)/\mathrm{d}\omega^2$ with $\Delta\tau$ for the laser pulse energy of 5 nJ and laser pulse width of 15 ps. Also shown in the figure is the gain $G(L)$ experienced by the laser pulse when it passes through the gain medium. When $\mathrm{d}^2\Phi(\omega_0)/\mathrm{d}\omega^2 > 0$, the pulse is up-chirped, and the pulse is down-chirped when $\mathrm{d}^2\Phi(\omega_0)/\mathrm{d}\omega^2 < 0$. For a laser to reach the lasing threshold, certain cavity gain is needed to overcome the cavity loss. For

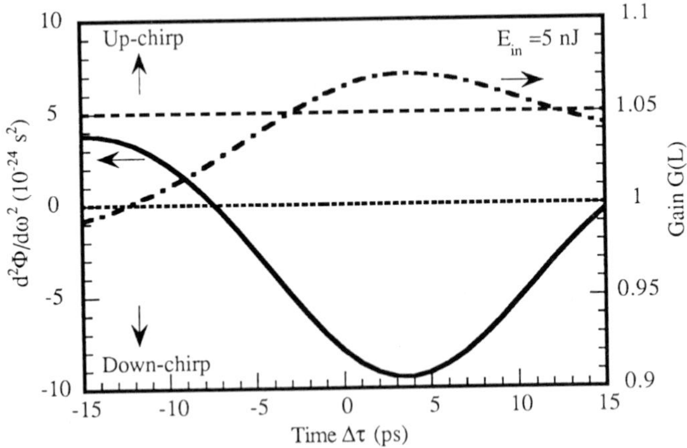

Figure 6.18. The dependence of $\mathrm{d}^2\Phi(\omega_0)/\mathrm{d}\omega^2$ on $\Delta\tau$ after the laser pulse passing through a 120 GaAs/AlGaAs MQW sample (solid line). Also shown in the figure are the gain $G(L)$ experienced by the laser pulse after its pass through the sample (dash-dotted line). The laser pulse energy is 5 nJ and the pulse width is 15 ps. The pump pulse width is 10 ps and the pump power is 150 mW at a repetition rate of 80 MHz. The short dashed line specifies the gain level of 1 and the chirp level of 0. The long dashed line specifies the gain level of 1.05.

example, if the round-trip cavity loss is 5%, a round-trip gain of 1.05 will be needed to sustain the laser operation. Because of this extra gain requirement, the allowance of $\Delta\tau$ will be limited. For the example illustrated here, the laser will be able to operate at certain range of $\Delta\tau > 0$ so that $G(L) > 1.05$. The laser pulse in this region will be down-chirped with $\mathrm{d}^2\Phi(\omega_0)/\mathrm{d}\omega^2$ between $-9 \times 10^{-24} \sim -5 \times 10^{-24}\mathrm{s}^2$. **This value is four orders of magnitude larger than the GVD caused by the intracavity material dispersion or the semiconductor DBR mirror and the dielectric mirrors.**

The magnitude of $\mathrm{d}^2\Phi(\omega_0)/\mathrm{d}\omega^2$ induced by the gain saturation and the pulsed optical pumping may be compared with that generated by a grating-pair. The dispersion from the grating-pair can be expressed by [154]

$$\frac{1}{2}\frac{\mathrm{d}^2\Phi(\omega_0)}{\mathrm{d}\omega^2} = -\frac{4\pi^2 c b_0}{\omega_0^3 a_\mathrm{c}^2 \cos^2\theta_{\mathrm{r}0}} \tag{6.24}$$

where c is the speed of light, b_0 is the center to center distance between the gratings, a_c is the grating constant (line spacing). The diffraction angle $\theta_{\mathrm{r}0}$ is related to the incident angle θ_i by

$$\sin\theta_{\mathrm{r}0} = \frac{2\pi c}{\omega a} - \sin\theta_\mathrm{i} \tag{6.25}$$

Both $\theta_{\mathrm{r}0}$ and θ_i are assumed to be on the same side of the grating normal. If assuming that the grating has 1200 lines/mm, λ is 0.88 μm, and $\cos\theta_{\mathrm{r}0} \sim 1$, $\mathrm{d}^2\Phi(\omega_0)/\mathrm{d}\omega^2 \sim -5 \times 10^{-24}\mathrm{s}^2$ will be equivalent to a b_0 of ~ 1 m. Since the dispersion obtainable from a prism-pair is much smaller than that from a grating-pair, such a large equivalent intracavity dispersion would be difficult to compensate by the usual intracavity prism-pair dispersion compensation method [174] widely adopted in the colliding pulse modelocking (CPM) [175] and Ti:sapphire laser [11] systems, while the high loss of grating-pair compressors [154, 155] prevents them from being used within the VCSEL cavity.

The third order derivative of phase to the frequency $\mathrm{d}^3\Phi(\omega_0)/\mathrm{d}\omega^3$ reveals the nonlinearity of the chirp. Figure 6.19 shows the variation of $\mathrm{d}^3\Phi(\omega_0)/\mathrm{d}\omega^3$ with $\Delta\tau$ for the laser pulse energy of 5 nJ and pulse width of 15 ps. For comparison, $\mathrm{d}^2\Phi(\omega_0)/\mathrm{d}\omega^2$ is also shown in the same figure. Notice that $\mathrm{d}^3\Phi(\omega_0)/\mathrm{d}\omega^3 = 0$ at the position of $\Delta\tau = 1.7$ ps. Physically it means that the chirp around the peak of the laser pulse has the best linearity at this position. The consequence is that the chirp of the output pulse can thus be ultimately compensated by a grating-pair that has the linear GVD.

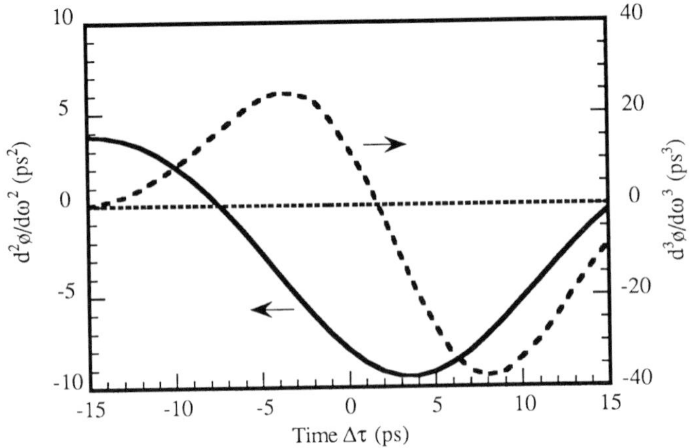

Figure 6.19. The variation of chirp nonlinearity $d^3\Phi(\omega_0)/d\omega^3$ with $\Delta\tau$ for the laser pulse energy of 5 nJ and pulse width of 15 ps (dashed line). The pump pulse width is 10 ps and the pump power is 150 mW at a repetition rate of 80 MHz. The correspondent $d^2\Phi(\omega_0)/d\omega^2$ (solid line) is shown for comparison.

6.3.2 *Gain dispersion*

The chirp due to the SPM and the XPM will broaden the spectrum of the pulse. It will eventually reach such a width that the finite gain shape of the gain material or other magnitude filtering elements may limit its further broadening. The gain saturation and the linewidth enhancement factor α can be ignored during the discussion of the gain dispersion since their role related to the pulse chirping has been included in the discussion of the SPM and XPM. This procedure is similar to the split-step Fourier method [171] that has been used in the analysis of light transmitting in the optical fiber.

The gain shape of the semiconductor materials is assumed Lorentzian. The exact gain shape of the semiconductor materials is somewhat different in that the gain shape is closer to the Gaussian shape on the higher frequency side [176]. After a single pass through the gain medium of length L, the electric field $\tilde{A}(z,\omega)$ can be written in terms of the input $\tilde{A}(0,\omega)$, as given by [151]

$$\tilde{A}(L,\omega) = \tilde{A}(\omega)\exp[i\Phi(\omega)] \tag{6.26}$$

with the real amplitude

$$\tilde{A}(\omega) = \tilde{A}(0,\omega)\exp\left(\frac{\frac{1}{2}g_{\mathrm{p}}L}{1 + T_2^2(\omega - \omega_0)^2}\right) \tag{6.27}$$

and the phase

$$\Phi(\omega) = \frac{\frac{1}{2}g_{\mathrm{p}}LT_2(\omega - \omega_0)}{1 + T_2^2(\omega - \omega_0)^2} \tag{6.28}$$

The peak power gain coefficient g_{p} is expressed by

$$g_{\mathrm{p}} = a(N_{\mathrm{p}0} - N_0) \tag{6.29}$$

and T_2 is a bandwidth parameter. From Equation (6.28), the maximum value of the second derivative of the phase, $\Phi''_{\max}(\omega)$, can be expressed by $\Phi''_{\max} = 0.73 g_{\mathrm{p}} L T_2^2$. Using the data in Table 6.1, $\Phi''_{\max}(\omega)$ is $\sim 1 \times 10^{-27}\mathrm{s}^2$. **This is three orders of magnitude smaller than the $\Phi''(\omega)$ caused by the gain saturation and the pulsed pumping.** In other words, the phase dispersion caused by the finite gain shape will have little effect on the formation of the laser pulse chirping in an modelocked VCSEL. The main effect of the finite gain shape on the laser pulse is its amplitude modulation to the spectrum of the pulses.

6.3.3 *Discussion of laser pulse evolution*

Laser pulse evolution in a laser cavity involves pulse shortening mechanisms and pulse broadening mechanisms. In the frequency domain, it is a balance between spectral broadening and spectral narrowing. In a synchronously pumped laser system, the leading edge of the laser pulse is shortened by the rising of the gain, the trailing edge is confined by saturation of the gain. This gain modulation mechanism causes pulse shortening. In the frequency domain, pulse shortening is represented by spectral broadening. For the modelocked VCSEL, strong SPM and XPM will cause extra spectral broadening, thus the time–bandwidth product will be larger than the Fourier transform limit. With the shortening of the pulse width and the broadening of the spectrum, the dispersive elements in the laser cavity start to act on the laser pulse. These dispersive elements include material dispersion, phase dispersion of gain, DBR structure related dispersion, and other intentionally introduced filters. The dispersion can be a pulse shortening factor or pulse broadening factor, depending on the relative sign between the pulse chirping and the dispersion. The phase dispersion factor is small, however, compared with the SPM and XPM effects in the modelocked VCSEL. One of the dominant pulse broadening or spectral narrowing mechanisms is thus the amplitude modulation to the pulse spectrum from a spectral filter with a finite filter bandwidth within the laser cavity. Carrier transport due to the nonuniform carrier distribution within the gain medium also plays an important role in limiting the laser pulse shortening, as discussed in Section 6.2.2 and Section 6.4.

If an initially unchirped pulse of 15 ps is launched into a GaAs VCSEL, for example, the pulse will become down-chirped due to the gain saturation and the pulsed optical pumping during the first round trip traveling in the cavity. The down-chirped input pulse becomes more down-chirped at the output during the second round trip. This process will continue so that the laser pulse will be shortened and the laser spectrum will be broadened. Interestingly, the down-chirp acquired by the laser pulse during the following round trips in the laser cavity tends to decrease after several passes due to the pulse shortening (Figure 6.17). Two important pulse buildup processes are examined here.

In the first case, the filter bandwidth is not very wide, but the spectrum of the laser pulses has become as wide as that of the spectral width of a Fourier transform limited subpicosecond pulse. The actual pulse width is still more than 8 ps since the pulse is strongly chirped. Further spectral broadening is suppressed by the amplitude modulation of the filter within the cavity. On each additional pass through the gain medium, the laser pulse will acquire extra bandwidth because of the gain modulation, the SPM and the XPM. The acquired extra bandwidth will be curtailed by the filter, so that after a round trip the pulse spectrum remains unchanged. In the time domain, this will be the case that the pulse shortening is balanced by the pulse broadening. The output pulse will be down-chirped. This is the case of the modelocked GaAs VCSEL in Section 6.2.1, and the XPM is a dominant factor.

In the second case, the filter bandwidth is very wide and the gain modulation induced pulse shortening is strong enough to obtain a pulse width shorter than 8 ps. The sign of the laser pulse chirp acquired during each additional pass through the gain medium will thus be positive, as shown in Figure 6.17. This will counteract the down-chirp acquired by the laser pulse in the former round trips so that the laser pulse will soon become up-chirped. As the laser pulse becomes more and more up-chirped, the spectral broadening caused by the gain modulation and the phase modulation will be offset by the filter. The laser pulse width will thus be stabilized, and the stabilized laser pulse will be up-chirped. This case is not able to be reconstructed experimentally with the pump pulse width of 10 ps or shorter. It can be done, however, with a pump pulse width over 100 ps [162]. In that case, a laser pulse of shorter than 30 ps acquires a positive chirp when passing through the gain medium, rather than the down-chirp in the case of the pump pulse width of 10 ps (Figure 6.20). The laser pulse will thus have an up-chirp when it is stabilized in the laser cavity and the SPM is a dominant factor.

According to the above discussion, the laser pulse can be made unchirped with the insertion of a matched filter in the cavity to limit extra bandwidth. Another possibility will be to utilize the down-chirp characteristics of a saturable absorber (Figure 6.15). This approach involves choosing parameters appropriate for the saturable absorber and is only appropriate to compensate up-chirped pulses.

It should be pointed out that in some situations the intracavity filter effect may not be the mechanism in limiting the laser pulse shortening, but the finite carrier transport time within the gain medium. Such a difference, however, does not affect the basic idea of the pulse formation discussion. It is also worth mentioning that the laser chirp is linearly proportional to the linewidth enhancement factor α. The optimum way to decrease the laser chirp is to have a smaller α, which is possible by tuning the laser λ_0.

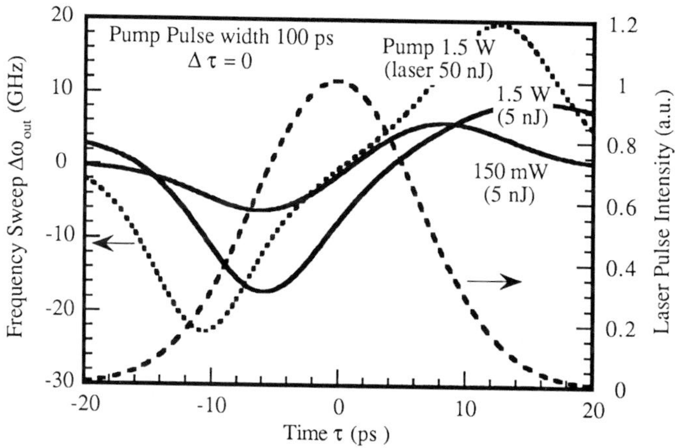

Figure 6.20. The instantaneous frequency sweep $\Delta\omega_{\text{out}}$ of a 15 ps pulse with a pulse energy of 60 nJ or 5 nJ after a single pass through a 120 GaAs/AlGaAs MQW sample at $\Delta\tau = 0$. The pump pulse width is 100 ps and the pump power is 1.5 W, or 150 mW at a repetition rate of 80 MHz. The SPM due to the gain saturation dominates and the laser pulse acquires up-chirp.

6.4 Carrier transport effect on modelocked VCSELs

It was demonstrated in Section 6.2.2 that limiting carrier transport effects with the use of a periodic gain structure was effective at reducing pulse width from a few picoseconds to under 200 fs. In this Section, these carrier and photon dynamics are analyzed.

Examine the standing wave pattern established in the gain medium during the lasing action, as shown in Figure 6.8. Spatial hole-burning occurs because of this standing wave pattern. Those spatial holes are caused by the rapid carrier depletion due to the strong field at those locations. The carriers outside of the high field regions can be depleted by transporting into these spatial holes through thermionic emission, diffusion and tunneling, or by a slow process of spontaneous emission. As an approximation to simplify the simulation process, the gain medium is separated into two parts (Figure 6.21). One part accounts for the gain medium seeing the lasing field, the other part accounts for the gain medium seeing no lasing field. Assuming the volumes of the two regions are equal, the rate equations for the carrier densities in the two regions are given by

$$\begin{aligned}\frac{\partial N_1(z,\tau)}{\partial \tau} = \frac{\alpha_{\mathrm{p}1}(z,\tau)}{\hbar\omega_{\mathrm{p}}\sigma} P_{\mathrm{p}}(z,\tau+\Delta\tau) - \frac{g_1(z,\tau)}{\hbar\omega_0\sigma}|A(z,\tau)|^2 \\ - \frac{N_1(z,\tau)}{\tau_{\mathrm{c}}} + \frac{N_2(z,\tau)-N_1(z,\tau)}{\tau_{\mathrm{d}}}\end{aligned} \tag{6.30a}$$

$$\begin{aligned}\frac{\partial N_2(z,\tau)}{\partial \tau} = \frac{\alpha_{\mathrm{p}2}(z,\tau)}{\hbar\omega_{\mathrm{p}}\sigma} P_{\mathrm{p}}(z,\tau+\Delta\tau) \\ - \frac{N_2(z,\tau)}{\tau_{\mathrm{c}}} - \frac{N_2(z,\tau)-N_1(z,\tau)}{\tau_{\mathrm{d}}}\end{aligned} \tag{6.30b}$$

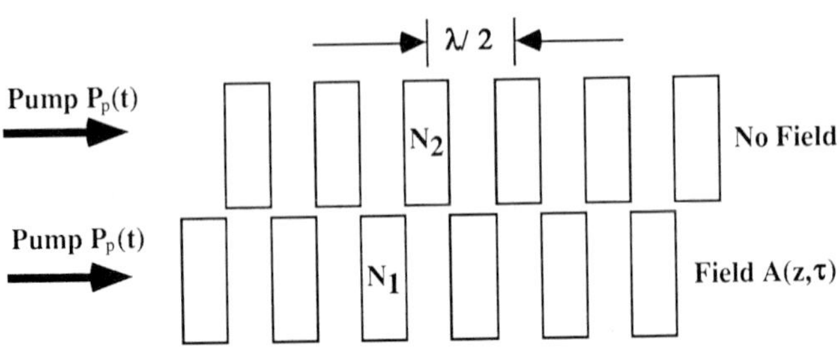

Figure 6.21. The model used to describe the gain in the simulation to include the carrier transport effect due to the spatial hole-burning.

where $N_1(z,\tau)$ is the carrier density in the gain region seeing the lasing field, $N_2(z,\tau)$ is the carrier density in the gain region seeing no lasing field, the gain coefficient $g_1(z,\tau)$ is assumed to be linear with carrier density $N_1(z,\tau)$ as shown by (6.4), $\alpha_{p1}(z,\tau)$ and $\alpha_{p2}(z,\tau)$ are the absorption coefficients to the pump at frequency ω_p in the two regions, respectively, and their relations with the carrier density $N_1(z,\tau)$ and $N_2(z,\tau)$ are given by (6.5), $P_p(z,\tau+\Delta\tau)$ is the pump power as given by (6.3), σ is the active cross-sectional area, τ_c is the carrier lifetime, and ω_0 is the optical central frequency of the lasing field. $A(z,\tau)$ is the normalized lasing field so that $|A(z,\tau)|^2$ stands for the instantaneous laser power. τ_d stands for the effective carrier transport time from the dark field region to the bright field region in the gain medium. Equations (6.30) can be rewritten as

$$\begin{aligned}\frac{\partial g_1(z,\tau)}{\partial\tau} = {} & \frac{\alpha_{p1}(z,\tau)}{\gamma E_{psat}}P_p(z,\tau+\Delta\tau) - \frac{g_1(z,\tau)}{E_{sat}}|A(z,\tau)|^2 - \frac{g_1(z,\tau)}{\tau_c} \\ & - \frac{aN_0}{\tau_c} + \frac{g_2(z,\tau)-g_1(z,\tau)}{\tau_d}\end{aligned} \tag{6.31a}$$

$$\begin{aligned}\frac{\partial g_2(z,\tau)}{\partial\tau} = {} & \frac{\alpha_{p2}(z,\tau)}{\gamma E_{psat}}P_p(z,\tau+\Delta\tau) - \frac{g_2(z,\tau)}{\tau_c} - \frac{aN_0}{\tau_c} \\ & - \frac{g_2(z,\tau)-g_1(z,\tau)}{\tau_d}\end{aligned} \tag{6.31b}$$

where the gain saturation energy E_{sat} and E_{psat} are given by (6.8) and (6.9), γ is defined by (6.10), and the gain coefficient $g_2(z,\tau)$ is related to $N_2(z,\tau)$ as shown by (6.4). Following Section 6.3.1, Equation (6.31) can be integrated over length L, as shown by

$$\begin{aligned}\frac{dG_1(\tau)}{d\tau} = {} & \frac{P_p(\tau+\Delta\tau)}{\gamma E_{psat}}\{1-\exp[\gamma G_1(\tau) - b(N_{p0}-N_0)L]\} \\ & - \frac{P_{in}(\tau)}{E_{sat}}\{\exp[G_1(\tau)]-1\} - \frac{G_1(\tau)}{\tau_c} - \frac{aN_0L}{\tau_c} \\ & + \frac{G_2(\tau)-G_1(\tau)}{\tau_d}\end{aligned} \tag{6.32a}$$

$$\frac{dG_2(\tau)}{d\tau} = \frac{P_p(\tau + \Delta\tau)}{\gamma E_{psat}} \{1 - \exp[\gamma G_2(\tau) - b(N_{p0} - N_0)L]\}$$

$$- \frac{G_2(\tau)}{\tau_c} - \frac{aN_0L}{\tau_c} - \frac{G_2(\tau) - G_1(\tau)}{\tau_d} \tag{6.32b}$$

where $G_1(\tau)$ and $G_2(\tau)$ are defined as in (6.14), L is the length of the gain in the two regions, and $P_{in}(\tau)$ is the input laser power. The initial conditions to solve Equation (6.32) are

$$G_1(-\infty) = -aN_0L \tag{6.33a}$$

$$G_2(-\infty) = -aN_0L \tag{6.33b}$$

The fourth-order Runge-Kutta method is used to integrate Equation (6.32) and to obtain $G_1(\tau)$. The output laser power $P_{out}(\tau)$ and corresponding phase of the field $\phi_{out}(\tau)$ can be expressed by the input power $P_{in}(\tau)$ and phase $\phi_{in}(\tau)$ using $G_1(\tau)$ following Section 6.3.1, as shown by

$$P_{out}(\tau) = P_{in}(\tau)e^{G_1(\tau)} \tag{6.34a}$$

$$\phi_{out}(\tau) = \phi_{in}(\tau) - \frac{1}{2}\alpha G_1(\tau) \tag{6.34b}$$

The laser pulse buildup in the cavity follows the route shown in Figure 6.22. The laser pulse starts from a field $E_1(\tau)$. On the laser pulse traveling route, it is amplified by the gain medium and the relation between $E_2(\tau)$ and $E_1(\tau)$ are governed by Equation (6.34) in the form of their amplitude and their phase. The linear loss comes from the output coupler and whatever optical components are within the laser cavity. If the net intensity loss is Λ,

$$E_3(\tau) = \sqrt{1 - \Lambda}E_2(\tau) \tag{6.35}$$

The gain dispersion is due to the finite gain bandwidth and the relation between $E_4(\tau)$ and $E_3(\tau)$ are governed by Equation (6.26) in the frequency domain. If there is any filter in the cavity, it can be expressed in the frequency domain by $\exp[-(\omega - \omega_{f0})^2/\Delta\omega_f^2]$ in terms of the intensity transmission so that

$$\tilde{E}_5(\omega) = \tilde{E}_4(\omega)\exp\left[-(\omega - \omega_{f0})^2 / (2\Delta\omega_f^2)\right] \tag{6.36}$$

where ω_{f0} is the center frequency of the filter and $\Delta\omega_f$ is the filter bandwidth [172]. The influence of the filter on the pulse phase is neglected.

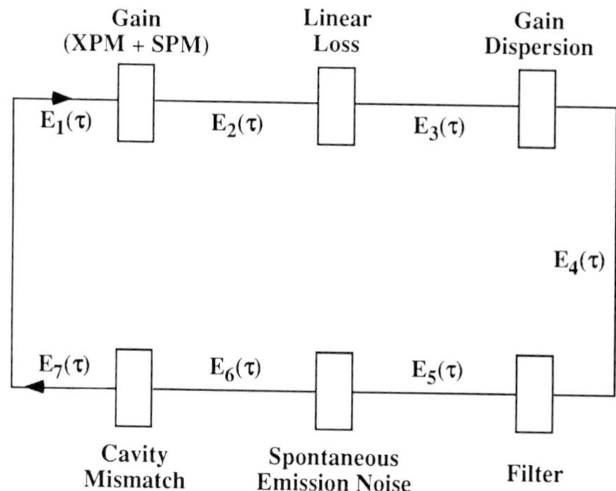

Figure 6.22. Laser pulse traveling route used in the pulse build up modeling, which represents the VCSEL configuration as shown in Figure 6.2.

Spontaneous emission noise usually exists among semiconductor lasers and is taken into account during the simulation [177]. The spontaneous noise can be obtained using a noise generator for the random phase generation, while the amplitude observes the Gaussian distribution with a bandwidth equal to the gain bandwidth in the frequency domain. $\tilde{E}_6(\omega)$ is obtained by adding this noise to $\tilde{E}_5(\omega)$. The VCSEL cavity and the pump laser cavity will not always match in length. The difference in cavity lengths will introduce a time-shift $\Delta\tau$ so that

$$E_7(\tau) = E_6(\tau - \Delta\tau) \tag{6.37}$$

where $\Delta\tau > 0$ if the VCSEL cavity is longer than the pump laser cavity. This concludes one of the many pulse cycles within the laser cavity. The following cycle will start with the output pulse $E_7(\tau)$ to replace the input $E_1(\tau)$ until a steady state is reached.

The modelocked VCSEL starts initially from the spontaneous noise. It takes many round trips for the noise signal to evolve into a laser pulse. If the change in the intensity in one round trip is small,

$$\sqrt{\left|\frac{\sum_j I_j^2(n+1) - \sum_j I_j^2(n)}{\sum_j I_j^2(n)}\right|} \le \delta \tag{6.38}$$

the modelocked VCSEL is considered to have reached the steady state, where $I_j(n)$ is the laser pulse intensity at the jth sampling point after the nth round trip in the cavity. The number of required round trips to reach the steady state depends on the choice of δ. For example, it may take around 350 round trips to have a pulse converge if δ is taken as 0.5%, but it will take around 2950 round trips if δ is 0.1%. Even though the intensity condition in Equation (6.38) is satisfied, the laser pulse may include fine structures that keep changing, as well as its phase. Since the experimentally measurable values are the intensity autocorrelation traces of laser pulses and the power spectra that average over many pulses, only those autocorrelation traces and their power spectra will be examined in the simulation to compare with the experiment. During the simulation, the pump pulse is 200 fs at a wavelength of 800 nm. The average pump power is 200 mW at 89 MHz. Other parameters used for the simulation are listed in Table 6.1. Both the intensity autocorrelation traces and the power spectra of the pulses are obtained by averaging the corresponding pulses over the 1000th to the 1500th round trips. The averaging corresponds to an integration time of 5.6 μs for a laser cavity of 89 MHz.

Although the model used for simulation is simplified, some featured predictions are verified by the experiment. Figure 6.23(a) shows the average autocorrelation traces at a carrier transport time τ_d of 4 ps for several different cavity mismatch times $\Delta\tau$. No intracavity filter is included during this simulation. Most of the pulse energy is contained in the side-lobes or pedestals if the pulse width is narrow. The problem is not so severe for a gain medium without carrier transport, as shown by Figure 6.23(b) with an infinite carrier transport time τ_d. Pulses are generally narrower in Figure 6.23(b) than in Figure 6.23(a). With carrier transport effects, pulses are also noisier, as shown by the corresponding power spectra in Figure 6.24(a), than those pulses without carrier transport, as shown by the corresponding power spectra in Figure 6.24(b). Experiments using a periodic-gain structure in a modelocked VCSEL have supported those predictions [156, 157].

The pulse energy is predicted larger for the case with carrier transport than without carrier transport (Figure 6.25). This is understandable as those carriers transported into the area seeing the strong lasing field will contribute to the gain. The average output power in the experiment for the case with carrier transport is 13 mW [156, 157], which converts to an intracavity pulse energy of 2.9 nJ for a cavity frequency of 89 MHz and an output coupler of 95% reflectivity. This pulse energy is consistent with the simulation in Figure 6.25. The average output power in the experi-

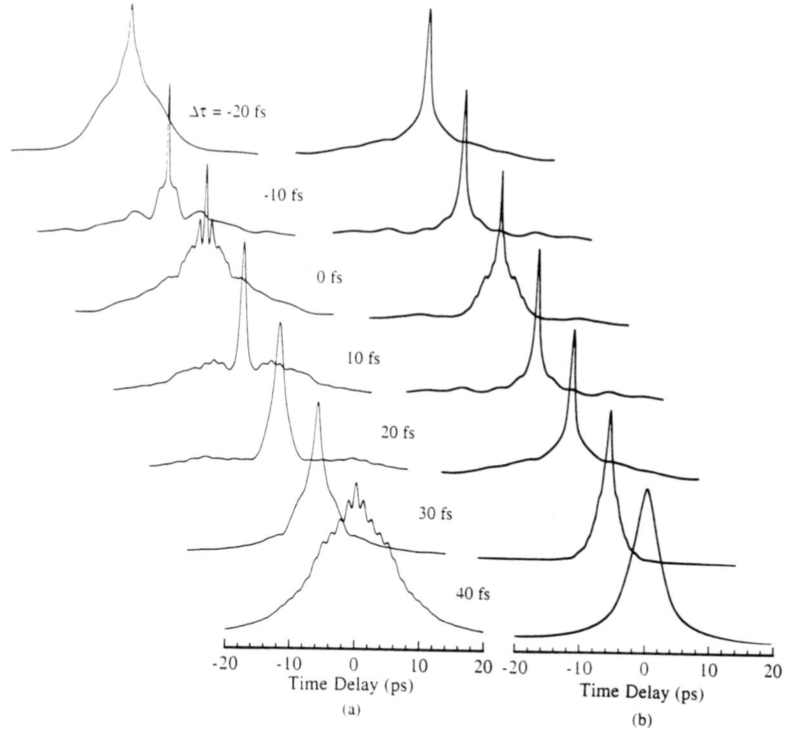

Figure 6.23. Average intensity autocorrelation traces of pulses obtained from simulation for different cavity mismatch times $\Delta\tau$ and without any filter in the cavity (a) at a carrier transport time of 4 ps, and (b) at an infinite carrier transport time.

ment for the case without carrier transport is 6 mW, which converts to an intracavity pulse energy of 2.2 nJ for a cavity frequency of 89 MHz and an output coupler of 97% reflectivity. The pulse energy of 2.2 nJ is slightly larger than the predicted value in Figure 6.25, partly due to the residual carrier transport since the periodic gain structure is not perfect in completely eliminating the carrier transport effect. It may also be due to the fact that the bulk material is used for the periodic gain structure in the experiment instead of the MQW. A bulk semiconductor gain medium has a smaller differential gain, and thus a larger gain saturation energy. The maximum available power from such a laser system will in principle be larger. Even so, the experiment has verified the tendency of a lower pulse energy at a longer carrier transport time.

To examine how effective the model is in simulating the laser pulse formation, it is necessary to examine if the experimental results can be

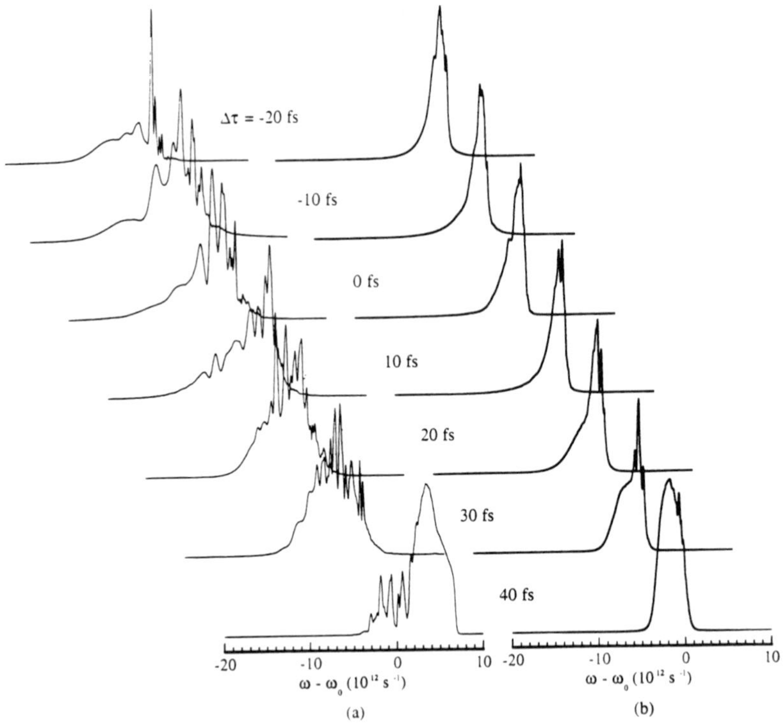

Figure 6.24. Average power spectra of pulses from simulation for different cavity mismatch times $\Delta\tau$ and without any filter in the cavity (a) at a carrier transport time of 4 ps and (b) at an infinite carrier transport time.

reproduced using the model. The comparison of the intensity autocorrelations of pulses between the experiment and the simulation is a straightforward approach. The modelocked laser with a MQW gain medium when pumped by 200 fs pulses from a Ti:sapphire laser was unstable during the experiment. This is also demonstrated in the simulation, as shown by the spectra in Figure 6.24(a). No extensive data were therefore available for comparison in this case. The laser became stable when a birefringent filter with a thickness of 0.5 mm was mounted into the laser cavity to replace a Brewster-angle-placed glass plate. A series of intensity autocorrelation traces of pulses was obtained by varying the laser cavity length and by rotating the filter, as shown by the solid curves in Figure 6.26. Those curves are reproduced in simulation with the filter included. The 0.5 mm filter corresponds to a filter bandwidth $\Delta\omega_f$ of about $48 \times 10^{12}\,\mathrm{s}^{-1}$. The central frequency of the filter can be tuned around

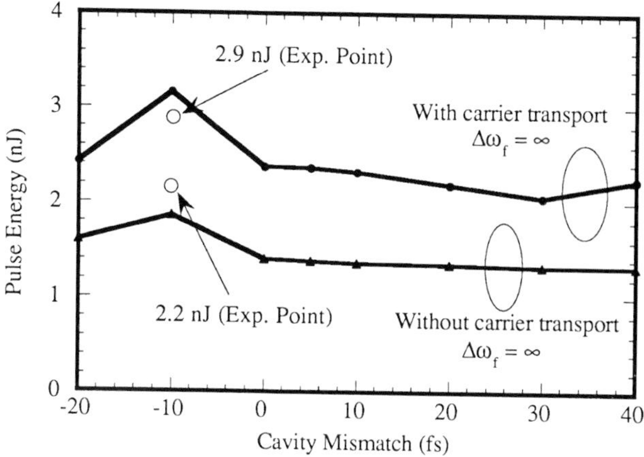

Figure 6.25. Pulse energy varying with cavity mismatch time from simulation for a gain with a carrier transport time of 4 ps and a gain without carrier transport. $\Delta\omega_f$ stands for the filter bandwidth.

the gain peak. To reduce the computing time, only cases involving the filter central frequency of 0 and ± 1.6 THz away from the gain peak are examined, which are shown by the dashed lines in Figure 6.26. In the simulation, every 10 fs cavity mismatch time corresponds to a cavity mismatch length of 1.5 μm. The central frequency of the filter is tuned 1.6 THz larger than the gain peak in Figure 6.26(a). In Figure 6.26(b), the central frequency of the filter is aligned to the gain peak, and the central frequency of the filter is 1.6 THz smaller than the gain peak in Figure 6.26(c). The results from the simulation agree well with the experiment in terms of the pulse width. The difference in the details of the pulse sub-structure is also within the acceptable range. Better agreement can be expected if finer filter tuning is used in simulation.

It has to be pointed out that the model used for the simulation has neglected any effects of two-photon absorption, spectral hole-burning and carrier heating. The time constant for those effects is typically within 300 fs. Those effects must be taken into account to predict the laser pulse behavior more precisely [178] when the laser pulse width falls into that regime. The model thus needs to be improved to fully reproduce the experimental results when using the periodic-gain structure in the modelocked VCSEL to generate pulses with a pulse width of shorter than 300 fs. Nevertheless, the agreement between measured pulses and calculated results is generally very good. This model can be used for designing optimized VCSELs.

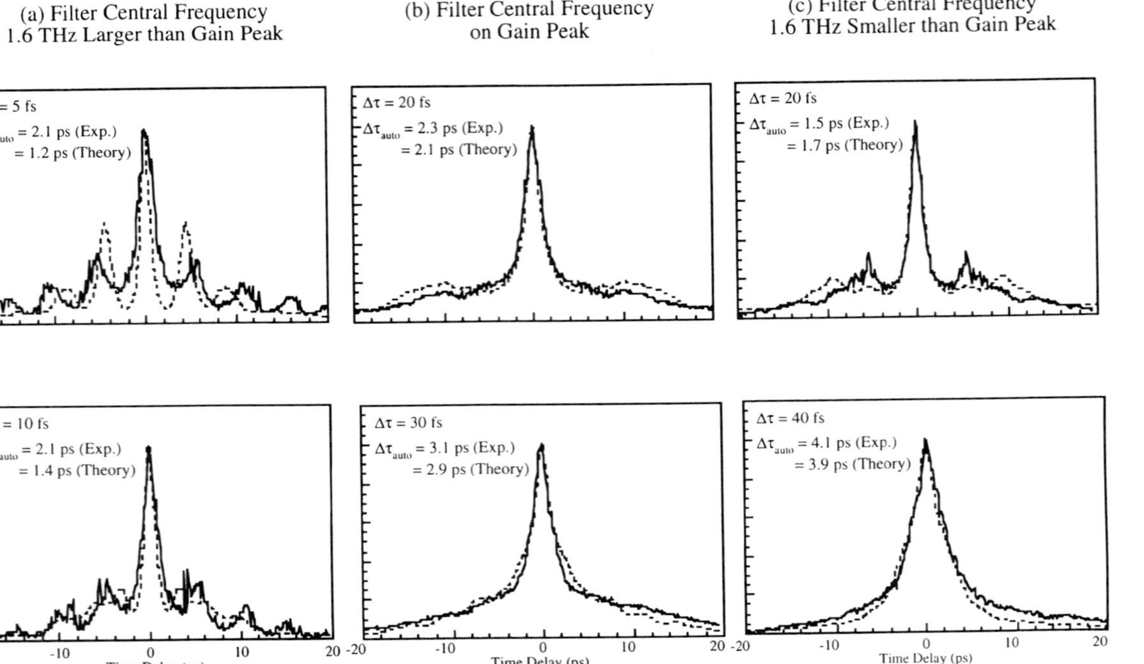

Figure 6.26. Intensity autocorrelation traces obtained from the experiment (solid lines) using a MQW gain medium in the modelocked VCSEL and a birefringent filter of 0.5 mm in the cavity for the power spectrum control. The dashed lines are obtained using the model including a carrier transport time of 4 ps. The filter central frequency is (a) 1.6 THz larger than the gain peak, (b) aligned with the gain peak, and (c) 1.6 THz smaller than the gain peak.

6.5 Electrically pumped modelocked semiconductor lasers

Optically pumped VCSELs have been modelocked to generate subpicosecond laser pulses with peak powers over 100 W. The requirement for a compact short pulse source makes it important that a modelocked VCSEL be electrically pumped. This section will discuss the first actively modelocked electrically pumped VCSEL [179].

6.5.1 *Gain consideration*

The MQW gain structure, as well as the periodic gain structure designed for optical pumping, is appropriate for electrical pumping only if the cavity loss is low and the required number of quantum wells is small, less than 20. It is difficult for injected carriers to cross over hundreds of quantum wells, as in the optically pumped lasers. Since the external cavity configuration of the modelocked VCSEL requires a certain gain thickness to overcome the intracavity loss, bulk semiconductor materials are used as the active medium. As the first attempt for the experiment in the electrically pumped modelocked VCSEL, the gain thickness was chosen to be 1 μm.

To determine the doping of the active layer, broad area stripe lasers were fabricated out of those samples with different dopants. The lowest threshold current density and the highest differential quantum efficiency were observed from the stripe lasers with n-doped active layers.

Figure 6.27 schematically shows the completed device structure. A distributed Bragg reflector (DBR) mirror alternated with quarter wavelength thick n-doped ($1 \times 10^{18}\ cm^{-3}$) AlAs and $Ga_{0.9}Al_{0.1}As$ is grown on an n-GaAs substrate. The interface of the two materials is linearly graded and selectively heavily doped to smooth the heterojunction, and thus to decrease the series resistance [180]. The measurement shows that the DBR has a reflectivity centered around 825 nm with a bandwidth of 60 nm, which is good for the device to operate at 100 K. On top of the DBR mirror is a $Ga_{0.7}Al_{0.3}As$ n-cladding of 0.5 μm thick with doping concentration of $1 \times 10^{18}\ cm^{-3}$. The active layer is 1 μm thick with n-doping of $2 \times 10^{17}\ cm^{-3}$ to ease current spreading. In order that holes can be uniformly injected into the active region transversely under the ring p-contact, a p^+p^-p structure is used as the p-cladding. The p-layer of $Ga_{0.7}Al_{0.3}As$ is 0.5 μm thick and doped at $1 \times 10^{18} cm^{-3}$. The p^--layer and p^+-layer of $Ga_{0.7}Al_{0.3}As$ are both 0.7 μm thick and doped at $2 \times 10^{17}\ cm^{-3}$ and $2 \times 10^{18}\ cm^{-3}$, respectively. Holes can thus uniformly

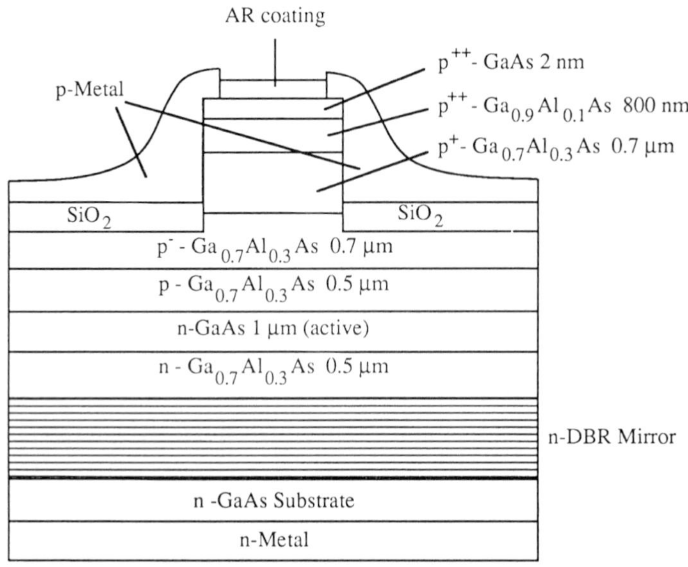

Figure 6.27. The schematic diagram of completed device structure for an electrically pumped modelocked VCSEL.

spread within the p^+ layer because of the higher resistive p^- layer. An observation of an almost circular uniform near-field of the spontaneous emission shows that the p^+p^-p structure for p-cladding does work to uniformly spread the current transversely in the active region in spite of the ring p-metal contact. The p-contact layer of $Ga_{0.9}Al_{0.1}As$ is heavily doped at 2×10^{19} cm^{-3} with a thickness of 800 nm and a 2 nm GaAs cap layer is grown to protect the layer of $Ga_{0.9}Al_{0.1}As$ from oxidizing. The I-V characteristics measurement shows that the completed devices have typically a contact resistance of 65 Ω.

6.5.2 *Modelocked vertical cavity lasers*

An external cavity configuration is used for modelocking. The device is mounted on a copper cold finger that is placed in a dewar maintained at liquid N_2 temperature. The temperature of the cold finger is −172°C during the laser operation. Both sides of the dewar window are AR coated. The monolithic DBR mirror forms one end of the laser. The external cavity laser is finished by an output coupler and an intracavity 19-mm-focal-length lens that has the same AR coating as that of the dewar window. The laser cavity round-trip length is 31.25 cm, which corresponds to a modelocking frequency of 0.96 GHz. This is the twelfth harmonic of the trigger rate of 80 MHz of the synchronous-scan streak camera. When a square pulsed current is applied to the device, the laser starts to lase at the current of 15 mA with a 1% output coupler. Figure 6.28 shows the L–I curves of this laser with output couplers of three different reflectivities from 95% to 99%. The current pulse width is 2 μs and the repetition rate of the current source can be increased to 100 kHz without degrading the laser operation. Device heating occurs when the repetition rate goes over 100 kHz or the current pulse width increases, which prohibits the CW operation of this device. The laser wavelength is between 825 nm and 830 nm, depending on the injected current level. The laser beam has a circular TEM_{00} transverse mode for all three cases. The roll-off in curve (c) is due to the device heating. The maximum external differential quantum efficiency is 9.2 %.

Active modulation on top of the DC bias is used to actively modelock the electrically pumped GaAs VCSEL. An amplified RF signal at 0.96 GHz from a synthesized RF signal generator is used to drive a step recovery diode (SRD). The negative electrical pulse of 200 ps from the SRD is sent into a broadband pulse inverter and is then applied onto the VCSEL device through a microwave bias-tee. The modelocked pulse width from the VCSEL is measured with a synchronous-scan streak camera that is triggered at 80 MHz. The quasi-DC bias has a pulse width of 2 μm at the repetition rate variable from 100 Hz to 100 kHz.

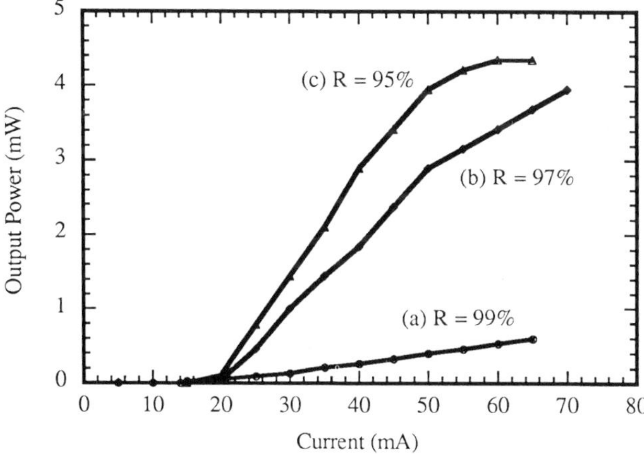

Figure 6.28. L–I characteristics of the external-cavity VCSEL for three different output couplers with reflectivities from 95% to 99%.

Figure 6.29 shows a typical streak camera pulse width measurement, which is conducted with the streak camera integration time at 100 ms and a device quasi-DC bias of 30 mA at the repetition rate of 50 kHz. The pulse width is 122 ps. A wider pulse width value is measured if the streak camera integration time is set longer than 100 ms. For example, the measured pulse width is 137 ps at an integration time of 193 ms. This is because the measurement averages over many pulses during one scan. Timing jitter between pulses is included during the averaging. More pulses are averaged during the measurement for longer streak camera integration time, and thus more timing jitter influence is added onto the measured pulse width. The severe timing jitter is partly due to the fact that the quasi-DC bias pulses and the modulation pulses are not synchronized.

To decrease the timing jitter influence on the pulse width measurement, we need to decrease the streak camera integration time, so that the number of pulses to be averaged per measurement can be decreased. Single-shot streak camera measurements would be preferred [167]. Since the specified minimum integration time of the synchronous-scan streak camera is 100 ms, the repetition rate of the quasi-DC bias may be decreased to further decrease the number of pulses to be averaged per measurement. For example, the measured pulse width is 81 ps with the repetition rate of the quasi-DC bias at 1 kHz. The corresponding power spectrum shows a spectral width of 0.5 nm. Less timing jitter influence in the measurement contributes to the decreased pulse width. Note that

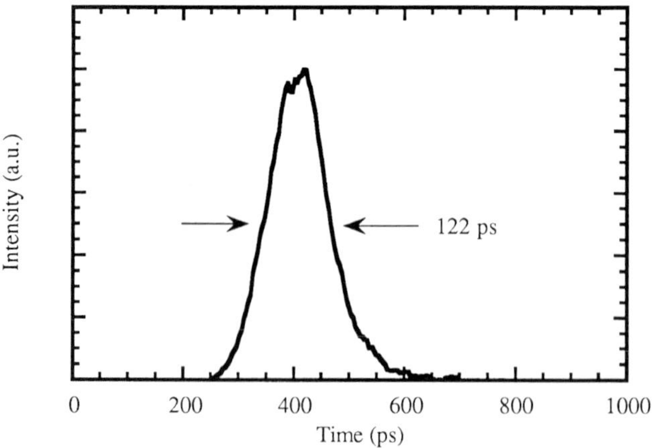

Figure 6.29. Pulses at a quasi-DC bias of 50 kHz measured with a synchronous-scan streak camera at an integration time of 100 ms.

the modelocked pulse build-up time is around 1 μs, and the quasi-DC bias has a pulse width of 2 μs. The synchronous-scan streak camera measurement also averages all the pulses including the initial portion of the pulses. For this reason, the measured pulse width of 81 ps is still wider than the actual stabilized laser pulse width. A single-shot measurement should be able to identify this process and to determine the stabilized pulse width more precisely. To fundamentally solve the problem, however, it is necessary to run the laser CW modelocked, which is prohibited by device heating at present.

The laser pulse width is dependent on the quasi-DC bias level. When the quasi-DC bias is increased, the pulse width is increased, and the pulse energy is increased as well. Figure 6.30 shows the pulse widths and the corresponding output pulse energy for several different quasi-DC biases at 50 kHz, measured with the shortest possible streak camera integration time. The maximum output pulse energy is 4.6 pJ at a bias of 60 mA. The output coupler has a reflectivity of 95%, thus the maximum intracavity pulse energy is 92 pJ. The gain saturation energy is 840 pJ for a gain diameter of 15 μm with a Γ of unity. Since the ultimate limitation to the intracavity pulse energy is the gain saturation energy E_{sat}, at least a factor of nine increase in the intracavity pulse energy should be expected from this modelocked VCSEL.

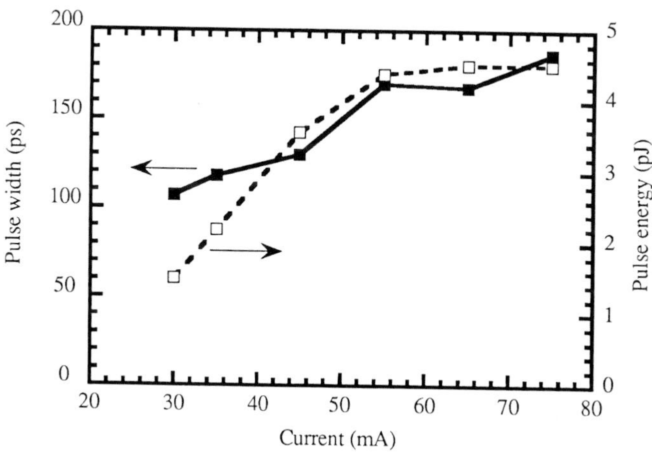

Figure 6.30. Pulse width as a function of the quasi-DC bias at a repetition rate of 50 kHz (solid line) and the corresponding output laser pulse energy (dashed line).

6.6 Conclusions

Vertical cavity surface emitting semiconductor lasers have been modelocked with synchronous optical pumping. A GaAs/AlGaAs MQW surface emitting laser is modelocked to generate 10 ps pulses at room temperature when pumped by a modelocked dye laser. Pulses generated from the modelocked VCSEL are severely chirped due to the XPM caused by pulsed optical pumping and the SPM caused by the gain saturation. After chirp compensation using a grating-pair-telescope configuration outside of the laser cavity, the pulse width is reduced to 320 fs at a peak power of 64 W. An intracavity birefringent filter is effective in limiting the time–bandwidth product of the pulses directly from the laser oscillator. When the laser with the MQW gain is pumped by a modelocked Ti:sapphire laser, the output pulse width is reduced to around 1 ps. Carrier transport is responsible in preventing shorter pulse generation from the modelocked VCSEL. A periodic gain structure is used to eliminate the source of carrier transport. The periodic gain structure also acts as a spectral filter to curb the pulse chirping and increases the effective differential gain. The laser pulse width with its intensity autocorrelation FWHM of 300 fs is obtained, which is one of the shortest reported in any modelocked semiconductor laser. This pulse width is ultimately limited by the carrier heating and spectral hole-burning in the gain medium if only an active modelocking technique is used. Generation of shorter pulses directly from the modelocked VCSELs relies on the implementation of passive modelocking techniques.

Vertical cavity surface emitting lasers with InGaAs/InP gain have been modelocked at 77 K when synchronously pumped by a modelocked 1.32 μm YAG laser. The output laser pulse width is around 5–30 ps before chirp compensation. The pulse width is reduced to 150 fs at a peak power over 3 kW after a single stage chirp compensation. The second stage pulse compression reduces the pulse width to 21 fs.

The results from the modelocked VCSELs have clearly demonstrated that the surface emitting lasers can be modelocked to generate picosecond or even subpicosecond pulses at much higher output power than that of edge-emitting in-plane semiconductor lasers. Another advantage of the modelocked VCSELs is the good laser beam quality.

An optically pumped modelocked VCSEL works fine as a lab tool, but is not very cost effective compared with an electrically pumped in-plane semiconductor laser. It is possible that a future modelocked VCSEL will be pumped by an arrayed edge-emitting diode laser. The laser will then be passively modelocked with a semiconductor saturable absorber. In some sense, a modelocked VCSEL will work like a coherent beam converter that converts a transverse-mode uncorrelated laser beam from the arrayed laser into a correlated beam. Since semiconductor materials can be tailored to satisfy different wavelength requirements, the modelocking method will be an efficient way to build a compact short pulse source for various applications.

One of the advantages of semiconductor lasers is the capability of being electrically pumped. An electrically pumped actively modelocked vertical cavity surface emitting laser has been successfully demonstrated. To make the device work CW modelocked at room temperature, the device structure design needs to be improved for better heat dissipation and higher quantum efficiency. Reports on monolithic VCSELs have shown that high power-conversion efficiency of 14.9% can be achieved with a better p-DBR mirror design [117]. Greater than 100 mW CW output power at room temperature has also been achieved with a p-side down device mounted on a diamond heat sink for better heat dissipation [115]. An electrically pumped CW modelocked VCSEL operating at room temperature should thus be promising if those techniques are implemented. The excellent beam quality and the potential high output power from such a laser should make it an attractive candidate for a compact short laser pulse source.

References

[1] L. E. Hargrove, R. L. Fork and M. A. Pollack, 'Locking of He-Ne laser modes induced by synchronous intracavity modulation,' *Appl. Phys. Lett.*, **5**, 4 (1964).

[2] H. W. Mocker and R. J. Collins, 'Mode competition and self-locking effects in a Q-switched ruby laser,' *Appl. Phys. Lett.*, **7**, 270 (1965).

[3] A. J. De Maria, D. A. Stetser and H. Heynau, 'Self mode-locking of lasers with saturable absorbers,' *Appl. Phys. Lett.*, **8**, 174 (1966).

[4] C. V. Shank and E. P. Ippen, 'Subpicosecond kilowatt pulses from a mode-locked cw dye laser,' *Appl. Phys. Lett.*, **24**, 373 (1974).

[5] I. S. Ruddock and D. J. Bradley, 'Bandwidth-limited subpicosecond pulse generation in mode-locked cw dye lasers,' *Appl. Phys. Lett.*, **29**, 296 (1976).

[6] J. C. Diels, E. W. Van Stryland and G. Benedict, 'Generation and measurement of pulses of 0.2 psec duration,' *Opt. Comm.*, **25**, 93 (1978).

[7] R. L. Fork, B. I. Greene and C. V. Shank, 'Generation of optical pulses shorter than 0.1 ps by colliding pulse mode-locking,' *Appl. Phys. Lett.*, **38**, 671 (1981).
[8] L. F. Mollenauer and R. S. Stolen, 'The soliton laser,' *Opt. Lett.*, **9**, 13 (1984).
[9] K. J. Blow and B. P. Nelson, 'Improved mode locking of an F-center laser with a nonlinear nonsoliton external cavity,' *Opt. Lett.*, **13**, 1026 (1988).
[10] D. E. Spence, P. N. Kean and W. Sibbett, '60-fsec pulse generation from a self-mode-locked Ti-sapphire laser,' *Opt. Lett.*, **16**, 42 (1991).
[11] N. Sarukura, Y. Ishida and H. Nakano, 'Generation of 50-fsec pulses from a pulse-compressed, cw, passively mode-locked Ti-sapphire laser,' *Opt. Lett.*, **16**, 153 (1991).
[12] M. T. Asaki, C. P. Huang, D. Garvey, J. P. Zhou, H. C. Kapteyn and M. M. Murnane, 'Generation of 11-fs pulses from a self-mode-locked Ti-sapphire laser,' *Opt. Lett.*, **18**, 977 (1993).
[13] R. N. Hall, G. E. Fenner, J. D. Kingsley, T. J. Soltys and R. O. Carlson, 'Coherent light emission from GaAs junctions,' *Phys. Rev. Lett.*, **9**, 366 (1962).
[14] M. I. Nathan, W. P. Dumke, G. Burns, F. H. Dills and G. Lasher, 'Stimulated emission of radiation from GaAs p-n junctions,' *Appl. Phys. Lett.*, **1**, 62 (1962).
[15] N. Holonyak and S. F. Bevacqua, 'Coherent (visible) light emission from $Ga(As_{1-x}P_x)As$ junctions,' *Appl. Phys. Lett.*, **1**, 82 (1962).
[16] T. M. Quist, R. J. Keyes, W. E. Krag, B. Lax, A. L. McWhorter, R. H. Rediker and H. J. Zeiger, 'Semiconductor maser of GaAs,' *Appl. Phys. Lett.*, **1**, 91 (1962).
[17] T. H. Maiman, 'Stimulated optical radiation in ruby,' *Nature*, **187**, 493 (1960).
[18] E. P. Harris, 'Spiking in current-modulated CW GaAs external cavity lasers,' *J. Appl. Phys.*, **42**, 892 (1971).
[19] V. N. Morozov, V. V. Nikitin, and A. A. Sheronov, 'Self-synchronization of modes in a GaAs semiconductor injection laser,' *JETP Lett.* **7**, 256 (1968).
[20] P. T. Ho, L. A. Glasser, E. P. Ippen and H. A. Haus, 'Picosecond pulse generation with a CW GaAlAs laser diode,' *Appl. Phys. Lett.*, **33**, 242 (1978).
[21] L. A. Glasser, 'CW mode locking of a GaInAsP diode laser,' *Electron. Lett.*, **14**, 725 (1978).
[22] H. Ito, H. Yokoyama, S. Murata and H. Inaba, 'Picosecond optical pulse generation from an R. F. modulated AlGaAs D. H. diode laser,' *Electron. Lett.*, **15**, 738 (1979).
[23] J. P. van der Ziel and R. M. Mikulyak, 'Mode locking of strip buried heterostructure (AlGa)As lasers using an external cavity,' *J. Appl. Phys.*, **51**, 3033 (1980).
[24] E. P. Ippen, D. J. Eilenberger and R. W. Dixon, 'Picosecond pulse generation by passive mode locking of diode lasers,' *Appl. Phys. Lett.*,

37, 267 (1980); 'Picosecond pulse generation with diode lasers,' *Picosecond Phenomena II*, p. 21 (Ed. R. M. Hochstrasser, W. Kaiser, and C. V. Shank, Springer-Verlag, 1980).

[25] H. Yokoyama, H. Ito and H. Inaba, 'Generation of subpicosecond coherent optical pulses by passive mode locking of an AlGaAs diode laser,' *Appl. Phys. Lett.*, **40**, 105 (1982).

[26] P. T. Ho, 'Coherent pulse generation with a GaAlAs laser by active mode locking,' *Electron. Lett.*, **15**, 527 (1979).

[27] H. Ito, H. Yokoyama and H. Inaba, 'Bandwidth limited picosecond optical pulse generation from actively mode-locked AlGaAs diode laser,' *Electron. Lett.*, **16**, 620 (1980).

[28] J. P. van der Ziel, 'Active mode locking of double heterostructure lasers in an external cavity,' *J. Appl. Phys.*, **52**, 4435 (1981).

[29] J. C. AuYeung and A. R. Johnston, 'Picosecond pulse generation from a synchronously pumped mode-locked semiconductor laser diode,' *Appl. Phys. Lett.*, **40**, 112 (1982).

[30] A. Olsson and C. L. Tang, 'Active mode locking of linear and ring external-cavity semiconductor lasers,' *IEEE J. Quantum Electron.*, **QE-17**, 1977 (1981).

[31] J. P. van der Ziel, W. T. Tsang, R. A. Logan, R. M. Mikulyak and W. M. Augustyniak, 'Subpicosecond pulses from passively mode-locked GaAs buried optical guide semiconductor lasers,' *App. Phys. Lett.*, **39**, 525 (1981).

[32] P. A. Morton, R. J. Helkey and J. E. Bowers, 'Dynamic detuning in actively mode locked semiconductor lasers,' *IEEE J. Quantum Electron.*, **25**, 2621 (1989).

[33] J. Chesmoy, M. C. Klein, L. Chusseau and J. M. Lourtioz, 'Period doubling and period quadrupling for an actively mode locked laser diode with extended cavity,' *J. Appl. Phys.*, **67**, 7615 (1990).

[34] M. B. Holbrook, W. E. Sleat and D. J. Bradley, 'Bandwidth-limited picosecond pulse generation in an actively mode-locked GaAlAs diode laser,' *Appl. Phys. Lett.*, **37**, 59 (1980).

[35] D. J. Bradley, M. B. Holbrook and W. E. Sleat, 'Bandwidth-limited picosecond pulses from an actively mode-locked GaAlAs diode laser,' *IEEE J. Quantum Electron.*, **QE-17**, 658 (1981).

[36] J. Chen, W. Sibbett and J. I. Vukusic, 'Tunable mode-locked semiconductor lasers incorporating Brewster-angled diodes,' *Opt. Comm.*, **48**, 427 (1984).

[37] J. T. K. Chang and J. I. Vukusic, 'Active mode locking of InGaAsP Brewster angled semiconductor lasers,' *IEEE J. Quantum Electron.*, **QE-23**, 1329 (1987).

[38] A. Mar, D. J. Derickson, R. J. Helkey, J. E. Bowers, R. T. Huang and D. Wolf, 'Actively mode-locked external-cavity semiconductor lasers with transform-limited single pulse output,' *Opt. Lett.*, **17**, 868 (1992).

[39] R. J. Helkey, W. X. Zou, A. Mar, D. B. Young and J. E. Bowers, 'Curved and tapered waveguide mode-locked InGaAs/AlGaAs

semiconductor lasers fabricated by impurity induced disordering,' *51st Annual Device Research Conference, IIB-4* (Santa Barbara, CA, 1993).

[40] D. J. Derickson, R. J. Helkey, A. Mar, J. R. Karin, J. E. Bowers and R. L. Thornton, 'Suppression of multiple pulse formation in external cavity mode locked semiconductor lasers using intrawaveguide saturable absorbers,' *IEEE Photon. Tech. Lett.*, **4**, 333 (1992).

[41] A. J. Lowery and I. W. Marshall, 'Stabilization of mode locked pulses using traveling wave semiconductor laser amplifier,' *Electron. Lett.*, **26**, 104 (1990).

[42] M. Serenyi, J. Kuhl and E. O. Gobel, 'Pulse shortening of actively mode-locked diode lasers by wavelength tuning,' *Appl. Phys. Lett.*, **50**, 1213 (1987).

[43] Yu. L. Bessonov, A. P. Bogatov, P. P. Vasil'ev, V. N. Morozov and A. B. Sergeev, 'Generation of picosecond pulses in an injection laser with an external selective resonator,' *Sov. J. Quantum Electron.*, **12**, 1510 (1982).

[44] S. Lundqvist, T. Andersson and S. T. Eng, 'Generation of tunable single-mode picosecond pulses from an AlGaAs semiconductor laser with grating feedback,' *Appl. Phys. Lett.*, **43**, 715 (1983).

[45] A. P. Bogatov, P. P. Vasil'ev, V. N. Morozov and A. B. Sergeev, 'Direct detection of picosecond pulses emitted by an injection laser with active mode locking,' *Sov. J. Quantum Electron.*, **13**, 1303 (1984).

[46] J. E. Epler, G. S. Jackson, N. Holonyak, Jr., M. Weinstein, R. D. Burnham and T. L. Paoli, 'Mode-locked coupled-stripe quantum well laser operation ($\lambda \sim 7350$Å) in a tunable ($\Delta\hbar\omega \sim 37$meV $> kT$) external grating cavity,' *Appl. Phys. Lett.*, **47**, 1022 (1985).

[47] P. P. Vasil'ev, V. N. Morozov, G. T. Pak, Yu. M. Popov and A. B. Sergeev, 'Measurement of the frequency shift of a picosecond pulse from a mode-locked injection laser,' *Sov. J. Quantum Electron.*, **15**, 859 (1985).

[48] P. Kempf and B. K. Garside, 'Dynamics of mode-locked laser diodes employing a repetitive short pulse drive current,' *Appl. Opt.*, **26**, 4522 (1987).

[49] D. M. Bird, R. M. Fatah, M. K. Cox, P. D. Constantine, J. C. Regnault and K. H. Cameron, 'Miniature packaged actively mode locked semiconductor laser with tunable 20 ps transform limited pulses,' *Electron. Lett.*, **26**, 2086 (1990).

[50] A. S. Hou, R. S. Tucker, and G. Eisenstein, 'Pulse compression of an actively modelocked diode laser using linear dispersion in fiber,' *IEEE Photon. Tech. Lett.*, **2**, 322 (1990).

[51] M. S. Ozyazici and M. S. Demokan, 'Actively mode locked 1.3 and 1.55 μm InGaAsP diode lasers,' *Opt. and Quantum Electron.*, **23**, 1169 (1991).

[52] J. M. Wiesenfeld, M. Kuznetsov and A. S. Hou, 'Tunable, picosecond pulse generation using a compressed, modelocked laser diode source,' *IEEE Photon. Tech. Lett.*, **2**, 319 (1990).

[53] R. S. Tucker, G. Eisenstein and I. P. Kaminow, '10 GHz active mode-locking of a 1.3 μm ridge-waveguide laser in an optical-fiber cavity,' *Electron. Lett.*, **19**, 552 (1983).

[54] R. S. Tucker, S. K. Korotky, G. Eisenstein, U. Koren, L. W. Stulz and J. J. Veselka, '20 GHz active mode-locking of a 1.55 μm InGaAsP laser,' *Electron. Lett.*, **21**, 239 (1985).

[55] G. Eisenstein, R. S. Tucker, U. Koren and S. K. Korotky, 'Active mode-locking characteristics of InGaAsP-single mode fiber composite cavity lasers,' *IEEE J. Quantum Electron.*, **QE-22**, 142 (1986).

[56] M. Kuznetsov, D. Z. Tsang, J. N. Walpole, Z. L. Liau and E. P. Ippen, 'Multistable mode locking of InGaAsP semiconductor lasers,' *Appl. Phys. Lett.*, **51**, 895 (1987).

[57] S. W. Corzine, J. E. Bowers, G. Przybylek, U. Koren, B. I. Miller and C. E. Soccolich, 'Actively mode-locked GaInAsP laser with subpicosecond output,' *Appl. Phys. Lett.*, **52**, 348 (1988).

[58] J. E. Bowers, P. A. Morton, A. Mar and S. W. Corzine, 'Actively mode locked semiconductor lasers,' *IEEE J. Quantum Electron.*, **QE-25**, 1426 (1989).

[59] N. Onodera, Z. Ahmed, R. S. Tucker and A. J. Lowery, 'Stability of harmonically driven mode locked semiconductor lasers,' *Appl. Phys. Lett.*, **59**, 3527 (1991).

[60] S. Sanders, A. Yariv, J. Paslaski, J. E. Ungar and H. A. Zarem, 'Passive mode locking of a two section multiple quantum well laser at harmonics of the cavity round-trip frequency,' *Appl. Phys. Lett.*, **58**, 681 (1991).

[61] S. Akiba, G. E. Williams and H. A. Haus, 'High rate pulse generation from InGaAsP laser in Selfoc lens external resonator,' *Electron. Lett.*, **17**, 527 (1981).

[62] Y. Tada, H. Yokoyama, H. Ito and H. Inaba, 'Ultrashort optical pulse generation from microwave modulated AlGaAs diode laser with Selfoc rod resonator,' *Opt. Comm.*, **47**, 187 (1983).

[63] T. E. Dimmick, P. T. Ho and G. L. Burdge, 'Coherent pulse generation by active modelocking of a GaAlAs laser in a Selfoc lens extended resonator,' *Electron. Lett.*, **20**, 831 (1984).

[64] G. Eisenstein, R. S. Tucker, I. P. Kaminow, T. P. Lee and C. A. Burrus, 'InGaAsP 1.3 μm optical amplifier modulator integrated with a fibre-resonator mode-locked laser,' *Electron. Lett.*, **20**, 624 (1984).

[65] G. Eisenstein, R. S. Tucker, S. K. Korotky, U. Koren, J. J. Veselka, L. W. Stulz, R. M. Jopson and K. L. Hall, 'Active mode-locking of an InGaAsP 1.55 μm laser in a fibre resonator with an integrated single-mode-fibre output port,' *Electron. Lett.*, **21**, 173 (1985).

[66] K. Y. Lau and A. Yariv, 'Direct modulation and active mode locking of ultrahigh speed GaAlAs laser at frequencies up to 18 GHz,' *Appl. Phys. Lett.*, **46**, 326 (1985).

[67] C. Harder, J. S. Smith, K. Y. Lau and A. Yariv, 'Passive mode locking of buried heterostructure lasers with nonuniform current injection,' *Appl. Phys. Lett.*, **42**, 772 (1983).

[68] K. Y. Lau, I. Ury and A. Yariv, 'Passive and active mode locking of a semiconductor laser without an external cavity,' *Appl. Phys. Lett.*, **46**, 1117 (1985).

[69] K. Y. Lau, 'Efficient narrow band direct modulation of semiconductor injection lasers at millimeter wave frequencies of 100 GHz and beyond,' *Appl. Phys. Lett.*, **52**, 2214 (1988).

[70] K. Y. Lau, 'Short pulse and high frequency signal generation in semiconductor lasers,' *J. Lightwave Tech.*, **7**, 400 (1989).

[71] K. Y. Lau, 'Narrow band modulation of semiconductor lasers at millimeter wave frequencies (> 100 GHz) by mode locking,' *IEEE J. Quantum Electron.*, **26**, 250 (1990).

[72] S. Sanders, L. Eng, J. Paslaski and A. Yariv, '108 GHz passive mode locking of a multiple quantum well semiconductor laser with an intracavity absorber,' *Appl. Phys. Lett.*, **56**, 310 (1989).

[73] R. S. Tucker, U. Koren, G. Raybon, C.A. Burrus, B. I. Miller, T. L. Koch and G. Eisenstein, '40 GHz active mode locking in a 1.5 μm monolithic extended cavity laser,' *Electron. Lett.*, **25**, 621 (1989).

[74] P. P. Vasil'ev and A. B. Sergeev, 'Generation of bandwidth limited 2 ps pulses with 100 GHz repetition rate from multisegmented injection laser,' *Electron. Lett.*, **25**, 1049 (1989).

[75] P. A. Morton, J. E. Bowers, L. A. Koszi, M. Soler, J. Lopata and D. P. Wilt, 'Monolithic hybrid mode locked 1.3 μm semiconductor lasers,' *Appl. Phys. Lett.*, **56**, 111 (1990).

[76] S. Sanders, L. Eng and A. Yariv, 'Passive mode locking of monolithic InGaAs/AlGaAs double quantum well lasers at 42 GHz repetition rate,' *Electron. Lett.*, **26**, 1087 (1990).

[77] S. Sanders, L. Eng, J. Paslaski and A. Yariv, '108 GHz passive mode locking of a multiple quantum well semiconductor laser with an intracavity absorber,' *Appl. Phys. Lett.*, **56**, 310 (1990).

[78] M. C. Wu, Y. K. Chen, T. Tanbun-Ek, R. A. Logan, M. A. Chin and G. Raybon, 'Transform-limited 1.4 ps optical pulses from a monolithic colliding-pulse mode-locked quantum well laser,' *Appl. Phys. Lett.*, **57**, 759 (1990).

[79] Y. K. Chen, M. C. Wu, T. Tanbun-Ek, R. A. Logan and M. A. Chin, 'Subpicosecond monolithic colliding-pulse mode-locked multiple quantum well lasers,' *Appl. Phys. Lett.*, **58**, 1253 (1991).

[80] P. G. May and M. Bierbaum, 'Monolithic mode locking of long cavity GaAs-AlGaAs semiconductor lasers,' *IEEE Photon. Tech. Lett.*, **3**, 296 (1991).

[81] J. H. Zarrabi, E. L. Portnoi and A. V. Chelnokov, 'Passive mode locking of a multistripe single quantum well GaAs laser diode with an intracavity saturable absorber,' *Appl. Phys. Lett.*, **59**, 1526 (1991).

[82] D. J. Derickson, R. J. Helkey, A. Mar, J. R. Karin, J. G. Wasserbauer and J. E. Bowers, 'Short pulse generation using multisegment mode locked semiconductor lasers,' *IEEE J. Quantum Electron.*, **QE-28**, 2186 (1992).

[83] P. B. Hansen, G. Raybon, M. D. Chien, U. Doren, B. I. Miller, M. G. Young, J. M. Verdiell and C. A. Burrus, 'A 1.54 μm monolithic semiconductor ring laser: cw and mode locked operation,' *IEEE Photon. Tech. Lett.*, **4**, 411 (1992).

[84] D. J. Derickson, P.A. Morton, J. E. Bowers and R. L. Thornton, 'Comparison of timing jitter in external and monolithic cavity mode locked semiconductor lasers,' *Appl. Phys. Lett.*, **59**, 3372 (1991).

[85] R. J. Helkey, D. J. Derickson, A. Mar, J. G. Wasserbauer, J. E. Bowers, and R. L. Thornton, 'Repetition frequency stabilisation of passively mode-locked semiconductor lasers,' *Electron. Lett.*, **28**, 1920 (1992).

[86] J. P. Ryan, L. S. Goldberg and D. J. Bradley, 'Comparison of synchronous pumping and passive mode-locking of cw dye lasers for the generation of picosecond and subpicosecond light pulses,' *Opt. Comm.*, **27**, 127 (1978).

[87] Y. Ishida, T. Yajima and K. Naganuma, 'Generation of broadly tunable subpicosecond light pulses from a synchronously and passively mode-locked cw dye laser,' *Japan. J. Appl. Phys.*, **19**, L717 (1980).

[88] J. P. van der Ziel, R. A. Logan and R. M. Mikulyak, 'Generation of subpicosecond pulses from an actively mode locked GaAs laser in an external ring cavity,' *Appl. Phys. Lett.*, **39**, 867 (1981).

[89] P. P. Vasil'ev, V. N. Morzov, Y. M. Popov and A. B. Sergeev, 'Subpicosecond pulse generation by a tandem-type AlGaAs DH laser with colliding pulse mode locking,' *IEEE J. Quantum Electron.*, **QE-22**, 149 (1986).

[90] J. P. van der Ziel and R. A. Logan, 'Dispersion of the group velocity refractive index in GaAs double heterostructure lasers,' *IEEE J. Quantum Electron.*, **QE-19**, 164 (1983).

[91] Y. Silberberg and P. W. Smith, 'Subpicosecond pulses from a mode-locked semiconductor laser,' *IEEE J. Quantum Electron.*, **QE-22** , 759 (1986).

[92] J. Kuhl, M. Serenyi and E. O. Göbel, 'Bandwidth-limited picosecond pulse generation in an actively mode-locked GaAs laser with intracavity chirp compensation,' *Opt. Lett.*, **12**, 334 (1987).

[93] G. M. Carter, L. Zheng and K. Y. Huang, 'Compression of pulses from a mode locked GaAs laser diode in an extended cavity with a fiber grating reflector,' *Appl. Phys. Lett.*, **61**, 379 (1992).

[94] C. F. Lin and C. L. Tang, 'Colliding pulse mode locking of a semiconductor laser in an external ring cavity,' *Appl. Phys. Lett.*, **62**, 1053 (1993).

[95] J. R. Andrews and R. D. Burnham, 'High peak power and gateable picosecond optical pulses from a diode array traveling-wave amplifier and a mode-locked diode laser,' *Appl. Phys. Lett.*, **49**, 1004 (1986).

[96] P. J. Delfyett, C. H. Lee, G. A. Alphonse and J. C. Connolly, 'High peak power picosecond pulse generation from AlGaAs external cavity

mode locked semiconductor laser and traveling-wave amplifier,' *Appl. Phys. Lett.*, **57**, 971 (1990).

[97] P. J. Delfyett, C. H. Lee, L. T. Florez, N. G. Stoffel, T. J. Gmitter, N. C. Andredakis, G. A. Alphonse and J. C. Connolly, 'Generation of subpicosecond high-power optical pulses from a hybrid mode-locked semiconductor laser,' *Opt. Lett.*, **15**, 1371 (1990).

[98] P. J. Delfyett, L. Florez, N. Stoffel, T. Gmitter, N. Andreadakis, G. Alphonse and W. Ceislik, '200-fs optical pulse generation and intracavity pulse evolution in a hybrid mode-locked semiconductor diode-laser/amplifier system,' *Opt. Lett.*, **17**, 670 (1992).

[99] J. P. van der Ziel, H. Temkin, R. D. Dupuis and R. M. Mikulyak, 'Mode-locked picosecond pulse generation from high power phase-locked GaAs laser arrays,' *Appl. Phys. Lett.*, **44**, 357 (1984).

[100] H. Masuda and A. Takada, 'Picosecond optical pulse generation from mode locked phased laser diode array,' *Electron. Lett.*, **25**, 1418 (1989).

[101] J. C. Kuo, C. S. Chang and C. L. Pan, 'Buildup of steady state picosecond pulses in an actively mode locked laser diode array,' *Opt. Lett.*, **16**, 1328 (1991).

[102] L. Y. Pang, J. G. Fujimoto and E. S. Kintzer, 'Ultrashort pulse generation from high power diode arrays by using intracavity optical nonlinearities,' *Opt. Lett.*, **17**, 1599 (1992).

[103] A. Mar, R. Helkey, T. Reynolds and J. Bowers, 'Mode-locked multi-segment resonant-optical-waveguide diode laser arrays,' *IEEE/LEOS Summer Topical Meeting on Optical Microwave Interactions*, PD2 (Santa Barbara, CA, 1993).

[104] L. F. Mollenauer, B. M. Nyman, M. J. Neubelt, G. Raybon and S. G. Evangelides, 'Demonstration of soliton transmission at 2.4 Gbit/s over 12000 km,' *Electron. Lett.*, **27**, 178 (1991).

[105] A. J. Taylor, J. M. Wiesenfeld, G. Eisenstein, R. S. Tucker, J. R. Talman and U. Koren, 'Electro-optic sampling of fast electrical signals using an InGaAsP injection laser,' *Electron. Lett.*, **22**, 61 (1986).

[106] P. A. Morton, R. J. Helkey, A. Mar, D. J. Derickson and J. E. Bowers, 'Monolithic mode locked laser arrays in optical computing,' *SPIE Digital Optical Computing II*, **1215**, 94 (1990).

[107] W. R. Christian and M. J. Rosker, 'Picosecond pulsed diode ring-laser gyroscope,' *Opt. Lett.*, **16**, 1587 (1991).

[108] H. Soda, K. Iga, C. Kitahara and Y. Suematsu, 'GaInAsP/InP surface emitting injection lasers,' *Japan. J. Appl. Phys.*, **18**, 2329 (1979).

[109] K. Iga, F. Koyama and S. Kinoshita, 'Surface emitting semiconductor lasers,' *IEEE J. Quantum Electron.*, **QE-24**, 1845 (1988).

[110] M. Orenstein, A. C. Von Lehmen, C. Chang-Hasnain, N. G. Stoffel, J. P. Harbison and L. T. Florez, 'Matrix addressable vertical cavity surface emitting laser array,' *Electron. Lett.*, **27**, 437 (1991).

[111] D. B. Young, J. W. Scott, F. H. Peters, B. J. Thibeault, S. W. Corzine, M. G. Peters, S. L. Lee and L. A. Coldren, 'High-power temperature-insensitive gain-offset InGaAs/GaAs vertical-cavity surface-emitting lasers,' *IEEE Photon. Tech. Lett.*, **5**, 130 (1993).

[112] K. Tai, G. Hasnain, J. D. Wynn, R. J. Fischer, Y. H. Wang, B. Weir, J. Gamelin and A. Y. Cho, '90% coupling of top surface emitting GaAs/AlGaAs quantum well laser output into 8 μm diameter core silica fibre,' *Electron. Lett.*, **26**, 1628 (1990).

[113] K. Iga, S. Ishikawa, S. Ohkouchi and T. Nishimura, 'Room temperature pulsed oscillation of GaAlAs/GaAs surface emitting laser,' *Appl. Phys. Lett.*, **45**, 348 (1984).

[114] F. Koyama, S. Kinoshita and K. Iga, 'Room-temperature cw operation of GaAs vertical cavity surface emitting laser,' *Trans. IEICE*, **E71**, 1089 (1988).

[115] F. H. Peters, M. G. Peters, D. B. Young, J. W. Scott, B. J. Thibeault, S. W. Corzine and L. A. Coldren, 'High power vertical cavity surface emitting lasers,' *Electron. Lett.*, **29**, 200 (1993).

[116] R. S. Geels and L. A. Coldren, 'Submilliamp threshold current vertical-cavity laser diodes,' *Appl. Phys. Lett.*, **57**, 1605 (1990).

[117] M. G. Peters, F. H. Peters, D. B. Young, J. W. Scott, B. J. Thibeault and L. A. Coldren, 'High wallplug efficiency vertical-cavity surface-emitting lasers using lower barrier DBR mirrors,' *Electron. Lett.*, **29**, 170 (1993).

[118] K. Iga, 'Surface emitting lasers,' *Opt. Quantum Electron.*, **24**, S97 (1992).

[119] T. Baba, Y. Yogo, K. Suzuki, F. Koyama and K. Iga, 'Near room temperature continuous wave lasing characteristics of GaInAsP/InP surface emitting laser,' *Electron. Lett.*, **29**, 913 (1993).

[120] J. J. Dudley, D. I. Babic, R. Mirin, R. J. Ram, T. Reynolds, E. L. Hu, J. E. Bowers, L. Yang and B. I. Miller, 'Low threshold, electrically injected InGaAsP (1.3 μm) vertical cavity lasers on GaAs substrates,' *51st Annual Device Research Conference*, IIIB-8 (Santa Barbara, CA, 1993).

[121] S. Uchiyama and K. Iga, 'Two-dimensional array of GaInAsP/InP surface-emitting lasers,' *Electron. Lett.*, **21**, 162 (1985).

[122] E. Ho, F. Koyama and K. Iga, 'Effective reflectivity from self-imaging in a talbot cavity and on the threshold of a finite 2-D surface emitting laser array,' *Appl. Opt.*, **29**, 5080 (1990).

[123] J. Stone, C.A. Burrus and J. C. Campbell, 'Laser action in photopumped GaAs ribbon whiskers,' *J. Appl. Phys.*, **51**, 3038 (1980).

[124] M. A. Duguay, T. C. Damen, J. Stone, J. M. Wiesenfeld and C.A. Burrus, 'Picosecond pulses from an optically pumped ribbon-whisker laser,' *Appl. Phys. Lett.*, **37**, 369 (1980).

[125] T. C. Damen, M. A. Duguay, J. Shah, J. Stone, J. M. Wiesenfeld and R. A. Logan, 'Broadband tunable picosecond semiconductor lasers,' *Appl. Phys. Lett.*, **39**, 142 (1981).

[126] J. Stone, J. M. Wiesenfeld, A. G. Dentai, T. C. Damen, M. A. Duguay, T. Y. Chang and E. A. Caridi, 'Optically pumped ultrashort cavity $In_{1-x}Ga_xAs_yP_{1-y}$ lasers: picosecond operation between 0.83 and 1.59 μm,' *Opt. Lett.*, **6**, 534 (1981).

[127] J. M. Wiesenfeld and J. Stone, 'Picosecond pulse generation in optically pumped, ultrashort-cavity, InGaAsP, InP, and InGaAs film lasers,' *IEEE J. Quantum Electron.*, **QE-22**, 119 (1986).

[128] J. M. Wiesenfeld and J. Stone, 'Chirp in picosecond film lasers and pulse compression by linear dispersion in optical fibers,' *Opt. Lett.*, **8**, 262 (1983).

[129] D. Marcuse and J. M. Wiesenfeld, 'Chirped picosecond pulses: evaluation of the time-dependent wavelength for semiconductor film lasers,' *Appl. Opt.*, **23**, 74 (1984).

[130] J. R. Karin, L. G. Melcer, R. Nagarajan, J. E. Bowers, S. W. Corzine, P.A. Morton, R. S. Geels and L. A. Coldren, 'Generation of picosecond pulses with a gain-switched GaAs surface-emitting laser,' *Appl. Phys. Lett.*, **57**, 963 (1990).

[131] L. G. Melcer, J. R. Karin, R. Nagarajan and J. E. Bowers, 'Picosecond dynamics of optical gain switching in vertical cavity surface emitting lasers,' *IEEE J. Quantum Electron.*, **27**, 1417 (1991).

[132] A. Mukherjee, M. Mahbobzadeh, C.F. Schaus, and S. R. J. Brueck, 'Ultrafast operation of optically pumped resonant periodic gain GaAs surface emitting lasers,' *IEEE Photon. Tech. Lett.*, **2**, 857 (1990).

[133] J. Lin, J. K. Gamelin, S. Wang, M. Hong and J. P. Mannaerts, 'Short pulse generation by electrical gain switching of vertical cavity surface emitting laser,' *Electron. Lett.*, **27**, 1957 (1991).

[134] J. Lin, J. K. Gamelin, K. Y. Lau and S. Wang, 'Ultrafast (up to 39 GHz) relaxation oscillation of vertical cavity surface emitting laser,' *Appl. Phys. Lett.*, **60**, 15 (1992).

[135] J. L. Jewell, Y. H. Lee, A. Scherer, S. L. McCall, N. A. Olsson, J. P. Harbison and L. T. Florez, 'Surface-emitting microlasers for photonic switching and interchip connections,' *Opt. Eng.*, **29**, 210 (1990).

[136] F. S. Choa, Y. H. Lee, T. L. Koch, C. A. Burrus, B. Tell, J. L. Jewell and R. E. Leibenguth, 'High-speed modulation of vertical-cavity surface-emitting lasers,' *IEEE Photon. Tech. Lett.*, **3**, 697 (1991).

[137] G. Hasnain, K. Tai, N. K. Dutta, Y. H. Wang, J. D. Wynn, B. E. Weir and A. Y. Cho, 'High temperature and high frequency performance of gain-guided surface emitting lasers,' *Electron. Lett.*, **27**, 915 (1991).

[138] P. Pepeljugoski, J. Lin, J. Gamelin, M. Hong and K. Y. Lau, 'Ultralow timing jitter in electrically gain-switched vertical cavity surface emitting lasers,' *Appl. Phys. Lett.*, **62**, 1588 (1993).

[139] G. Hasnain, J. M. Wiesenfeld, T. C. Damen, J. Shah, J. D. Wynn, Y. H. Wang and A. Y. Cho, 'Electrically gain-switched vertical-cavity surface-emitting lasers,' *IEEE Photon. Tech. Lett.*, **4**, 6 (1992).

[140] D. Tauber, G. Wang, R. S. Geels, J. E. Bowers and L. A. Coldren, 'Large and small signal dynamics of vertical cavity surface emitting lasers,' *Appl. Phys. Lett.*, **62**, 325 (1993).

[141] G. Wang, R. Nagarajan, D. Tauber and J. Bowers, 'Reduction of damping in high-speed semiconductor lasers,' *IEEE Photon. Tech. Lett.*, **5**, 642 (1993).

[142] W. L. Cao, A. M. Vacher and C. H. Lee, 'Synchronously pumped mode-locked GaAs laser,' *Appl. Phys. Lett.*, **38**, 653 (1981).

[143] A. M. Vaucher, W. L. Cao, J. D. Ling and C. H. Lee, 'Generation of tunable picosecond pulses from a bulk GaAs laser,' *IEEE J. Quantum Electron.*, **QE-18**, 187 (1982).

[144] C. B. Roxlo and M. M. Salour, 'Synchronously pumped mode-locked CdS platelet laser,' *Appl. Phys. Lett.*, **38**, 738 (1981).

[145] C. B. Roxlo and M. M. Salour, 'Dewar design for optically pumped semiconductor lasers,' *Rev. Sci. Instrum.*, **53**, 458 (1982).

[146] C. B. Roxlo, R. S. Putnam and M. M. Salour, 'Optically pumped semiconductor platelet lasers,' *IEEE J. Quantum Electron.*, **QE-18**, 338 (1982).

[147] J. Yorsz, S. G. Shevel and E. P. Ippen, 'Optically pumped Zn_xCdS platelet lasers,' *Opt. Comm.*, **48**, 139 (1983).

[148] R. S. Putnam and M. M. Salour, 'Modelocked picosecond pulses from 490 nm to 2 μm with optically pumped semiconductor lasers,' *Proc. SPIE*, **439**, 66 (1983).

[149] W. B. Jiang, R. Mirin and J. E. Bowers, 'Mode-locked GaAs vertical cavity surface emitting lasers,' *Appl. Phys. Lett.*, **60**, 677 (1992).

[150] D. C. Sun, S. R. Friberg, K. Watanabe, S. Machida, Y. Horikoshi and Y. Yamamoto, 'High power and high efficiency vertical cavity surface emitting GaAs laser,' *Appl. Phys. Lett.*, **61**, 1502 (1992).

[151] W. B. Jiang, D. J. Derickson and J. E. Bowers, 'Analysis of laser pulse chirping in mode-locked vertical cavity surface emitting lasers,' *J. Quantum Electron.*, **QE-29**, 1309 (1993).

[152] J. A. Giordmaine, M. A. Duguay and J. W. Hansen, 'Compression of optical pulses,' *IEEE J. Quantum Electron.*, **QE-4**, 252 (1968).

[153] R. A. Fisher and J. A. Fleck, 'On the phase characteristics and compression of picosecond pulses,' *Appl. Phys. Lett.*, **15**, 287 (1969); R. A. Fisher, P. L. Kelley and T. K. Gustafson, 'Subpicosecond pulse generation using the optical Kerr effect,' *Appl. Phys. Lett.*, **14**, 140 (1969).

[154] E. B. Treacy, 'Optical pulse compression with diffraction gratings,' *IEEE J. Quantum Electron.*, **QE-5**, 454 (1969).

[155] O. E. Martinez, '3000 times grating compressor with positive group velocity dispersion: application to fiber compensation in 1.3–1.6 μm region,' *IEEE J. Quantum Electron.*, **QE-23**, 59 (1987).

[156] W. B. Jiang, M. Shimizu, R. P. Mirin, T. E. Reynolds and J. E. Bowers, 'Femtosecond periodic gain vertical-cavity lasers,' *Photo. Tech. Lett.*, **5**, 23 (1993).

[157] W. B. Jiang, M. Shimizu, R. P. Mirin, T. E. Reynolds and J. E. Bowers, 'Femtosecond periodic gain vertical-cavity semiconductor lasers,' *OSA Proc. Ultrafast Electronics and Optoelectronics*, **14**, 21 (1993).

[158] R. Nagarajan, M. Ishikawa, T. Fukushima, R. S. Geels and J. E. Bowers, 'High speed quantum-well lasers and carrier transport effects,' *IEEE J. Quantum Electron.*, **28**, 1990 (1992).

[159] S. W. Corzine, R. S. Geels, J. W. Scott, R. H. Yan and L. A. Coldren, 'Design of Fabry–Perot surface-emitting lasers with a periodic gain structure,' *IEEE J. Quantum Electron.*, **25**, 1513 (1989).

[160] W. B. Jiang, S. R. Friberg, H. Iwamura and Y. Yamamoto, 'High powers and subpicosecond pulses from an external-cavity surface-emitting InGaAs/InP multiple quantum well laser,' *Appl. Phys. Lett.* **58**, 807 (1991).

[161] J. F. Pinto, L. W. Stratton and C. R. Pollock, 'Stable color-center laser in K-doped NaCl tunable from 1.42 to 1.76 μm,' *Opt. Lett.*, **10**, 384 (1985).

[162] W. H. Xiang, S. R. Friberg, K. Watanabe, S. Machida, W. B. Jiang, H. Iwamura and Y. Yamamoto, 'Femtosecond external-cavity surface-emitting InGaAs/InP multiple-quantum-well laser,' *Opt. Lett.*, **16**, 1394 (1991).

[163] J. P. Heritage, R. N. Thurston, W. J. Tomlinson, A. M. Weiner and R. H. Stolen, 'Spectral windowing of frequency-modulated optical pulses in a grating compressor,' *Appl. Phys. Lett.*, **47**, 87 (1985).

[164] L. F. Mollenauer, R. H. Stolen and J. P. Gordon, 'Experimental observation of picosecond pulse narrowing and solitons in optical fibers,' *Phys. Rev. Lett.*, **45**, 1095 (1980).

[165] W. H. Xiang, S. R. Friberg, K. Watanabe, S. Machida, Y. Sakai, H. Iwamura and Y. Yamamoto, 'Sub-100 femtosecond pulses from an external-cavity surface-emitting InGaAs/InP multiple quantum well laser with soliton-effect compression,' *Appl. Phys. Lett.*, **59**, 2076 (1991).

[166] D. C. Sun and Y. Yamamoto, 'Passive stabilization of a synchronously pumped external-cavity surface-emitting InGaAs/InP multiple quantum well laser by a coherent photon-seeding technique,' *Appl. Phys. Lett.*, **60**, 1286 (1992).

[167] K. Watanabe, H. Iwamura and Y. Yamamoto, 'Effect of additive-pulse mode locking on an external-cavity surface-emitting InGaAs semiconductor laser,' *Opt. Lett.*, **18**, 1642 (1993).

[168] S. de Silvestri, P. Laporta and O. Svelto, 'The role of cavity dispersion in cw mode-locked lasers,' *IEEE J. Quantum Electron.*, **QE-20**, 533 (1984).

[169] P. Laporta and V. Magni, 'Dispersive effects in the reflection of femtosecond optical pulses from broadband dielectric mirrors,' *Appl. Opt.*, **24**, 2014 (1985).

[170] A. N. Pikhtin and A. D. Yas'kov, 'Dispersion of the refractive index of semiconductors with diamond and zinc-blende structures,' *Sov. Phys. Semicond.*, **12**, 622 (1978).

[171] G. P. Agrawal, *Nonlinear Fiber Optics*, Chapters 1 & 2. (Academic Press, 1989).

[172] D. M. Kim, J. Kuhl, R. Lambrich and D. von der Linde, 'Characteristics of picosecond pulses generated from synchronously pumped cw dye laser system,' *Opt. Comm.*, **27**, 123 (1978).

[173] G. P. Agrawal and N. A. Olsson, 'Self-phase modulation and spectral broadening of optical pulses in semiconductor laser amplifiers,' *IEEE J. Quantum Electron.*, **QE-25**, 2297 (1989).

[174] R. L. Fork, O. E. Martinez and J. P. Gordon, 'Negative dispersion using pairs of prisms,' *Opt. Lett.*, **9**, 150 (1984).

[175] W. H. Xiang, W. B. Jiang and Y. Ishida, 'Femtosecond pulses generated from non-colliding pulse mode-locked ring dye lasers,' *Opt. Comm.*, **86**, 70 (1991).

[176] A. Yariv, *Quantum Electronics*, Chapter 21. (John Wiley & Sons, 3rd Edn., 1989).

[177] T. Urisu and Y. Mizushima, 'Transient pulse buildup mechanisms in a synchronously-pumped mode-locked dye laser,' *J. Appl. Phys.*, **57**, 1518 (1985).

[178] A. Dienes, J. P. Heritage, M. Y. Hong and Y. H. Chang, 'Time-domain and spectral-domain evolution of subpicosecond pulses in semiconductor optical amplifiers,' *Opt. Lett.*, **17**, 1602 (1992).

[179] W. B. Jiang, M. Shimizu, R. P. Mirin, T. E. Reynolds and J. E. Bowers, 'Electrically pumped mode-locked vertical cavity semiconductor lasers,' *Opt. Lett.*, **18**, 1937 (1993).

[180] R. S. Geels and L. A. Coldren, 'Low threshold, high power, vertical-cavity surface-emitting lasers,' *Electron. Lett.*, **27**, 1984 (1991).

7

High power ultrafast semiconductor injection diode lasers

PETER J. DELFYETT

7.1 Introduction

The generation of ultrafast optical pulses from semiconductor diode lasers is extremely attractive owing to the compact and efficient properties of these devices. Applications of these devices range from photonic switching (Smith, 1984), electro-optic sampling (Valdmanis *et al.*, 1982; Valdmanis *et al.*, 1983), optical computing (Miller, 1987; Miller, 1989), optical clocking (Delfyett *et al.*, 1991), applied nonlinear optics (Miller *et al.*, 1983), and other areas of ultrafast laser technology (Delfyett *et al.*, 1991). There have been many recent advances in ultrafast pulse generation from diode lasers in the past few years, with many researchers concentrating on device fabrication, device physics, theoretical modeling and systems applications.

In the past, picosecond optical pulse generation from diode lasers has been accomplished by modelocking and gain switching techniques (Ho *et al.*, 1978; Ito *et al.*, 1979). Several researchers have used degraded semiconductor lasers, which rely on defects in the semiconductor to provide a mechanism for passive modelocking (Ippen *et al.*, 1980; van der Ziel *et al.*, 1981; Yokoyama *et al.*, 1982). The pulses produced from these systems ranged from a few picoseconds to just under a picosecond. The drawback with utilizing degraded diodes is that the lifetime of these devices is limited, typically several hours. Proton implanted multiple quantum well structures (MQW) have been used to passively modelock semiconductor lasers, producing pulses of 1.6 ps in duration which were then compressed to 0.83 ps (Silberberg *et al.*, 1984; Smith *et al.*, 1985; Silberberg and Smith, 1986). The advantage of utilizing MQW structures in an external cavity is that this removes the damaged material from the semiconductor chip, thus increasing the lifetime and reliability of the system. Active modelocking techniques have also been used to generate single ultrashort optical pulses with a wide range of success (Ho, 1979; van der Ziel, 1981). The typical optical pulses obtained range from 5 to 30 ps; the shorter pulses obtained by utilizing intracavity optical elements which compensate for the group velocity dispersion (Kuhl *et al.*, 1987; Putnam and Salour, 1983). The shortest optical pulses directly generated from an actively modelocked semiconductor laser are 0.58 ps. However, the output in this case is multi-pulse (Corzine *et al.*, 1988; Bowers *et al.*, 1989). Theoretical treatments of active modelocking with diode lasers in this limit are given in Morton *et al.* (1989) and Schell *et al.* (1991). Most recently, monolithic hybrid modelocked structures are being employed to generate ultrahigh repetition rate optical pulse trains with pulse durations of ~1 ps (Morton *et al.*, 1990; Wu *et al.*, 1990) (see Chapters 8 and 9). Other devices, such as surface emitting laser diodes, also show potential as sources of ultrafast optical pulses (Xiang *et al.*, 1991) (see Chapter 6).

Gain switching can also be utilized to generate picosecond optical pulses. Typical output pulse durations are generally on the order of twenty to thirty picoseconds by direct gain switching. Grating compressors, as used by Kuznetsov *et al.* (1990), and fiber compression techniques have been employed by Liu *et al.* (1989, 1991) to compress the chirped optical pulses generated from gain switched diodes. Most recently, a gain switched distributed feedback diode laser and erbium fiber amplifiers have exploited the soliton compression effect to produce optical pulses of ~100 fs (Nakazawa *et al.*, 1990a, 1990b). Limitations in these schemes are the increased temporal jitter of the generated optical pulses, low output power, and additional complexity required for shorter pulses.

With the advent of high power semiconductor laser devices, experimental results have shown the potential for generating relatively high peak power optical pulses from a semiconductor diode laser system (Andrews and Burnham, 1986; van der Ziel *et al.*, 1984; Delfyett *et al.*, 1990). In this chapter, the techniques and underlying physics involved in generating high power ultrashort optical pulses from semiconductor optical amplifiers will be covered. The main concepts that are to be stressed experimentally are: (1) the elimination of the residual facet reflectivity which causes multiple pulse outputs by using angled striped semiconductor traveling wave optical amplifiers, (2) generation of short pulses by using intracavity MQW saturable absorbers, (3) exploitation of the frequency chirp impressed on the pulse by performing pulse 'compression' techniques, and (4) creation of high output powers by amplification techniques.

7.2 Active modelocking

The generation of ultrashort optical pulses from semiconductor diode lasers via optical modelocking techniques typically requires an external cavity oscillator (Aspin and Carroll, 1979; Haus 1980). One main difficulty of constructing an external cavity with a semiconductor diode laser is the stringent requirements of the refractive index and thickness of the antireflecting coating deposited on the cleaved facet which faces the rear reflector. The typical requirements for the Fabry–Perot laser structure is 5% fluctuation of the index of refraction and 6 nm thickness variation of the antireflection coating. These stringent requirements are necessary to provide a residual facet reflectivity of less than 10^{-4}, owing to the large single pass gain of diode laser structures (Marcuse, 1989). Incomplete elimination of the Fabry–Perot (FP) modes associated with the diode chip only allows modelocking within the mode clusters of the diode chip. The consequence of this is the production of optical pulses with a high degree of temporal substructure within the pulse envelope. This temporal substructure manifests itself as coherent spikes in the autocorrelation trace. The coherent spikes are temporally separated by the round-trip time of the laser diode chip, nominally 6 ps for a 250 μm device length.

In order to overcome the detrimental effects of residual facet reflectivity, spectral selection of a single mode cluster can be performed. This has an effect of eliminating other mode clusters and allows the production of a clean modelocked optical pulse. The drawback of this technique is that it limits the spectral width of the modelocked optical pulse, and as a result limits the minimum pulse duration obtainable with this method.

An alternative method of overcoming the problem of residual facet reflectivity is to use an angled stripe semiconductor traveling wave amplifier device (Alphonse *et al.*, 1988). The main reason for using this type of device is that this geometry eliminates the coupling of reflected light back into the gain medium and ensures the production of a clean modelocked optical pulse (Holbrook *et al.*, 1980; Bradley *et al.*, 1981; Chen *et al.*, 1984). The elimination of the FP modes can now allow for complete modelocking over a wider spectral range, thus allowing for much shorter pulse widths. In addition, the device can also be used in a master oscillator-power amplifier configuration, if high output powers are desired. These concepts are illustrated schematically in Figure 7.1.

The devices which are used in these experiments are gain guide angled stripe semiconductor traveling wave optical amplifiers (TWA). The

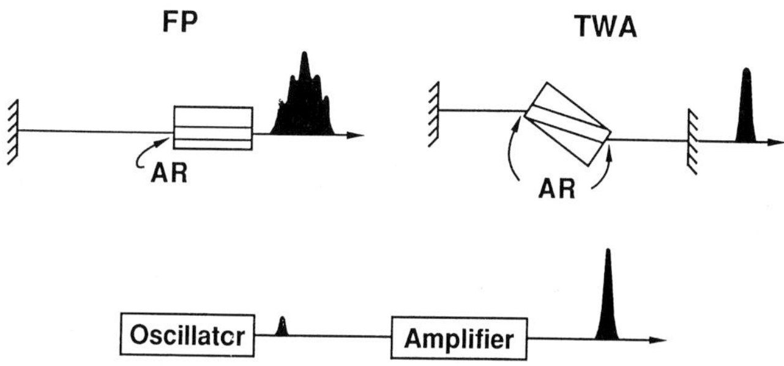

Figure 7.1. Schematic diagram illustrating the advantages of a traveling wave amplifier external cavity modelocked laser with a Fabry–Perot external cavity modelocked laser.

device structure is depicted schematically in Figure 7.2. The key feature of this device is the thin active region which is ~80 nm. The thin active region and the gain guided nature of the device are the attributes of these devices which allow the generation of high output energies. These devices can typically amplify optical pulses to over 100 pJ per pulse. This can be understood by noting that the output saturation power of semiconductor optical amplifiers is given by

$$P_s = \frac{h\nu A}{\Gamma a \tau} \tag{7.1}$$

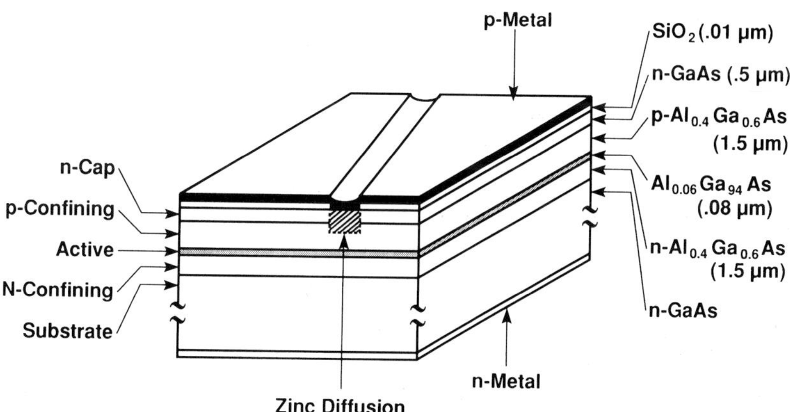

Figure 7.2. Schematic diagram of the semiconductor traveling wave optical amplifier used in the experiments.

where $h\nu$ is the photon energy, A is the area of the active region, Γ is the optical confinement factor, a is the differential gain, and τ is the gain recovery time. From this, it can be seen that these devices will have large A due to the gain guiding effect and small Γ due to the thin active region, resulting in a large output saturation power.

In addition to the angled stripe geometry of the device, antireflection coatings are deposited on both facets. The effect of the coatings, in addition to suppressing any residual Fabry–Perot effect, is to increase the output power by reducing the normal 30% reflection losses owing to the semiconductor–air interface. The device performance characteristics are summarized in Figure 7.3(a,b). In Figure 7.3(a), the spectral distribution of the spontaneous emission and the L-I curve are presented. This plot shows a spectrally broad output, without the usual FP modes associated with semiconductor lasers. In addition, the L-I curve, shown in Figure 7.3(b), displays an exponentially increasing output power with increasing injection current, which is characteristic of spontaneous emission. In addition, the normal knee associated with the onset of lasing in diodes is absent.

The generalized experimental arrangement is schematically represented in Figure 7.4. A low power oscillator, which is comprised of a semiconductor traveling wave optical amplifier, is used to generate optical pulses via active, passive or hybrid modelocking. An identical semiconductor TWA is used as a post amplifier to increase the output power of the generated pulse train. In addition, chirp compensatory techniques can then be employed to reduce the pulse width, due to the frequency sweep which is impressed upon the optical pulse, if desired.

The detailed experimental setup is depicted schematically in Figure 7.5 for the active modelocking experiments. The device structure is the angled stripe AlGaAs semiconductor traveling wave optical amplifier described above. The device is located inside the external cavity, with either a 600 l/mm reflective diffraction grating or a plane high reflector and lens in the cat's eye geometry as the rear reflector and a 20% output coupler defining the external cavity. Lenses are used to collect and collimate the output light from the diode, and additional focusing lenses are used in the cat's eye geometry to increase the cavity stability. In addition, an intracavity slit is incorporated to control the transverse mode profile. The incorporation of a reflective diffraction grating is a convenient method for limiting the spectral width of the modelocked laser. In addition, tuning can be achieved without altering the cavity alignment, which can be encountered when using some types of intracavity etalons. The

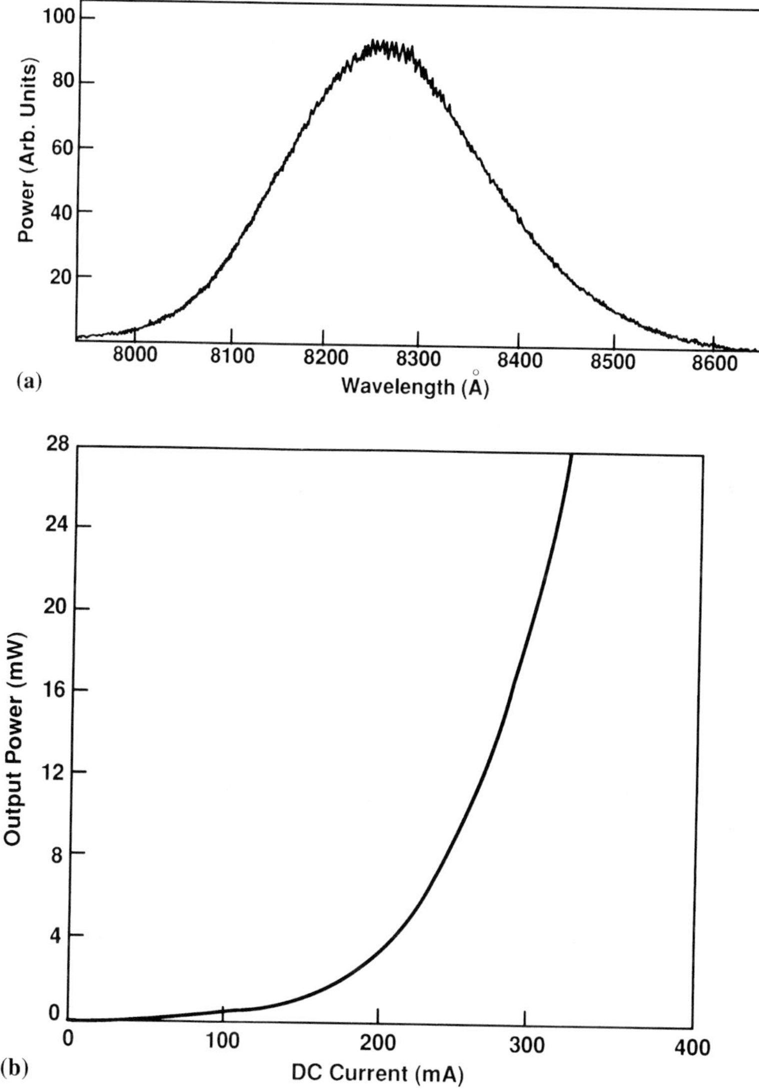

Figure 7.3. Device characteristics. (a) Spontaneous emission output; (b) L–I curve for the TWA.

laser diode is thermo-electrically cooled with a temperature stabilization control circuit which stabilized the temperature to within a tenth of a degree Celsius. The output of the modelocked laser is detected with a fast photodiode and monitored with a sampling oscilloscope and an RF spectrum analyzer. Optical autocorrelation techniques are used to measure

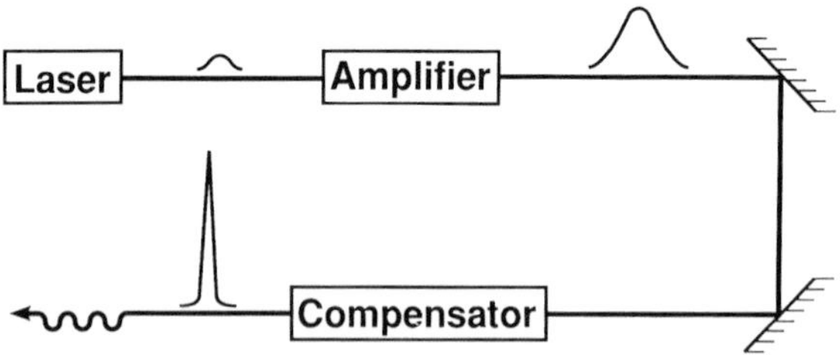

Figure 7.4. Schematic diagram of the method used to generate a high power ultrafast optical pulse.

pulse widths below the resolution of the electronic measurement system, ~25 ps. A small amount of the modelocked output is directed into a 0.25 m spectrometer with a reticon for the optical spectral analysis. Both DC and RF currents are applied to the diode through a bias tee. Optimum modelocking is achieved by tuning the modulating current into resonance with the external cavity and observing the modelocked pulses on the sampling oscilloscope, and by observing the electrical spectrum from the photodiode. The DC bias current is then systematically varied until the optimum pulse widths are measured by optical autocorrelation techniques. For optimum pulse widths, the DC bias current is 185 mA with ~1 W of RF power, using a 960 MHz modulating current. The DC current is ~35 mA below the CW lasing threshold current.

An illustration of the residual facet reflectivity effects which are encountered in external cavity modelocked diode lasers is shown in Figure 7.6(a–d) for both the FP and TWA external cavities. Figure 7.6(a) is an autocorrelation trace of an actively modelocked optical pulse train emitted from an external cavity FP semiconductor diode

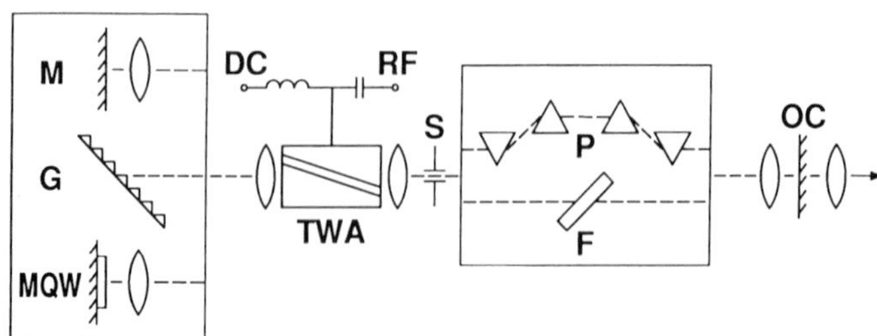

Figure 7.5. Schematic diagram of the experimental setup. Legend: M – mirror; OC – output coupler; MQW – multiple quantum well saturable absorber; TWA – traveling wave amplifier; S – slit; F – filter; G – grating; P – prism sequence.

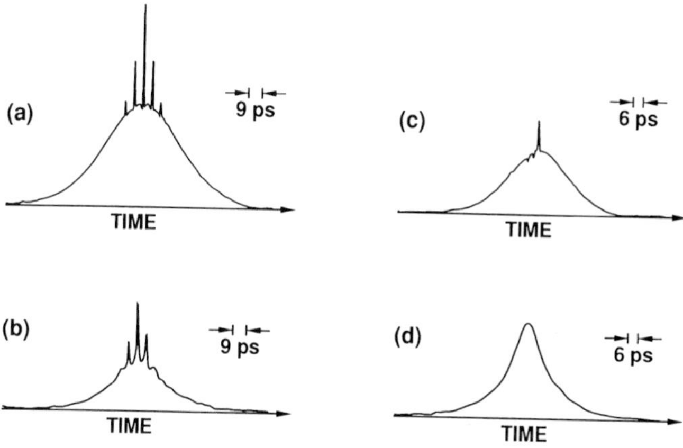

Figure 7.6. Autocorrelation traces from two different actively modelocked laser systems, illustrating the effect of residual facet reflectivity. (a) FP diode without bandwidth limitation. (b) FP diode with bandwidth limitation. (c) TWA without bandwidth limitation. (d) TWA with bandwidth limitation.

laser, with a residual facet reflectivity owing to an imperfect antireflection coating on the facet that faces the rear reflector. The rear reflector in this case is the plane reflecting mirror and lens combination. The coherent spikes associated with the modelocking of mode clusters are clearly observable and are separated by the round-trip time of the diode, corresponding to a diode length of 400 μm. Figure 7.6(b) is an autocorrelation trace of a spectrally filtered modelocked optical pulse train from the same external cavity laser. The spectral filtering in this case is accomplished by utilizing an intracavity etalon or a reflective diffraction grating to select only one mode cluster. The coherent spikes observed in Figure 7.6(b) are noticeably suppressed, giving a cleaner optical pulse.

Figure 7.6(c) is an autocorrelation trace of an actively modelocked optical pulse train emitted from an external cavity TWA semiconductor diode laser without spectral selection. The salient feature in the autocorrelation trace is the production of a single coherent spike centered at the peak of the optical pulse. This coherent signal is an indication that the multiple reflections normally encountered with Fabry–Perot external cavity semiconductor lasers have been suppressed owing to the angle stripe geometry of the device. Figure 7.6(d) is an autocorrelation trace from the actively modelocked external cavity TWA laser, incorporating a 600 l/mm rear reflective diffraction grating as the spectrally limiting optical element. This autocorrelation trace shows an optical pulse width of ~15 ps without any coherent spike superimposed on the generated optical pulse. These results clearly show the importance of eliminating the residual facet reflectivity of diode lasers in order to achieve clean optical pulses from external cavity modelocking techniques.

In order to obtain high peak output power from a diode laser, an amplification step is required. Utilizing an identical semiconductor traveling wave amplifier device, which was used as the gain medium in the external cavity modelocked laser, provides an efficient and compact single pass amplifier (Wiesenfeld *et al.*, 1988; Eisenstein *et al.*, 1988). Experiments were performed with the active modelocked laser described above, which emits pulses of ~15 ps in duration with 1 mW of average output power at 870 nm with a 960 MHz repetition rate. These pulses were injected into the TWA to measure the maximum amplified output power obtainable from these devices and to determine the linearity, i.e., the saturation characteristics of the TWA.

In Figure 7.7(a), a plot of the amplified output power is shown as a function of input bias current, with the injected optical power held constant. This result shows that the amplifier reaches a region where the gain is unity at approximately 125 mA of DC bias current. The output power increases to nearly 30 mW at a bias current of 275 mA. In Figure 7.7(b), a

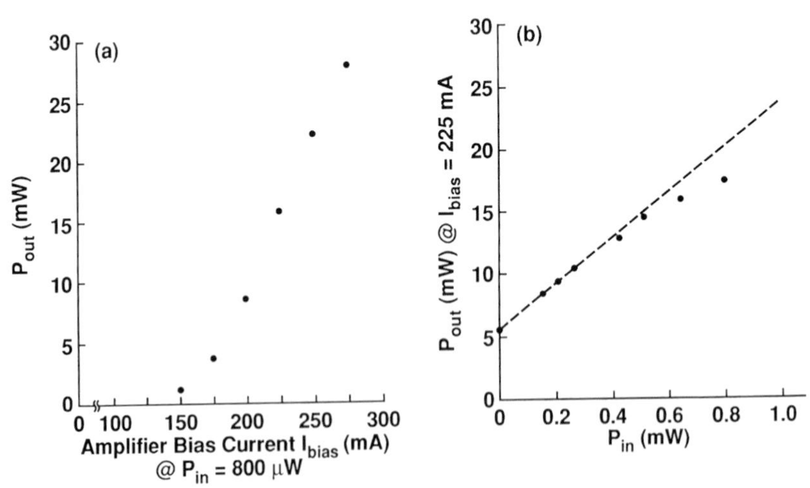

Figure 7.7. Amplification characteristics of the semiconductor traveling wave optical amplifier. (a) Output power versus injection current with input optical power held constant. (b) Output power versus input optical power with injection current held constant.

plot of the amplified output power is shown as a function of injected power, with the bias current held constant. The salient feature of this graph shows that the output power increases linearly with increasing input optical power. At the highest input power of 800 μW, the gain is reduced by ~0.5 dB from the dashed line, showing that the amplifier is entering the saturation region. Extrapolation of the data yields an input power of −3 dB gain compression to be ~2 mW. The slope of the dashed line is less than unity owing to the spontaneous emission at low input power levels.

The amplified pulse widths were measured using standard autocorrelation techniques. Autocorrelation traces of the amplified pulses were recorded with varying values of input optical power and amplifier bias currents. Typical autocorrelation traces comparing the injected mode-locked laser pulse to the amplified optical pulse are shown in Figure 7.8(a–d). For an amplifier bias current of 225 mA and an injected optical

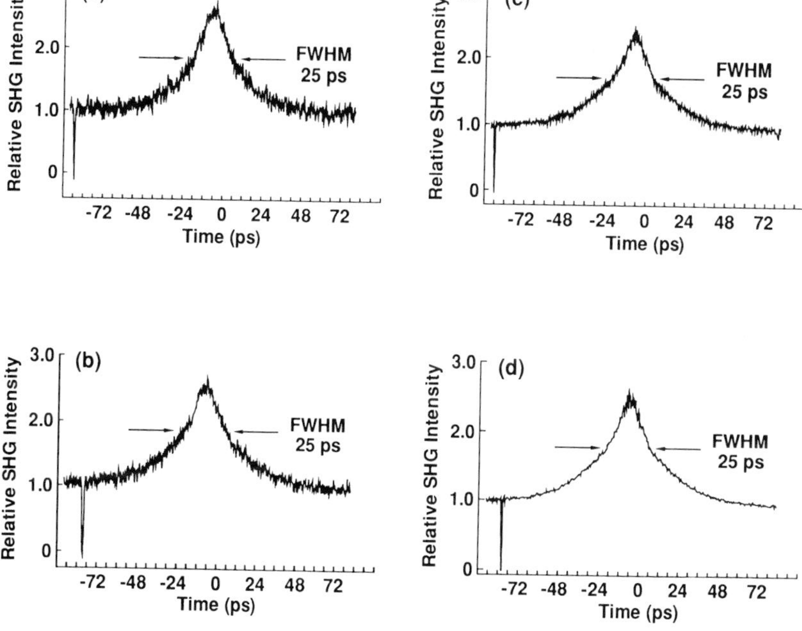

Figure 7.8. Autocorrelation traces of the amplified active modelocked optical pulses at different output peak power levels. (a) input optical pulse, (b) output optical pulse with 1 W peak power, (c) output optical pulse with 2 W peak power, (d) output optical pulse with 3 W peak power.

power of 480 μW, the average power after subtracting the spontaneous emission was measured to be 14 mW, translating to a peak power of 1 W. The autocorrelation trace of this pulse is shown in Figure 7.8(b), with the injected optical pulse shown in Figure 7.8(a). Increasing the bias current to 275 mA increased the average output power to 26 mW, implying a peak power of 2 W. The autocorrelation trace of this pulse is shown in Figure 7.8(c) for comparison.

In order to obtain the highest output power from the amplifier, an RF current with ~1 W of power was applied to the diode amplifier. The phase of the RF current was adjusted so the injected optical pulse experiences the maximum gain. The DC bias current in this case was 275 mA. Under these conditions, CW average output power from the amplifier in excess of 38 mW was achieved. The autocorrelation trace for these experimental conditions is shown in Figure 7.8(d). Increasing further the bias current to 300 mA increased the modelocked amplified output power to 56 mW, with 10 mW of background spontaneous emission. This amplifier output power translates to a peak power over 3 W, assuming a 15 ps pulse duration.

From these results, the potential for generating ultrashort high power optical pulses from semiconductor diode lasers is clearly apparent. Below, we will show how higher peak output powers can be achieved by reducing the optical pulse width, utilizing passive/hybrid modelocking techniques and chirp compensatory techniques.

7.3 Passive and hybrid modelocking with multiple quantum well saturable absorbers

The spectral limiting performed in actively modelocked external cavity semiconductor lasers is important for the production of clean modelocked optical pulses. However, with this technique the spectral limit imposed on the lasing spectrum inhibits the production of subpicosecond and femtosecond optical pulses due to the transform limit. A method to circumvent this problem is to initiate a pulse which has the proper phase relation of the longitudinal modes of the external cavity laser. This is typically accomplished by introducing a saturable absorber inside of the external cavity laser. In this case, an ultrashort optical pulse is built up from the large optical noise transients in the spontaneous emission. However, with this technique the random fluctuations in the spontaneous emission and cavity length influence the pulse shaping mechanisms. This ultimately leads to instabilities in the pulse shape and the pulse-to-pulse timing stability.

The first observations of passive modelocking were in degraded diode lasers (Ippen *et al.*, 1980). Defects associated with aging processes were found to play a major role as saturable absorbing centers. Subsequently, absorbing centers were controllably introduced by proton implantation techniques. In this case, implantation is performed on one facet of the laser diode. With appropriate diode biasing conditions, external cavity lasers constructed with these devices will pulsate at a repetition rate inversely proportion to the round-trip time of the external cavity, producing optical pulses on the order of a few picoseconds. Shorter pulses (~0.6 ps) have been generated from these devices, however these pulses were not single optical pulses and rather occurred in a pulse burst (van der Ziel *et al.*, 1981; Yokoyama *et al.*, 1982). Multi-contact diodes with nonuniform current injection or a reversed bias section accomplish a similar task (Harder *et al.*, 1983; Lau *et al.*, 1985; Lau, 1989; Vasil'ev, 1988), and this method has recently been shown to be extremely attractive for monolithic modelocked laser diodes (Morton *et al.*, 1989; Wu *et al.*, 1990). An alternative method is to incorporate a saturable absorber which is spatially removed from the laser diode, such as a laser diode with a controllable amount of absorption (Stallard and Bradley, 1983; McInerney *et al.*, 1985), or rely on the excitonic absorption properties of multiple quantum well semiconductor structures. The advantage of utilizing the excitonic absorption of MQWs as the saturable mechanism is the approximately ten times lower saturation intensity as compared with the direct band transition (Haus, 1975).

Passively modelocked operation of a semiconductor diode laser utilizing multiple quantum wells is achieved by designing the MQW such that the excitonic absorption peak of the MQW closely matches the peak of the spontaneous emission of the laser device. This assures that it will be possible to force the laser to operate within the excitonic absorption region. In Figure 7.9(a,b), spectra are shown of the traveling wave amplifier emission and of the multiple quantum well absorber transmission. The absorber transmission is obtained by using the spontaneous emission of the TWA as a broadband probe, thus allowing for easy identification of the excitonic absorption maximum with respect to the spontaneous emission output. The peak emission wavelength of the TWA devices used in the passive modelocked experiments occurs at ~830 nm. The absorber is designed to have 100 periods of 7 nm GaAs wells separated by 10 nm AlGaAs barriers (30% aluminum) (Chemla *et al.*, 1984). The absorber is

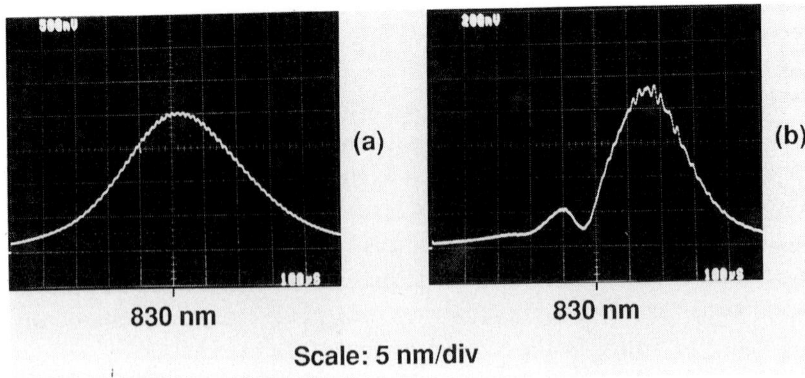

Figure 7.9. Emission spectra of the TWA; (b) transmission spectra of the MQW absorber.

typically grown on 50 nm of AlAs, so that lift-off technology can be used to remove the absorber from the substrate and attach it to a high reflection mirror (Yablonovitch *et al.*, 1987). The key advantage of utilizing the saturable absorber attached to the rear reflector is that the position of the absorber in this case mimics the absorber position in the classic colliding pulse modelocked dye laser. The absorbers are proton implanted with a single dose of 200 keV protons with a density of $\sim 10^{13}$ cm^{-2} to reduce the absorption recovery time (Smith *et al.*, 1985).

The experimental arrangement used for generating passively modelocked optical pulses resembles the active modelocked system with the exception of the rear reflector being replaced by the MQW saturable absorber, and only a DC current is applied to the diode. The modelocked laser was formed by placing the traveling wave amplifier inside of an external cavity. The transverse mode profile of the laser is controlled by an adjustable slit, and the operating wavelength is controlled by an intracavity filter, which has a 10 nm passband as measured with a broadband source. The multiple quantum well structure was designed to have the excitonic absorption occur at ~825 nm, which is approximately 5 nm shifted towards the shorter wavelength region from the spontaneous emission peak of the TWA used in these experiments.

Passive modelocking could be achieved by tuning the laser to operate in the excitonic absorption region. By monitoring the spectrum which was reflected from the MQW-mirror reflector, a substantial broadening of the lasing spectrum would occur when the laser was tuned to operate in the excitonic absorption region. The typical DC bias threshold current in this case was 146 mA, as compared with 117 mA for the laser without the

MQW absorber. Under these conditions the laser would emit pulses of approximately 5 ps in duration at 828 nm, with a repetition frequency equal to the longitudinal mode spacing of the external cavity, which in this case was 302 MHz. The typical average output power from the oscillator was 500 μW. Noticeable satellite pulses were apparent in the autocorrelation traces when the modelocking was adjusted for the robust generation of the shortest pulses, consistent with previous experimental observations (Silberberg *et al.*, 1984). Figure 7.10(a) is an autocorrelation trace of the passively modelocked laser pulse, showing a 5 ps pulse width and the existence of satellite pulses. Comparing this result with the active modelocking result of Figure 7.6(d), the pulse shortening mechanism owing to the saturable absorber becomes apparent.

In order to stabilize the output pulse, a modulated injection current can be applied to the laser through the bias-tee, in a similar fashion to the active modelocked system. In this case, hybrid modelocking is performed. The advantage of utilizing a hybrid modelocking scheme is that it takes advantage of the stability which is offered by the active modelocked system, and also takes advantage of the additional pulse shortening mechanisms provided by the saturable absorber. Experimentally, applying ~500 mW of RF current to the laser diode, where the RF frequency matches the longitudinal mode spacing of the external cavity, the satellite pulses generated from the passive modelocked laser could be suppressed, emitting single 5 ps pulses with 800 μW of average output power (Figure 7.10(b)). The output pulse train was emitted at a repetition rate of 302 MHz, with an operating wavelength of 828 nm. The pulse duration is sensitive to the operation wavelength, thus eliminating the potential for wide tunability in this laser system. The spectral width of the optical

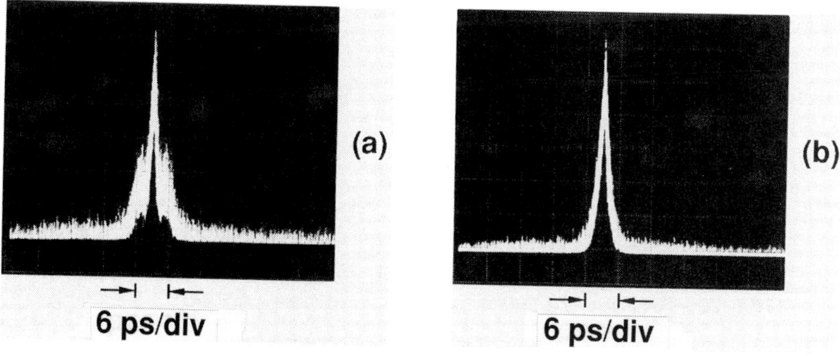

Figure 7.10. Autocorrelation traces of (a) a passive modelocked optical pulse, and (b) a hybrid modelocked optical pulse.

pulses under these conditions is typically 2.5 nm. This clearly shows that the hybrid system takes advantage of the pulse shortening mechanisms provided by passive modelocking, and provides additional stability to the pulse shaping mechanisms by removing the satellite pulses.

Differences in the stability of passive modelocked lasers and hybrid modelocked lasers can also be observed utilizing complementary frequency domain techniques (von der Linde, 1986; Keller *et al.*, 1989). This is accomplished by measuring the power spectrum of the generated optical pulse train with a high speed photodetector and analyzing the detected signal with an RF spectrum analyzer or phase noise analyzer. In Figure 7.11, single components of the RF spectra of both the passive and hybrid modelocked laser systems are shown. The salient feature of these RF spectra point out the differences in the timing stability of the two modelocking processes. In Figure 7.11(a), an individual component of the RF spectrum of the passively modelocked laser is shown. The poorly defined value of the repetition rate, or jitter, is seen to give rise to broadened wings of the RF spectral components. Figure 7.11(b) is the same RF spectral component of the hybrid modelocked laser. In this case, the RF power is applied to the diode without any further attempt to optimize the operation parameters of the laser. In this figure, a well defined position of the RF component is observed. This is an indication that the application of the active driving signal in hybrid modelocking imparts an additional temporal stability to the operation of the semiconductor laser source. The timing jitter in both of these cases can be estimated from

$$\frac{\Delta T}{T} = \frac{1}{2\pi n}\left[\frac{P_S}{P_C}\frac{\Delta\nu}{B}\right]^{\frac{1}{2}} \tag{7.2}$$

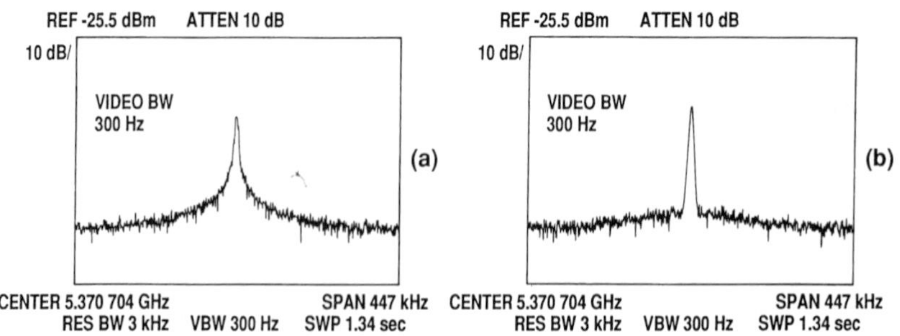

Figure 7.11. RF spectra of the passive modelocked laser (a), and the hybrid modelocked laser (b), illustrating the additional temporal stability obtained from an active drive signal.

where T is the pulser period, ΔT is the change in pulse period, n is the harmonic number of the power spectrum at which the measurement is being performed, P_S and P_C are the power in the phase noise sideband and the carrier, respectively, $\Delta\nu$ is the -3 dB bandwidth of the phase noise sideband, and B is the resolution bandwidth of the spectrum analyzer. From the figures, the timing jitter can be estimated to be 6 ps and 2 ps for the passive and hybrid modelocked lasers, respectively. Further optimization of the hybrid modelocked laser can reduce the residual phase noise sidebands present in Figure 7.11(b) below the noise floor of the spectrum analyzer. Utilizing a low noise floor RF spectrum analyzer can now reveal the phase-noise sidebands associated with the optimized hybrid modelocked laser. In Figure 7.12(a,b), the phase noise characteristics of the hybrid modelocked laser are shown. A typical component of the RF power spectrum is shown in Figure 7.12(a). In this figure we show the 41st harmonic of the power spectrum which shows the existence of the phase noise sidebands. The power ratio P_S/P_C is measured and plotted versus harmonic number, with the resulting data shown in Figure 7.12(b). In this figure the frequency-squared dependence of the power ratio is observed, which assures the correct interpretation of the existence of the noise sidebands to be due to timing jitter. From this data, the relative timing jitter can

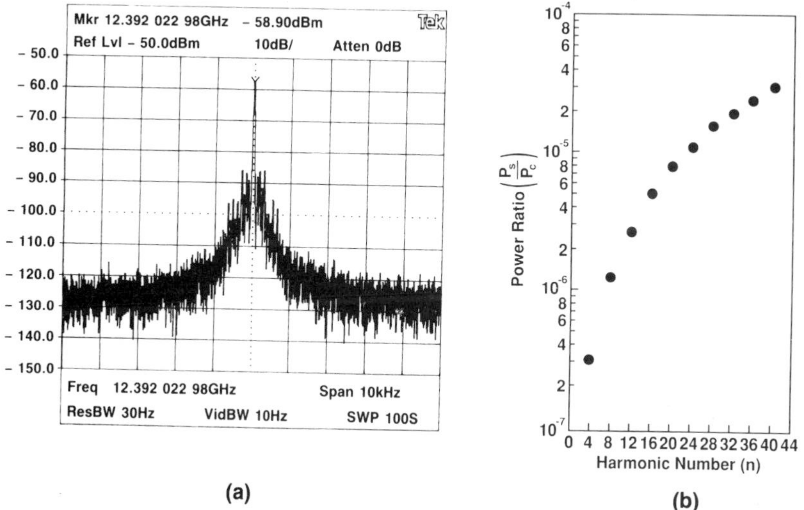

Figure 7.12. RF spectra of the hybrid modelocked laser (a). The frequency squared increase in the phase noise sidebands is verified and shown in (b).

be obtained by using Equation 7.2, and extracts a relative timing jitter of ~400 fs, over a measurement interval of 100 s. This value of jitter is extremely low as compared with other standard laser systems, such as modelocked argon and YAG systems (von der Linde, 1986; Rodwell *et al.*, 1989), but is typical of modelocked semiconductor laser systems (Taylor *et al.*, 1986; Derickson *et al.*, 1991). The low jitter characteristics of the laser system make it an ideal candidate for applications in optical timing applications and electro-optic sampling. It should be noted that the locking range of the RF drive signal in pure active modelocking is typically 10^{-3} of the drive frequency, while for hybrid modelocking, the locking range is $\sim 10^{-5}$.

In order to obtain high peak power optical pulses, the generated pulse train was amplified by an identical semiconductor traveling wave optical amplifier. The light from TWA 1 contained a total of 50 mW of output power, with 32 mW of amplified signal power and 18 mW of spontaneous emission. The DC bias current to TWA 1 in this case was ~300 mA, with an additional 1 W of RF power, where the RF frequency matched the modelocking frequency and was phase adjusted to yield the maximum gain. The ratio of the amplified signal power to the spontaneous emission power in these experiments was less than the 80% achieved in the active modelocked experiments described above due to the lower repetition rate used here. The amplified signal power of 32 mW translates to 20 W of peak power and also shows that the diode is capable of producing pulses with more than 100 pJ of energy per pulse.

From the pulse width measurements and the optical spectrum measurements shown in Figure 7.12(a,b), the time–bandwidth product can be calculated from

$$\text{Time–bandwidth product} = c\Delta\tau\frac{\Delta\lambda}{\lambda^2} = \Delta\tau\Delta f \tag{7.3}$$

From the measurements in Figures 7.13(a,b), the time–bandwidth product is 5.44, which is approximately 12 times the transform limit assuming Gaussian shaped pulses. The shape of the hybrid modelocked autocorrelation trace and the large time–bandwidth product suggests that there possibly exists a large frequency sweep, or chirp, impressed on the modelocked pulse. In order to take advantage of this frequency sweep, a standard dual grating pulse compressor was constructed and employed to compress the modelocked pulse (Martinez, 1987).

The pulse compressor utilized two blazed gratings with 1800 grooves/mm, with two 15 cm focal length lenses. The total loss from the compen-

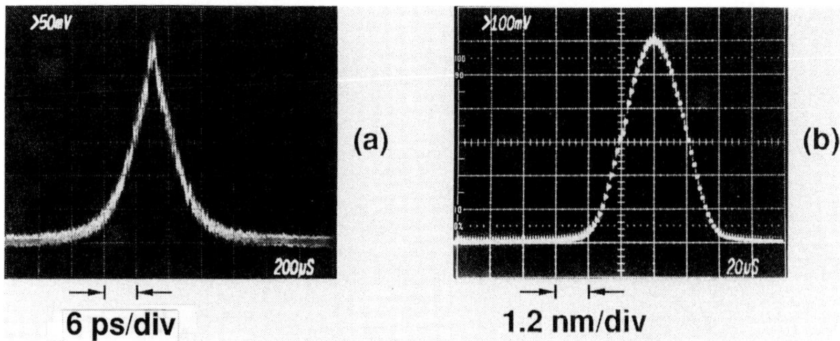

Figure 7.13. (a) Autocorrelation trace of the hybrid modelocked laser pulse after amplification through TWA 1, displaying a measured FWHM of 7.2 ps which leads to a deconvolved pulse of 5.1 ps. (b) The corresponding optical spectrum is displayed with a FWHM bandwidth of 2.5 nm. The time–bandwidth product is 5.44, which is twelve times the transform limit.

sator was 4 dB. An adjustable slit was located at the Fourier plane of the telescope inside of the compressor, thus acting as a spectral filter (Heritage *et al.*, 1985; Weiner and Heritage, 1987). The spectral filter could be adjusted to eliminate unwanted spontaneous emission, yielding high contrast amplified pulses, and also could be used to select the spectral region of the optical pulse which contains the most linear chirp. The compressor was constructed in the double pass geometry. The compressor length was then adjusted to yield the shortest autocorrelation trace after amplification. The position of the second grating in the pulse compressor was varied in both the positive and negative dispersion regimes to achieve the maximum compression ratio. From our experiments, the maximum pulse compression occurs by having the grating in the negative dispersion regime, approximately 7 cm away from the zero dispersion position. The sign of the chirp in our case is identical to that which was obtained in previous pure passive modelocking experiments (Silberberg and Smith, 1986), but opposite to that obtained in pure active modelocking experiments (Wiesenfeld *et al.*, 1990). This shows that the inclusion of the saturable absorber greatly influences the chirping mechanism in the hybrid modelocked system as compared with pure active modelocking.

Using this compressor, optical pulses could be compressed down to 0.46 ps, with 10 mW of amplified signal power directly after the compressor. The spectral window in the compressor filters out the spontaneous emission giving only 1 mW of additional spontaneous emission

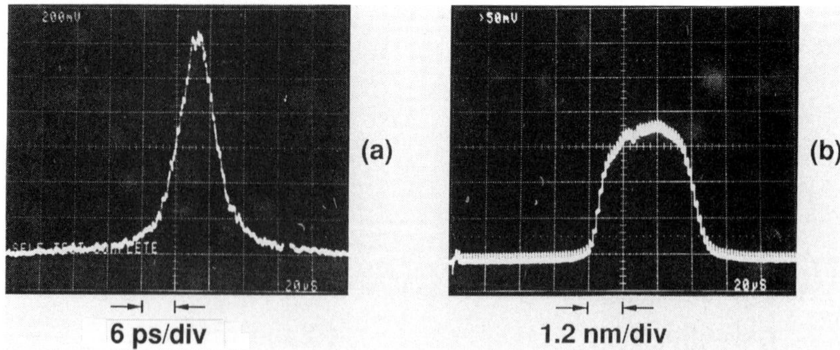

Figure 7.14. (a) Autocorrelation trace of the compressed pulse with an optimized oscillator to yield the shortest compressed pulse. The autocorrelation FWHM is 0.72 ps which gives a deconvolved pulse width of 0.46 ps assuming a sech^2 pulse shape. (b) The corresponding spectrally windowed optical spectrum, with a bandwidth of 3.2 nm.

power. Figure 7.14(a,b) shows the autocorrelation trace and windowed spectra of the optical pulse for this experimental configuration. The pulse after the compressor, with 10 mW of amplified average signal power imply a peak power of 72 W.

By adjusting the slit width located at the Fourier plane of the compressor, the pulse can be arbitrarily broadened, while maintaining transform limited pulses. Figure 7.15(a,b) shows the autocorrelation trace and windowed spectrum of a pulse with a time–bandwidth product of ~0.44. This method of arbitrary pulse width control is attractive for investigating the dynamics of systems which are dependent on the input pulse width.

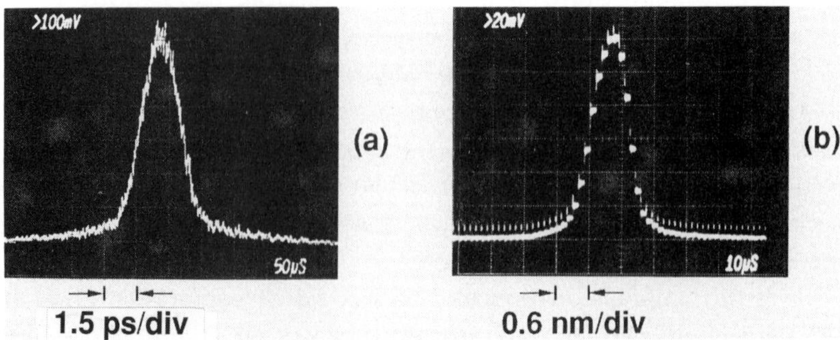

Figure 7.15. Autocorrelation trace and windowed spectrum of the spectrally windowed pulse.

7.4 Cubic phase compensation

In the above section, it was shown how the gain dynamics associated with carrier depletion and gain recovery in semiconductor traveling wave amplifiers can impart a nonlinear frequency sweep onto an ultrashort optical pulse. The nonlinear chirp arises directly from the integrating nonlinearity associated with gain depletion, the partial gain recovery if present, and the resulting time varying refractive index owing to the plasma effect. This nonlinear frequency sweep can prevent the complete compensation of the chirp if only quadratic phase compensation techniques are used. In this section, a method for independently compensating the cubic phase distortion is presented. The cubic phase distortion is the next dominant term in the Taylor series expansion for the nonlinear chirp. Thus, by independently compensating for both the quadratic and cubic phase distortion, one can take advantage of the nonlinear frequency sweep which is impressed on optical pulses generated from semiconductor lasers.

The importance of residual cubic phase distortion of ultrafast optical pulses was pointed out in pulse compression experiments with colliding-pulse modelocked dye lasers and fiber grating pulse compression techniques (Fork *et al.*, 1987). It was demonstrated that cubic phase distortion could be controlled by incorporating a four prism sequence in conjunction with the standard grating compressors, and balancing the relative quadratic and cubic phase compensation provided by each compressor. The drawback with this method is that there is not an independent control of the quadratic and cubic compensation.

The cubic compensator used in this experiment is schematically represented in Figure 7.16. The optical pulses are obtained from the external cavity hybrid modelocked semiconductor diode laser system. The pulses are amplified with a TWA amplifier to boost the power, and directed into the dual grating compensator to compensate for the quadratic phase distortion which leads to linear frequency sweep. The compressed optical pulses are then sent to an autocorrelator for temporal diagnostics. Once optimal compression has been obtained using only the quadratic compressor, the resultant optical pulses are then injected into the pure cubic compensator. This compensator is composed of a 1800 l/mm reflective diffraction grating, a 75 cm achromatic lens placed one focal length away from the grating, and a deformable mirror which provides for cubic compensation placed one focal length from the lens. The optical pulses are sent into the autocorrelator for

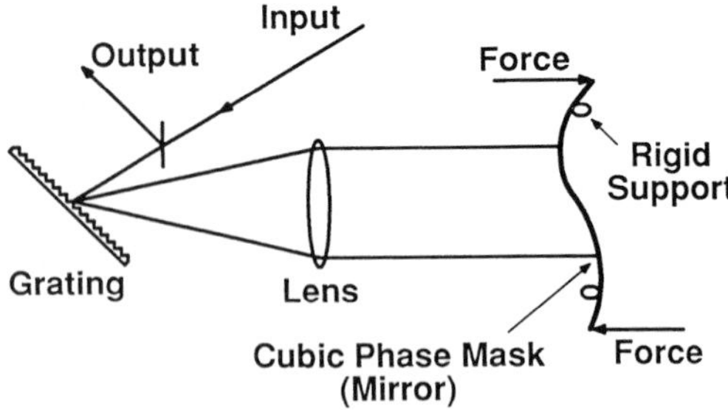

Figure 7.16. Experimental setup showing a schematic diagram of the cubic phase compensator.

temporal diagnostics. The cubic compensator is then adjusted for maximal pulse compression.

The result of the experiment is summarized in Figure 7.17. The laser oscillator is aligned so that it operates with the widest possible spectrum. In this mode, the oscillator emits pulses with a spectrum sufficiently wide to support optical pulses of ~200 fs in duration. Figure 7.17(a) is an autocorrelation trace of the optical pulses after the linear frequency chirp has been compensated by the quadratic phase compensator. The FWHM pulse duration is ~410 fs. The large wings observed on the autocorrelation trace are a clear indication that higher order phase distortion is contained in the optical pulse. Figure 7.17(b) is an autocorrelation trace of the optical pulses after cubic phase distortion has been removed by the new compensator. The FWHM of the optical pulse in this case is 290 fs, showing that the cubic compensator has compressed the optical pulses after the quadratic distortion has been removed. More

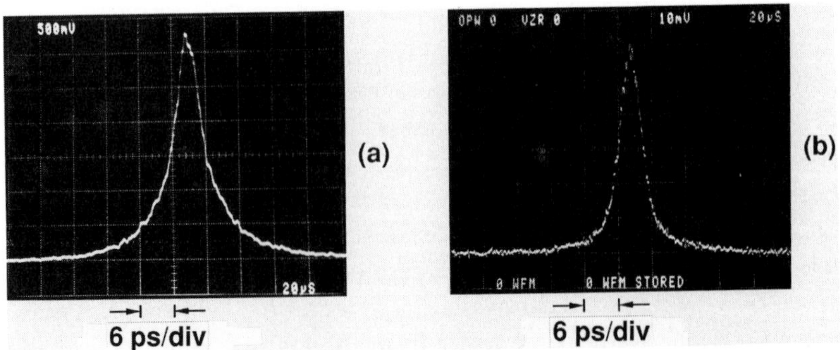

Figure 7.17. Autocorrelation trace of the optical pulse: (a) before cubic phase compensation and (b) after cubic phase compensation.

importantly, the extremely large wings which were apparent in the autocorrelation trace of Figure 7.17(a) are completely absent in the autocorrelation of Figure 7.17(b). It should be noted that the physical mechanism responsible for the cubic phase distortion is unresolved. The cubic term is most likely impressed on the pulse during the modelocking process; smaller amounts of cubic phase distortion may also arise from the external traveling wave amplifier. It should also be noted that the conditions for good cubic phase compensation are highly dependent on the operation point of the laser system, i.e., bias current, RF frequency, and laser wavelength.

In order to quantify the results presented above, a simulation of the cubic compression was performed. The nonlinear phase of the optical pulse can be expanded in the form

$$\begin{aligned}\Phi(\omega) =&\Phi(\omega_0) + \Phi'(\omega_0)(\omega - \omega_0)\\ &+\frac{1}{2}\Phi''(\omega_0)(\omega - \omega_0)^2+\frac{1}{6}\Phi'''(\omega_0)(\omega - \omega_0)^3\end{aligned} \tag{7.4}$$

where Φ is the phase, the prime denotes differentiation, and ω and ω_0 are the optical frequency and the center carrier frequency, respectively. Figure 7.18(a and b) are two computed autocorrelation traces of optical pulses which have varying amounts of cubic phase distortion, without any quadratic phase distortion. This is appropriate for the experiment, since the quadratic phase distortion has been removed by the dual grating compressor. In Figure 7.18(a), the optical pulse is computed for a hyperbolic secant-squared pulse shape having a spectral width corresponding to the experimentally measured width of ~4 nm. The cubic phase, Φ''', is computed to be 0.036 ps^3, such that the simulated and measured auto-

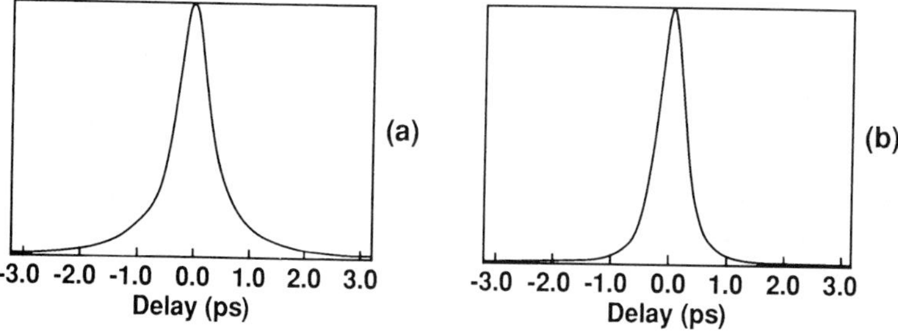

Figure 7.18. Autocorrelation traces of optical pulses with varying amounts of residual cubic phase distortion. (a) Computed optical pulse corresponding to the experimental input pulse. (b) Computed optical pulse corresponding to the compressed output pulse.

correlation traces match. The compressed pulse is computed by reducing the cubic phase term. The final computed autocorrelation trace giving the best fit to the experimental result is shown in Figure 7.18(b). In this case, Φ''' is calculated to be 0.015 ps^3. The residual amount of cubic phase distortion, according to the calculation, is a result of the higher order phase terms in the expansion not being identically equal to zero, and these higher order terms are grouped into a residual cubic term.

7.5 Intracavity dynamics

Thus far, it has been shown that the nonlinear effects in semiconductor traveling wave amplifiers are large and may significantly contribute to the pulse shaping dynamics in external cavity diode lasers. Previous work has investigated the transient behavior of an active modelocked diode laser in the 'turn on' and 'turn off' conditions (AuYeung *et al.*, 1982). In experiments described in this section, the intracavity spectral and temporal characteristics of the optical modelocked pulse are measured, to determine the nature of the pulse shaping mechanisms in this particular type of hybrid modelocked laser.

The layout for the laser oscillator is shown in Figure 7.5. The external cavity semiconductor laser is constructed from the AlGaAs angled striped semiconductor traveling wave optical amplifier, a modified multiple quantum well saturable absorber, a 50% output coupler, and a four prism sequence to compensate for intracavity group velocity dispersion (Valdmanis *et al.*, 1985; Valdmanis *et al.*, 1986). The multiple quantum well saturable absorber incorporates a modified design as compared with the type which was used in the passive and hybrid modelocking experiments described above. The new design utilizes wells of dimension 6 nm through 9 nm where each well is separated by a 10 nm barrier. This design creates a broader absorption region as compared with the one well dimension absorber. The wider absorption region creates the possibility for the support of a wider modelocked spectrum and tunability. The absorption recovery was reduced to ~150 ps by 200 keV proton implantation with a density of 10^{13} cm^{-2}. The prisms are composed of the glass type SF18 (index of refraction 1.72), with a prism separation of 33 cm. The external cavity has a fundamental longitudinal mode spacing of 111.8 MHz. The semiconductor traveling wave optical amplifier chip is placed 15 cm from the rear reflector. Under passively modelocked operation, the laser oscillator operates at a repetition rate of 335.4 MHz, due to the $1/e$ gain recovery time of the traveling wave amplifier chip, which was measured to be 1.1 ns. In order to achieve hybrid modelocked operation, an RF signal with ~0.5 W of power is applied to the TWA through a bias-tee network. During hybrid modelocked operation, the RF frequency is chosen to match the third harmonic of the fundamental frequency, i.e. 335.4 MHz. The application of the RF drive signal broadens the passively modelocked spectrum significantly, giving a lasing bandwidth of ~7.2 nm, at a center wavelength of 838 nm. The optical pulses produced from the oscillator however were highly chirped, with a pulse duration of ~10 ps and a time–bandwidth product of 30. These results are contradictory to most ultrafast laser systems with GVD compensation, since the inclusion of the four prism sequence is to balance the residual group velocity dispersion and chirp impressed upon the pulse during the modelocking process.

In order to obtain a better understanding of the modelocking dynamics, intracavity spectra were measured at three important locations. These locations are (1) at the output coupler, (2) before the saturable absorber, and (3) after the saturable absorber. The results of these measurements are summarized in Figure 7.19.

In Figure 7.19(e), the spectrum of the pulse is shown after exiting the gain medium, before the pulse enters the saturable absorber. The salient feature of this pulse spectrum is the predominant red peak, which is due to self phase modulation resulting from rapid depletion of the gain. The measured gain experienced by the pulse is >30. In Figure 7.19(f), the pulse spectrum is shown after the pulse exits the multiple quantum well saturable absorber. A drastic modification of the optical pulse spectrum is observed, giving evidence that the optical nonlinearity associated with the saturable absorber is changing the optical pulse characteristics. The output spectrum, shown in Figure 7.19(d), is also modified with respect to the pulse spectrum shown in Figure 7.19(f). In this case, the reflected pulse enters the gain medium during a time when the gain is low. As a result, the gain medium immediately enters saturation as the pulse is injected into the device. This is confirmed by observing that the gain experienced by the pulse is ~5. This is understandable due to the fact that the gain medium is located close to the rear saturable absorber reflector, and as a result, the gain does not have a sufficient time to recover completely.

The large changes in the optical pulse spectra imply that there are accompanying modifications in the optical pulse shape. In Figure 7.19(a–c), the temporal pulse shapes of the corresponding spectra are shown. The pulse shapes were recorded with a Hamamatsu streak camera

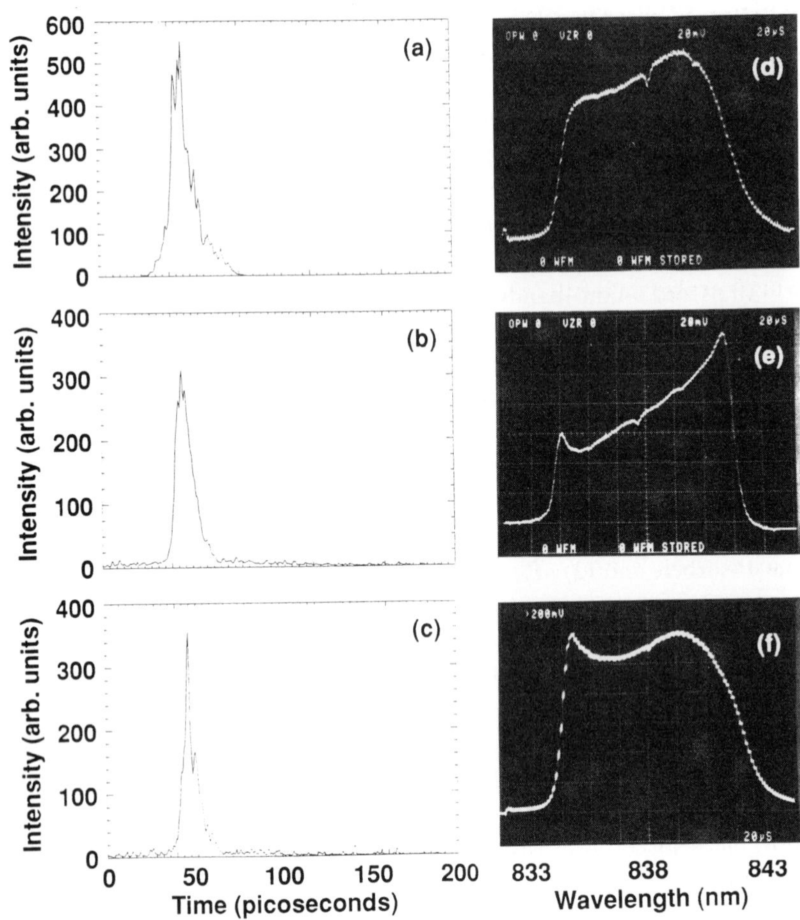

Figure 7.19. Temporal pulse shapes and corresponding optical spectra at different locations in the cavity. (a) Output pulse shape, (b) pulse shape after the gain medium (before saturable absorber), (c) pulse shape after the saturable absorber, (d) output spectrum, (e) spectrum after the gain medium, and (f) spectrum after the saturable absorber.

system, with a temporal resolution of 2 ps. Several single-shot events were recorded for each location, and the resulting temporal profiles were averaged to eliminate the noise fluctuations in the streak camera output. In Figure 7.19(a), the temporal shape of the output optical pulse is shown. The pulse has an asymmetric shape, with a FWHM of 10 ps. In Figure 7.19(b), the temporal shape of the pulse is shown after returning from the output coupler and propagating through the gain medium. The salient difference between the two pulse shapes is the reduced fall time of the pulse in Figure 7.19(b), due to the saturation of the gain. The pulse width in this case is 8 ps. In Figure 7.19(c), the temporal shape of the pulse is shown after reflecting from the saturable absorber mirror. The pulse shape at this location exhibits a fast leading edge of the pulse, which is due to the saturation of the absorption of the multiple quantum well structure, giving the optical pulse a duration of 4 ps. The optical pulse is then broadened as it passes through the gain medium a second time, due to the large gain compression experienced by the pulse (Lowery, 1988; Tohyama *et al.*, 1991). This broadening is explained by the fact that the injected power of the pulse is high and the gain is low during the passage of the pulse through the amplifier the second time. As the pulse enter the gain medium, the gain immediately starts to drop and saturates before the peak of the pulse. As a result, the output is a broadened pulse, with the leading edge slightly advanced as compared with the input pulse. Group velocity dispersion in the laser diode does not play a role in the broadening process for a bandwidth of 7 nm; broadening due to GVD is only ~200 fs (van der Ziel *et al.*, 1983), which is much less than the broadening experienced by the optical pulse.

These results show changes in the optical modelocked pulse shape of more than 50% within one round trip. Conventional theories of modelocking using quantum well saturable absorbers show that fractional changes in the modelocked pulse are on the order of a few percent (Haus and Silberberg, 1985), leading to the conclusion that the hybrid modelocking method used here imparts additional optical nonlinearities into the modelocking process. The modifications in the spectral and temporal characteristics of the modelocked optical pulse also lead to the conclusion that the optical nonlinearities experienced by an optical pulse in a hybrid modelocked semiconductor laser are sufficiently large that dispersion compensation with prisms provides insufficient compensation for the chirp which is induced by the gain dynamics. It should be noted that the laser system without the prism sequence or MQW absorber could not support the optical bandwidth obtained in these experiments.

Even though the intracavity prism sequence cannot compensate for the frequency sweep induced during the modelocking process, femtosecond optical pulses can still be generated from this laser system. This is achieved by injecting the optical pulse train into a standard dual grating compressor arranged in a double pass geometry. This optical arrangement provides for a larger dispersion and thus can compensate the linear frequency sweep impressed on the optical pulse, thus allowing for the compression of an optical pulse. In the experimental configuration described above, the optical pulse is injected into a semiconductor traveling wave optical amplifier prior to compression in order to obtain high output power. This is analogous to chirped pulse amplification schemes (Maine *et al.*, 1988). The advantages of amplifying the optical pulse prior to compression is two fold: (1) the effect of the pulse-width dependent gain saturation due to carrier heating and cooling is avoided (Lai *et al.*, 1990), thus allowing for the generation of high output power optical pulses, and (2) spectral filtering techniques can be employed in the Fourier plane of the compressor to eliminate the spontaneous emission from the amplifier, thus creating high contrast amplified pulses. It should be noted that the optical pulse shape and spectrum are negligibly affected by the external amplification stage. This is easily understood by noting that the spectral distortion is inversely proportional to the pulse width (below in Equation 7.8). For a 10 ps pulse, spectral distortion of the order of 0.4 angstroms may occur. This will have little effect on the pulse shape, and has been confirmed by independent spectral and cross-correlation measurements of the pulse after amplification.

Experimentally, amplification of the hybrid modelocked optical pulse train was accomplished by utilizing an identical semiconductor traveling optical amplifier. With an injection power of 1–2 mW, amplified average output powers of over 30 mW were obtained, corresponding to over 100 pJ of energy per pulse. It should be noted that the injected optical power contained $\sim$75 μW of total background spontaneous emission distributed over the entire output emission spectrum. The amplified pulse train was then injected into the dual grating compressor in a double pass geometry. The compressed pulse was then measured using a standard rotating mirror autocorrelator. Figure 7.20 shows the compressed output of the amplified modelocked optical pulse train generated from the hybrid modelocked laser system. The autocorrelation trace shows an optical pulse width of 207 fs in duration, assuming a hyperbolic secant-squared optical pulse shape. The throughput power from the compressor yielded 11.5 mW of amplified output power, with less than 100 μW of background spontaneous emission, owing to the spectral filter. These results imply a peak power of 165 W, making these optical pulses both the shortest and most intense ever generated from an all semiconductor diode laser system.

7.6 Amplification characteristics/dynamics

The modelocking and amplification results described above show that the TWA devices used in these experiments are sufficient for providing high peak power optical pulses which can be used for a variety

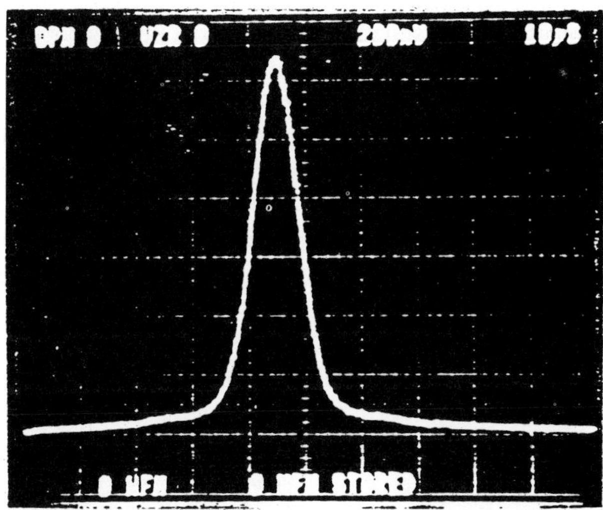

Figure 7.20. Autocorrelation of the compressed pulse, showing a FWHM pulse duration of 207 fs, assuming a hyperbolic secant-squared pulse profile. The average power after compression is 11.5 mW, implying a peak power of 165 W. Scale: 320 fs/div.

of photonic applications. Additional important information is contained in the temporal dynamics of these devices. These dynamics can potentially influence the performance and usage of the TWA.

In this section, we show that large optical nonlinear effects exist in semiconductor traveling wave amplifiers. These effects are due to nonlinear gain depletion and subsequent hot carrier thermalization (Stix *et al.*, 1986; Kesler and Ippen, 1987; Hall *et al.*, 1990), which can be observed as a rapid gain depletion accompanied by a fast (~1 ps) partial recovery of the gain of a semiconductor traveling wave amplifier after femtosecond optical pulse injection. Theoretical treatment of this phenomena is given in Willatsen *et al.* (1991) and Gomatam and DeFonzo (1990).

The optical pulses employed in the experiment utilized the hybrid mode-locked external cavity semiconductor laser system described above. Two types of experiment were performed to observe and verify the optical nonlinearity induced by carrier heating. The first type employed a single beam in which the input optical pulse width could be varied from 460 fs to 5 ps, either by non-optimal pulse compression or by spectral filtering techniques. These pulses were then injected into a semiconductor traveling wave amplifier operating in the gain regime. The corresponding amplified pulse spectra were then measured with weak and strong optical injection into the traveling wave optical amplifier. The second type of experiment employed dual beams in a standard time resolved pump-probe geometry, where the time-resolved gain dynamics and time-resolved spectra could be measured.

The results of the single beam experiments are summarized in Figure 7.21. In Figure 7.21(a), the input optical pulse spectrum is shown. With low optical injection (<1 mW), the amplified optical spectrum replicates the input spectrum independent of input pulse duration. With high optical injection, the amplified pulse spectrum is severely distorted. In Figure 7.21(b), the output amplified optical pulse spectrum is shown for the case of an input pulse duration of 460 fs and an input average power of 5 mW to the traveling wave amplifier being studied. There are clearly two distinct spectral peaks which occur on both the high and low energy sides of the amplified pulse spectrum.

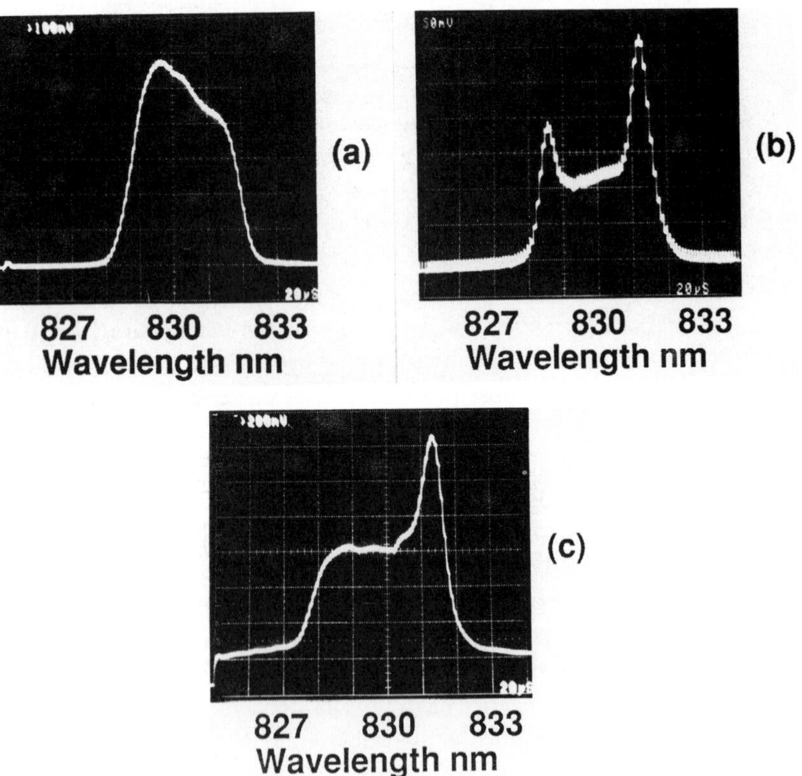

Figure 7.21. (a) Input optical spectrum to the traveling wave amplifier. (b) Output optical spectrum from the TWA with a 460 fs input pulse. (c) Output optical pulse spectrum from the TWA with a 2 ps input pulse.

In Figure 7.21(c), the output amplified optical pulse spectrum is shown for the case when the input optical pulse duration is 2 ps with an average power of 5 mW. The pulse broadening was accomplished by non-optimal pulse compression, resulting in a chirped injected pulse. This spectrum shows a distinct peak on the low energy side of the spectrum, while the high energy side is completely absent of any spectral peak. It should be noted that this effect was observed, independent of the sign of the residual chirp on the injected pulse, i.e., the effect is intensity dependent. Similar effects were observed with broadened pulses generated by the spectral filtering technique, however the spectral distortion is less dramatic owing to the narrow spectral width of the injected beam. Summarizing the single beam experiments, it is observed that under high injection power, the amplified optical pulse spectrum is severely distorted, where the spectral distortion is highly pulse-width dependent. The two regimes can be classified as spectral distortion which occurs when the input pulse duration is less 1 ps, and spectral distortion which occurs when the input pulse duration is greater than 1 ps.

In order to obtain a better understanding of the pulse-width dependent spectral distortion process, the time-resolved gain dynamics were measured with short pulses (~500 fs) and then with long pulses (~2 ps). The experimental setup used for these experiments is shown in Figure 7.22. Standard pump-probe techniques were used, accompanied by lock-in

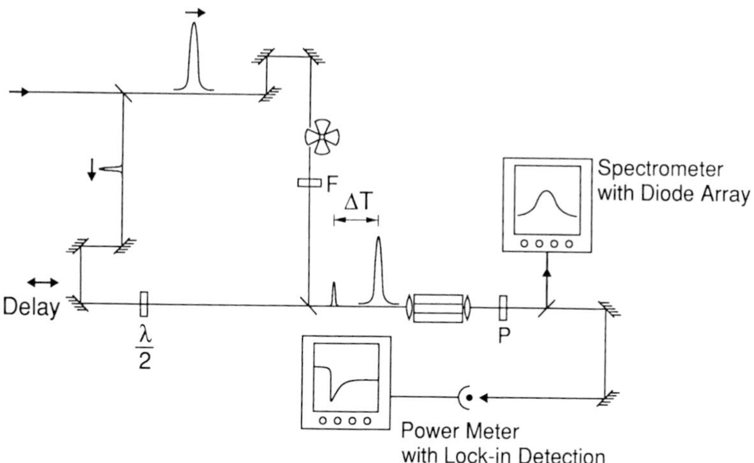

Figure 7.22. Schematic diagram of the pump-probe setup used for investigating the time-resolved gain dynamics and time-resolved spectral measurements.

detection methods. The input pump and probe beams were crossed polarized to facilitate selection of either beam at the output of the amplifier. The results of these experiments are summarized in Figure 7.23. Figure 7.23(a) shows a plot of the gain of the probe pulse with respect to the time delay of the pump pulse for the case of an input pulse duration of 0.5 ps. The salient features of the gain dynamics in this case are the rapid depletion of the probe gain near $t = 0$, the fast (~1.5 ps) partial recovery of the gain, and then the slow (~1.1 ns) gain recovery. The fast partial recovery is due to the dynamic carrier cooling effect, while the slow recovery is the normal carrier recovery. It should be noted that these measurements are consistent with previously measured values in AlGaAs lasers (Kesler and Ippen, 1987; Hultgren and Ippen, 1991); faster time constants have been measured in InGaAsP laser structures (Hall *et al.*, 1990).

In Figure 7.23(b), the gain dynamics of the traveling wave amplifier are shown for the case when the input optical pulse is the non-optimal compressed pulse with a duration of 2 ps. The salient feature of the gain dynamics in this case is the step-like response of the probe gain, which is the gain depletion and long carrier recovery. The dynamic carrier cooling effect does not play a major role in the gain dynamics for pulses longer than 2 ps. The gain dynamics in this case sharply contrast with the results which are obtained when the input pulse duration is 0.5 ps.

The spectral distortion observed in the single beam experiments can be directly related to the ultrafast gain dynamics once the effects of self-phase modulation are considered (Olsson and Agrawal, 1989; Agrawal and Olsson, 1989; Alfano, 1989; Shen, 1984). The time varying gain of the TWA leads to a temporally varying refractive index due to carrier depletion. From self-phase modulation theory, there will be an accompanying

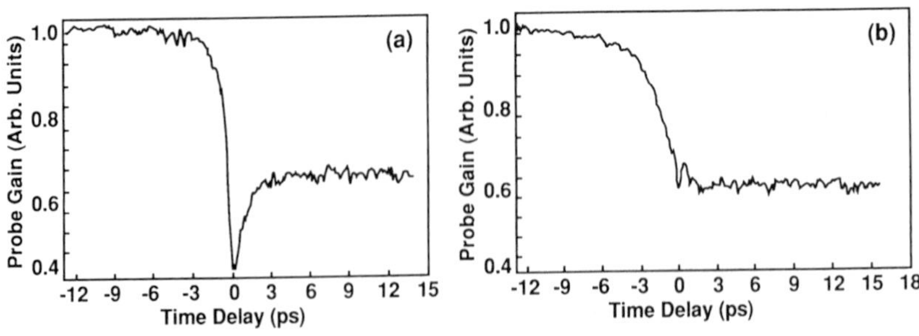

Figure 7.23. Gain dynamics obtained from time-resolved pump-probe measurements for (a) input pulse widths of 500 fs, and (b) input pulse widths of 2 ps.

frequency shift due to the rapidly changing refractive index. The instantaneous frequency follows

$$\omega_{\text{inst}} = \omega_0 - \frac{\omega_0 L}{c} \frac{\partial n(t)}{\partial t} \tag{7.5}$$

i.e., rapid increase in the refractive index lowers the carrier frequency and vice versa. In the above equation, ω_{inst} is the instantaneous frequency, ω_0 is the optical carrier frequency, L is the interaction length and $n(t)$ is the time varying refractive index.

The signal measured in the time-resolved gain experiments is proportional to

$$G(t) = \int_{-\infty}^{t} h(t-\tau) I^2(\tau) \mathrm{d}\tau \tag{7.6}$$

where $G(t)$ is the time varying gain, $h(t)$ is the system impulse response, and $I^2(\tau)$ is the intensity autocorrelation function of the optical pulses used in the experiment. For semiconductor amplifiers, it is important to realize that reductions in the gain lead to an increase in refractive index; increases in gain lower the refractive index. This is directly due to the change in carrier concentration, which changes the refractive index via the plasma effect (Thompson, 1980). Additional changes in the refractive index are also caused by the change in gain via the Kramers–Kroenig relation (Yariv, 1989). The salient feature is that the sign of the index change is the same for both mechanisms, when the optical frequency is at the center of the gain spectrum.

In the experiments described, the impulse response is pulse-width dependent. The most general form of the impulse response can phenomenologically be written as

$$h(t) = [a_1 \exp(t/\tau_1) + a_2(t/\tau_2)]u_{-1}(t) \tag{7.7}$$

where the as are constants, u_{-1} is the negative unit step response, and τ_1 and τ_2 are the characteristic gain recovery times due to the thermalization of hot carriers and normal gain recovery, respectively.

From our experiments, it is observed that when $\tau_1 < \tau_p < \tau_2$, the effect of carrier cooling is not observed, thus leaving only the second term in Equation 7.6. In this case, the time-resolved gain is just the integral of the pulse intensity, which ultimately leads to an instantaneous frequency shift which is directly proportional to the intensity envelope of the input optical pulse. In this case, the instantaneous frequency experiences a red shift which has its maximum deviation at the peak of the optical pulse. The carrier then recovers to its initial frequency as the pulse intensity decreases. For the case when $\tau_p \approx \tau_1$, the impulse response takes the form which is given by Equation 7.6. The instantaneous frequency shift of the input pulse is now modified by the carrier cooling effect. In this case, the gain depletion causes a red shift of the carrier frequency, however, the carrier cooling effect partially replenishes the gain, causing a rapid reduction in the refractive index which ultimately leads to a blue shift of the carrier frequency. These effects are shown in Figure 7.24(a–f), which plot the time dependent gain for the two pulse-width regimes, the resulting change in refractive index and the corresponding instantaneous frequency shift.

In order to confirm this model, experiments were performed to time-resolve the spectral distortion of a probe pulse. The setup is identical to the setup employed for the pump-probe gain dynamics (see Figure 7.22), with the spectrum of the probe beam being analyzed by a spectrometer with a diode array readout. Lock-in detection methods were not required for these experiments.

In Figures 7.25 and 7.26, plots of the time-resolved probe spectra and peak wavelength shift are shown for the short pulse and long pulse regimes, respectively. In these figures, the effects of phase modulation

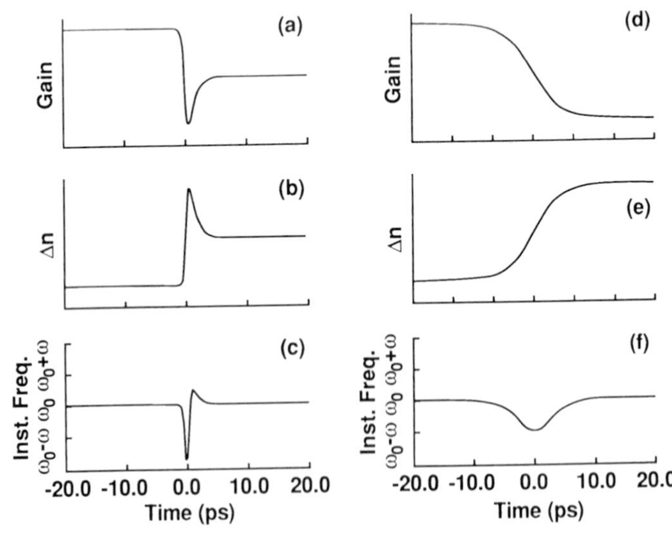

Figure 7.24. Simulation of the time-resolved gain dynamics for the short (~0.5 ps) and long (~2 ps) regime. (a–c) Gain, index, and instantaneous frequency for the short pulse regime. (d–f) Gain, index and instantaneous frequency for the long pulse regime.

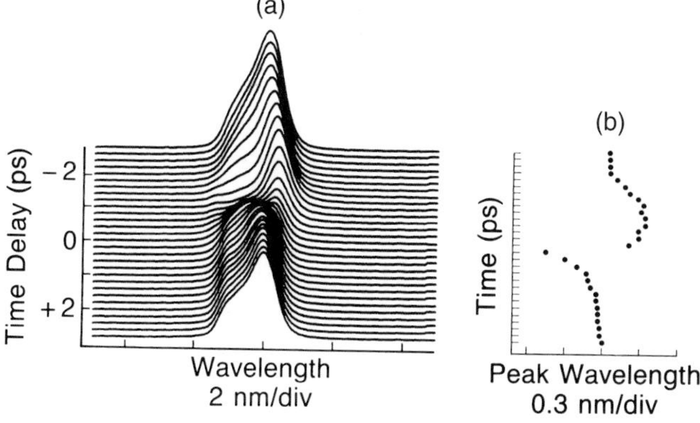

Figure 7.25. (a) Time-resolved spectra of the probe pulse for input pulses of 500 fs. (b) Plot of the peak wavelength shift versus time.

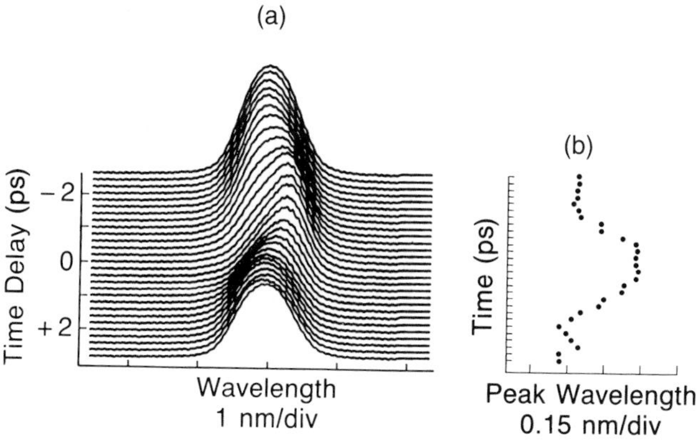

Figure 7.26. (a) Time-resolved spectra of the probe pulse for input pulses of 2 ps. (b) Plot of the peak wavelength shift versus time.

due to gain depletion and hot carrier thermalization are clearly observed. In Figure 7.25(a,b), for the case of an input pulse duration of ~0.5 ps, one observes a red shift at early times accompanied by a simultaneous reduction of the intensity of the probe spectra. This is due to the effects of index change and gain depletion. At maximum gain depletion near $t = 0$, the peak of the probe spectra sweeps through the original carrier frequency towards the high energy side of the spectrum. This blue peak is then observed to increase in intensity while simultaneously returning to the original carrier frequency as the gain recovers. This blue shift is a direct manifestation of the refractive index change due to the partial recovery of the gain by the thermalization of hot carriers in the conduction band of the amplifier. The carrier thermalization time can be obtained directly from Figure 7.25(b), noting that the instantaneous frequency shift is proportional to the derivative of the time varying index. From Figure 7.25(b), one obtains a carrier thermalization time of 1.5 ps, in excellent agreement with the value obtained from the time-resolved gain measurement of Figure 7.23(a).

In Figure 7.26(a), for an input pulse duration of 2 ps, a red shift is observed in the amplified probe pulse spectra as the gain is depleted. The maximum peak wavelength shift occurs at $t = 0$ in this case. At maximum gain depletion, the peak wavelength returns to its original position. One also observes that the shift of the optical pulse spectrum follows the intensity autocorrelation function, which is clearly observed in Figure 7.26(b), showing a 'pulse width' of ~2 ps. It should be stressed that the pulse broadening was accomplished by spectral filtering in this experiment to eliminate artifacts caused by the residual time-dependent frequency of the test pulse.

The maximum frequency deviation obtained in these experiments can be estimated from the maximum gain depletion. The single pass small signal gain $\exp(gL)$ of these devices is ~100. The maximum phase change can be obtained from the maximum change in gain divided by four, where a factor of two is included for the electric field gain and a second factor of two is included owing to the relative strengths of the real and imaginary parts of the susceptibility, χ_r and χ_i, according to the Kramers–Kroenig relation, i.e., $\chi_{r\,max} = \chi_{i\,max/2}$ (Yariv, 1989). Thus, the maximum frequency deviation Δf can be given as

$$\Delta f = \frac{1}{2\pi}\frac{gL}{4\tau_p} \tag{7.8}$$

with τ_p as the pulse width. For pulses of 1 ps in duration, this gives a maximum frequency deviation of 180 GHz, corresponding to 0.4 nm, which is in good agreement with the observed frequency shifts and other experimental observations (Agrawal and Olsson, 1989; Olsson and Agrawal, 1989; Hultgren and Ippen, 1991).

7.7 Applications of modelocked semiconductor laser diodes in synchronous optical networks

In this section, several applications of high power ultrafast semiconductor laser diodes will be discussed, as they pertain to synchronous optical networks. Future high speed networks may incorporate photonic techniques to provide a variety of functions, e.g., switching, processing and computing, and transmission, owing to the vast amounts of bandwidth that these technologies may provide. One example in particular is the timing or synchronization which may be required in some photonic network architectures. In these high data rate networks, precise timing information, or clocks, are required to assure the fidelity of the network and to allow the network to operate optimally at the data rates for which the network has been designed. Present electronic methods of performing these timing functions such as clock distribution, clock recovery and synchronization become difficult at high data rates. It has been demonstrated above that the timing jitter associated with the modelocked optical pulse train output is sufficiently low, such that this source may become an attractive alternative for providing precisely timed trains of optical pulses in synchronous high data rate networks. In the following sections, three scenarios which require synchronous timing information for photonic network applications will be discussed. These three scenarios are optical clock distribution, optical clock recovery and optical clock synchronization.

7.7.1 *Optical clock distribution*

The distribution of the master clock in high speed digital switching machines is a long-standing problem. Achieving appropriate clock phasing without distortion in a multi-chip high speed digital switching system is often the single most difficult obstacle to attaining theoretical switching speed limits. Design difficulties emerge when typical distances between synchronously clocked integrated circuits become comparable with vt, where v is the speed of signal propagation and t is the clock pulse rise time. Additional difficulties arise when signal speeds approach the point where typical integrated circuit gate delays are a significant fraction of $T/4$, where T is the clock period. Unfortunately, both of these situations usually occur concurrently. Under these circumstances, transmission lines are essential for the distribution of clock and data signals to each printed circuit board. In this case, impedance matching issues become extremely important. Owing to the limitations in electronic fanout capability, clock distribution is typically achieved by cascading many buffer amplifiers, and finally shipping the output of the last buffer via a point-to-point transmission line to the destination printed circuit board. In many applications, such as future generation telecommunications switching machines or high speed supercomputers, as many as 2000 printed circuit boards require a master clock. Usually, the fundamental hardware limit to the attainable system switching speed is not the switching limits of the transitors in the machine, but the clock skew limit imposed by the distribution technique employed (Hartman, 1986).

Photonics offers new insight into the solutions of these problems. The fiber medium is a low loss, low dispersion channel. Over short links, the channel is virtually bandwidth unlimited. The transport of baseband information over a fiber can be accurately thought of as the modulation of a carrier oscillating at approximately 2×10^{14} Hz. For this type of modulation the carrier need not be coherent, since direct detection methods are usually employed. With a 2×10^{14} Hz carrier, even extremely wideband digital signals of ~10 Gb/s convert to narrow band transmission within the fiber. Even over a few inches of fiber, millions of carrier wavelengths are traversed. Therefore, transmission line matching problems reduce to problems analogous to narrowband microwave matching.

The use of optics as an alternative in clock distribution has been demonstrated in the past (Hartman, 1986). Dynamic clock jitter as small as 50 ps has been reported for fanouts of 10. Most of this jitter is due to laser diode turn-on delay. The incorporation of a modelocked laser in the distribution scenario virtually eliminates this source of jitter. Furthermore, the use of laser amplifiers can significantly enhance the fanout capability if the output power of the laser amplifier is high enough. It should be mentioned that static clock skew, due to different path lengths, etc., is an additional system concern but is not specifically addressed in this section.

In this section, we describe an experiment in which the high power modelocked femtosecond semiconductor laser system is used as a source of extremely low jitter intense optical pulses. The experimental arrangement utilized in these experiments is diagrammed schematically in Figure 7.27. The master clock is the hybrid modelocked semiconductor laser system operating at 830 nm, with 0.5 ps optical pulses with 10 mW of average power. The optical pulse train produced by this system was distributed to 1024 ports using an optical fiber splitter. The fiber splitter consisted of a Corning 1:16 fiber splitter connected to a Canstar 64 port fiber splitter. Both fiber splitters are multimode, with 50 μm cores. The typical splitting loss at each port is 1 dB. The light was coupled to the input of the fiber splitter using a commercially available laser diode collimating lens. The typical input coupling efficiency to the fiber splitter was approximately 40%. At the output of the fiber splitter, the optical

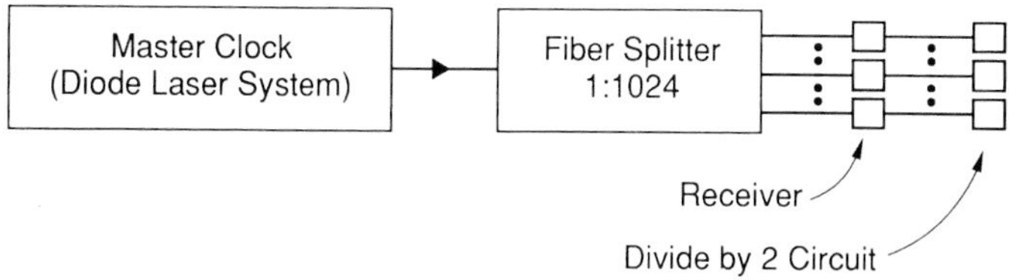

Figure 7.27. Schematic diagram of the optically distributed clocking network.

pulse train was detected and amplified using a commercial (Tektronix PG 701) 700 MHz OE receiver, with a 1V/mW sensitivity at 850 nm. The output of the receiver was then used to drive a divide-by-two ECL logic circuit to generate the square wave clocking waveform.

The pulse-to-pulse timing jitter of the modelocked pulse train was discussed earlier and measured to be 400 fs. The extremely low timing jitter and high power contained in the optical pulse train (over 70 W peak) are the key features that enable this optical source to be used as a master clock in an optically distributed clocking network.

Since, in this scheme, the idea is to distribute a well defined clock signal to many separate ports, or receivers, by splitting a single pulse, it is immediately noted that this superimposes several requirements on the photonic network. Since this is a single pulse splitting techique, it is seen that the port-to-port timing jitter is determined only by the transmission path and detection electronics. Being that the pulse is split many times, a high power optical source or many independent optical amplifiers are required to meet the large fanout requirement. In addition, this technique is not readily applicable to multiwavelength or wavelength division multiplexed networks, since, in this technique, a single pulse is providing the timing information.

Figure 7.28(a,b) shows a trace of the received optical clock and of the generated square wave clock signal. The clock signal is the output of a divide-by-two digital circuit, with the received pulse train shown in the top trace as the input. The received clock signal has $\sim 1\ \mu$W incident on the detector and a repetition rate of 302 MHz, with the generated clock frequency at 151 MHz. Both the received optical clock and the generated square wave clocking wave form are seen to be extremely stable and robust.

In Figure 7.29, an expanded view of the received clock signal, as displayed on the sampling oscilloscope, is shown. The sampled signal is extremely stable, suggesting that the received clock signal has not been

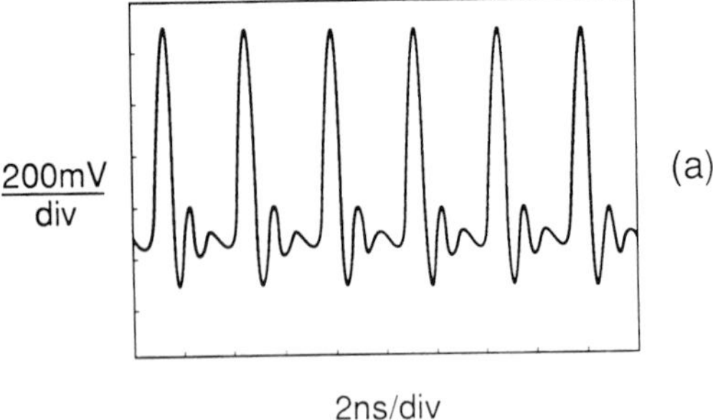

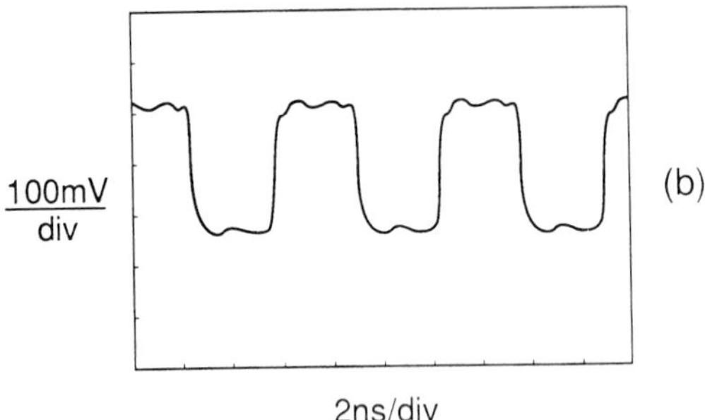

Figure 7.28. (a) Output of the OE receiver, showing the received optical clock after being split by a 1:1024 fiber splitter network. (b) Resulting square wave clocking waveform generated by the received clock and the divide-by-two circuit.

degraded by timing jitter. Unfortunately, frequency domain techniques to measure the timing jitter on this waveform are not suitable. This is due to the low bandwidth of the receiver, which does not allow for a sufficient number of harmonics to be analyzed in the power spectrum. In addition, it is likely that, as the clock signal is passively split, an additional jitter component may be added. This component, due to optical reflections within the fiber splitter structure, or even to modal noise, would add to the total jitter seen at a single port. Furthermore, since the jitter component represents a correlated (or colored) noise signal, it may not add in quadrature to the uncorrelated jitter sources. An experiment to measure this jitter contribution was performed.

Figure 7.30 shows the experimental arrangement which is used to measure the timing jitter. The output of the correlator $V_C(t)$ is proportional to

$$V_C(t) = V_1(t + \Delta T) \times V_2(t + \delta t) \tag{7.9}$$

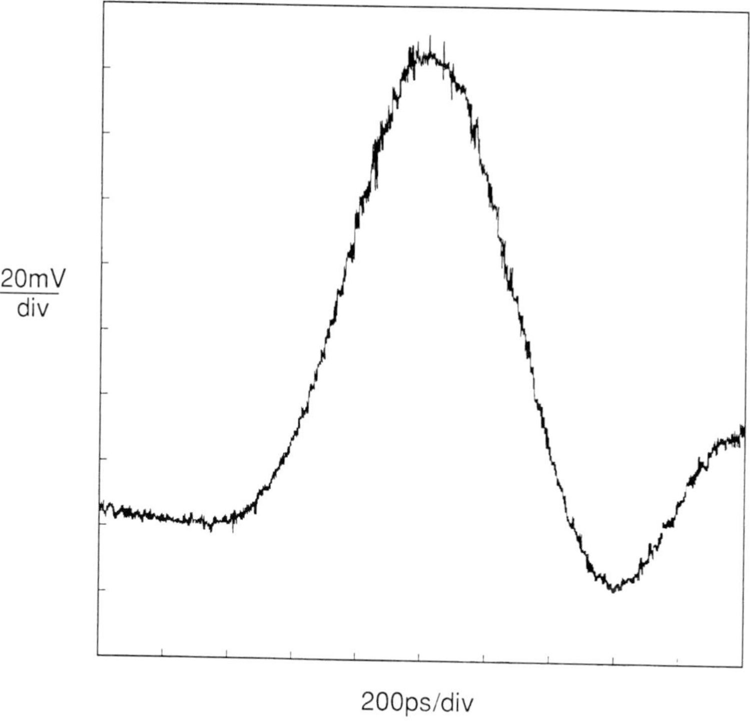

Figure 7.29. Expanded view of the received optical clock as seen on a 20 GHz sampling oscilloscope.

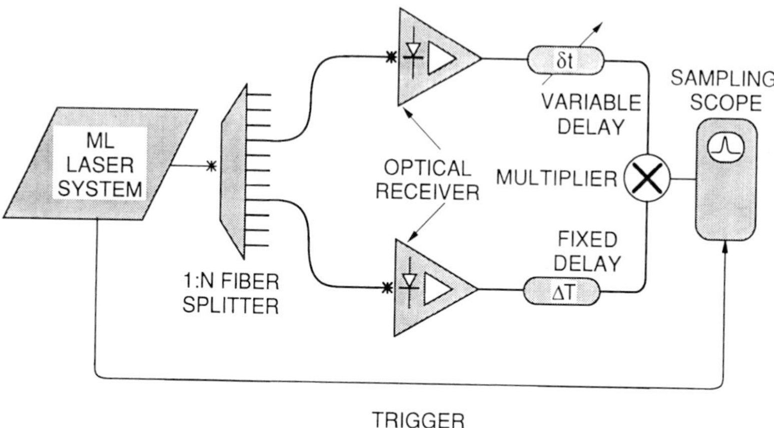

Figure 7.30. Schematic diagram of the correlated jitter measurement.

where V_1 and V_2 are the receiver output voltages of port 1 and port 2, respectively. Under steady-state conditions, but with a jitterless clock source present, the correlator is initially arranged so that the output of the microwave mixer yields a correlation peak. Next, δt is adjusted so that the output of the microwave mixer is not perfectly correlated. Under these circumstances, the height of the correlation peak is a sensitive function of the relative delay between inputs 1 and 2. Using a broadband amplifier as the filter in the correlator, one can measure small changes in the relative delay between two ports on a fast time scale. In our experiments, an equivalent 20 GHz bandwidth filter (sampling oscilloscope) was used as the band limiting element. By measuring the correlation peak voltage at various different time delays δt, one obtains a calibration of correlation voltage change per unit time delay. Under these circumstances, we measured a correlation jitter sensitivity of approximately 0.6 ps/mV. By using a maximum/minimum excursion feature on the sampling oscilloscope in our experiment, we were able to investigate the total dynamic clock arrival time variation (jitter) between any two outputs over a period of time. In Figure 7.31, the maximum/minimum excursion

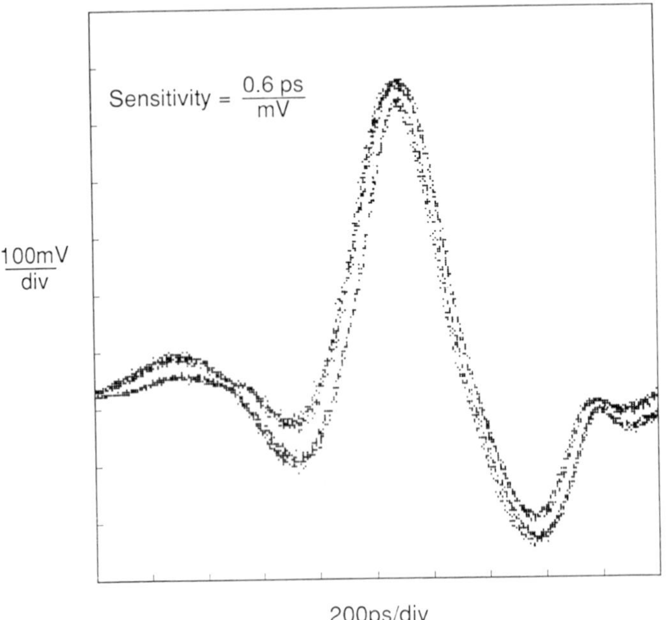

Figure 7.31. Output signal of the correlator showing the maximum and minimum excursions, yielding a jitter of 12 ps.

voltage of the correlator is displayed, which shows a correlation peak voltage excursion of approximately 20 mV. Measurements of this kind on any two output ports yielded a total pulse-to-pulse jitter of approximately 12 ps over measurement periods of approximately 1 hour. This 12 ps result represents total excursion ($>6\sigma$) and includes all sources of jitter, correlated and uncorrelated, rapidly varying and slowly varying, including optical reflections, possibilities of modal noise, and temporal and thermal variations of the electronics. It is emphasized that in spite of the fact that multimode optics were employed, no modal noise effects were observed. Even in our case, where a 1:16 splitter was connected via multimode connectors to a 1:64 splitter of a different variety (different vendor and different design type), no evidence of modal noise effects could be found. The temporal jitter obtained in these measurements is more than an order of magnitude improvement over what can be achieved by conventional electrical or direct optical modulation techniques. One can estimate the maximum fanout capability of the present clock network by examining the power budget. The maximum number of output ports N_{max} to be driven can be given by

$$N_{max} = (N_{ph})(hc/\lambda)(B/P_{min})L^{-1} \tag{7.10}$$

Here, N_{ph} is the number of photons per bit, L is the loss factor, P_{min} is the minimum detection power, B is the bit-rate, and h, c, and λ are Planck's constant, the speed of light, and the operating wavelength, respectively. In the limiting case of no optical loss in the coupling and splitting of the master optical clock signal ($L = 1$), and with a minimum detectable power of 1 μW, one could expect the maximum fanout capability to be $\sim$7000, with a clock rate of 300 MHz. Higher clocking rates will result in a proportionally smaller split ratio, due to the finite number of photons that one can extract from the laser amplifier used in the experiment, i.e., doubling the repetition rate reduces the split factor by two since the number of photons per pulse have been reduced by a factor of two. Additional reductions in the fanout capabilities also arise at higher repetition rates owing to the larger power required by the receivers.

In this section, it was demonstrated how a wideband electrical clock signal could be transformed into a narrowband optical signal and distributed with optical fibers to provide timing information in a photonic network. Below, we give an example where a narrowband electrical clock signal is distributed before it is converted to an optical timing signal.

7.7.2 *Synchronization of hybrid modelocked ultrafast semiconductor laser diode systems*

In certain applications, such as network synchronization, photonic switching, or other ultrafast multibeam optical techniques, precise temporal generation of ultrashort optical pulses from several individual sources may be required. In this section, two hybrid modelocked semiconductor laser systems are synchronized by a single coherent electrical driving signal. The advantage of this technique is that direct modulation techniques of each diode can be performed, thus providing the synchronization, without any sophisticated control electronics to control the relative timing jitter. In addition, the distribution of a narrowband electrical signal is far more simple than distributing a wideband electrical signal, since dispersive effects over the transmission media are eliminated. Unfortunately, a narrowband electrical timing signal, such as a sine wave, may not be the optimum choice for a clock signal because the rise-time of a sine wave is slow compared to a square wave of the same frequency. This shallow slope of the sine wave signal may transfer small amplitude fluctuations into timing fluctuations from electronic circuits which are triggered at well defined voltage levels. This is a classic example of how amplitude fluctuations are converted into phase or timing fluctuations of a signal. An attractive feature, however, is that the narrowband signal may be globally broadcast, as with a geosynchronous satellite, to many independant locations, providing clocking information over an extremely wide area. Since the narrowband electrical signal is used to drive independant modelocked diode lasers, this technique would allow for synchronization between photonic networks operating at different wavelengths, e.g., 850 nm, 1310 nm or 1550 nm, or for several closely spaced wavelengths that may be encountered in wavelength division multiplexed networks.

The experimental arrangement is shown in Figure 7.32. Two individual external cavity hybrid modelocked semiconductor diode laser systems are constructed from the AlGaAs angled stripe traveling wave optical amplifiers. Each laser system is initially passively modelocked with AlGaAs multiple quantum wells relying on the excitonic absorption to provide the saturable absorption. The cavity length, bias current, RF drive and wavelength of each laser are adjusted for an identical operating repetition rate of ~335 MHz. The pulses from each laser system are amplified with identical semiconductor traveling wave amplifiers to increase the peak power. Dual grating compressors are then encorporated to compensate

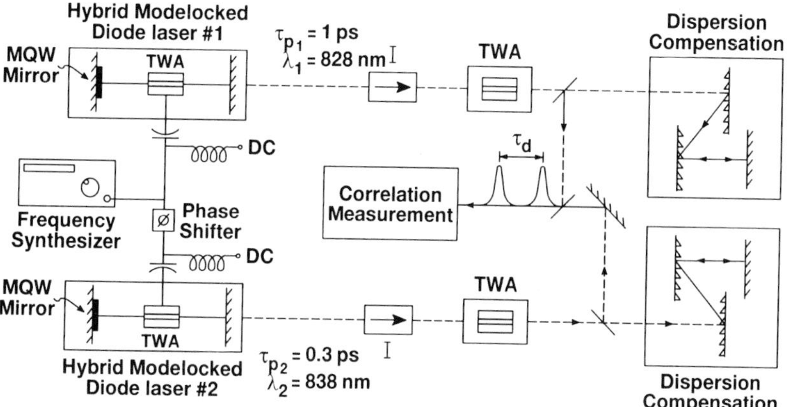

Figure 7.32 Schematic diagram of the experimental setup showing two hybrid modelocked laser systems driven by a single oscillator. Each system contains a modelocked oscillator, an external optical amplifier, and a dispersion compensator.

linear chirp impressed on the optical pulses during modelocking. The output characteristics from laser system 1 are: wavelength 828 nm, pulse duration 0.5 ps, peak power 70 W. Laser system 2 emits 0.2 ps pulses at 838 nm, with 165 W peak power. Both systems operate at 335 MHz. The peak power obtained from these laser systems are sufficient for inducing fundamental nonlinear optical phenomena required for applications such as photonic switching.

The relative timing jitter between the two diode laser systems can be obtained by utilizing nonlinear optical cross-correlation methods. The measurement method is illustrated schematically in Figure 7.33. The two beams from each laser system are directed into a commercially available rotating mirror autocorrelator, with a small relative time delay, τ_d, between the pulses in each beam. The measured signal is the autocorrelation of this two-pulse input sequence. With the correlation of two pulses I_1 and I_2 given as

$$R_{12}(\tau) = \int_{-\infty}^{\infty} I_1(t) I_2(t+\tau)\,\mathrm{d}t \tag{7.11}$$

it is easy to show that the measured signal from the autocorrelator gives

$$S(\tau) = R_{11}(\tau) + R_{22}(\tau) + R_{12}(\tau - \tau_d) + R_{21}(\tau + \tau_d), \tag{7.12}$$

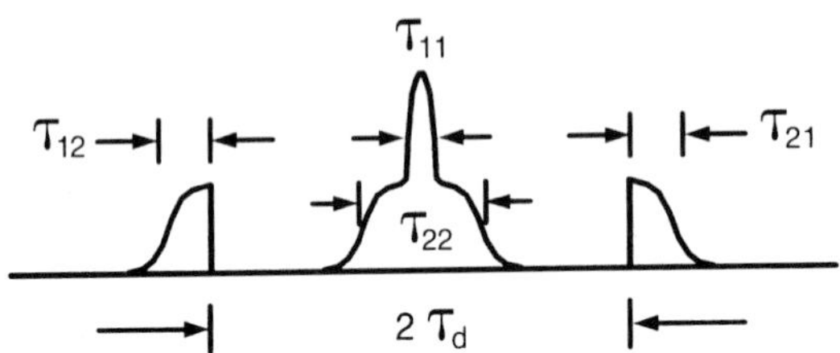

Figure 7.33. Schematic diagram of the two-pulse correlation measurement technique.

where R_{ii}, $i = 1, 2$, is the autocorrelation function and R_{ij} is the cross-correlation function between signals 1 and 2. The signal maintains all the properties associated with conventional correlations, e.g., $R_{ij}(\tau) = R_{ji}(-\tau)$.

In the limiting case of no temporal variation of the time delay τ_d between the two pulses, i.e., no relative timing jitter, the measured signal gives information about the individual pulse widths, and cross-correlational pulse widths. If one pulse is substantially shorter than the other, i.e. $\tau_1 << \tau_2$, then the cross-correlation signal gives information about the intensity temporal shape of pulse 2.

To show these features, a measurement was performed with a 12 ps uncompressed pulse and a 0.2 ps compressed pulse from one of the diode laser systems. Figure 7.34 shows the signal generated from the double pulse correlation measurement. The central peak is the sum of the autocorrelation signals from both the compressed and uncompressed pulses, yielding a narrow signal superimposed on a broad base. The side bands show the cross-correlation between the uncompressed and compressed pulses, showing a fast rising edge with a slow trailing edge. This type of pulse shape is commonly encountered from modelocked diode lasers and has been previously measured using streak camera diagnostics, as shown in Figure 7.19.

The second case is encountered when the relative timing jitter is greater than the maximum autocorrelation width. In this case, the cross-

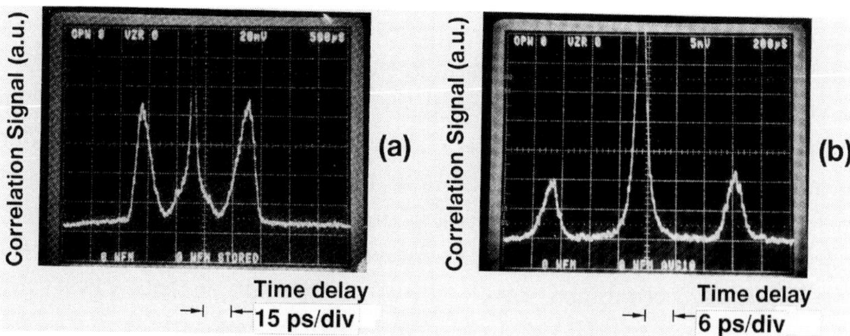

Figure 7.34. (a) Displayed correlation signal, showing the result obtained with a pulse pair from one laser system. The input pulse parameters are 12 ps and 0.3 ps. The result shows the sum of the individual autocorrelation terms located at $\tau = 0$, and the cross-correlation terms located at $\pm\tau$ show the exponential like pulse shape generated from the modelocked oscillator before dispersion compensation. (b) Displayed correlation signal, showing a 3 ps relative timing jitter between the laser systems, as shown in the correlation side peak signals. The calculated jitter is 2.8 ps with the input pulse widths of 1 and 0.3 ps from the individual laser.

correlation signal is a measure of the temporal distribution of the timing jitter, and is proportional to R_{ij} convolved with the jitter distribution $J(t)$,

$$R'_{ij}(\tau) = \int_{-\infty}^{\infty} J(t)R_{ij}(t+\tau)\mathrm{d}t. \tag{7.13}$$

Relying on these features, the relative timing jitter between the two laser systems can be experimentally measured. The resultant correlation signal is displayed in Figure 7.34(b). The correlation signal displays the expected $2n - 1$ peaked signal ($n = 2$). The center peak is the sum of the autocorrelations of both laser pulse 1 and laser pulse 2, i.e., R_{11} and R_{22}. The two side peaks show R'_{ij} which contains information about the cross-correlation pulse width R_{ij} and the relative timing jitter distribution $J(t)$. The pulse widths used in this experiment are 1 ps and 0.3 ps. The FWHM of the cross-correlation peak is 3 ps, which in our case is a measure of the relative timing jitter. This measurement was obtained by averaging the correlation signal for > 2 sec. Assuming a Gaussian distribution for the timing jitter $J(t)$, the jitter can be calculated from

$$\tau_{12}^2 = \tau_{11}^2 + \tau_{22}^2 + \tau_{\mathrm{J}}^2. \tag{7.14}$$

From this one calculates a relative timing jitter of 2.8 ps. This shows that high power modelocked semiconductor diode laser systems can be highly synchronized without any external feedback control mechanism, thus allowing for cost-effective synchronization techniques at independent remote locations. The results demonstrated in these experiments may prove useful for future high speed add-drop multiplexers, network synchronization utilizing the building integrated timing supply architecture (BITS), and in combined wavelength division multiplexed (WDM) time division multiplexed (TDM) lightwave networks. It should be noted that the 2.8 ps result obtained in this experiment represents an upper bound on the relative timing jitter, i.e., the maximum relative timing jitter can be as low as 3 ps for diode laser systems without external feedback control mechanisms.

The relative timing jitter measured in Figure 7.34(b) has several distinct components. These sources may be minimized with appropriate laser system design. The main sources of timing jitter are cavity length fluctuations, spontaneous emission fluctuations, and fluctuations in the RF driving source. Spontaneous emission fluctuations can be suppressed by using microcavity laser structures, such as vertical cavity surface emitting lasers and microdisk lasers, owing to their low spontaneous emission properties. Fluctuations in the DC drive source contribute to spontaneous emission fluctuations, hence low noise current sources will assist in minimizing this effect. Phase noise fluctuations encountered in the RF drive signal may still be the most influential mechanism associated with the timing jitter. Low noise crystal oscillators designed for a specific frequency of operation will be superior to standard frequency synthesizers which are made to operate over extremely wide ranges. Additional sources of jitter can be introduced by the external optical amplifier stage and by fluctuations in the chirp impressed on the optical pulse during modelocking. The additional jitter introduced by the external amplifier is expected to be minimal owing to the fact that the optical pulse is modified by the amplifier on a single pass basis only, thus limiting the amount of timing variation which can occur on the pulse. The latter case of temporal variations on the phase of the optical pulse can lead to larger effects, owing to the temporal shift which can be imparted by the dispersion compensator, i.e., phase shifts in frequency transform to temporal shifts in time.

In summary, two high power hybrid modelocked semiconductor laser systems were synchronized utilizing only a common RF driving source. The resultant timing jitter was measured using a correlational method, displaying a relative timing jitter of 3 ps measured over a 2 second interval. The measured jitter contains all sources, from external driving sources, environmental fluctuations, and from sources concerned with the detailed physics involved with pulse production, amplification and compression.

7.7.3. *Optical clock recovery*

The third technique of generating a synchronized optical clocking signal relies on the optical injection of one hybrid modelocked laser to synchronize a separate passive modelocked laser, in a master–slave configuration, as illustrated schematically in Figure 7.35. This technique is appropriate for providing an all-optically recovered clocking signal from an intensity modulated optical pulse train. This technique may prove useful in applications of all-optical signal regeneration, all-optical signal processing, photonic switching, all-optical add-drop multiplexers, or other types of all-optical multi-pulse processing techniques.

The salient features of this technique are that it provides a synchronized control pulse at a receiver location, without any *a priori* knowledge of the information contained within the optical data train. An additional advantage of this technique is that large phase changes, which may occur owing to different times of arrival of optical data packets from different master lasers, can be tracked by the slave laser. This technique, however, requires one to utilize a small portion of the optical data signal and as a result, is immediately most suitable for single wavelength networks.

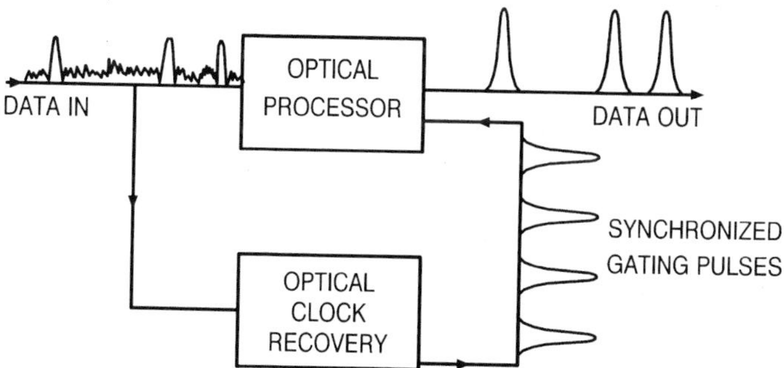

Figure 7.35. Schematic diagram of all-optical clock recovery oscillator. Potential applications include photonic switching, add-drop multiplexors, CDMA and holographic pulse processing.

The lasers used in the experiment are identical to those described earlier. They are comprised of three main components; a low power hybrid modelocked master oscillator, a high power single pass semiconductor traveling wave optical amplifier and a grating compensator for pulse compression. The wavelength and duration of the pulses generated from these laser systems are 830 nm and ~500 fs, respectively, with ~10–12 mW of average output power.

As mentioned earlier, the method used for synchronizing the master hybrid modelocked laser with a slave passive modelocked laser requires an injection seed pulse from the master laser into the slave laser. Both the master and slave lasers must have similar spectral characteristics. The free running slave laser shows considerable phase noise sidebands associated with pulse-to-pulse timing jitter, which is characteristic of passive modelocked lasers (Figure 7.37(a)). Since the slave laser is not synchronized, the optical pulses will appear randomly on a sampling oscilloscope (Figure 7.36(a). With ~100 μW of optically injected power from the master laser into the slave laser, the optical pulses became synchronized

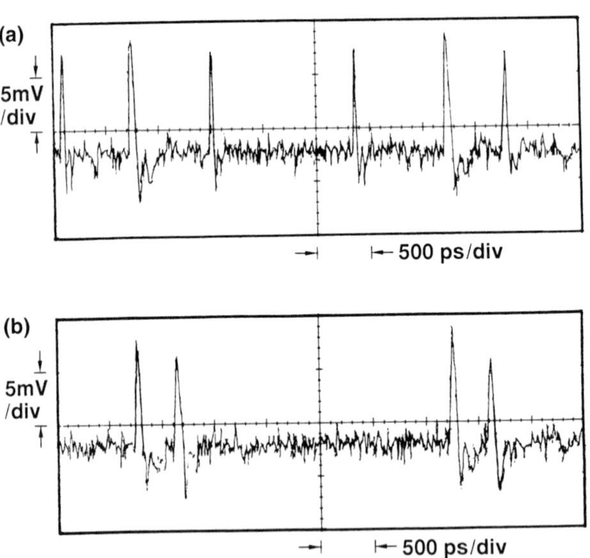

Figure 7.36. Time domain results of the optical clock recovery oscillator. (a) Displayed signals of the master laser and the slave laser without optical injection. (b) Same as in (a), except with optical injection into the slave laser. The salient feature is the reproduction of the clock signal from the slave laser as compared with the free running mode without optical injection.

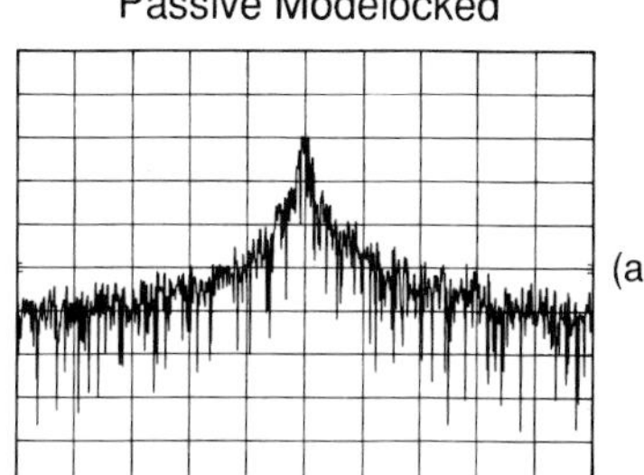

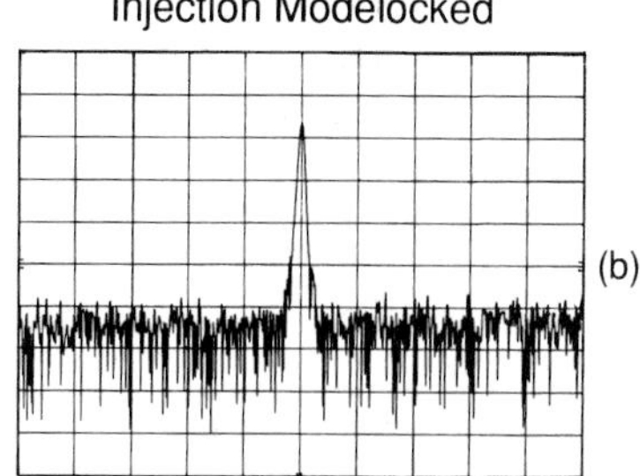

Figure 7.37. Frequency domain results of the optical clock recover oscillator. (a) Displayed power spectrum of the free running clock, showing large phase noise sidebands, which are attributed to timing jitter. (b) Same as in (a) except with optical injection into the slave laser. The salient feature is the dramatic reduction of the phase noise sidebands, showing the increased robustness of the recovered optical clock signal.

with the master laser, as observed on the sampling oscilloscope and by the elimination of the phase noise sidebands of the slave laser (Figure 7.37(a,b)). The resultant pulse-to-pulse timing jitter on the slave laser was measured to be below $\sim$3 ps.

In practice, the system issues which need to be considered are (1) the injected power required to initiate a recovered clock signal, (2) the spectral mismatch which can be tolerated by the slave oscillator, (3) the cavity mismatch of the slave laser, and (4) the tolerance of the slave oscillator to a large number of 'zeros' in the data stream and the required number of 'ones' in order to recover the clock signal.

7.8 Conclusion and future directions

In conclusion, several techniques useful for generating high power ultrafast optical pulses from diode lasers have been described. These techniques are active modelocking, passive/hybrid modelocking, quadratic phase compensation, and cubic phase compensation. Insight to the limiting mechanisms for high power ultrashort pulse generation has also been provided by time-resolved gain dynamics and time-resolved spectra, and also by investigating the intracavity pulse shaping mechanisms. Intracavity chirp compensation has proven to be challenging for the generation of femtosecond optical pulses in external cavity diode lasers, however, alternative methods for intracavity compensation of frequency chirp which allows for larger dispersion, such as grating compressors, may prove fruitful for this task. In this case, the detrimental effects of the pulse-width dependent gain saturation and spectral distortion must be avoided if high output peak powers are desired. It should also be noted that the type of intracavity dispersion required for semiconductor lasers is *dynamic*, i.e., the chirp impressed on the optical pulse during modelocking is greatly influenced by the different time scales involved in the gain dynamics. As a result, the compensation technique must change as the gain dynamics change owing to the compression of the injected optical pulses from the compensator.

The techniques presented in this chapter can be extended to yield potentially even shorter and higher output peak powers. The emission bandwidth of the TWA can be broadened by utilizing quantum well traveling wave amplifiers, and the absorber design can be extended to cover a larger spectral region. Higher output powers can be achieved by modifying the external amplifier, e.g., utilizing multistripe or broad area devices. In addition, pulse stretching prior to amplification may prove useful (Hansen *et al.*, 1989; Lee and Delfyett, 1991); experimentally, the average amplifier output powers were higher in the active modelocked system, where 15 ps optical pulses were used at 960 MHz, as compared with the 5–10 ps pulses at 335 MHz.

From the experimental techniques described in this study, it is possible to generate optical pulses of 200 fs in duration with over 165 W of peak power, making these pulses both the shortest and most intense ever generated from an all-semiconductor injection diode laser system. The pulses generated from this system are sufficiently powerful for inducing nonlinear phenomena in waveguide structures, which may influence photonic switching applications.

In addition, several applications have been highlighted to show how modelocked semiconductor diode lasers may be utilized in future photonic networks, in terms of providing an easy method for generating accurately timed trains of optical pulses for timing information. These applications were optical clock distribution, optical clock synchronization, and optical clock recovery. Areas for future research would necessarily include novel packaging such that optical bench research prototypes can be incorporated into a meaningfully useful marketable product. The results in this work show the potential impact that compact diode laser sources may have in future lightwave and photonic applications.

Acknowledgments

The work described in this Chapter was performed at Bell Communications Research, Red Bank, New Jersey, 07701. The author would like to acknowledge the collaborative interactions with several colleagues: Gerard Alphonse, Nick Andreadakis, Leigh Florez, Tom Gmitter, Davis Hartman, John Heritage, Chang-Hee Lee, Yaron Silberberg, Peter Smith, Ned Stoffel, and Andrew Weiner.

References

Agrawal, G. P. and Olsson, N. A. (1989) *IEEE J. Quantum Electron.*, **25**, 2297.

Alfano, R. R. (ed.) (1989) *The Supercontinuum Laser Source* (Springer Verlag, New York).

Alphonse, G. A., Gilbert, D. B., Harvey, M. G. and Ettenberg, M. (1988) *IEEE J. Quantum Electron.*, **24**, 2454.

Andrews, J. R. and Burnham, R. D. (1986) *Appl. Phys. Lett.*, **49**, 1004–6.

Aspin, G. J. and Carroll, J. E. (1979) *Solid State and Electr. Dev.*, **3**, 220–3.

AuYeung, J. C., Bergman, L. A. and Johnston, A. R. (1982) *Appl. Phys. Lett.*, **41**, 124–6.

Bowers, J. E., Morton, P. A., Mar, A. and Corzine, S. W. (1989) *IEEE J. Quantum Electron.*, **QE-25**, 1426–39.

Bradley, D. J., Holbrook, M. B. and Sleat, W. E. (1981) *IEEE J. Quantum Electron.*, **QE-17**, 658-662.

Chemla, D. S., Miller, D. A. B., Smith, P. W., Gossard, A. C. and Wiegmann, W. (1984) *IEEE J. Quantum Electron.*, **QE-20**, 265–75.

Chen, J., Sibbett, W. and Vukusic, J. I. (1984) *Opt. Commun.*, **48**, 427–31.

Corzine, S. W., Bowers, J. E., Przybylek, G., Koren, U., Miller, B. I. and Soccolich, C. E. (1988) *Appl. Phys. Lett.*, **52**, 348–50.

Delfyett, P. J., Lee, C.-H., Alphonse, G. A. and Connolly, J. C. (1990) *Appl. Phys. Lett.*, **57**, 971.

Delfyett, P. J., Hartman, D. H. and Ahmad, S. Z. (1991a) *J. Light. Tech.*, **9**, 1646–9.

Delfyett, P. J., Gayen, S. K. and Alfano, R. R. (1991b) in *Ultrafast Laser Technology*, R. A. Meyers (ed.), *Encyclopedia of Lasers and Optical Technology* (Academic Press, San Diego, CA).

Derickson, D. J., Morton, P. A. and Bowers, J. E. (1991) *Appl. Phys. Lett.*, **59**, 3372–4.

Eisenstein, G., Hansen, P. B., Weisenfeld, J. M., Tucker, R. S. and Raybon, G. (1988) *Appl. Phys. Lett.,* **53**, 1239–41.

Fork, R. L., Brito Cruz, C. H., Becker, P. C. and Shank, C. V. (1987) *Opt. Lett.,* **12**, 483–5.

Gomatam, B. N. and DeFonzo, A. P. (1990) *IEEE J. Quantum Electron.,* **26**, 1689–1704.

Hall, K. L., Mark, J., Ippen, E. P. and Eisenstein, G. (1990) *Appl. Phys. Lett.,* **56**, 1740–2.

Hansen, P. B., Weisenfeld, J. M., Eisenstein, G., Tucker, R. S. and Raybon, G. (1989) *IEEE J. Quantum Electron.,* **25**, 2611–20.

Harder, C., Smith, J. S., Lau, K. Y. and Yariv, A. (1983) *Appl. Phys. Lett.,* **42**, 772–4.

Hartman, D. H. (1986) *Opt. Eng.*, **25**, 1086–1102.

Haus, H. (1975) *IEEE J. Quantum Electron.,* **QE-11**, 736-46.

Haus, H. A. (1980) *IEE Proc.*, **127**, 323–29.

Haus, H. and Silberberg, Y. (1985) *J. Opt. Soc. Am.,* **B2**, 1237–43.

Heritage, J. P., Thurston, R. N., Tomlinson, W. J., Weiner, A. M. and Stolen, R. H. (1985) *Appl. Phys. Lett.*, **47**, 87–9.

Ho, P.-T. (1979) *Electron. Lett.*, **15**, 526–7.

Ho, P.-T., Glasser, L. A., Ippen, E. P. and Haus, H. A. (1978) *Appl. Phys. Lett.*, **33**, 241–2.

Holbrook, M. B., Sleat, W. E. and Bradley, D. J. (1980) *Appl. Phys. Lett.*, **37**, 59–61.

Hultgren, C. T. and Ippen, E. P. (1991) *Appl. Phys. Lett.*, **59**, 635–7.

Ippen, E. P., Eilenberger, D. J. and Dixon, R. W. (1980) *Appl. Phys. Lett.*, **37**, 267–9.

Ito, H., Yokoyama, H., Murata, S. and Inaba, H. (1979) *Electron. Lett.*, **15**, 738–40.

Keller, U., Li, K. D., Rodwell, M. and Bloom, D. M. (1989) *IEEE J. Quantum Electron.* **25**, 280–7.

Kesler, M. P. and Ippen, E. P. (1987) *Appl. Phys. Lett.*, **51**, 1765.

Kuhl, J., Serenyi, M. and Gobel, E. O. (1987) *Opt. Lett.*, **12**, 334.

Kuznetsov, M., Weisenfeld, J. M. and Radzihovsky, L. R. (1990) *Opt. Lett.,* **15**, 180–2.

Lai, Y., Hall, K. L. and Ippen, E. P. (1990) *IEEE Phot. Tech. Lett.*, **2**, 711–13.

Lau, K. Y. (1989) *IEEE J. Light. Tech.*, **7**, 400–19.

Lau, K. Y., Ury, I. and Yariv, A. (1985) *Appl. Phys. Lett.*, **46**, 1117–19.

Lee, C.-H. and Delfyett, P. J. (1991) *IEEE J. Quantum Electron.,* **27**, 1110–14.

Liu, H.-F., Fukazawa, M., Kawai, Y. and Kamiya, T. (1989) *IEEE J. Quantum Electron.,* **QE-25**, 1417–25.

Liu, H.-F., Ogawa, Y. and Oshiba, S. (1991) *Appl. Phys. Lett.,* **59**, 1284–6.

Lowery, A. J. (1988) *Electron. Lett.,* **24**, 1125.

Maine, P., Strickland, D., Bado, P., Pessot, M. and Mourou, G. (1988) *IEEE J. Quantum Electron.,* **QE-24**, 398–403.

Marcuse, D. (1989) *IEEE J. Light. Tech.,* **7**, 336–9.

Martinez, O. E. (1987) *IEEE J. Quantum Electron.,* **QE-23**, 59.

McInerney, J., Reekie, L. and Bradley, D. J. (1985) *Electron. Lett.,* **21**, 117–18.

Miller, D. A. B. (1987) *Opt. Eng.,* **26**, 368–72.

Miller, D. A. B. (1989) *Opt. Lett.,* **14**, 146–8.

Miller, D. A. B., Chemla, D. S., Smith, P. W., Gossard, A. C. and Wiegmann, W. (1983) *Opt. Lett.,* **8**, 477–9.

Morton, P. A., Helkey, R. J. and Bowers, J. E. (1989) *IEEE J. Quantum Electron.,* **25**, 2621–3.

Morton, P. A., Bowers, J. E., Koszi, L. A., Soler, M., Lopata, J. and Wilt, D. P. (1990) *Appl. Phys. Lett.*, **56**, 111–13.

Nakazawa, M., Suzuki, K. and Yamada, E. (1990a) *Electron. Lett.,* **26**, 2038–40.

Nakazawa, M., Kurokawa, K., Kubota, H., Suzuki, K. and Kimura, Y. (1990b) *Appl. Phys. Lett.,* **57**, 653–5.

Olsson, N. A. and Agrawal, G. P. (1989) *Appl. Phys. Lett.,* **55**, 13.

Putnam, R. S. and Salour, M. M. (1983) *Proceedings of Society of Photo-Optical Instrumentation Engineers,* **439**, 66.

Rodwell, M. J. W., Bloom, D. M. and Weingarten, K. J. (1989) *IEEE J. Quantum Electron.,* **25**, 817–27.

Schell, M., Weber, A. G., Bottcher, E. H., Scholl, E. and Bimberg, D. (1991) *IEEE J. Quantum Electron.,* **27**, 402–9.

Shen, Y. R. (1984) *The Principles of Nonlinear Optics* (Wiley, New York).

Silberberg, Y. and Smith, P. W. (1986) *IEEE J. Quantum Electron.,* **QE-22**, 759–61.

Silberberg, Y., Smith, P. W., Eilenberger, D. J., Miller, D. A. B., Gossard, A. C. and Wiegmann, W. (1984) *Opt. Lett.,* **9**, 507–9.

Smith, P. W. (1984) *Phil. Trans. R. Soc. Lond. A,* **313**, 349–55.

Smith, P. W., Silberberg, Y. and Miller, D. A. B. (1985) *J. Opt. Soc. Am. B,* **2**, 1228–36.

Stallard, W. A. and Bradley, D. J. (1983) *Appl. Phys. Lett.,* **43**, 626–8.

Stix, M. S., Kesler, M. P. and Ippen, E. P. (1986) *Appl. Phys. Lett.,* **48**, 1722–4.

Taylor, A. J., Weisenfeld, J. M., Eisenstein, G. and Tucker, R. S. (1986) *Appl. Phys. Lett.,* **49**, 681–3.

Thompson, G. H. B. (1980) *Physics of Semiconductor Laser Devices* (J. Wiley Inc., New York).

Tohyama, M., Takahashi, R. and Kamiya, T. (1991) *IEEE J. Quantum Electron.,* **QE-27**, 2201–9.

Valdmanis, J. A. and Fork, R. L. (1986) *IEEE J. Quantum Electron.,* **QE-22**, 112–18.

Valdmanis, J. A., Mourou, G. and Gabel, C. W. (1982) *Appl. Phys. Lett.,* **41**, 211–12.

Valdmanis, J. A., Mourou, G. A. and Gabel, C. W. (1983) *IEEE J. Quantum Electron.,* **19**, 664–7.

Valdmanis, J. A., Fork, R. L. and Gordon, J. P. (1985) *Opt. Lett.,* **10**, 131.

van der Ziel, J. P. (1981) *J. Appl. Phys.,* **52**, 4435–46.

van der Ziel, J. P. and Logan, R. A. (1983) *IEEE J. Quantum Electron.,* **QE-19**, 164–8.

van der Ziel, J. P., Tsang, W. T., Logan, R. A., Mikulyak, R. M. and Augustyniak, W. M. (1981) *Appl. Phys. Lett.,* **39**, 525–7.

van der Ziel, J. P., Temkin, H., Dupuis, R. D. and Mikulyak, R. M. (1984) *Appl. Phys. Lett.,* **44**, 357–9.

Vasil'ev, P. P. (1988) *IEEE J. Quantum Electron.,* **QE-24**, 2386–91.

von der Linde, D. (1986) *Appl. Phys.,* **B39**, 201–17.

Weiner, A. M. and Heritage, J. P. (1987) *Revue Phys. Appl.,* **22**, 1619–28.

Wiesenfeld, J. M., Eisenstein, G., Tucker, R. S., Raybon, G. and Hansen, P. B. (1988) *Appl. Phys. Lett.,* **53**, 1239–41.

Wiesenfeld, J. M., Kuznetsov, M. and Hou, A. S. (1990) *IEEE Phot. Tech. Lett.,* **2**, 319.

Willatsen, M., Uskov, A., Olesen, M. H., Tromborg, B. and Jauho, A.-P. (1991) *IEEE Phot. Tech. Lett.,* **3**, 606–9.

Wu, M. C., Chen, Y. K., Tanbun-Ek, T., Logan, R. A., Chin, M. A. and Raybon, G. (1990) *Appl. Phys. Lett.,* **57**, 759–61.

Xiang, W. H., Friberg, S. R., Wantanabe, K., Sakai, Y., Iwamura, H. and Yamamoto, Y. (1991) *Appl. Phys. Lett.,* **59**, 2076–8.

Yablonovitch, E., Gmitter, T., Harbison, J. P. and Bhat, R. (1987) *Appl. Phys. Lett.,* **51**, 2222–4.

Yariv, A. (1989) *Quantum Electronics*, 3rd edn. (Wiley, New York).

Yokoyama, H., Ito, H. and Inaba, H. (1982) *Appl. Phys. Lett,* **40**, 105–7.

8

The hybrid soliton pulse source

PAUL A. MORTON

8.1 Introduction

The hybrid soliton pulse source (HSPS) described in this chapter is being developed at AT&T Bell Laboratories as a possible source for ultra-long distance soliton communication systems. The HSPS integrates a semiconductor laser gain section and a fiber Bragg reflector in a hybrid device, to capitalize on the inherent advantages each possesses. The semiconductor section allows simple and efficient pumping of the device, and the fiber Bragg reflector provides excellent wavelength and optical bandwidth control. Together these components provide a source which is both simple and inexpensive, whilst satisfying the exacting performance requirements for ultra-long distance soliton transmission systems. Practical soliton transmission systems take advantage of the gain from erbium doped fiber amplifiers to provide lossless transmission, coupled with the nonlinearity in optical fibers to balance dispersion and provide dispersionless transmission (see Mollenauer *et al.,* 1986; 1992; 1993; Gordon and Haus, 1986; Gordon and Mollenauer, 1991; Kodama and Hasegawa, 1987; 1992; Haus, 1991; Mecozzi *et al.*, 1991). Current records for soliton transmission, which offer the largest product of bit-rate multiplied by the error free transmission distance, are 20 GBit/s over 13 000 km and 10 GBit/s over 20 000 km, by Mollenauer *et al.* (1993). In order for soliton transmission systems to become a reality, a practical source must be produced which can meet the stringent performance characteristics these systems require. The HSPS described in this chapter is a simple and elegant device which meets these requirements.

While the performance requirements for a pulse source for soliton transmission are severe, a source which can meet them can therefore quite easily provide the necessary performance required by many other applications. Examples of other applications include clock/strobe signals in opto-electronic integrated circuits (OEICs) and other optical processing elements. A clock source can be used for high speed interconnects and local area networks, and in communications systems where optical pulses have many applications. Optical pulses are particularly useful at higher bit-rates (Andrekson *et al.*, 1992), when optical multiplexing/demultiplexing can overcome the bottleneck due to slower electronic components. Short pulse sources can also be used for measurement and metrology applications. Modelocked semiconductor lasers are ideal sources of short pulses with low timing jitter (Derickson *et al.*, 1991), which is a major concern in many of these applications.

This chapter starts with a description of the general requirements of a pulse source for soliton transmission in Section 8.2, as this is the major driver for the development of the hybrid source. It then goes through the development of the HSPS, outlining how different parts of the design allow the device to meet the performance criteria set out in Section 8.2. Early work on the HSPS using silicon optical bench (SIOB) technology to provide the external reflector is described in Section 8.3. The optimum version of the hybrid pulse source, which uses an optical fiber with an integrated Bragg reflector, is described in Section 8.4, together with initial results. The results of these first devices showed an instability, which only worsened as the quality of Bragg reflectors improved. A description of where this instability originates, and how it can be overcome, is shown in Section 8.4.1. The method for overcoming the instability can be expanded to provide a source with an extremely wide operating frequency range, due to a novel wavelength self-tuning mechanism. Results of this wide operating frequency range are shown in Section 8.4.2. The same device which operates stably as a pulse source can be used as a high power, single-longitudinal mode CW laser source with narrow linewidth and low noise. Results for CW operation are described in Section 8.4.3. The HSPS so far described can produce transform limited pulses at a well controlled wavelength and with a wide variation of the operating frequency. Shorter pulse widths can be obtained from the device by using a semiconductor gain section with higher differential gain, or by broadening the bandwidth of the grating. Results from varying these two parameters are described in Section 8.4.4. The ability to tune the operating wavelength of the device is discussed in Section 8.4.5. In Section 8.4.6, a novel arrangement utilizing a two-section semiconductor laser is described. This arrangement is optimized to produce very high peak output power at the expense of the fidelity of the pulse source. A fully packaged HSPS device is described in Section 8.5, together with system measurements taken on a soliton transmission loop incorporating sliding-frequency guiding filters. The final Section, 8.6, is a brief discussion of future prospects.

8.2 Pulse source requirements for soliton transmission systems

Ultra-long distance soliton transmission systems discussed in this chapter refer to systems operating with lengths up to 10 000 km, which are sufficiently long to provide transoceanic links between most continents. A simple schematic diagram of a soliton system is shown in Figure 8.1. Data from network connections is multiplexed up to the operating bit-rate, for example 2.5 GBit/s, then the data is impressed upon a series of optical pulses by a modulator. The pulse source is required to provide this train of optical pulses. The modulated pulses are then transmitted through many spans of optical fiber separated by optical amplifiers. The optical amplifiers provide for lossless transmission, while the system is designed so that nonlinear effects in the fiber counteract dispersion to allow dispersionless transmission. The soliton pulses themselves can be viewed as the natural bits of an optical transmission system, providing the only stable waveform for transmission over long distances. The

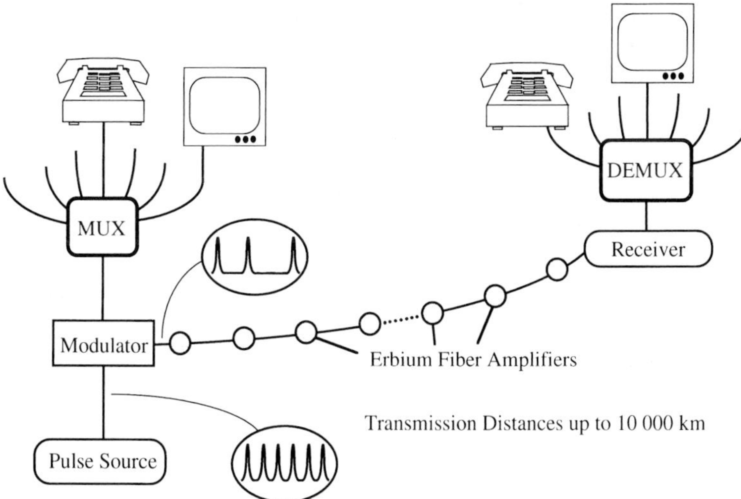

Figure 8.1. Schematic diagram of an ultra-long distance soliton transmission system.

requirements for a practical optical pulse source for use in soliton transmission systems can be split into two major categories. The first, and most important, is the performance, which sets very stringent requirements on all major operating parameters of the source. The second, which has a large bearing on the possible commercialization of soliton based communications systems, is the cost and reliability of a source.

The major parameters of a pulse source are its operating wavelength, repetition frequency, pulse width, pulse shape and time–bandwidth product. Control of the operating wavelength is critical to allow soliton propagation, particularly in systems employing wavelength division multiplexing, where multiple channels are placed close together and the required channel spacing must be kept constant. The use of sliding guiding optical filters to control soliton transmission (Mollenauer *et al.,* 1993) also requires specific operating wavelengths in multiple wavelength systems. The repetition rate of the source must be exactly the same as the clock rate in the transmission system, for example 2.48832 GHz or twice and four times this rate for standard Sonet applications. This represents a severe restriction for standard modelocked lasers, which have a fixed cavity length and typically only operate over a very small frequency range. The fixed cavity length determines the operating frequency, which is close to the cavity resonance frequency (1/round-trip time). Small deviations from this frequency take the device off resonance and stop the modelocking action. It is also important when using a modelocked laser as the source that the device operates at its fundamental cavity frequency, to maintain coherence between all pulses in the output waveform. When operating at a harmonic of the fundamental cavity frequency a more complex laser design may be necessary, as described by Harvey and Mollenauer (1993). The required pulse width depends on the operating bit-rate of the transmission system. Shorter pulse widths require higher average power levels for soliton propagation, while longer pulses tend to interact with neighboring pulses. Pulse widths are typically chosen to occupy approximately one fifth of the operating bit period, giving a range from 20 ps to 80 ps, for systems operating from 10 GBit/s to 2.5 GBit/s. The optimum pulse shape for a standard soliton transmission system would be sech^2 in shape, to match the soliton shape, however it has been shown that other symmetrical pulse shapes, such as Gaussian shaped pulses, will also work well (Harvey and Mollenauer, 1993). In systems with sliding guiding filters, the optimum pulse shape is more complex due to the interaction between the pulses and filters, and it is unlikely that a pulse source can produce precisely the optimum shape.

One major restriction on the pulse source is that the output waveform must be transform limited, or close to transform limited, to allow optimum system operation. This restriction is reduced somewhat in systems employing sliding-frequency guiding filters, which are more tolerant to lower fidelity pulse waveforms.

The pulse source must meet all of the stringent performance criteria described above, and continue to do so over long periods of time. The reliability of the source will depend on its complexity, and how many ways it can fail. The cost of the source must also be included in the decision as to whether a soliton based system is economical to produce. The HSPS is in principle a very simple device that could be produced fairly inexpensively, and with correct packaging should be very reliable. This source should therefore be considered a major contender in the race for a simple and inexpensive pulse source for soliton transmission systems, as well as a general pulse source for many other applications.

8.3 Hybrid soliton pulse source with a silicon optical bench reflector

The realization of long distance soliton based transmission systems requires a reliable, stable source of transform limited pulses at 1.55 μm with a pulse width of 20–80 ps, at repetition rates of 2.5–10.0 GHz. The relatively long pulse widths, compared with more typical short pulse results from modelocked semiconductor lasers, require a very narrow optical bandwidth to obtain transform limited pulses. The need for wavelength division multiplexing of solitons also requires accurate wavelength control for each source. External cavity lasers utilizing air cavities require the alignment of bulk optics, together with a diffraction grating for wavelength control and an etalon for bandwidth control. Integrated structures provide an attractive alternative, with monolithic modelocked lasers (Tucker *et al.*, 1989; Morton *et al.*, 1990; Wu *et al.*, 1990; Hansen *et al.*, 1992) providing short pulses with a compact, reliable device. However, the monolithic devices are constrained by their physical length to relatively high repetition frequencies, and at present do not have the bandwidth control necessary to produce longer pulses that are transform limited.

The hybrid soliton pulse source (HSPS) using silicon optical bench (SIOB) technology is a good candidate as a practical source for soliton transmission systems (Raybon *et al.*, 1988; Morton *et al.*, 1991). A low loss silica waveguide external cavity allows long cavity lengths and therefore low repetition rates. A high level of fabrication control, due to the use of mature silicon processing technology, allows precise control of operating wavelength and extremely narrow Bragg reflectors (Henry *et al.,* 1989; Adar *et al.*, 1991; 1992). Initial attempts to produce a modelocked hybrid pulse source were carried out using a Si_3N_4 SIOB Bragg reflector to provide the external cavity for a semiconductor laser. In 1988, Raybon *et al.* actively modelocked the laser to produce pulses of 27 ps at repetition frequencies of 1.8, 3.6, 5.4 and 7.2 GHz, with a minimum time–bandwidth product of 2.0. In this early experiment, problems with butt-coupling the laser diode to the Si_3N_4 waveguide limited the range of measurements taken.

In 1991, Morton *et al.* overcame some of the initial problems by first attaching a short fiber stub to the silica waveguide and then aligning the lensed fiber to the semiconductor laser. In this work the SIOB waveguides used were also different from those of the earlier experiments, providing a lower waveguide loss and also a narrower Bragg reflection spectrum. A schematic diagram of this HSPS device is shown in Figure 8.2. The experiments used a 1.55 μm strained quantum well laser (Tanbun-Ek *et al.*, 1990) with an anti-reflection (AR) coating on one facet, and the other facet left as cleaved to provide a second output as a monitor. The laser diode had a threshold of 12 mA before coating, and after coating did not lase for bias currents up to 100 mA. The output from the AR coated facet was coupled into the silica waveguide using a

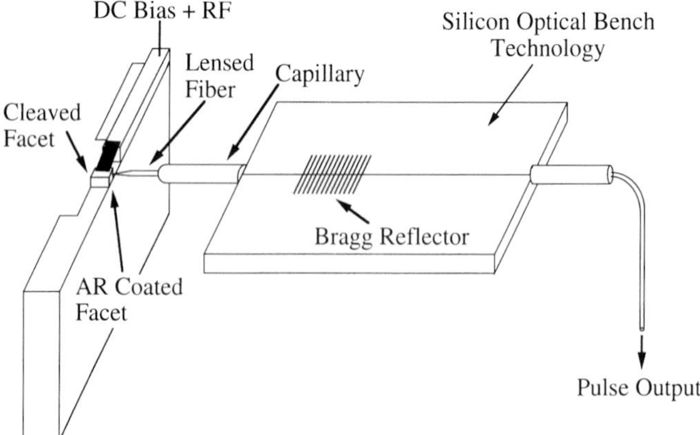

Figure 8.2. Schematic diagram of the hybrid soliton pulse source with silicon optical bench Bragg reflector.

short lensed fiber stub attached to the waveguide. This allowed easier and more stable alignment to the laser diode than butt-coupling (used by Raybon *et al.,* 1988) and was also expected to improve coupling efficiency. The device output was taken from a fiber attached to the other end of the silica waveguide, allowing for simplicity of alignment, with only one critical fiber alignment to the laser diode necessary. The output power was 50% less from this output fiber than from a fiber coupled to the cleaved facet of the laser diode, although by applying a high reflectivity coating to the facet, and optimizing the reflectivity of the Bragg reflector, it would be possible to maximize this output. The silica waveguide was a 'pglass' waveguide (Adar *et al.,* 1991), which on other samples have been measured to have a loss of only 0.03 dB/cm. A first-order Bragg reflector was formed by etching 0.5 μm into the core of the silica waveguide (Adar *et al.,* 1992). The Bragg reflector had a measured reflectivity full width at half-maximum (FWHM) of 0.17 nm (21 GHz at 1.55 μm), a maximum reflectivity of 77% and a physical length of 7 mm.

The experimental setup, which is shown in Figure 8.3, includes a microwave synthesizer and amplifier to modulate the laser diode. The output from the silica waveguide is used to measure pulse width using a 34 GHz bandwidth detector and 50 GHz sampling oscilloscope. The pulse width is deconvolved from the measurements using a simple sum of squares approximation. The results described in this chapter were taken with two high speed detectors, the second proving a higher bandwidth and less ringing. The ringing seen in some results of pulse shape is due to a

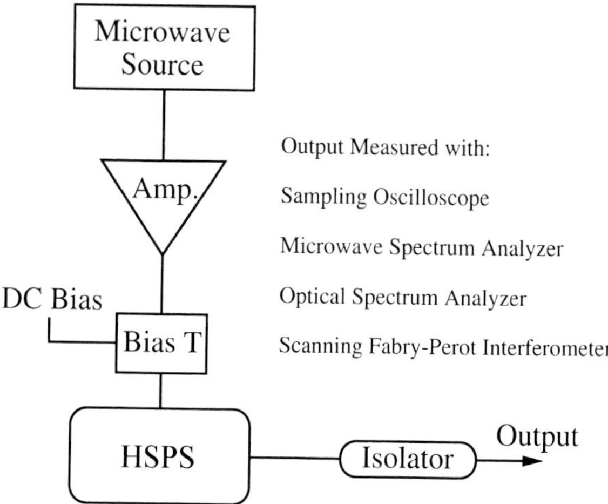

Figure 8.3. Experimental setup to drive and measure the HSPS.

combination of the detector and sampling oscilloscope. This observation has been confirmed by measurements on a streak camera. The monitor output from the cleaved facet of the laser diode is used to measure optical spectra on a scanning Fabry–Perot interferometer and optical spectrum analyzer, together with the microwave spectra on a 22 GHz microwave spectrum analyzer. The measured comb lines on the microwave spectrum analyzer indicate how well the device is modelocked.

When operated with a DC bias current, the lowest device threshold was 35 mA. This is a large rise from the original laser threshold of 12 mA, and shows that there is still a large coupling loss between the laser diode and silica waveguide. Improvements in fabrication techniques for connecting the fiber stub to the silica waveguide should improve the coupling and threshold current. The maximum power in the output fiber was 300 μW. The device lased within the bandwidth of the Bragg reflector, with a suppression of the laser diode modes of 40 dB. Under DC operation the device usually oscillated in a single longitudinal mode, although as many as 5 modes were sometimes seen. This multiple mode operation gave rise to mode hopping, and output fluctuations. When modelocked, from 3 to 5 main modes oscillated and the average output power became stable.

Active modelocking was performed by modulating the drive current of the laser diode at frequencies close to the fundamental cavity frequency (1/cavity round-trip time). The optimum modelocking frequency was 3.84 GHz, corresponding to a cavity length of 2.6 cm. For a bias level of 60 mA, and an RF power of 20.4 dBm at 3.84 GHz, modelocked pulses with a FWHM of 51 ps and an optical bandwidth of 7 GHz were obtained (Figure 8.4), giving a time–bandwidth product of 0.36. This is close to the theoretical minimum time–bandwidth product of 0.31 for sech^2 shaped pulses. The average power from the output fiber was 52.5 μW, giving a peak power of 260 μW. The use of an erbium fiber amplifier could easily boost this power up to the level required for an $N = 1$ soliton (1–3 mW) for pulse propagation.

The frequency dependance of the device is a very important property of a practical modelocked pulse source. Figure 8.5 shows the effect of detuning the modulation frequency away from the optimum frequency, with the bias current fixed at 60 mA and the RF optimized for best performance. Pulse widths around 55 ps with a time–bandwidth product of less than 0.5 were found for a bandwidth greater than 60 MHz. This unusually large useful frequency range is due to a novel mechanism of the HSPS caused by the spatially distributed nature of the Bragg reflector.

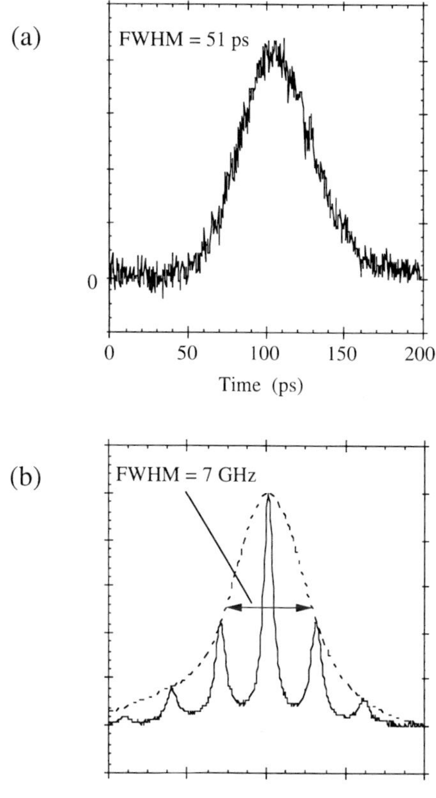

Figure 8.4. (a) Measured pulse shape and (b) optical spectrum for 60 mA bias current and 20.4 dBm RF modulation at 3.84 GHz.

The distance to the effective reflection position within the reflector depends on wavelength, being shortest for the Bragg wavelength and increasing for wavelengths on either side of it. Because of this effect, the effective cavity length of the device can vary if the operating wavelength is changed. As the modulation frequency is varied, the device changes its effective cavity length by shifting the center wavelength, so as to keep the device on resonance. This effect is shown in Figure 8.5(d). At modulation frequencies above 4.1 GHz the device center wavelength is 1.5475 μm, the same wavelength the device exhibits under DC operation. As the frequency is reduced, the center wavelength reduces and so the effective cavity length increases. The results show a maximum shift in wavelength of over 0.1 nm for a modulation frequency change of 300 MHz. This result is also confirmed by spectral results from the scanning

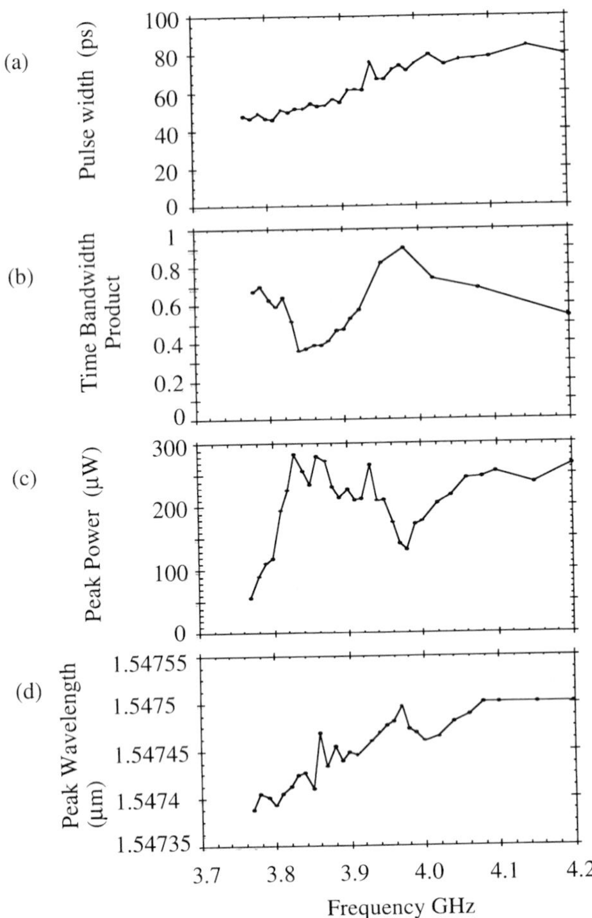

Figure 8.5. (a) Pulsewidth, (b) time–bandwidth product, (c) peak power and (d) center wavelength versus modulation frequency (60 mA bias, optimum RF power).

Fabry–Perot interferometer in which the mode spacing is seen to follow the modulation frequency over the range from 3.78 to 4.0 GHz, where modelocking occurs. Above this frequency the device tends towards its (multimode) DC operating spectra and is just modulated rather than modelocked. The large detuning at 3.78 GHz (0.1 nm) pushes the operating wavelength well away from the Bragg reflector reflectivity peak, and as a result the average output power drops considerably (Figure 8.5(c)). The stability of this source, together with more discussion of the large tuning range, is addressed in Sections 8.4.1 and 8.4.2.

To summarize these results, in this first demonstration of an HSPS, the device produced near transform limited pulses at 1.55 μm, with a pulse width of 51 ps at a 3.85 GHz repetition rate and a time–bandwidth product of 0.36. The HSPS also has a large useful modulation frequency range of over 60 MHz (with time–bandwidth product <0.5), due to a novel mechanism in which the effective cavity length self-tunes with the modulation frequency to keep on resonance. This wide operating bandwidth allows flexibility in device fabrication when a specific operating frequency is required. The excellent accuracy and reliability of SIOB technology in fabricating silica based waveguides and Bragg reflectors allows both absolute wavelength control and very narrow reflection bandwidths. This control lends itself to wavelength division multiplexing, in which an array of lasers could be coupled to Bragg reflectors with offset center wavelengths, to provide multi-channels of transform limited pulses.

8.4 Hybrid soliton pulse source with a fiber Bragg reflector

Although the initial results from the HSPS using SIOB technology were promising, the creation of a new class of devices based on Bragg reflectors written directly into the core of optical fibers provided a more practical platform for producing the HSPS devices (Morton *et al.*, 1992a). The low coupling loss, and simplicity of packaging a fiber external cavity with integrated Bragg reflector (Bird *et al.*, 1991) make it preferable to the initial SIOB embodiment.

The Bragg reflectors are manufactured using holographic techniques, which allows great control over the Bragg wavelength and intensity profile. The reflector is produced by exposing an optical fiber to a two beam interference pattern, at a wavelength near the peak of a defect absorption band at 242 nm (Meltz *et al.*, 1989; Mizrahi and Sipe, 1993). The absorbed UV light permanently modifies the index of refraction in the core, transferring the interference pattern to the core index profile, and therefore forming a grating, or Bragg reflector. The grating intensity profile follows that of the laser beams, and is therefore approximately Gaussian. This provides for a very clean reflection spectrum free from any significant side-lobes. Some of the reflectors described in this chapter use highly sensitized optical fibers in their fabrication. These gratings use a standard communications fiber which has been sensitized by soaking in hydrogen gas (Lemaire *et al.*, 1993). Once the gratings are written in sensitized fibers, they are heated to displace the residual hydrogen.

A schematic diagram of the experimental arrangement is shown in Figure 8.6. A 1.55 μm laser diode is used, with one facet high reflectivity coated for improved cavity Q, and the other AR coated to allow coupling to the external cavity and suppress Fabry–Perot modes. The external cavity is composed of an AR coated lensed fiber stub fusion spliced to a section of photo-sensitive fiber with an integrated Bragg reflector. This arrangement allows efficient coupling from the laser diode to the fiber. The grating has an approximately Gaussian taper with a 5 mm FWHM, which results in a Bragg reflector with a reflectivity FWHM of 0.2 nm. The peak reflectivity, which is determined by exposure conditions, is 63%, with the remaining 37% transmission forming the output of the laser. The reflection spectrum in this case was slightly asymmetric, with a small shoulder on the short wavelength side. We believe that this contributed to the stability of the laser, as discussed in Section 8.4.1. The effective cavity length of the device depends on the position of the fiber grating and the operating wavelength. A DC bias and microwave drive are applied to the laser diode, with the modulation period being close to the fundamental modelocking frequency (inverse of cavity round-trip time). In the results presented in this section, a step recovery diode is used to provide short (40–50 ps) electrical pulses to the laser diode at the modelocking frequency. The output of the device is passed through an optical isolator, and then split between the different measurement instruments.

The DC light/current characteristics of the fiber HSPS are shown in Figure 8.7. The threshold is low (11 mA) and the output power comparable with a standard laser diode (> 5 mW at 100 mA). This is particularly good when it is noted that this power is already coupled into a single-mode fiber. The operating wavelength without modulation is 1.53265 μm,

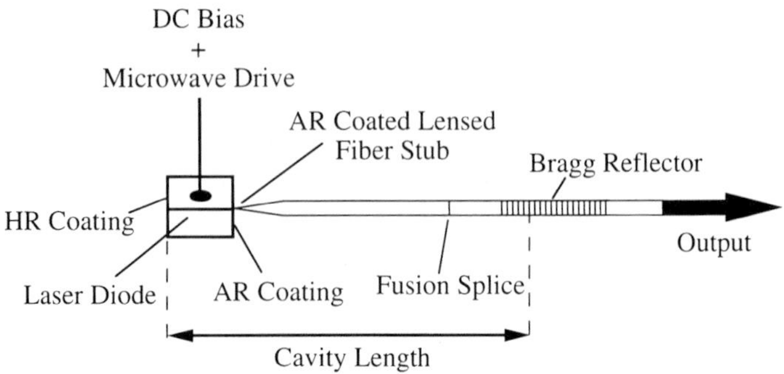

Figure 8.6. Schematic diagram of an HSPS with fiber Bragg reflector.

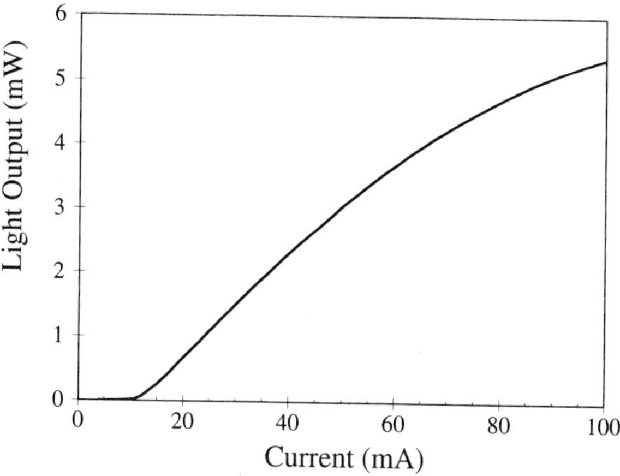

Figure 8.7. Typical DC light/current characteristics.

as determined by the grating. The optimum modelocking conditions were found to be a DC bias current of 36 mA and an RF frequency of 2.37 GHz with approximately +27 dBm power into the step recovery diode. The measured optical pulse shape is shown in Figure 8.8(a), as seen on the sampling oscilloscope. The pulse width in this case is 18.5 ps. The peak power level is calculated from the average power measured under modelocked conditions which is 2.1 mW (before the isolator), giving a peak power of 49 mW. The optical spectrum under these conditions is shown in Figure 8.8(b). The FWHM of the optical spectrum is 0.13 nm, giving a time–bandwidth product of 0.31, which is transform limited if a sech^2 pulse shape is assumed. The stability of the optical pulses was confirmed by the microwave spectrum analyzer, which showed sharp comb lines at harmonics of the modulation frequency, with no noise components at intermediate frequencies. Under other conditions (41 mA bias, +30 dBm at 2.38 GHz) peak powers of up to 70 mW were found, with a small degradation of time–bandwidth product to 0.32. However, spectral anomalies were sometimes seen in these nonoptimized cases, producing a twin lobed optical spectrum. The spectral anomalies of this device are believed to be due to the spatially distributed Bragg reflector properties, and are described in Section 8.4.1.

In this section we have described the first results of an HSPS with a fiber external cavity and integrated Bragg reflector. The device produces time–bandwidth limited pulses with a pulse width of 18.5 ps and high peak output power of 49 mW. The operating wavelength is 1.53 μm which is compatible with fiber optical amplifiers for use in amplified

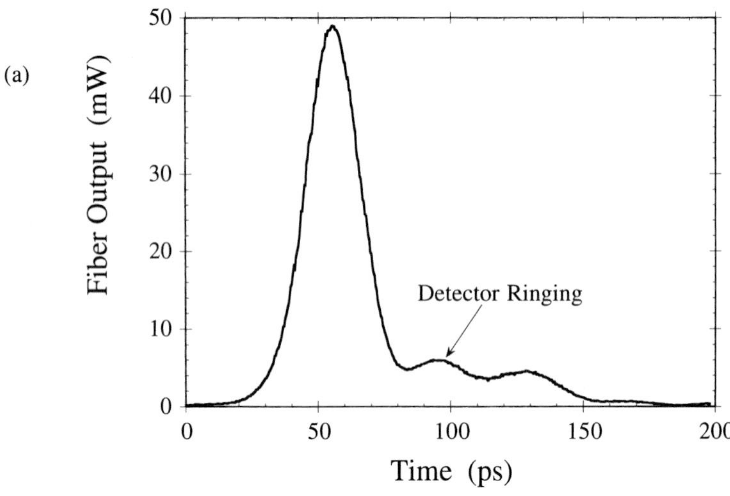

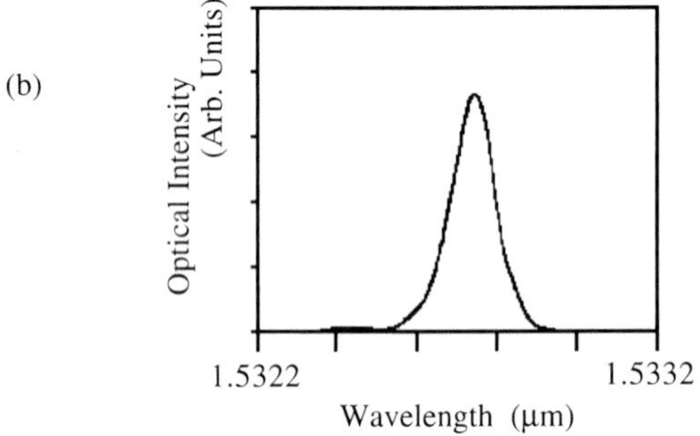

Figure 8.8. (a) Measured optical pulse shape and (b) optical spectrum, under optimum modelocked conditions at a repetition rate of 2.37 GHz.

long distance soliton transmission systems. While producing stable operation under specific driving conditions, we found that in general the device output was not too stable, and often produced a twin lobed output spectrum. In order to overcome this stability problem a more complicated reflector design is necessary.

8.5 Spectral instabilities: cause and solution

In order to understand the spectral instabilities described in Section 8.4, it is first important to understand the Bragg reflector characteristics. The Bragg reflectors used in these experiments can be characterized for pulsed operation as having a penetration depth L_{eff}, where L_{eff} describes the distance to some effective reflection plane (Figure 8.9). In this case L_{eff} is wavelength dependent. A simplified model for an unchirped Bragg reflector (modeled using Laser Matrix (Schatz, 1994)) designed to give 50% reflectivity and a bandwidth of about 0.2 nm was used to give the plots of reflectivity and penetration depth shown in Figure 8.10. The shortest penetration depth occurs at the Bragg wavelength, where the feedback is at a maximum. This depth increases for wavelengths on either side of the Bragg wavelength. The penetration depth translates into a time delay for an optical pulse reflecting off the grating, the slope of this curve is therefore related to the dispersion of the grating $\delta\tau/\delta\nu$. The unchirped reflector has fairly constant dispersion on the two sides of its reflection peak, however around the peak in reflectivity the dispersion changes rapidly from one sign to the other.

The effect of this passive reflector in a modelocked cavity with a semiconductor gain section must now be considered. For stable modelocking to occur, the dynamics of the semiconductor section must counteract the dispersion of the grating, so that a pulse can replicate itself after one round trip of the cavity. As an optical pulse travels through the semiconductor section and is amplified, it will modify the carrier density in the laser, which in turn modifies the refractive index. Self-phase modulation occurs due to the changing refractive index, which chirps the frequency of the optical pulse. This chirp can be considered to be approximately linear across the center of the pulse, and can have either sign (or be zero) depending on whether carrier depletion from stimulated emission is larger or smaller than the injected carrier density. If the laser diode produces a chirp which is approximately linear, then it can only counteract a grating reflection with constant dispersion. In the unchirped reflector case, this means that stable modelocked operation can be achieved with

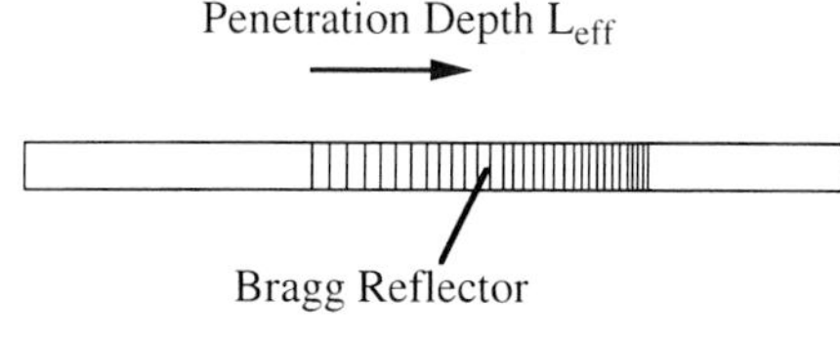

Figure 8.9. Schematic diagram of a Bragg reflector showing penetration depth L_{eff}.

(a)

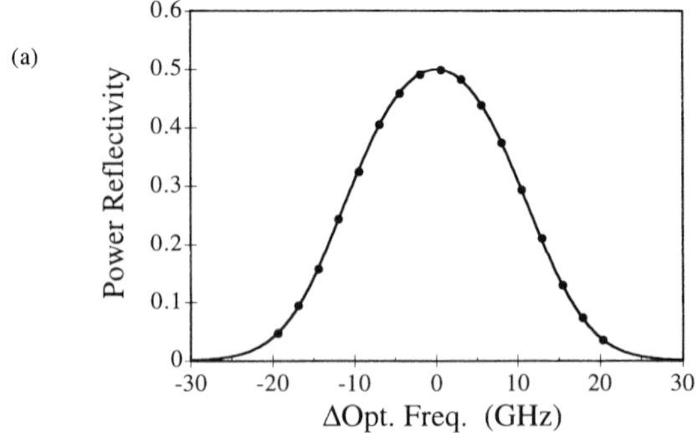

(b)

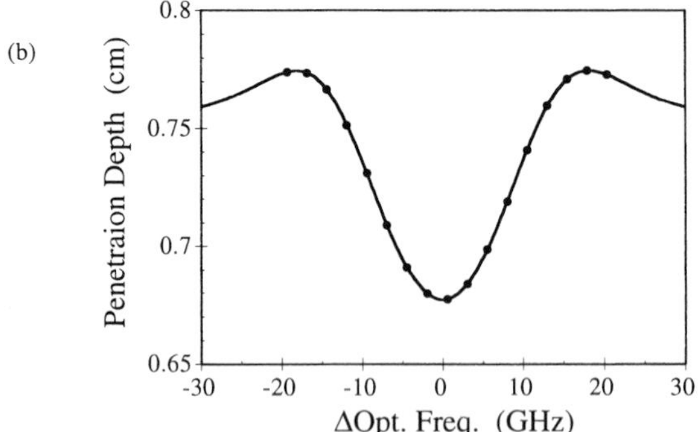

Figure 8.10. (a) Reflectivity and (b) penetration depth versus frequency offset from Bragg frequency, modeled to produce 50% reflectivity with 0.2 nm bandwidth.

the device operating on one side or other of the grating, but not at the Bragg wavelength where the dispersion is changing rapidly. This provides us with two problems. The first is that the device will not operate at the Bragg wavelength where the feedback is maximum. The second problem, which is even more important, is that the device output will be unstable, as there are two degenerate states of operation. The first modelocked state has the set of locked modes on one side of the Bragg wavelength, the other state is when the modes are on the other side of the Bragg wavelength. As there is no mechanism to choose which set of modes should oscillate, the device output can jump from one set to the other. This conclusion is confirmed by experimental results using different unchirped fiber Bragg reflectors, in which it is seen that the devices would not modelock with the spectrum centered on the Bragg wavelength, but always on one side or the other of the peak. An example of

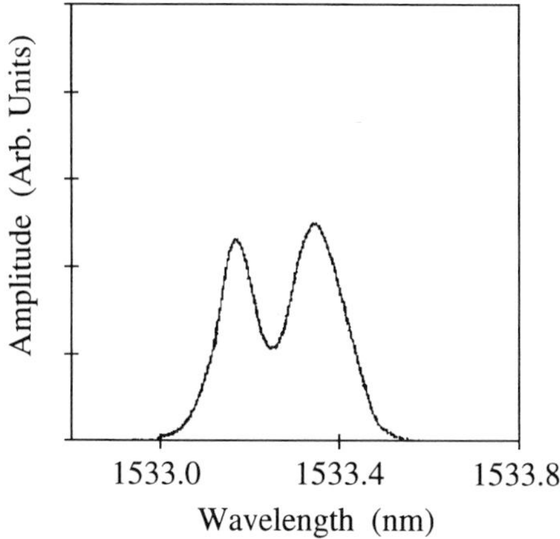

Figure 8.11. Time averaged optical spectrum from a device with an unchirped Bragg reflector.

the time averaged optical spectrum from a device with an unchirped Bragg reflector is shown in Figure 8.11. In this case the two lobes indicate where two separate sets of locked modes are operating. The time averaging process loses the information about whether the two sets of modes are oscillating together or if the output jumps between the two sets, and at what rate this occurs.

The stable modelocked results shown in Section 8.4 rely on an imperfection in the Bragg reflector which breaks the symmetry of the two sets of modes, allowing only one set of modes to operate, but only under specific driving conditions. However, as mentioned in that section, if a cleaner reflector is used, the unstable output nature of the device using an unchirped Bragg reflector occurs for all operating conditions.

It is possible to confirm the idea of how the reflectivity of the grating varies with effective cavity length by looking at the relative intensity noise (RIN) spectrum of the device, which is shown in Figure 8.12 for a device with an unchirped Bragg reflector. The noise peaks occur at the cavity resonance frequency (1/round-trip time) and its harmonics. By looking closely at the fundamental noise peak an asymmetry in the peak is noticeable. The maximum feedback occurs at the Bragg wavelength, which from the simple model occurs at the minimum effective cavity length. This minimum cavity length corresponds to the maximum cavity frequency at which feedback occurs. For lower frequencies (longer cavity lengths) feedback occurs from the wavelengths on either side of the Bragg wavelength. For higher frequencies there should be very little feedback

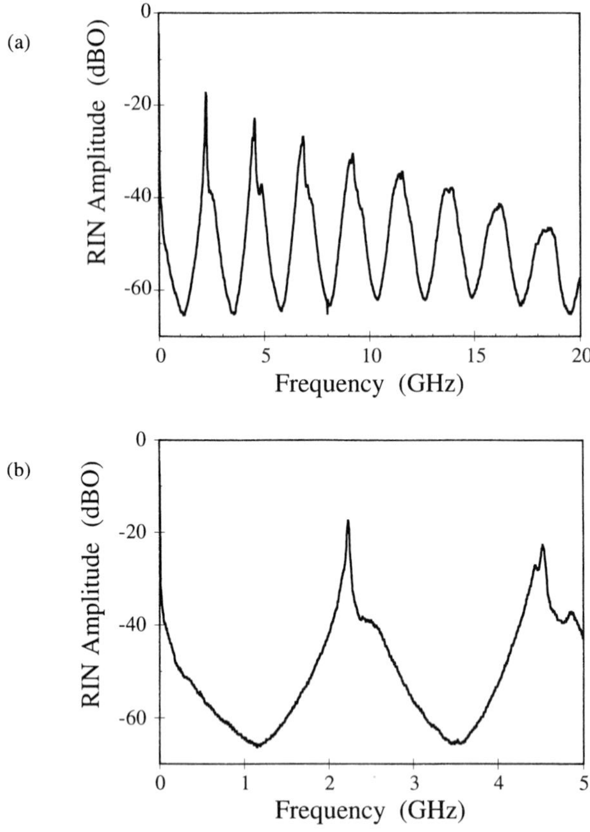

Figure 8.12. Relative intensity noise spectrum from a device with an unchirped Bragg reflector, (a) is the spectrum from 0 to 20 GHz, (b) is a closer view from 0 to 5 GHz.

from the grating. This argument predicts the asymmetric noise peak shown in the RIN spectrum, with the noise level falling quickly on the higher frequency side of the noise peak.

In order to remove the degeneracy seen for an HSPS with an unchirped Bragg reflector, and allow the device to operate stably with a single set of locked modes, a linearly chirped Bragg reflector can be used (Morton *et al.,* 1993). The linear chirp can provide a linearly varying penetration depth versus wavelength, as shown in Figure 8.13, which is calculated for an orientation where the grating period is reduced when moving away from the semiconductor section. This grating produces an almost constant dispersion, whose penetration depth allows only one stable operat-

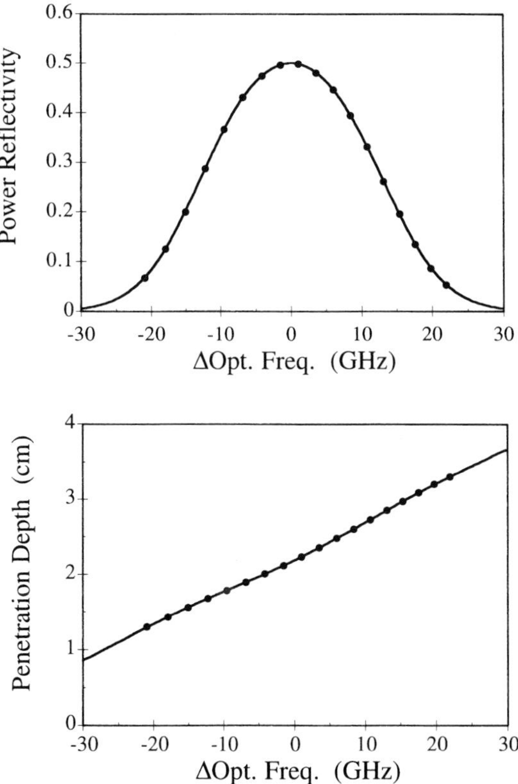

Figure 8.13. (a) Reflectivity and (b) penetration depth versus frequency offset from Bragg frequency for a chirped Bragg reflector, modeled to produce 50% reflectivity with 0.2 nm bandwidth.

ing point for a specific effective cavity length. When actively modelocked, the modulation frequency applied to the laser diode specifies the required effective cavity length, and for any value of this cavity length there is only one possible operating wavelength. The chirped reflector therefore stabilizes the output of the HSPS. Another advantage of having a penetration depth which varies over a large spatial length is that the device can have a large variation in effective cavity lengths, and therefore operate over a large modulation frequency range. The novel wavelength self-tuning mechanism that allows this large frequency range is described in more detail in Section 8.4.2.

The results in Section 8.3 for the HSPS using a SIOB reflector did not show any instabilities in the output spectrum, and operated quite stably, but only on one side of the Bragg wavelength (the short wavelength side).

This indicates that the reflector used may have an intrinsic asymmetry causing the short wavelength set of locked modes to oscillate in preference to the longer wavelength modes.

The orientation of a chirped Bragg reflector affects the sign of dispersion the grating produces. The semiconductor laser can, however, produce both signs of chirp to counteract the different signs of dispersion, by changing the operating conditions such as DC bias and RF power level. This has been tested experimentally, and it is found that the devices do in fact produce stable modelocking with both grating orientations. For CW operation of the hybrid laser, as described in Section 8.4.3, it is found that only one orientation of the chirped reflector allows stable operation. It is for this reason that the results presented for the HSPS generally have the grating oriented in one direction, this being the direction to allow stable CW operation, so that the devices can be used in both applications.

8.6 Wide operating frequency range using a chirped Bragg reflector

A practical soliton transmission system may be required to operate at the Sonet rate of 2.48832 GHz with a pulse width of around 50 ps. This tight specification on the operation frequency requires accurate control of the modelocked laser cavity length, while the relatively long pulse width requires a narrow optical bandwidth for transform limited pulses to be maintained. The HSPS utilizing a uniform grating Bragg reflector integrated into a fiber external cavity has been shown to operate close to the required frequency with transform limited pulses and very high output powers (Morton *et al.*, 1992a). However, this device worked well only under specific operating conditions, and showed spectral instabilities when operating parameters were changed. In this section we describe results using a Bragg reflector with a linearly chirped grating, which has overcome the spectral instability problem of the earlier device. By distributing the chirped grating over a fairly large physical length we have also dramatically improved the operating frequency range of the device over previous modelocked semiconductor lasers. The new device offers the advantages of the previous HSPS structure, including the simplicity of packaging this kind of device, together with much greater stability over all operating ranges, plus a very broad operating frequency range.

The experimental setup is the same as shown in Figure 8.6, except that the design of the reflector has been changed. The reflector has a linear chirp of 0.2 nm distributed over 1 cm of the fiber, and a peak reflectivity of 45%. This reflectivity, plus the high coupling efficiency of the lensed fiber (up to 70%), provides a low loss external cavity, which is important for good device operation. Lower feedback (see Section 8.3) produces similar results but with much lower power levels. The Bragg reflector is oriented with the reflections from longer wavelengths occurring from the section of grating nearest the laser diode. The center of the grating is positioned 41 mm from the laser diode to give a fundamental modelocking frequency of around 2.5 GHz. A DC bias and RF drive were applied to the laser diode. For these results, sinusoidal modulation was sufficient to produce good modelocked results, whereas previously (see Section 8.4) short electrical pulses were required.

The device has a DC threshold current of 15 mA, and at 100 mA bias produces 5.6 mW of output power through the Bragg reflector, which is used as the output. The threshold and power output are similar to those of the original uncoated laser diode, showing that the feedback from the external cavity is high. It is important when using quantum well lasers to keep the threshold down close to that of the original laser diode, to take advantage of the good characteristics of the device (i.e. high differential gain at low carrier densities). Typical results for the output pulse shape from a 50 GHz sampling oscilloscope, and the optical spectrum from a scanning Fabry–Perot interferometer, are shown in Figure 8.14. These results show a pulse width of 44 ps and an optical spectrum of 8.6 GHz, giving a time–bandwidth product of 0.38.

The frequency dependance of the device is very important for use in a practical system. For modelocking to occur, the optical cavity must be on resonance with the modulation frequency, that is, the reciprocal round-trip time of the cavity must equal or be very close to the modulation frequency. This sets a tight specification on the cavity length. The novel behavior of the HSPS is due to the spatially distributed nature of the Bragg reflector, in particular that of the chirped device. In a chirped Bragg reflector the distance within the reflector to an effective reflection plane depends on wavelength, giving a linear change in this distance as the wavelength is varied. Within the modelocked cavity this translates into a change in overall cavity length, and therefore the modelocking frequency, with a change in wavelength. In practice, for a particular modulation frequency the HSPS self-tunes its operating wavelength to keep the device on resonance, and therefore produce good modelocking. By using a long chirped reflector, large changes in modulation frequency can be accommodated. Standard modelocked semiconductor lasers in an

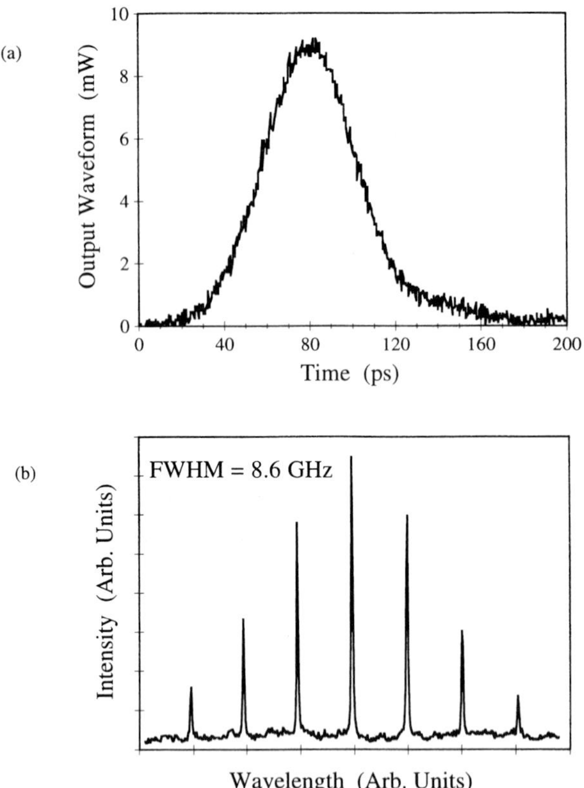

Figure 8.14. Typical results for (a) pulse shape and (b) optical spectrum, showing a time–bandwidth product of 0.38.

external cavity cannot change their cavity length by any appreciable amount, and therefore have a very limited frequency range over which modelocking will occur.

The self-tuning of the device wavelength is shown in Figure 8.15. The laser diode has a DC bias of 25 mA and the RF power is optimized for best operation at each modulation frequency (optimum RF power varied from 10–15 dBm). The modulation frequency is varied from 2 GHz to 2.9 GHz in 100 MHz steps, and the resulting envelopes of the optical spectra at each modulation frequency are shown. Good modelocking occurs for frequencies between 2.1 GHz and 2.8 GHz, giving an extremely wide modulation frequency range (700 MHz), or almost 30% of the center frequency. The center wavelength of the device varies by 0.3 nm for the 700 MHz change in frequency, utilizing almost the whole reflection spec-

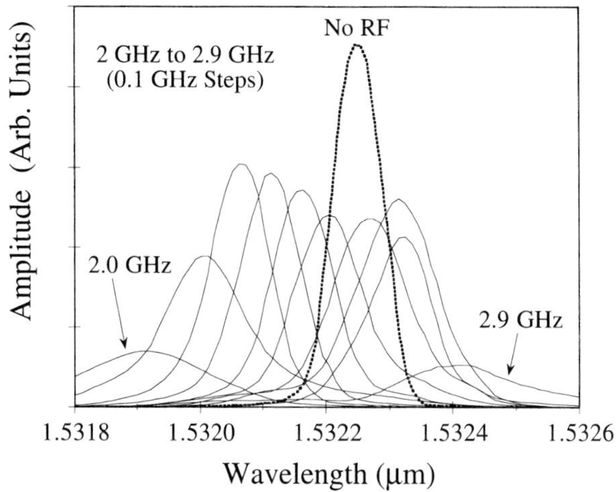

Figure 8.15. Envelopes of optical spectra for modulation frequencies of 2.0 GHz to 2.9 GHz in 100 MHz steps (spectra follow frequencies from left (low) to right).

trum of the chirped Bragg reflector. By using a longer Bragg reflector placed closer to the laser diode it should be possible to increase the frequency range of the device even more. For example, a 5 cm long chirped grating starting 5 mm from a laser diode could in principle operate over a modulation frequency range of 1.8 GHz to 17 GHz. There might, however, be problems in operating over such a large range as both fundamental and harmonic modelocking could occur at separate wavelengths.

The wavelength self-tuning gives rise to a small change in operating wavelength, depending on the required operating frequency. In some applications, independent control of the device wavelength may be required to tune the wavelength to a specific operating point. Ways to achieve this are described in Section 8.4.5.

The operating characteristics of the device are shown in Figure 8.16, as the modulation frequency is varied from 2.1 GHz to 2.8 GHz. Figure 8.16(a) shows the pulse width and peak power versus frequency. The pulse width is fairly constant at the center of the range with longer pulses occurring at lower modulation frequencies. The peak power of the pulses rises near the center of the range (2.5–2.6 GHz). This corresponds to the peak reflectivity of the Bragg reflector, as seen from the spectral results in Figure 8.15. These show the spectrum without modulation (therefore at the peak in reflectivity) coinciding with the spectra for modulation at

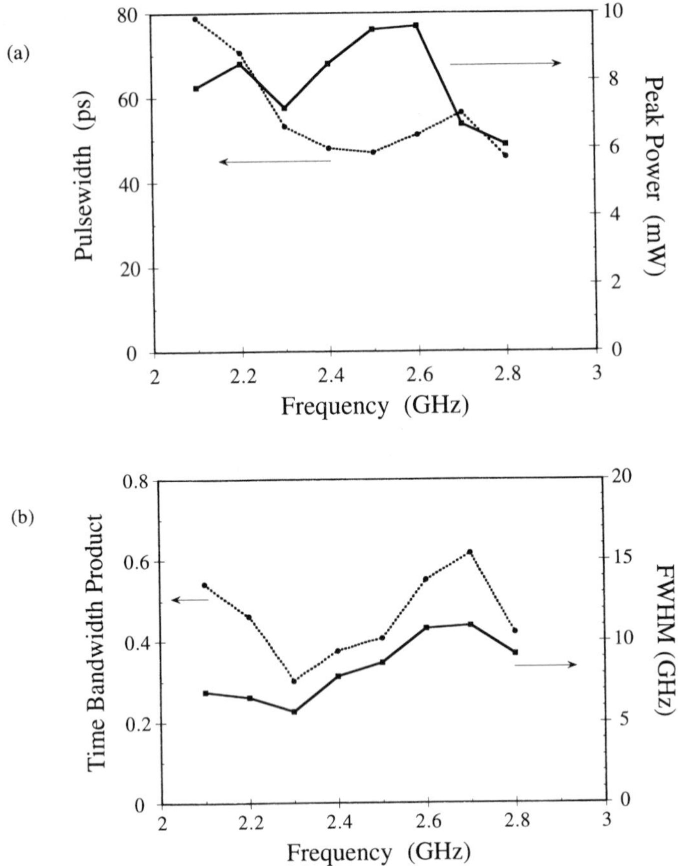

Figure 8.16. (a) Pulse width and peak power and (b) time–bandwidth product and spectral width versus modulation frequency.

2.5–2.6 GHz. Peak powers of up to 10 mW are observed from this device.

The time–bandwidth product and spectral width of the pulses are shown in Figure 8.16(b). The spectral width is lower at lower modulation frequencies and rises at the higher frequencies. The time–bandwidth product, which is very important for soliton transmission, is below 0.6 over the whole modelocking range, with a minimum of 0.3 at 2.3 GHz, which is close to the transform limit if a sech^2 pulse shape is assumed.

Due to the low frequency sinusoidal drive (~2.5 GHz), the laser diode effectively sees a long electrical drive pulse (200 ps), which tends to lead towards multi-pulse output from the modelocked laser (Morton *et al.*, 1989; Bowers *et al.*, 1989). In order to overcome this, the bias in these experiments was kept relatively low (25 mA). The RF power was also optimized for the pulse shape and time–bandwidth product, however the initial buildup of a trailing pulse was seen in some of the measurements. By increasing the RF power the pulse shape becomes cleaner with no sign of a secondary pulse, however at high RF power levels the optical spectrum also increases. Figure 8.17 shows the effect of RF power on the pulse width and spectral width of the pulses. At low RF powers the pulse width is long and a trailing pulse exists. As the RF is increased the pulse width reaches a minimum of just under 50 ps for 9 dBm RF input power to the laser, which corresponds to the minimum time–bandwidth product. The optical spectrum increases with RF power, and for higher RF levels the pulse width also increases. One way to eliminate the formation of a secondary pulse would be to use a multi-section laser diode which includes a saturable absorber section. A pulsed RF drive could also be used, but for future use of standard laser diode packaging we chose initially to stay with sinusoidal modulation. For higher modelocking frequencies (5 GHz, 10 GHz) the shorter effective electrical drive pulse will help inhibit multi-pulse buildup. If secondary pulses are suppressed by one of these means, the DC drive can be increased and the output power therefore increased.

The design of the chirp in the Bragg reflector is important as it establishes the performance characteristics of the HSPS. The reflector used in

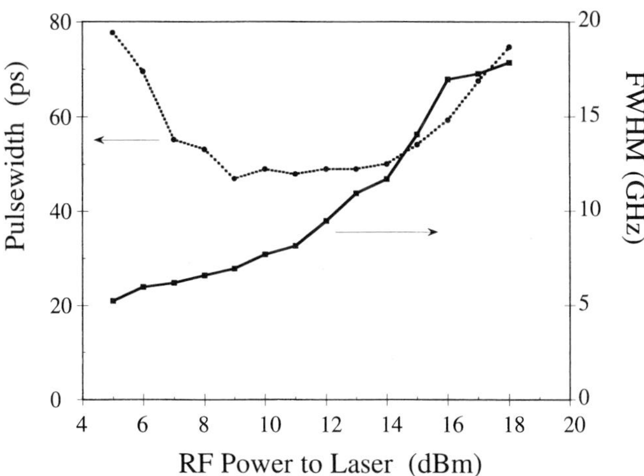

Figure 8.17. Pulse width and spectral width versus RF power level.

these experiments has a relatively small linear chirp (0.2 nm) over a long length (1 cm) which gives rise to a lot of dispersion (>300 ps/nm per pass). This dispersion spreads out the optical pulse, which counteracts the pulse shortening in the laser diode, and together these effects give the relatively long pulse widths required for soliton transmission. For shorter pulse widths a larger chirp/length ratio would be required.

8.7 CW operation with a chirped Bragg reflector

Modelocked operation of the hybrid laser has been described so far, however it has also been found that this device has attractive attributes for use as a CW source (Morton *et al.*, 1994b), which will be described in this section. High power, narrow linewidth single mode lasers with low noise have many applications in lightwave systems, particularly as sources in externally modulated communications systems. The ability to produce these devices with a tightly controlled operating wavelength is very important for many applications. Wavelength division multiplexed systems require separate lasers at fixed wavelengths. Long haul systems may require operation close to the dispersion minimum, or operation in the small passband of a chain of optical amplifiers.

The CW version of the hybrid device uses the same cavity setup described in Figure 8.6. The initial measurement of the transmission spectrum for the grating used in these experiments is shown in Figure 8.18, where the center wavelength is 1536.2 nm. After heating to displace

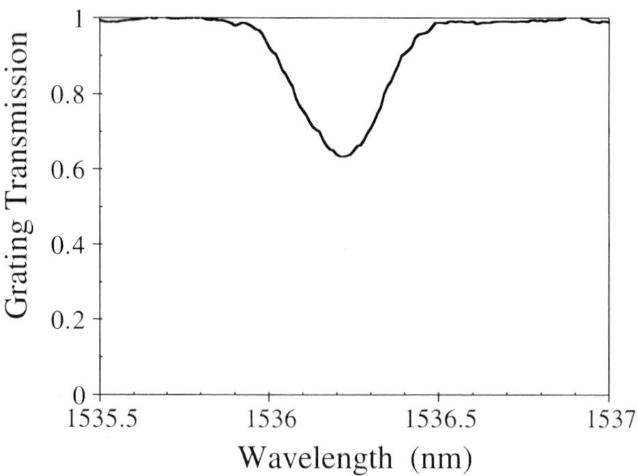

Figure 8.18. Transmission spectrum of a chirped Bragg reflector.

the residual hydrogen added in the sensitizing procedure, the wavelength shifts to 1535 nm where the device operates. The reflectivity of the grating is simply given by one minus the transmission, as the gratings used in this work are in the weak gratings regime and so have no significant loss (Mizrahi and Sipe, 1993). The reflector has a linear chirp distributed over 1 cm of fiber, giving an overall bandwidth of 0.28 nm, with a maximum reflectivity of 33%. This very narrow bandwidth, and also the ability to create gratings with precise wavelength control, is due to the uniformity of the fiber waveguide. Similar tolerances are not practical in semiconductor waveguides due to processing non-uniformities and the higher refractive indices. These fiber reflectors also provide a clean reflection spectrum (Mizrahi and Sipe, 1993), whereas semiconductor reflectors generally have side-lobes due to their abrupt amplitude profile. The hybrid approach therefore has an advantage over a monolithic semiconductor device such as a distributed Bragg reflector (DBR) laser (Koch *et al.,* 1988), due to the superior reflector characteristics and substantially lower optical loss. In this device the cavity length is 4.1 cm, which also allows it to be actively modelocked at a fundamental cavity frequency of 2.488 GHz. A shorter fiber cavity should lead to higher mode selectivity, although we have found this device to be very stable.

The use of a passive Bragg reflector eliminates the spatial hole-burning found in distributed feedback (DFB) laser diodes, where long-range fluctuations in the carrier density (over tens of microns) along the laser diode change the effective grating pitch and tend to destabilize single mode operation. The use of a semiconductor gain section reduces the spatial hole-burning common in solid-state lasers such as an erbium doped fiber laser (Mizrahi *et al.,* 1993), where hole-burning occurs on the scale of a wavelength, reducing the gain margin of the lasing mode and eventually leading to multi-mode operation at high pump power. Carrier diffusion in the semiconductor substantially reduces this effect as the diffusion length is far longer than the effective wavelength in the material.

The light/current characteristics of the device are shown in Figure 8.19. The threshold current is 7 mA, and the slope efficiency is 0.15. The low threshold is similar to that measured for the uncoated laser diode, and is achieved due to the high coupling efficiency between the laser diode and external cavity. A high coupling efficiency antireflection coated lensed fiber stub (Presby and Edwards, 1992; Edwards *et al.,* 1993) is used in this device. The output coupler (Bragg reflector) has a fairly low reflectivity of 33% to allow a high output power to be achieved. In order to obtain clean single mode operation it is essential that reflections from the AR

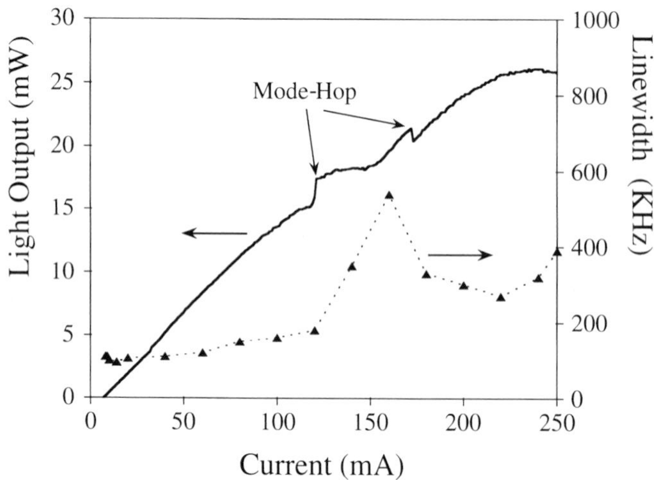

Figure 8.19. Light output and linewidth versus bias current.

coated laser facet and fiber lens are kept as small as possible. The very small residual Fabry–Perot modes in the output spectrum of this device clearly indicate that these reflections are insignificant. The maximum output power of 27.5 mW is achieved for a bias level of 250 mA. This high power level is useful in many practical systems, especially as the light is already coupled into a single mode fiber.

A scanning Fabry–Perot interferometer was used to monitor the fine spectral features of the device. Single longitudinal mode operation was maintained throughout the measured bias range. As the bias is varied the mode changes smoothly in wavelength due to the changing index in the semiconductor portion of the cavity. For sufficient change in cavity length the device will mode-hop to the next longitudinal mode (see Figure 8.19), however this hop is not accompanied by mode-beating or spurious noise. The change in wavelength with bias is small because the semiconductor waveguide only makes up a small part of a long cavity (about 1.6% of the optical length of the cavity is semiconductor). The device exhibits hysteresis in the mode behavior, going from one mode to the next as the bias current is increased, then staying with the new mode as the bias is reduced to original levels. This behavior is shown in the traces from a scanning Fabry–Perot interferometer shown in Figure 8.20. A feedback loop could be used to prohibit mode-hopping, as has been used in monolithic DBR lasers (Woodward *et al.,* 1992), to provide a robustly single mode device.

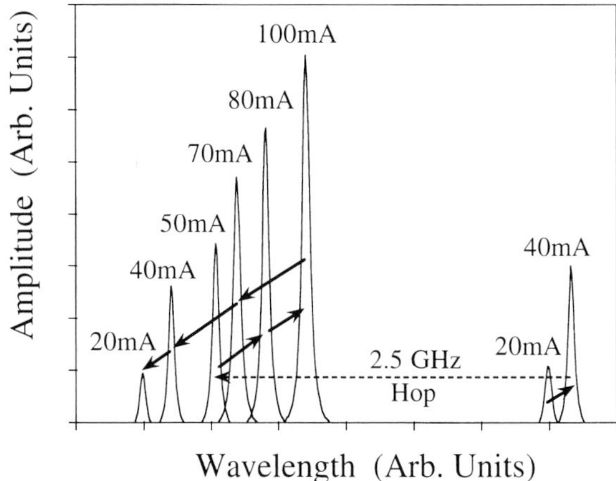

Figure 8.20. Scanning Fabry–Perot traces of the lasing mode behavior as the bias is first increased from 20 mA up to 100 mA, then reduced back to 20 mA.

The laser output remains in a stable single mode through the use of a chirped Bragg reflector. We find that the chirped reflector can either stabilize or destabilize single mode behavior depending on orientation. With the reflector oriented so that the grating period is reduced moving away from the semiconductor section, stable single longitudinal mode operation occurs. With the grating in the opposite direction a more complex dynamic occurs, leading to unstable mode behavior in our devices. The mechanism for this behavior can be revealed by considering the laser diode as a nonlinear element and the grating as a dispersive element. Over a time scale of several nanoseconds, the effective index of the laser can be considered to vary with the laser intensity I ($n = n_0 + n_2 I$). A slight rise and fall in intensity will be accompanied by a red then a blue shift in optical frequency. A grating in the correct orientation will tend to flatten out these intensity variations which will stabilize the output, whilst in the opposite orientation the variations will be increased, leading to unstable behavior. A rigorous description of this process, which includes gain saturation and the effects of electron and photon lifetimes, has been shown to agree with this simple picture (Sipe, 1993). In that case the problem is cast as a nonlinear Schrödinger equation and solved to find stability regions for various parameter values. Experimental evaluation of the mode behavior of this device has been carried out for gratings oriented in both directions, and stable single mode behavior found only in the direction described above. By utilizing a correctly oriented chirped

Bragg reflector the output can be guaranteed to stay in a single longitudinal mode regardless of cavity changes, although mode-hops will occasionally occur to reset the lasing mode close to the peak in the reflector.

The laser spectrum at threshold (7 mA), just above threshold (8 mA), and at the maximum output power of 27.5 mW (250 mA) are shown in Figure 8.21. At 1 mA above threshold the side-mode suppression ratio (SMSR) is already 35 dB, for only 0.1 mW output power. The SMSR rises quickly with the output power, with a maximum value of 57 dB at 200 mA bias, and for the maximum output power the SMSR is still over 55 dB. The side modes are residual Fabry–Perot modes of the laser diode cavity, and also from reflections of the fiber lens tip. The optical spectrum analyzer used to measure SMSR cannot resolve individual external cavity modes.

The linewidth of this device versus bias current, measured by a self homodyne technique using a 10 km delay length, is also shown in Figure 8.19. The minimum linewidth of 100 KHz is achieved close to threshold, and is fairly constant for power levels up to 5 mW. The narrow linewidth measurements are easily affected by acoustic pickup, and are probably limited by the experimental setup, as the fiber is held in free space and aligned to the laser diode. It is therefore expected that narrower linewidths will be obtained when the device is fully packaged. Further, a simple feedback circuit may also reduce the effect of acoustic pick-up.

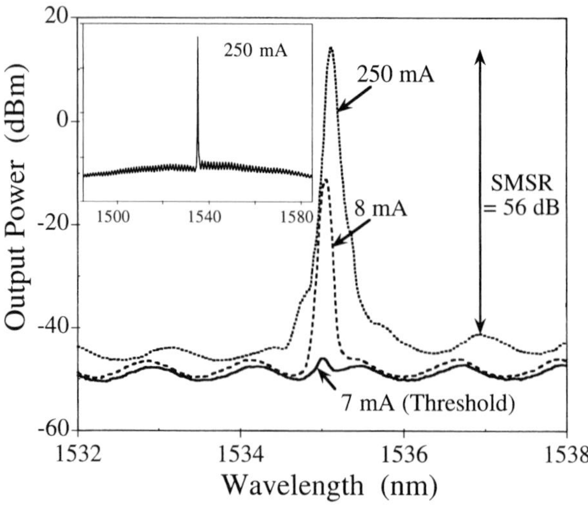

Figure 8.21. Optical spectra at 7 mA, 8 mA and 250 mA bias current.

The linewidth broadens as the bias current is increased, and shows a maximum of 540 KHz close to the region where the mode-hop occurs. The linewidth then decreases again above the mode-hop range. The broadening at higher currents may be due to the higher levels of spontaneous emission (see Figure 8.21).

A measurement of the RIN versus frequency is shown in Figure 8.22, for a bias current of 180 mA. The RIN is measured to be just under 160 dB/Hz up to 2 GHz, which is close to the measurement limit. This measurement shows the very low noise level of this device, providing a high signal/noise ratio source. A peak around 2.5 GHz and its harmonics shows the optical resonance due to the cavity round-trip time. This peak is fairly broad due to the long spatial extent of the Bragg reflector, which provides feedback at a range of effective cavity lengths. It is this phenomenon that allows an extremely large modelocking frequency range for this kind of device (Morton *et al.,* 1993). For CW operation a shorter cavity length would be optimum, moving the RIN peaks out to higher frequencies. The RIN spectrum for this device using a chirped Bragg reflector can be compared with the same measurement shown in Figure 8.12 for the unchirped reflector.

In this section we have described a hybrid laser that produces high output power in a stable single longitudinal mode, through the use of a correctly oriented chirped Bragg reflector. The output has a narrow linewidth, low noise, and well controlled wavelength. This device has applications in many lightwave systems. The device structure, using different

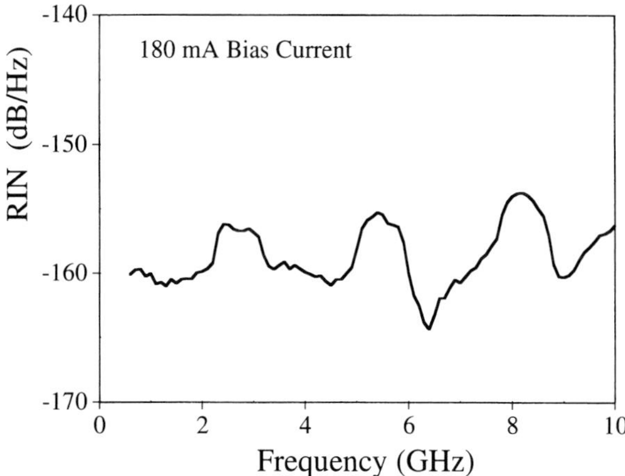

Figure 8.22. Relative intensity noise (RIN) spectrum for 180 mA bias current.

semiconductor materials, should allow similar operation from visible wavelengths out to 2 μm.

8.7.1 *Achieving short single pulse output*

The previous sections have described various aspects of the HSPS performance characteristics. One important parameter not discussed at length is how to achieve clean pulses, of the correct pulse width, which are symmetrical and transform limited. This section deals with some of these issues, and discusses ways of providing single, clean pulses, together with ways of producing shorter pulses than those described in Section 8.4.2.

The HSPS devices described so far have been designed to operate with a fundamental cavity frequency of 2.5 GHz. A soliton system operating at this bit-rate would require pulses of the order of 50 ps, similar to those obtained in Section 8.4.2. For operation at higher bit-rates, there is the question of whether to produce a source directly at these frequencies (5 GHz or 10 GHz), or to optically multiplex up to these bit-rates by simply splitting then re-combining the output pulse train with an appropriate time delay between the two arms. Similarly, optical demultiplexing at the receiving end can be easily accomplished (Andrekson *et al.,* 1992). HSPS devices can be made to operate at the higher bit-rates by shortening the cavity length, which can be accomplished by creating the lensed fiber end directly on the grating fiber. An HSPS device with a fundamental cavity frequency of 10 GHz was fabricated in this way, producing pulse widths of the order of 30 ps at a 10 GHz repetition rate, the pulse width being limited by the grating bandwidth. Working at the lower bit-rate of 2.5 GBit/s allows the use of conventional electronics at the transmitter and receiver, while working at the higher rates puts a strain on the availability of components and their performance. For these reasons it is useful to have a source operating at 2.5 GBit/s but with pulse widths compatible with higher bit-rates (i.e. 20 ps pulses). In order to achieve these shorter pulse widths the HSPS design must be modified.

When working with a 2.5 GHz HSPS, a sinusoidal RF drive signal will look to the laser like a fairly long electrical pulse, as the negative portion of the waveform is effectively removed by the diode characteristics of the semiconductor laser. The electrical pulse will be almost 200 ps in duration, which is far longer than the optical pulses produced by the device. This large mismatch in the pulse durations means that the drive pulse will generally continue injecting carriers into the laser section after the optical pulse has left the section. The carrier density can therefore rise after the pulse has left, which can produce sufficient gain for a secondary pulse to build up. This causes trailing shoulders on the output pulse, or secondary output pulses (Morton *et al.,* 1989). This phenomenon can be reduced by using shorter electrical drive pulses, using a laser section with higher differential gain, or using two section lasers.

Short electrical pulses can be produced by step recovery diodes (SRDs), which are commercially available components that are driven by a high power sinusoid (approximately 1 W) to produce short pulses at the drive frequency. SRDs with operating frequencies up to 2.5 GHz can be obtained commercially, with output pulse widths around 50 ps and peak voltages of around 5 V into 50 Ω. Using the short electrical pulse drive reduces the chance of secondary pulse output.

The use of a semiconductor gain section with a higher modulation efficiency can also reduce the possibility of secondary pulses, and at the same time produce shorter output pulses. The laser parameter of importance is the differential gain, dG/dN, i.e. the change in gain divided by the change in carrier density in the active section. This term controls the rate of stimulated emission, so that by increasing dG/dN for a given input optical pulse, the gain of the pulse will be higher and the carrier depletion much stronger. This will lead to shorter pulses, with fewer available carriers after the pulse to provide gain for a secondary pulse to build up.

The differential gain can be increased by using a compressively strained quantum well active region in the semiconductor laser section (Morton *et al.,* 1992b). An even larger increase can be obtained by P-doping this quantum well structure, as described in Morton *et al.* (1992c), where this effect was used to produce 1.55 μm Fabry–Perot lasers with a small signal modulation bandwidth of 25 GHz. By using devices with the same active region structure (P-doped compressively strained MQW), shorter, clean output pulses were obtained from an HSPS with a grating bandwidth of only 0.2 nm FWHM. Results from these experiments are shown in Figure 8.23. The laser DC bias is 14 mA, and an SRD was used to supply 50 ps electrical pulses to the laser at a repetition rate of 2.488 GHz. In these experiments the electrical pulse is passed through a variable attenuator before being applied to the laser. An attenuation setting of 9 dB is used for these results. Figure 8.23(a) shows the pulse shape, which is clean and symmetrical with a width of 25 ps. The spectrum is shown from both an optical spectrum analyzer (OSA) (Figure 8.23(b)), which shows the envelope of the modes, and a scanning Fabry–Perot interferometer (Figure 8.23(c))

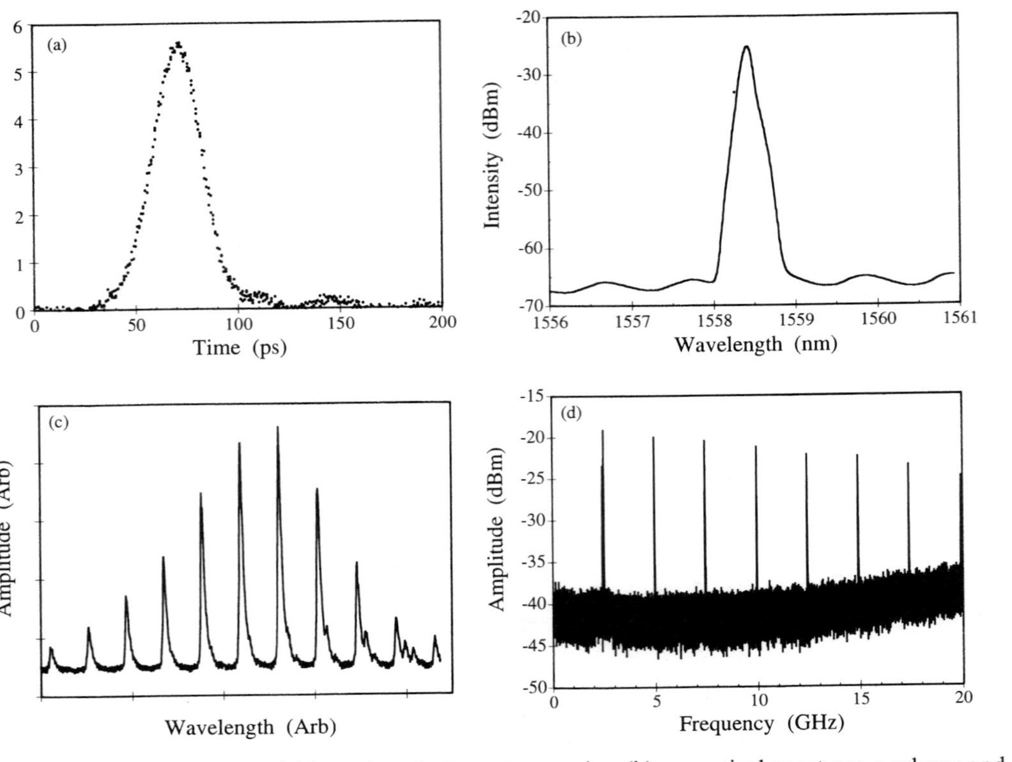

Figure 8.23. (a) Pulse width, and optical spectrum using (b) an optical spectrum analyzer and (c) a scanning Fabry–Perot interferometer; and (d) microwave spectrum, for an HSPS using a high differential gain laser diode.

which shows the actual locked modes. The OSA shows the high rejection of the residual Fabry–Perot modes of the laser diode, which are 40 dB below the lasing modes. The actual mode structure from the scanning Fabry–Perot interferometer has a clean shape. The final trace (Figure 8.23(d)) is from a microwave spectrum analyzer with an optical input, which shows the comb of modes at 2.488 GHz and its harmonics. This trace shows that there are no spurious signals or noise peaks in the output waveform. It is important that all these measurements are taken on an output waveform to ensure that the device is operating stably. The measured optical bandwidth is 11.5 GHz, giving a time–bandwidth product of 0.29, which is close to the transform limit if a sech^2 pulse shape is assumed.

The variations of pulse width and time–bandwidth product for this device as the amplitude of the driving electrical pulses is varied are shown in Figure 8.24. The pulse width can be varied from 30 ps down to 17.5 ps, with the time–bandwidth product remaining close to 0.3. This is a substantial tuning range of the pulse width over which clean, transform limited pulses are produced.

It is possible to use two-section laser diodes to help reduce the buildup of secondary pulses and also produce short pulses. In general, a shorter section is reverse biased, and the modulation applied to this section. The longer section is biased to provide gain to allow the device to oscillate. The major effect of the reverse bias is to reduce the average carrier

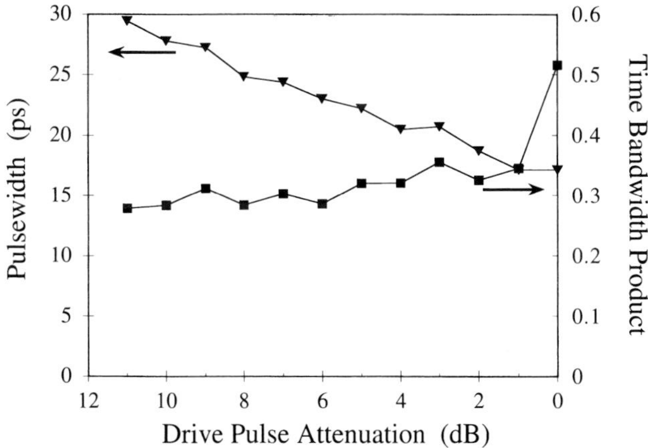

Figure 8.24. Pulse width and time–bandwidth product versus drive pulse attenuation for an HSPS using a high differential gain laser diode.

density in that section, which effectively increases the differential gain as this is larger at lower carrier densities. The higher differential gain of the short section has the same effect as the high differential gain laser described in previous paragraphs. Mar *et al.* (1992) used a two-section laser to produce single optical pulses of 1.4 ps at a repetition rate of 3 GHz using a special SRD which produced 29 ps electrical drive pulses with an amplitude of 7 V. In that work the authors used a long DC biased section to increase the round-trip time within the semiconductor cavity to be comparable with the electrical drive pulse width. This ensures that reflections of the 1.4 ps pulse from the AR coated facet of the semiconductor section do not see sufficient gain while returning through the semiconductor section to form secondary, reflection induced pulses (Morton *et al.*, 1989; Bowers *et al.*, 1989).

While it is possible to reduce the pulse width of a given HSPS by using high differential gain from a single or multi-section laser, or by using a short electrical drive waveform, it is simpler to actually modify the bandwidth of the Bragg reflector. By increasing the chirp rate in the chirped Bragg reflector a larger bandwidth can be easily achieved. These larger chirp rates produce a more linear variation of penetration depth versus wavelength, as they swamp the small quadratic chirp that is inherent in gratings with a Gaussian amplitude profile (Mizrahi and Sipe, 1993). An example of the transmission spectrum of a wide bandwidth chirped Bragg reflector is shown in Figure 8.25. This reflector has a bandwidth of 1.9

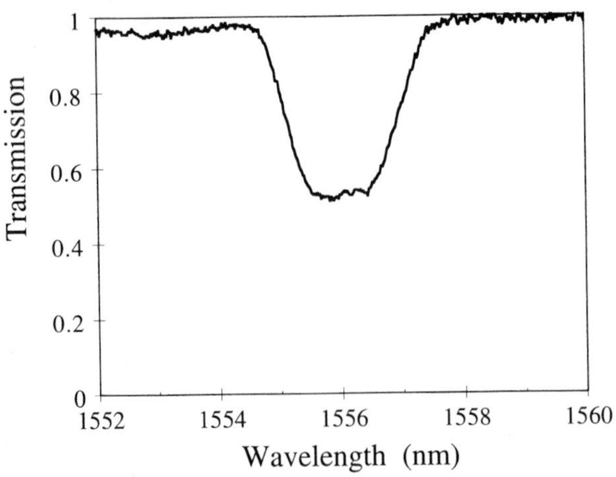

Figure 8.25. Transmission spectrum of a wide bandwidth (1.9 nm) chirped Bragg reflector.

nm FWHM spread along 1.2 cm of fiber length. The peak reflectivity of 48% is located at 1556 nm.

By combining the grating in Figure 8.25 with a high coupling efficiency AR coated microlensed fiber, an HSPS with a threshold current of 7 mA and maximum CW output power of 20 mW was created. Using a standard compressively strained multiple quantum well laser (Tanbun-Ek *et al.,* 1990), pulses of under 15 ps were obtained (close to the measurement resolution of 12.5 ps) for sinusoidal modulation at 2.488 GHz. An excellent suppression of the Fabry–Perot modes from the residual reflections of the AR coated laser facet and reflections from the lensed fiber end were achieved. These are generally dominated by reflections from the lensed fiber end. By using AR coated micro-machined lenses (Presby and Edwards, 1992; Edwards *et al.,* 1993), these reflections are made much smaller, giving rise to the suppression ratio of over 50 dB for these modes. The resulting pulse waveform and optical spectrum are shown in Figure 8.26. The output is a clean single pulse with a width of 14.6 ps, a peak power of 16.5 mW, and a time–bandwidth product of 0.35.

In this section we have shown how to optimize the pulse shape and reduce the pulse width for an HSPS using a specific Bragg reflector. Similar reductions in pulse width can be achieved by broadening the reflector bandwidth. This also allows the short pulse to be achieved with a sinusoidal drive, which is preferable for simple package design. When designing a device for use in a specific system, a compromise must be made, as the broader gratings allow shorter pulses to be obtained easily, while narrower gratings provide optimal bandwidth control and smaller variations in the device operating wavelength.

8.7.2 *Wavelength tuning the HSPS*

The center wavelength of operation of the HSPS is controlled by the integrated Bragg reflector. For the device described in Section 8.4.2, the reflector has a bandwidth of 0.2 nm, with a linear chirp spread out over 1 cm of fiber length. This narrow bandwidth provides for excellent control of the output wavelength, however in devices with a wider bandwidth some extra form of control may be necessary.

The wavelength self-tuning effect (Section 8.4.2) allows the device to operate over a wide modulation frequency range. However, the real need for this mechanism is to allow the device to be fabricated with some error in the exact placement of the grating in the cavity, and still be able to

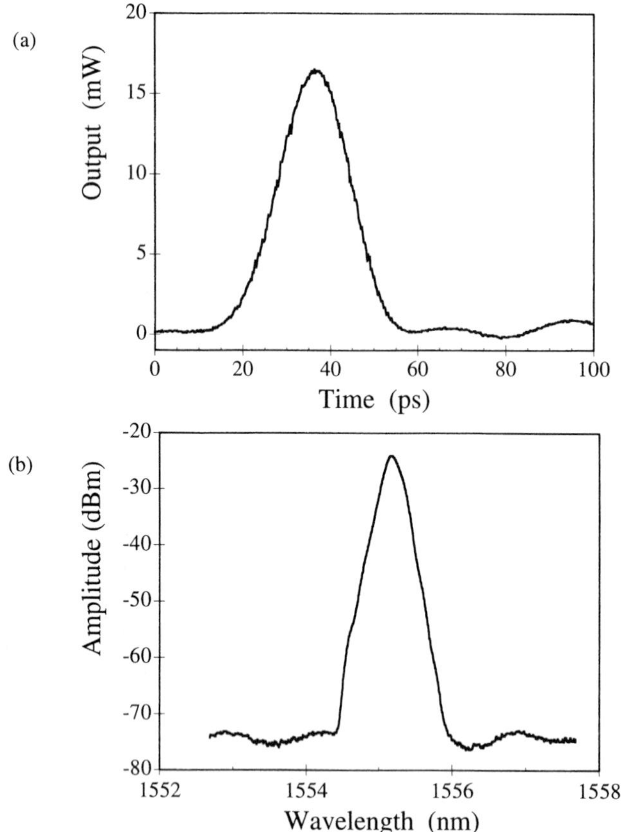

Figure 8.26. (a) Measured optical pulse shape and (b) optical spectrum for an HSPS using a wide bandwidth Bragg reflector.

operate at the required frequency. For a reasonable length error of ± 1 mm, the center wavelength of the device will be up to 0.02 nm away from the Bragg wavelength in the case of a 0.2 nm wide Bragg reflector. Broader gratings will produce larger errors. There will also be some small error in the actual Bragg wavelength of fabricated gratings due to manufacturing tolerances.

In order to have very fine control of the operating wavelength it is necessary to have an independent control of the wavelength, which can be used to overcome manufacturing errors in Bragg wavelength and wavelength shifts due to errors in cavity length. This independent control will need to produce wavelength changes of the order of a few angstroms to overcome these factors. In order to change the center wavelength, the

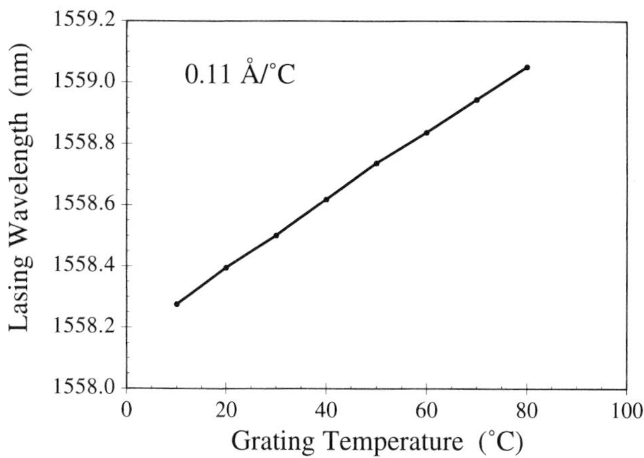

Figure 8.27. Center wavelength of the output spectrum for a modelocked HSPS versus the temperature of the Bragg reflector.

Bragg reflector effective period must be modified. Means to accomplish this have been characterized at great length by groups using these fiber gratings in sensor applications (Morey *et al.*, 1989). The two main parameters for modifying the grating are the temperature and the strain on the grating.

Temperature tuning of the grating produces a wavelength shift of approximately 0.01 nm/°C. This is small enough so that the grating does not need to be temperature stabilized with any high degree of accuracy, but large enough to allow several angstroms of tuning for reasonable temperature changes. The majority of this wavelength shift is due to the change in index of the material with temperature.

Experiments were carried out using an HSPS with a chirped Bragg reflector of around 0.2 nm FWHM, with a center wavelength of 1558 nm. The grating was placed on a peltier heat pump which allowed the temperature to be varied from 10 °C to 80 °C. The device was modelocked continuously as the temperature was varied, with some minor adjustments of drive levels to allow optimum operation at each temperature. The center wavelength of the output spectrum is plotted in Figure 8.27, versus the temperature of the grating. A linear variation of center wavelength with operating temperature was found, with a coefficient of 0.11 A/°C being extracted from the data. An overall tuning range of 0.78 nm was achieved for this temperature range.

The application of strain to a fiber grating can produce much larger changes in Bragg wavelength than those achieved from reasonable temperature changes. This may be useful if a large tuning range is required for the operating wavelength, however tighter control is necessary if fine control of the wavelength is also important. Strain can be applied by simply stretching the device mechanically, for example, by using piezoelectric transducers. Another possibility is to attach the grating to a metal with a high coefficient of expansion which will stretch the fiber when heated. In Morey *et al.* (1989), a wavelength shift of 1.75 nm was achieved with the application of 32 kpsi tension to a fiber. Higher levels of tension, which can be accommodated with properly re-coated fiber gratings, can provide wavelength shifts of 10 nm or more.

While strain can provide an easy way to tune the center wavelength of an HSPS, it can also cause problems. It is necessary to produce a strain free package to hold the fiber reflector in order to maintain the Bragg wavelength of the grating at any specific temperature. This must be considered in the packaging of all devices where fiber gratings are used as an accurate wavelength reference.

The results described in this section show that an independent control of the device operating wavelength can be achieved by temperature controlling the Bragg reflector. In this configuration 0.78 nm of temperature tuning were obtained for reasonable temperature changes. This allows the device to overcome variations in operating wavelength due to fabrication tolerances in the reflector Bragg wavelength and the HSPS cavity length.

8.7.3 *High power output using two-section laser diodes*

The HSPS fulfills all the requirements for simplicity, reliability and excellent operating characteristics necessary for soliton transmission systems, which put stringent requirements on a pulse source. For most other applications the specifications can be relaxed somewhat, which allows the device to be optimized for other particular attributes. One important parameter is the peak power of the modelocked pulses, which for many applications should be maximized even at the expense of the time–bandwidth product. In this section a hybrid pulse source utilizing a two-section laser diode in order to increase the attainable power levels of the modelocked pulses is described (Morton *et al.*, 1994a).

A schematic diagram of this source is shown in Figure 8.28. This device uses a two-section laser diode to allow operation with high average and peak power levels. The laser diode is a compressively strained multiple quantum well device with seven wells, which provides low loss and high differential gain (Tanbun-Ek *et al.*, 1990). The shorter section (75 μm) at

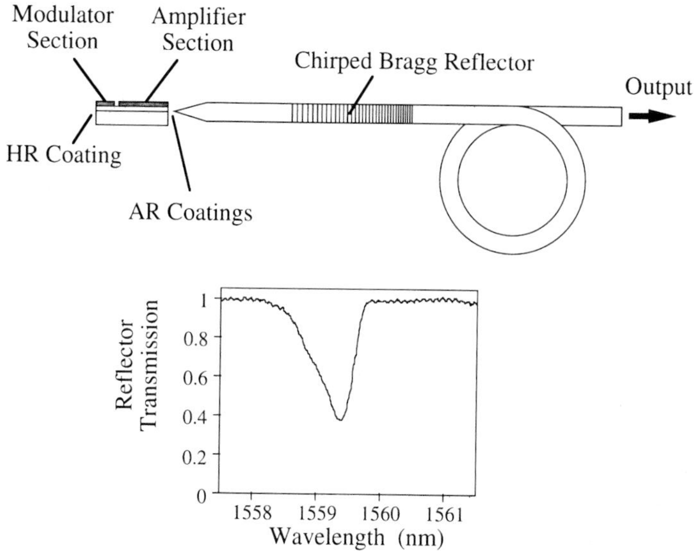

Figure 8.28. Schematic diagram of the hybrid pulse source with a two-section laser diode, and the transmission spectrum of the chirped Bragg reflector.

the end of the cavity is used to modulate the laser diode to actively mode-lock the device. The longer section (425 μm) is DC biased, and used as an amplifier section to allow high output powers. The two-section devices are produced by segmenting the usual stripe contact and etching between the contacts to increase the electrical isolation, while confining the etch to depths so as not to interfere with the optical waveguide of the device. The electrical isolation between sections in the device used is 270 Ω. Initial measurements of the Bragg reflector transmission spectrum are shown in Figure 8.28, where the central wavelength is 1559.3 nm. After heating to displace the residual hydrogen added in the sensitizing procedure, the Bragg wavelength shifts to 1557 nm. The chirped grating reflector has a peak reflectivity of 63% and a spectral full width at half-maximum (FWHM) of 0.71 nm. The peak reflectivity of 1557 nm is chosen so as to coincide with the gain in erbium doped fiber amplifiers. The chirp rate in the reflector is 16 A/cm, which gives a spatial FWHM of about 4.4 mm for this grating.

The modulation section of the laser diode is reverse biased with a voltage source, and a sinusoidal current then applied via a bias-tee. The reverse bias reduces the average carrier density in this section, and allows very efficient modulation of the gain by the drive signal. A modulation frequency around 2.5 GHz produces fundamental modelocking in this device. The modulation frequency can be varied substantially around this central frequency and modelocking still occurs, due to wavelength self-tuning (Morton *et al.,* 1993). The amplifier section provides gain to overcome cavity losses, and also provides the energy necessary to produce high power pulses. Under typical drive levels of 22 mA DC bias on the amplifier section, and −1.2 V on the modulation section with a 2.5 GHz RF signal of + 12 dBm, the device produces pulses of 24 ps with a peak power of 13.5 mW and a time–bandwidth product of 0.37. With the RF drive to the laser removed, the negative bias on the modulation section ensures that the device will not lase CW. This provides a very large suppression of the Fabry–Perot modes of the laser diode cavity, or modes due to reflections from the lensed fiber end. The energy supplied to the device can be increased by increasing the bias to the amplifier section. By simultaneously increasing the negative bias on the modulation section and increasing the RF power, the device can be kept in a stable modelocked configuration with the output power level of the device increased. By increasing the DC bias up to 160 mA, with a bias of –3.0 V on the modulation section and an RF drive of 15.6 dBm at 2.51 GHz , the results shown in Figure 8.29 are found. The pulse width is 23 ps, with an average output power of 7.8 mW, giving a peak power level of 137 mW. These power levels are an order of magnitude higher than those achieved from typical modelocked semiconductor lasers. The optical spectra measured on an optical spectral analyzer and with a scanning Fabry–Perot interferometer are also shown in Figure 8.29. These spectra show a large suppression of residual Fabry–Perot modes of 40 dB. The scanning Fabry–Perot interferometer shows that modes are locked together with an optical bandwidth of 40 GHz. This gives a time–bandwidth product of 0.88, which is sufficiently low for most practical applications. The HSPS described in Section 8.4.2 produced a time–bandwidth product of 0.3 (i.e. close to transform limited if a sech^2 pulse shape is assumed), but for a reduced peak power of 7 mW. The large number of longitudinal modes, and accompanying wide spectrum, may allow the pulses to extract more power from the semiconductor gain medium, due to the larger energy range and also reduced levels of spectral hole-burning (lower power per mode).

If even higher peak powers are required, for example if the output is to be split many times to be sent to different nodes, an erbium fiber amplifier can be used as a power amplifier. The high average power (7.8 mW) produced by the hybrid source is necessary to saturate the erbium fiber

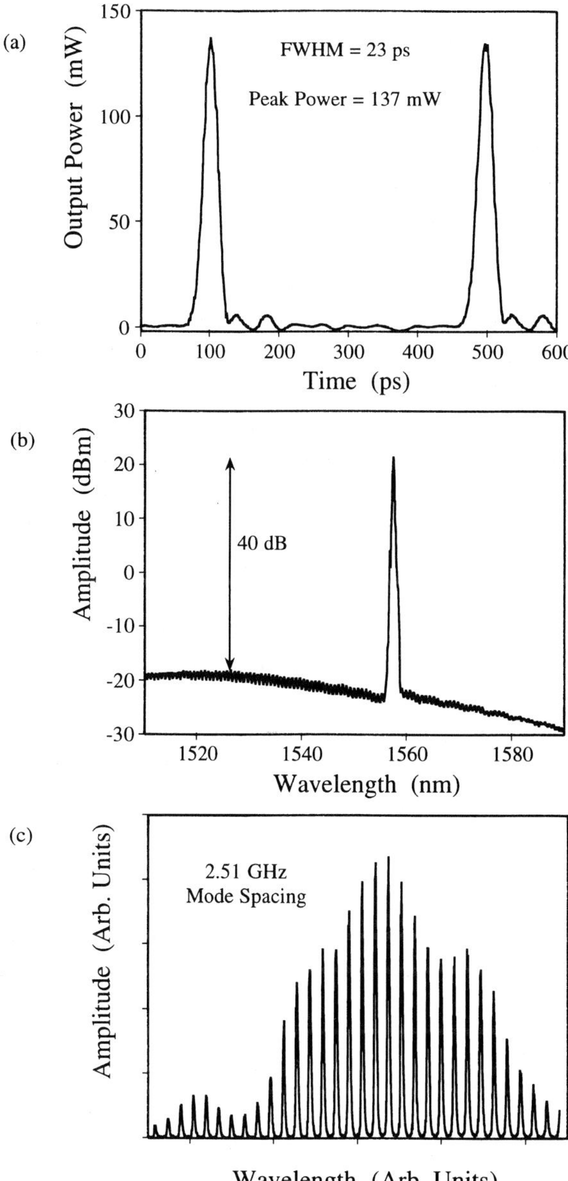

Figure 8.29. (a) Pulse output waveform, and optical spectra using (b) an optical spectrum analyzer and (c) a scanning Fabry–Perot interferometer, for 160 mA DC bias to the amplifier, −3.0 V bias and +15.6 dBm RF drive at 2.51 GHz to the modulator section.

amplifier and therefore reduce the level of noise added by the amplifier. This will help preserve the high signal/noise ratio produced by the modelocked hybrid pulse source. An erbium fiber power amplifier with a maximum average power of +20 dBm would amplify the pulses up to a peak power level of 1.7 W. Lower repetition rates would increase the peak power accordingly.

The peak power achieved at various DC bias levels is shown in Figure 8.30, together with the reverse bias applied to the modulation section. The trends show how the peak power rises with increasing amplifier bias, and saturates at higher bias levels. The absorber reverse bias is increased as the amplifier bias increases, to keep the device operating below threshold for CW operation. At the high bias levels the peak in the optical gain (see Figure 8.29(b)) moves towards shorter wavelengths due to band-filling in the gain region, and the spectral shape of the loss provided by the reverse biased section. Higher output powers would be possible if the gain peak of the material just above threshold was chosen to be at a longer wavelength than the Bragg reflector, so that under high bias the gain peak occurs at the Bragg reflector wavelength.

In this section we have seen that by using a two-section laser diode in the hybrid pulse source we are able to attain much higher power levels during modelocking. This is achieved by increasing the bias in the amplifier section, while reverse biasing the modulation section to prevent CW operation. Peak powers of 137 mW in the output fiber are achieved at a

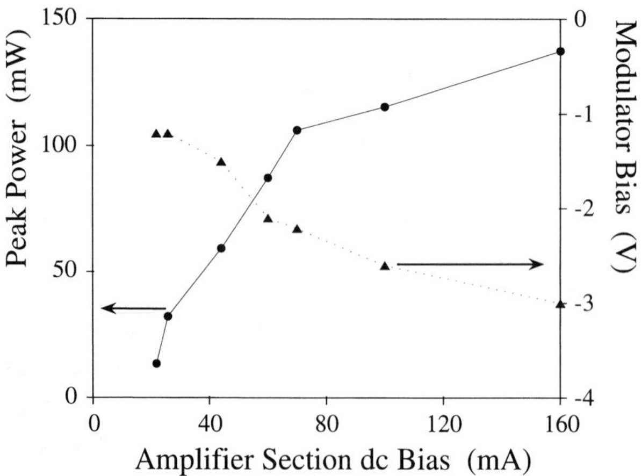

Figure 8.30. Peak power and required modulator bias versus the DC drive to the amplifier section.

repetition rate of 2.51 GHz and a wavelength of 1557 nm, with a time–bandwidth product of 0.88. The use of an erbium fiber amplifier would allow even higher pulse powers to be achieved, while still maintaining a high signal/noise ratio.

8.8 Packaged HSPS characteristics and soliton transmission results

This chapter has described the development of the HSPS from its early SIOB embodiment up to the present design using a fiber cavity with an integrated, chirped Bragg reflector. The use of a linearly chirped Bragg reflector has been shown to stabilize the output of the HSPS by eliminating the possibility of having two degenerate modelocked states. The linear chirp also allows the devices to be used over a large range of modulation frequencies, due to the wavelength self-tuning mechanism. This allows devices to be manufactured with a reasonable tolerance on the cavity length, and still be able to operate at the designed repetition frequency. The wavelength of the HSPS is tightly controlled by the Bragg reflector, due to its narrow bandwidth. Tuning to provide a fine control of the wavelength and overcome manufacturing variations in the operating wavelength can be accomplished by controlling the temperature or strain of the grating. By varying the temperature of the grating in an HSPS, 0.78 nm of wavelength tuning was shown for a reasonable (70°C) temperature range.

Ways to control the output waveform and provide single, clean transform limited pulses were described. Pulse widths from 15 to 50 ps were produced from devices with different reflector bandwidths. Devices with high differential gain were used to produce shorter pulses for a given grating bandwidth. The use of a pulsed electrical drive instead of a sinusoid was also shown to produce shorter optical pulses, with no secondary pulse buildup.

The HSPS has been shown experimentally to meet all the performance requirements for a source for ultra-long distance soliton transmission systems. The final requirement is that it must be possible to manufacture a stable and reliable source at a reasonable cost. To achieve this aim the HSPS must first be packaged to allow measurements of its stability when used in a transmission system. The packaged device must then meet the reliability requirements for a source in the terminal building of a transmission system. Initial packaging work has been carried out on the

HSPS, and a schematic diagram of the package layout and a photograph of a finished package are shown in Figure 8.31. To date only a small number of devices have been packaged, but results indicate that the packaged devices behave at least as well as the equivalent components measured on the bench. The package provides a much more robust connection from the laser to the fiber cavity, together with better immunity to mechanical vibrations.

(a)

Connections

1. DC + RF Input
2. Laser TE Cooler
3. Grating TE Cooler
4. Thermistor
5. Back Face Monitor
6. Thermistor

TE Coolers
(1)
Laser Submount
(2)
(3) (4) (5) (6)

(b)

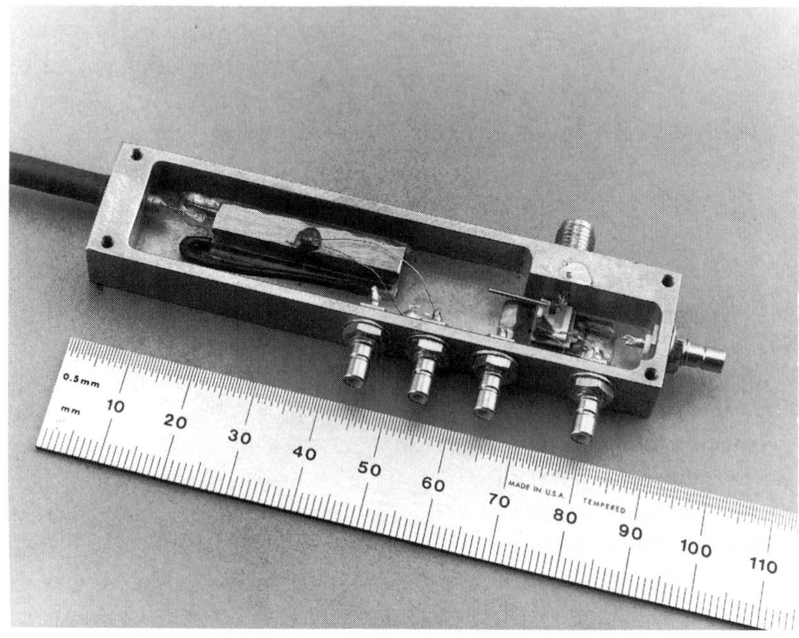

Figure 8.31. (a) Schematic diagram of the HSPS package and (b) photograph of a packaged HSPS.

In this section we describe the characteristics of a fully packaged HSPS over a wide range of operating conditions, together with initial system results using this source for ultra-long distance soliton transmission (Morton *et al.*, 1994c). The grating used in the package has a reflectivity of 43%, a FWHM of 3 A and is written with a chirp rate of 6.4 A/cm.

The device package is compact ($9 \times 2 \times 1$ cm) with the largest size being determined by the required length of the external cavity. In this case, a required modelocking frequency of 2.488 GHz sets the fiber cavity length at approximately 4 cm. The package includes temperature control for the laser diode and the fiber grating. The laser is kept at a constant temperature for stable operation, while the grating temperature can be varied to allow fine tuning of the operating wavelength. The laser DC bias and RF drive are applied through a coaxial microwave connector to the laser submount. The remaining connectors are used for temperature control and a back face monitor. The device is very simple, requiring just one alignment of a fiber to the laser diode, with the output taken through the Bragg reflector. Once packaged, the operation of the HSPS is extremely stable, results being taken over weeks of continuous operation.

The packaged device described in this section was designed in response to specifications from soliton system designers. This ability to meet design parameters set by others is of paramount importance for a practical pulse source. The HSPS specifications included an operating wavelength of 1557 nm, a modelocking frequency of 2.488 GHz (Sonet frequency) and a pulse width of 20 ps. The device operates CW in a single longitudinal mode, through correct orientation of the chirped Bragg reflector. The threshold current is 6 mA, which indicates that there is very high coupling between the laser diode and external cavity. A CW output power of 5.5 mW at 100 mA bias is achieved.

The device is very simple to use. Modelocking is accomplished by applying a DC current close to the threshold value plus a sinusoidal RF drive at the modelocking frequency. An example of the output from this device is shown in Figure 8.32. Pulses of 19.5 ps with peak powers of 5.5 mW are shown for drive conditions of 11 mA DC bias plus 21 dBm RF at 2.488 GHz. The optical spectrum in Figure 8.32(b) shows the large suppression of residual cavity modes of both the laser diode itself and modes from reflections off the fiber tip, these cavity modes being over 40 dB below the lasing modes. In this device the lasing modes are detuned almost 20 nm to the short wavelength side of the gain

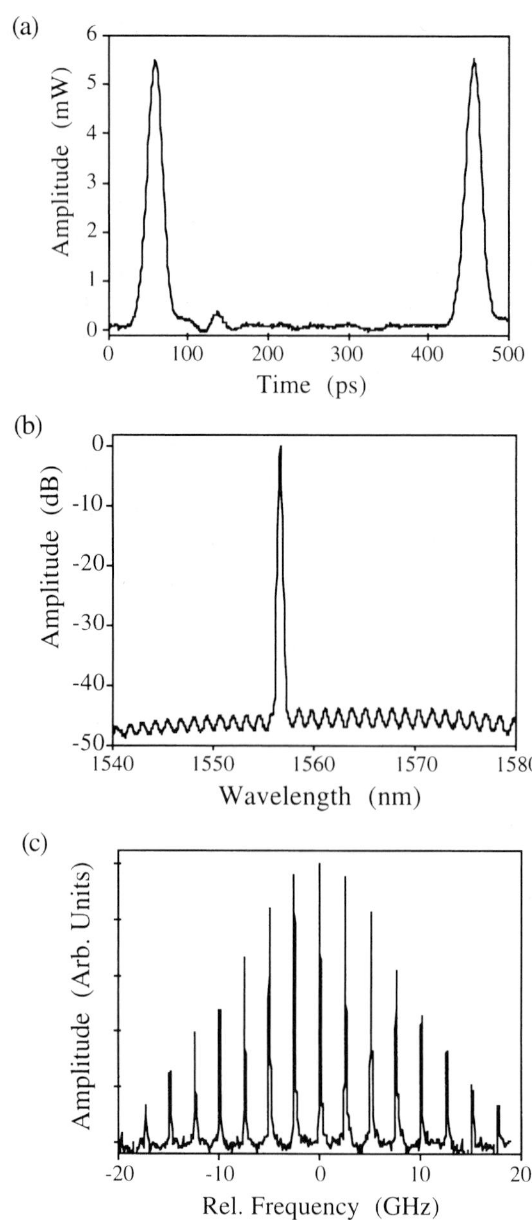

Figure 8.32. (a) Typical pulse waveform, and optical spectra using (b) an optical spectrum analyzer and (c) a scanning Fabry–Perot interferometer.

peak, to provide a higher differential gain and therefore produce the required short pulse width. The actual mode-structure is shown in Figure 8.32(c), the output from a scanning Fabry–Perot interferometer. This shows a very clean envelope of modes, with a FWHM of 19 GHz, giving a time–bandwidth product of 0.37.

One major attraction of this source is its stability over a large range of drive parameters. This is particularly important in developing a practical device for use in real systems, where inevitable production variations in cavity lengths, necessary drive levels etc. must not adversely affect the device operation. An example of this tolerance is shown in Figure 8.33, where the pulse width and time–bandwidth product are plotted for changes in RF power of over an order of magnitude. Over most of this range the pulse width remains fairly constant at around 20 ps. The time–bandwidth product slowly increases as the device is driven harder, but stays below 0.4 over most of the range. In this device the pulse width is set mainly by the dispersion in the grating, and the accompanying self-phase modulation in the laser diode section which must counteract it for stable modelocking to occur. The pulse width is also seen to be independent of DC bias level from 8 mA up to 18 mA, the range over which stable modelocking occurs.

Having shown that the pulse width can be well controlled over a large range of operating conditions, the other two important parameters for the pulse source are the modelocking frequency and the operating wavelength. Due to the wavelength self-tuning phenomena,

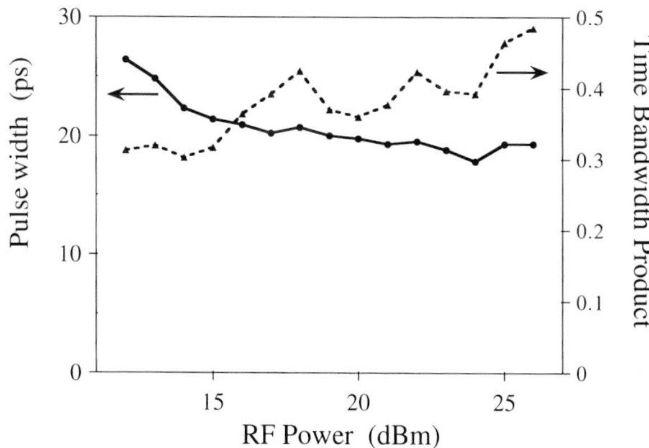

Figure 8.33. Pulse width and time–bandwidth product versus RF power level, with a DC bias of 11 mA and an RF frequency of 2.488 GHz.

the packaged HSPS has a very wide operating frequency range, with good modelocking achieved from 2.3 GHz to 2.95 GHz. The pulse width remains close to 20 ps as the modulation frequency is varied. For this variation in modulation frequency of 650 MHz, the wavelength shift is 5.2 A.

The overall wavelength variation from a group of HSPS devices will include the small shift due to wavelength self-tuning, plus several from fabrication and packaging of the gratings. It is then necessary to tune this operating wavelength to the precise system wavelength. The packaged device provides over 7 A tuning range for reasonable temperature changes (0.11 A/°C), and can therefore overcome variations in operating wavelength to set the wavelength to the system requirement.

Transmission experiments were carried out using the packaged HSPS. The 2.488 GHz pulse-train from the source was modulated with a 2^{14} word sequence from a pattern generator, and then optically multiplexed up to ~10 Gbit/s using the split/delay one arm/recombine technique (described in Mollenauer *et al.*, 1993). The 10 Gbit/s pulse-train was then sent into a recirculating loop, the schematic diagram of which is shown in Figure 8.34. This loop configuration includes three sections of transmission fiber, each followed by an erbium fiber amplifier and a sliding frequency guiding filter (Mollenauer *et al.*, 1993). The center frequencies of the filters are translated in time to provide a linear variation

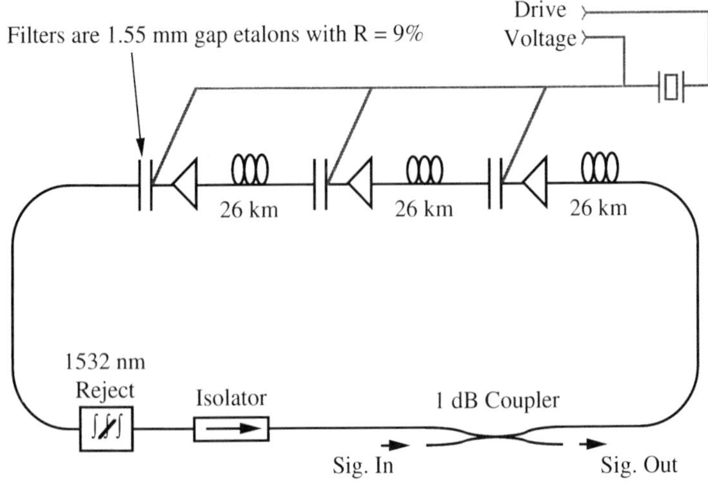

Figure 8.34. Schematic diagram of the soliton loop incorporating sliding-frequency guiding filters.

in center frequency versus transmission distance. In a straight line transmission system they would be replaced by selected fixed frequency filters. The pulse-train recirculates around the loop, and a portion is removed for bit error rate (BER) analysis. The 10 GBit/s bit stream is demultiplexed to 2.5 GBit/s using a nonlinear loop mirror (for example, see Andrekson *et al.*, 1992), and the BER measured versus transmission distance. The BER measurements are shown in Figure 8.35. The results show no error floor, and error free transmission (10^{-9} BER) is achieved for transmission of over 27 000 km. This corresponds to a bit rate $\times$ distance product of 270 Terabit-km/s.

In summary, we have described a fully packaged HSPS, which was designed to specifications, and produces 20 ps pulses at 2.488 GHz at a wavelength of 1557 nm. The source produces almost the same results for large variations in both DC bias current and RF drive level. The use of a linearly chirped Bragg reflector gives the device a large operating frequency range (2.3–2.95 GHz) due to wavelength self-tuning, which ensures the device will always operate at the design frequency. The operating wavelength is tuned over 7 A by temperature controlling the fiber grating, which allows any specified wavelength to be achieved. The HSPS was used in ultra-long distance soliton transmission experiments using a loop configuration including sliding-frequency guiding filters. Error free transmission at 10 GBit/s was achieved for a transmission distance of over 27 000 km.

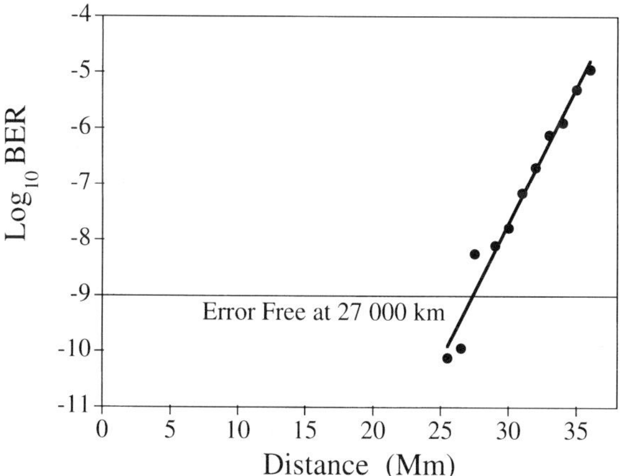

Figure 8.35. Measured bit error rate (BER) versus transmission distance.

8.9 Outlook

Looking to the future, there are many needs for short optical pulse sources, and many different devices which can fulfill these needs. The HSPS provides a stable, reliable pulse source for use in applications ranging from soliton transmission to measurement systems. Much work is still needed to fully exploit the capabilities of the HSPS, and to expand its use into different fields and applications. If this source is developed and produced in large numbers, its cost will fall to be comparable with standalone high performance lasers. This will allow the HSPS to be used in a larger range of applications than currently envisaged, and maybe provide the first short pulse laser system produced in large quantities.

Acknowledgments

The HSPS project is the result of a stimulating collaboration with Victor Mizrahi, Paul Lemaire and Turan Erdogan who together have pushed fiber grating technology to an artform. Tawee Tanbun-Ek and Ralph Logan have helped greatly in the development of semiconductor lasers used in all of these experiments. The author would like to thank Linn Mollenauer and Won Tsang for useful discussions and constant encouragement throughout this work, and also Renen Adar, Chuck Henry, Herman Presby, George Harvey, Dave Ackerman, Richard Schatz, Mike Sergent, Ken Wecht, Paul Sciortino and Jill Morton for help with this project.

References

Adar, R., Shani, Y., Henry, C. H., Kistler, R. C., Blonder, G. E. and Olsson, N. A. (1991) *Appl. Phys. Lett.*, **58**, 444.

Adar, R., Henry, C. H., Kistler, R. C. and Kazarinov, R. F. (1992) *Appl. Phys. Lett.*, **60**, 1779.

Andrekson, P. A., Olsson, N. A., Simpson, D. J., DiGiovanni, D. J., Morton, P. A., Tanbun-Ek, T., Logan, R. A. and Wecht, K. W. (1992) *IEEE Phot. Tech. Lett.*, **4**, 644.

Bird, D. M., Armitage, J. R., Kashyap, R., Fatah, R. M. A. and Cameron, K. H. (1991) *Electron. Lett.*, **27**, 1115.

Bowers, J. E., Morton, P. A., Mar, A. and Corzine, S. W. (1989) *IEEE J. Quantum Electron.*, **QE-25**, 1426.

Derickson, D. J., Morton, P. A., Bowers, J. E. and Thornton, R. L. (1991) *Appl. Phys. Lett.*, **59**, 3372.

Edwards, C. A., Presby, H. M. and Stulz, L. W. (1993) *Appl. Optics*, **32**, 2099.
Gordon, J. P. and Haus, H. A. (1986) *Opt. Lett.*, **11**, 665.
Gordon, J. P. and Mollenauer, L. F. (1991) *IEEE J. Light. Tech.*, **9**, 170.
Hansen, P. B., Raybon, G., Koren, U., Miller, B. I., Young, M. G., Chien, M., Burrus, C. A. and Alferness, R. C. (1992) *IEEE Phot. Tech. Lett.*, **4**, 215.
Harvey, G. T. and Mollenauer, L. F. (1993) *Opt. Lett.*, **18**, 107.
Haus, H. A. (1991) *J. Opt. Soc. Am. B*, **8**, 1122.
Henry, C. H., Blonder, G. E. and Kazarinov, R. F. (1989) *IEEE J. Light. Tech.*, **7**, 1530.
Koch, T. L., Koren, U., Gnall, R. P., Burrus, C. A. and Miller, B. I. (1988) *Electron. Lett.*, **24**, 1431.
Kodama, Y. and Hasegawa, A. (1987) *IEEE J. Quantum Electron.*, **QE-23**, 510.
Kodama, Y. and Hasegawa, A. (1992) *Opt. Lett.*, **17**, 31.
Lemaire, P. J., Atkins, R. M., Mizrahi, V. and Reed, W. A. (1993) *Electron Lett.*, **29**, 1191.
Mar, A., Derickson, D., Helkey, R. and Bowers, J. (1992) *Opt. Lett.*, **17**, 868.
Mecozzi, A., Moores, J. D., Haus, H. A. and Lai, Y. (1991) *Opt. Lett.*, **16**, 1841.
Meltz, G., Morey, W. W. and Glenn, W. H. (1989) *Opt. Lett.*, **14**, 823.
Mizrahi, V. and Sipe, J. E. (1993) *J. Lightwave Tech.*, in press.
Mizrahi, V., DiGiovanni, D. J., Atkins, R. M., Park, Y. K. and Delavaux, J. M. (1993) *J. Lightwave Tech.*, in press.
Mollenauer, L. F., Gordon, J. P. and Islam, M. N. (1986) *IEEE J. Quantum Electron.*, **QE-22**, 157.
Mollenauer, L. F., Gordon, J. P. and Evangelides, S. G. (1992) *Opt. Lett.*, **17**, 1575.
Mollenauer, L. F., Lichtman, E., Neubelt, M. J. and Harvey, G. T. (1993) *Electron. Lett.*, **29**, 910.
Morey, W. W., Meltz, G. and Glenn, W. H. (1989) *SPIE 1169, Fiber Optic and Laser Sensors VII.*
Morton, P. A., Helkey, R. J. and Bowers, J. E. (1989) *IEEE J. Quantum Electron.*, **QE-25**, 2621.
Morton, P. A., Bowers, J. E., Koszi, L. A., Soler, M., Lopata, J. and Wilt, D. P. (1990) *Appl. Phys. Lett.*, **56**, 111.
Morton, P. A., Adar, R., Kistler, R. C., Henry, C. H., Tanbun-Ek, T., Logan, R. A., Coblentz, D. L., Sergent, A. M. and Wecht, K. W. (1991) *Appl. Phys. Lett.*, **59**, 2944.
Morton, P. A., Mizrahi, V., Kosinski, S. G., Mollenauer, L. F., Tanbun-Ek, T., Logan, R. A., Coblentz, D. L., Sergent, A. M. and Wecht, K. W. (1992a) *Electron. Lett.*, **28**, 561.
Morton, P. A., Temkin, H., Coblentz, D. L., Logan, R. A. and Tanbun-Ek, T. (1992b) *Appl. Phys. Lett.*, **60**, 1812.
Morton, P. A., Logan, R. A., Tanbun-Ek, T., Sciortino Jr., P. F., Sergent, A. M., Montgomery, R. K. and Lee, B. T. (1992c) *Electron. Lett.*, **28**, 2156.
Morton, P. A., Mizrahi, V., Andrekson, P. A., Tanbun-Ek, T., Logan, R. A., Lemaire, P., Coblentz, D. L., Sergent, A. M., Wecht, K. W. and Sciortino Jr., P. F. (1993) *IELE Phot. Tech. Lett.*, **5**, 28.

Morton, P. A., Mizrahi, V., Tanbun-Ek, T., Logan, R. A., Lemaire, P., Erdogan, T., Sciortino Jr., P. F., Sergent, A. M. and Wecht, K. W. (1994a) *Opt. Lett.*, **19**.

Morton, P. A., Mizrahi, V., Tanbun-Ek, T., Logan, R. A., Lemaire, P., Presby, H. M., Erdogan, T., Woodward, S. L., Sipe, J. E., Phillips, M. R., Sergent, A. M. and Wecht, K. W. (1994b) *Appl. Phys. Lett.*, **64**.

Morton, P. A., Mizrahi, V., Harvey, G. T., Mollenauer, L. F., Tanbun-Ek, T., Logan, R. A., Presby, H. M., Erdogan, T., Sergent, A. M. and Wecht, K. W. (1994c) '270 Terabit.km/sec Soliton Transmission using a Packaged Hybrid Soliton Pulse Source', *IEEE Phot. Tech. Lett.*, to be published.

Presby, H. M. and Edwards, C. A. (1992) *Electron Lett.*, **28**, 582.

Raybon, G., Tucker, R. S., Eisenstein, G. and Henry, C. H. (1988) *Electron. Lett.*, **24**, 1563.

Schatz, J. R. (1994) *Laser Matrix*, part of Ph.D. dissertation.

Sipe, J. E. (1993) personal communication.

Tanbun-Ek, T., Logan, R. A., Chu, S. N. G., Sergent, A. M. and Wecht, K. W. (1990) *Appl. Phys. Lett.*, **57**, 2184.

Tucker, R. S., Koren, U., Raybon, G., Burrus, C. A., Miller, B. I., Koch, T. L. and Eisenstein, G. (1989) *Electron. Lett.*, **25**, 621.

Woodward, S. L., Koch, T. L. and Koren, U. (1992) *IEEE Phot. Tech. Lett.*, **4**, 417.

Wu, M. C., Chen, Y. K., Tanbun-Ek, T., Logan, R. A., Chin, M. A. and Raybon, G. (1990) *Appl. Phys. Lett.*, **57**, 759.

9

Monolithic colliding pulse modelocked diode lasers

MING C. WU AND YOUNG-KAI CHEN

9.1 Introduction

Compact and lightweight picosecond semiconductor laser sources are needed for high bit-rate time-division multiplexed communication systems (Andrekson *et al.*, 1992), 100 Gbit/s transmission systems (Kawanishi *et al.*, 1993), ultra-long distance soliton fiber transmission (Mollenauer *et al.*, 1991; 1992; Nakazawa *et al.*, 1991), picosecond optical logic gates (Nelson *et al.*, 1991), electro-optic sampling systems (Valdmanis and Mourou, 1986), and opto-electronic generation of millimeter and sub-millimeter waves (Scott *et al.*, 1992). Though there has been substantial progress in the area of ultrashort optical pulse generation using dye or solid state gain media (Ippen *et al.*, 1989; Krausz *et al.*, 1992), semiconductor optical sources are preferred for the above applications because of the following advantages. First, the repetition frequency of interest is usually above 1 GHz, thus substantially shorter cavity length is needed. Second, semiconductor lasers can be electrically pumped, eliminating the need of another pumping laser. Third, the power consumption is very low, typically of the order of 100 mW. No air or water cooling is required. They are also very energy efficient. Typical external quantum efficiencies of semiconductor lasers are between 30% and 90%.

More importantly, the semiconductor gain medium can be integrated with many other optical components using integrated optics techniques. In fact, the whole modelocked laser can be integrated monolithically on a single piece of semiconductor, completely eliminating optical alignment process. This is the main topic of this chapter.

There are two commonly used methods to generate short optical pulses using semiconductor lasers: gain switching (Downey *et al.*, 1987; Lau, 1988a) and modelocking (van der Ziel, 1985). Gain switching is achieved by injecting a short electrical pulse into the semiconductor laser. An optical pulse shorter than the electrical pulses can be generated by the nonlinear interaction of electrons and photons. The advantage of gain switching is the simplicity of the device and the variable repetition rate achievable. However, when the laser is suddenly switched from below to above threshold, there are significant fluctuations in carrier density and turn-on delay. These result in significant frequency chirp (Koch and Bowers, 1984), timing jitter and pulse energy fluctuation (Leep and Holm, 1992) for the gain-switched optical pulses.

Short optical pulses with pure spectral quality and low timing jitter can be achieved by modelocking. Early efforts of modelocked semiconductor lasers focused on external cavity configuration (for a review, see van der Ziel, 1985). The first successful modelocking of a semiconductor laser was reported by Ho *et al.* (1978). An external cavity was used to reduce the round-trip frequency to 5 GHz, at which the laser was modulated by a microwave oscillator. Such a modelocking scheme is called active modelocking. A pulse width of 23 ps was obtained. A shorter pulse width of 0.58 ps was later obtained by modulating the laser at higher harmonics (16 GHz) of the round-trip frequency (2 GHz) (Bowers *et al.*, 1989). Passive modelocking of a semiconductor laser was first demonstrated in 1980 (Ippen *et al.*, 1980). Optical pulses with duration of 1.3 ps were produced by aged laser diodes containing dark line defects in the active region, which provided the required saturable absorption. Later, 0.65-ps-long optical pulses were produced by van der Ziel *et al.* (1981) using a proton-bombarded section on the diode laser as the saturable absorber. One of the major issues with external cavity modelocked semiconductor lasers is multiple optical pulses within a period. These multiple pulses results from residue intracavity reflections from the laser facets.

9.2 Monolithic modelocked semiconductor lasers

There are many advantages to integrating the whole modelocked laser monolithically on a single piece of semiconductor: the ultrafast source becomes very compact (typically 500 μm wide, 100 μm thick, and a few hundred μm to several mm long). No optical alignment is needed since all the optical components are pre-aligned on the semiconductor substrate by optical lithography and microfabrication technologies. It is not necessary to operate the laser on optical tables. The monolithic ultrafast laser can be packaged in a similar fashion to standard semiconductor lasers. Since there are no moving parts, it is very stable thermally and mechanically. The monolithic cavities also minimize the undesired intracavity reflections which produced multiple optical pulses per period in external cavity modelocked lasers. Monolithic integration has made it possible to generate ultrashort optical pulses at hundreds of GHz. With monolithic cavities, it is possible to scale down the cavity length to only a few hundred micrometers. This has not been possible with external cavity modelocked lasers because the cavity lengths are usually limited by the finite thickness of bulk optical components. The photon round-trip time in monolithic semiconductor optical cavities is approximately 1.1 ps per 100 μm cavity length. Thus it is now possible to generate CW modelocked pulses at repetition frequencies of several hundred GHz. This is indeed the case, as will be discussed later where a CW modelocked laser operating at 350 GHz repetition frequency has been experimentally demonstrated.

The monolithic modelocked semiconductor laser has been an active research area recently. Several monolithic modelocking schemes have been reported, including active modelocking with integrated passive cavity (Tucker *et al.*, 1989; Raybon *et al.*, 1992; Hansen *et al.*, 1993a), active modelocking with ring cavity (Hansen *et al.*, 1992), hybrid modelocking with three contacts (Morton *et al.*, 1990; Brovelli and Jackel, 1991), hybrid monolithic colliding pulse modelocked (CPM) lasers (Wu *et al.*, 1990; 1993; Chen and Wu, 1992), passive modelocking with tandem-contacts (Lau, 1988b; 1990; Vasil'ev and Sergeev, 1989; Sanders *et al.*, 1990a; 1990b), passive monolithic CPM lasers (Chen, Y. K. *et al.*, 1991; Derickson *et al.*, 1992; Martins-Filho *et al.*, 1993) and a passive modelocked ring laser (Hohimer and Vawter, 1993). A very wide range of repetition rates was obtained: from 2.2 GHz (2-cm-long cavity, Hansen *et al*, 1993a) to 350 GHz (250-mm-long cavity, Chen, Y. K. *et al* ., 1991). The shortest pulse width obtained was 610 fs (Wu *et al.*, 1991a; Chen, Y. K. *et al.*, 1991; Chen, Y. K. and Wu, 1992). Comparison of various monolithic modelocking schemes will be discussed in Section 9.5.

Integrated wavelength-selective filters can also be incorporated into the cavities to control or tune the center wavelength of modelocked pulses. For example, active modelocked distributed Bragg reflector (DBR) lasers (Raybon *et al.*, 1991), passive modelocked DBR lasers (Arahira *et al.*, 1993), and an active modelocked laser with integrated vertical coupler filter (Raybon *et al.*, 1993) have been demonstrated. However, the pulse widths of these lasers are usually longer because of the additional bandwidth limitation. It has also been shown that temperature tuning is very effective in tuning the wavelength without broadening the pulse width (Wu *et al.*, 1991b).

Recent developments in the physics and technology of semiconductor lasers have made it possible to build monolithic modelocked semiconductor lasers. The material quality of semiconductor lasers has steadily improved over the past decade. The introduction of quantum well lasers (for a review, see Tsang, 1987; Arakawa and Yariv, 1986) has had a dramatic impact on the development of monolithic modelocked diode lasers. The threshold current densities of quantum well lasers are greatly reduced from that of the bulk semiconductor lasers, especially for lasers with long cavities (Choi and Wang, 1990; Thijs *et al.*, 1991). Threshold current densities below 100 A/cm^2 can now be routinely achieved. This has made it possible to make long-cavity lasers without excessive heating problems. A typical cavity length of a 10 GHz monolithic modelocked laser is around 4 mm. In addition to low threshold, quantum well lasers also have lower dispersion and lower frequency chirp (Arakawa and Yariv, 1986), both are advantageous for generating short optical pulses.

Another important development is the quantum well saturable absorber. An intracavity saturable absorber is needed for passive and hybrid modelocking. Most early attempts at passive modelocking in external cavity semiconductor lasers used saturable absorption created by optical damage. The damage was produced by aging (Ippen *et al.*, 1980) or proton bombardment (van der Ziel *et al.*, 1981). Though short optical pulses have been generated using those saturable absorbers, the reliability of such lasers remained a serious issue. It has been shown that a necessary condition for modelocking with a slow saturable absorber is that the absorber cross-section should be larger than the gain cross-section (Haus, 1975). Multiple quantum wells (MQW) have larger cross-section than the bulk semiconductors and are suitable for use as saturable absorbers. An MQW was utilized by Silberberg *et al.* (1984) to produce optical pulses with duration of 1.6 ps in external cavity configuration.

The MQW saturable absorbers also enable us to use the same quantum well materials as gain media as well as saturable absorbers, which is an important step towards monolithic integration of modelocked semiconductor lasers. Because of energy quantization, the density of states of the MQW is a step-like function. As a result, the maximum gain saturates with injected carrier (Arakawa and Yariv, 1986). The optical gain versus injected current density is illustrated in Figure 9.1. The differential gain (the slope of the curve in Figure 9.1) decreases with increasing bias. Since the absorber (or gain) cross-section is proportional to the differential

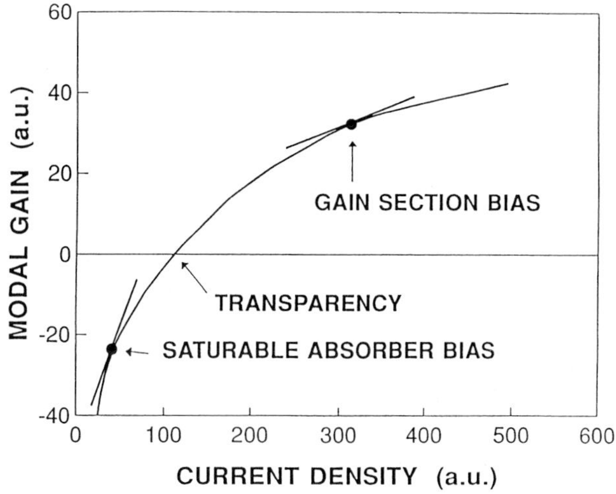

Figure 9.1. Optical gain versus injected carriers for multiple quantum wells. The absorber cross-section is larger than the gain cross-section, which satisfies the necessary condition for passive modelocking.

gain, the previously stated necessary condition for passive modelocking is satisfied if the MQW saturable absorber is biased at a lower level than the gain media comprising the same MQWs. Passive modelocking of monolithic semiconductor lasers using such an inhomogeneous pumping scheme was first attempted by Lau (1988b; 1990) where 10% optical modulation at 70 GHz was observed. Shorter optical pulses with ~ 2 ps pulse width and ~ 100 GHz repetition rate were obtained later using the same current pumping scheme (Vasil'ev and Sergeev, 1989; Sanders *et al.*, 1990a). It is noted that all the monolithic passive modelocked lasers reported to date (except Vasil'ev and Sergeev, 1989) have used the same MQW materials for both gain media and saturable absorbers.

9.3 Monolithic colliding pulse modelocked semiconductor lasers

The colliding pulse modelocking (CPM) scheme, first proposed by Fork, Greene and Shank (1981), was widely used to generate femtosecond optical pulses in dye lasers. The original CPM dye laser has a ring cavity with absorber dye jet located at 1/4 cavity length from the gain dye jet. Such an arrangement will create a pair of counterpropagating optical pulses in the ring cavity. These two pulses collide in the saturable absorber and create a transient grating that synchronizes, stabilizes and shortens the pulses (Shank, 1988). The coherent interaction of the optical pulses lowers the saturation energy and enhances the effective cross-section of the saturable absorber (Garmire and Yariv, 1967; Stix and Ippen, 1983). This enhancement of absorber cross-section is especially important for monolithic modelocked lasers to guard against low frequency self-pulsation, as will be discussed later (Wu *et al.*, 1993).

An external cavity semiconductor CPM laser has also been implemented using a semiconductor laser amplifier in a ring cavity. The saturable absorber was created by proton bombardment of one laser facet. The semiconductor laser amplifier was modulated at the round-trip frequency of 625 MHz. Optical pulses of 0.56 ps were produced, however, and a series of trailing pulses due to the undesired intracavity reflections at the other laser facet were also observed. A linear external cavity laser was also constructed to purposely use the intracavity reflections to create CPM effects (Vasil'ev *et al.*, 1986). Similar results (multiple pulses per cycle) were obtained.

In this section, we will describe a novel monolithic colliding pulse modelocked quantum well laser which produces transform-limited subpicosecond optical pulses at repetition frequencies from 32 GHz to 350 GHz.

9.3.1 *Device structure*

The schematic structure of the monolithic CPM quantum well laser is shown in Figure 9.2. It consists of a linear cavity with cleaved

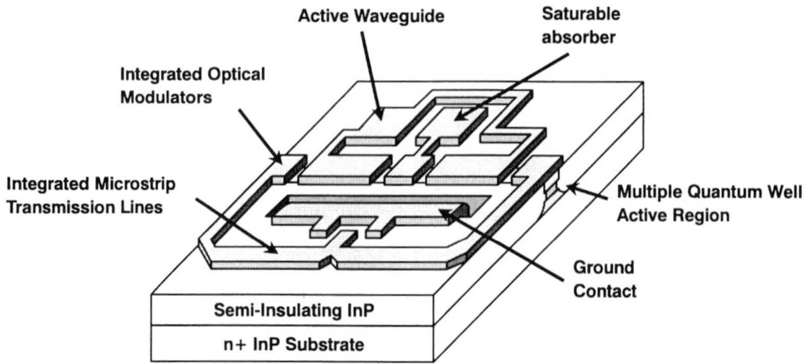

Figure 9.2. The schematic structure of the monolithic CPM diode laser.

facets as mirrors. The laser is divided into five sections by separate top electrodes. The active quantum wells extend continuously through all five sections. The center section of the quantum wells is reverse biased and functions as saturable absorber. For effective modelocking, the length of the saturable absorber section is designed to be shorter than the spatial duration of the pulse width. The two adjacent sections are connected together and forward biased by a current source. These two sections are called gain sections or active waveguides. The monolithic CPM laser also includes two optical modulators located near the cleaved facets. The modulators serve to synchronize the optical pulses with electrical clocks in hybrid modelocking schemes.

Though most of the CPM dye lasers use ring cavities, a linear cavity is used for our monolithic CPM laser. To use a linear cavity, the saturable absorber needs to be placed in the exact center of the cavity to within a fraction of pulse width (Stix and Ippen, 1983; Shank, 1988), which is 30 μm for a pulse duration of 100 fs. This is very difficult to achieve in practical positioning of the dye jets. On the other hand, in monolithic CPM semiconductor lasers, the saturable absorber is precisely defined by the photolithography process. Its relative position in the cavity is determined by the facet cleaving process which has an accuracy of a few microns. This is satisfactory for lasers with saturable absorbers longer than a few tens of microns, which is the case of our monolithic CPM lasers with repetition rates up to 350 GHz. For even higher repetition frequency (and thus shorter saturable absorber), mirrors defined by photolithography and dry etching might be required. The use of linear cavity configuration is also compatible with manufacturing technology for conventional semiconductor lasers.

A symmetric cavity is employed for our CPM laser so that the two colliding pulses will have equal amplitudes during collision. Therefore, the optical modulators needed for hybrid CPM are placed in pairs on both sides of the laser. It is important that the microwave signals sent to the two modulators be exactly in phase. A symmetric microwave transmission line has been incorporated monolithically to ensure the precise timing of the microwave distribution. The microwave signal is fed from the center of the transmission lines using a cascade-type probe.

The operation of the monolithic CPM laser is illustrated in Figure 9.3. The optical pulses traveling in the cavity and the microwave phases are shown at a time interval of $T/8$, where T is the photon round-trip time. The starting point of $t = 0$ is defined as the time the microwave peak reaches the modulators (this microwave phase is defined as zero). In the

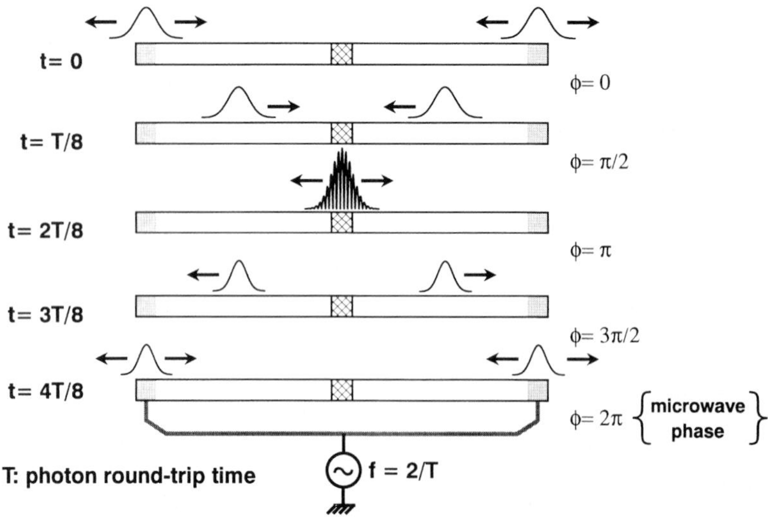

Figure 9.3. Schematic diagram illustrating the relative timing of optical pulse propagation and microwave modulation in monolithic CPM lasers.

steady state, the optical pulses are synchronized to the microwaves and a pair of optical pulses is formed near the facets. The optical pulses start to propagate towards each other, and at $t = T/8$ they travel to half the way between the modulators and the saturable absorber. The microwave phase is $\pi/2$. At $t = 2T/8$, the two optical pulses collide in the central saturable absorber and form a transient grating. The microwave phase at the modulator now reaches π. When the pulses travel to 3/4 of the cavity length, the microwave phase becomes $3\pi/2$. At $t = 4T/8$, the pulses reach the opposite modulator as the microwave phase reaches 2π. The microwave peak coincides with the optical pulses again. It is interesting to note that the microwave period is only half of the photon round-trip time T because the microwaves actually modulate both of the counterpropagating pulses. The repetition frequency of the output pulse train is also twice $1/T$ because both counterpropagating pulses are collected. Thus the fundamental repetition frequency of the linear cavity monolithic CPM laser is twice that of the same cavity with a single modelocked pulse in the cavity.

The photograph of a monolithic CPM laser is shown in Figure 9.4 (the third stripe from top). The cavity length is 2.5 mm, corresponding to a repetition rate of 32 GHz. The saturable absorber and the two modula-

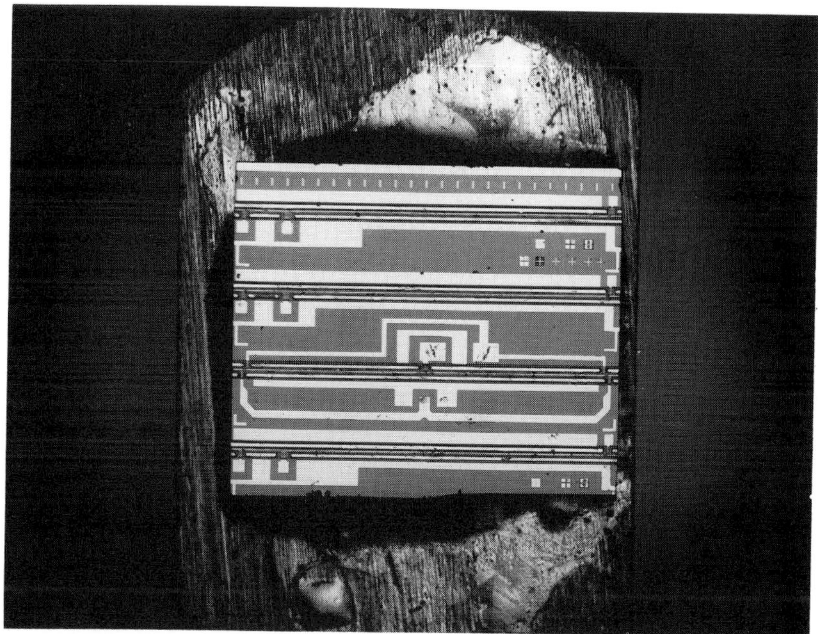

Figure 9.4. The photograph of a 2.54-mm-long monolithic CPM laser (third stripe from the top).

tors are each 70 μm long. Three other lasers with non-CPM configuration (saturable absorber on one side and modulator on the other side) are incorporated on the same chip for comparison. The results will be discussed later.

The semiconductor itself consists of a buried heterostructure (BH) InGaAs/InGaAsP/InP multiple quantum well laser (Tanbun-Ek *et al.*, 1989). The laser layers as described in Figure 9.5 are grown by organometallic vapor phase epitaxy (OMVPE). The MQW consists of five 5-nm-thick InGaAs quantum wells separated by four 22.5-nm-thick InGaAsP (with bandgap wavelength of $\lambda_g = 1.25$ μm) barrier layers. The lasing wavelength is 1.55 μm. The MQW is embedded in a graded index separate confinement heterostructure (GRIN-SCH). The lower GRIN-SCH comprises three 25-nm-thick InGaAsP layers with decreasing bandgaps ($\lambda_g = 1.08$ μm, 1.16 μm, and 1.25 μm, respectively), while the upper GRIN-SCH comprises the same layers with increasing bandgaps. The GRIN-SCH is sandwiched between two 2-μm-thick InP cladding layers and finally capped by a heavily doped InGaAsP contact layer. After the first OMVPE growth, 1-μm-wide laser waveguide strips are formed by

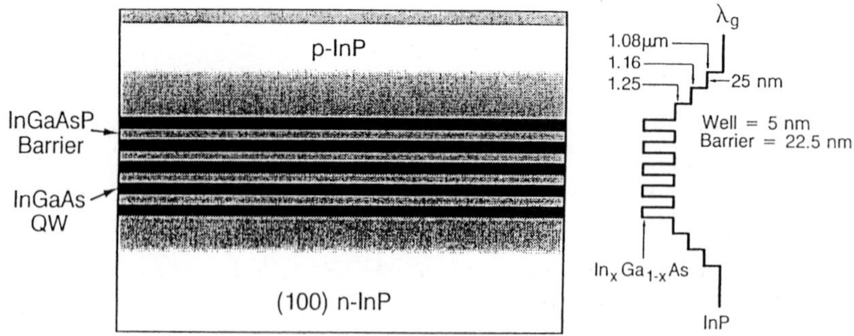

Figure 9.5. The epitaxial layer structure of monolithic CPM lasers.

conventional lithography and wet chemical etching. They are then surrounded by Fe-doped semi-insulating InP with a second OMVPE growth. Standard lithography, wet chemical etching, lapping and metallization processes are used to produce the final structure.

9.3.2 *DC and AC characteristics of the monolithic CPM lasers*

The finished monolithic CPM laser is first tested in the CW condition. The light versus current (L-I) curve of a 2.1-mm-long CPM laser is shown in Figure 9.6. The threshold current is 54 mA when the

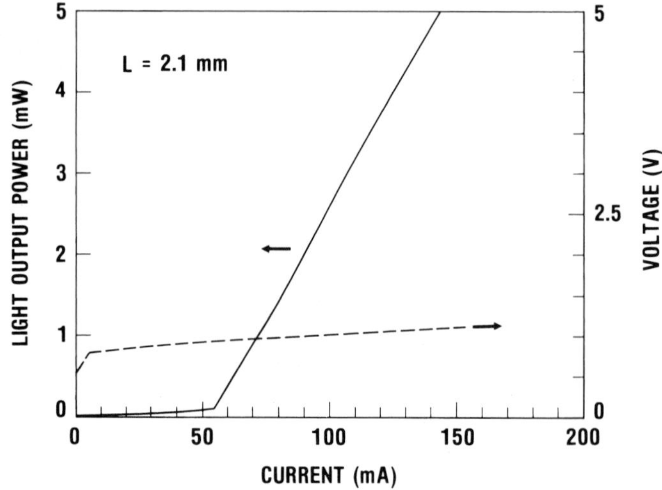

Figure 9.6. The light versus current and voltage versus current characteristics of monolithic CPM laser under uniform pumping.

CPM laser is uniformly pumped with all sections connected together. A linear L-I curve is observed. The voltage (I-V curve) across the laser is also shown in Figure 9.6. The laser has a sharp turn-on voltage at 0.8 V and low leakage current under the reverse bias, indicating a high quality p-n junction at the active quantum wells. Since our MQW saturable absorber does not need defects, lasers with high-quality optical and electrical characteristics are desired for modelocking experiments. This is confirmed by our CW and short pulse measurements. To investigate the effect of the saturable absorber, the L-I curves of the CPM laser are measured for various saturable absorber voltages from 0.2 V to −0.8 V with steps of −0.2 V, as shown in Figure 9.7 for bias voltages. With increasing reverse bias, the threshold gradually shifts to higher currents due to the increasing absorption in the absorber. Nonlinearity is also observed in the L-I curves for optical power above 1 mW, indicating the presence of strong interaction of the saturable absorber and the laser. As we will discuss later, optimum modelocking actually occurs in the linear regions below 1 mW.

The lasers are then tested for their AC response. The test setup is shown in Figure 9.8. The small signal frequency response of the CPM laser is characterized by an HP 8510 network analyzer and a high speed photodetector with a bandwidth of 32 GHz. The input RF signal is sent

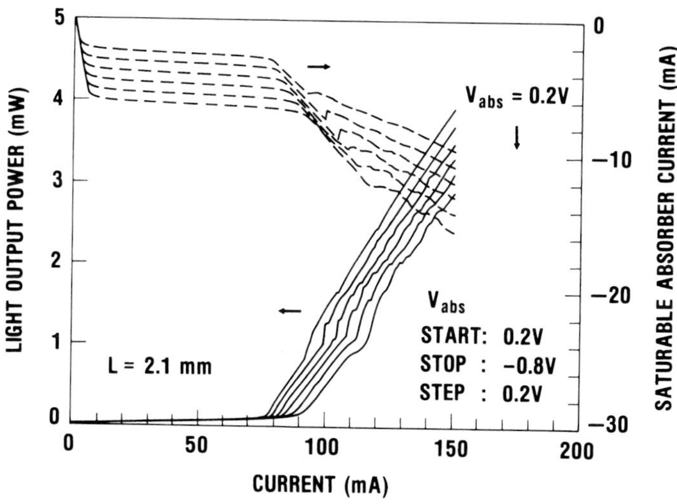

Figure 9.7. The light versus current characteristics of a monolithic CPM laser with various reverse biases for the saturable absorber. The photocurrent from the saturable absorber is also shown.

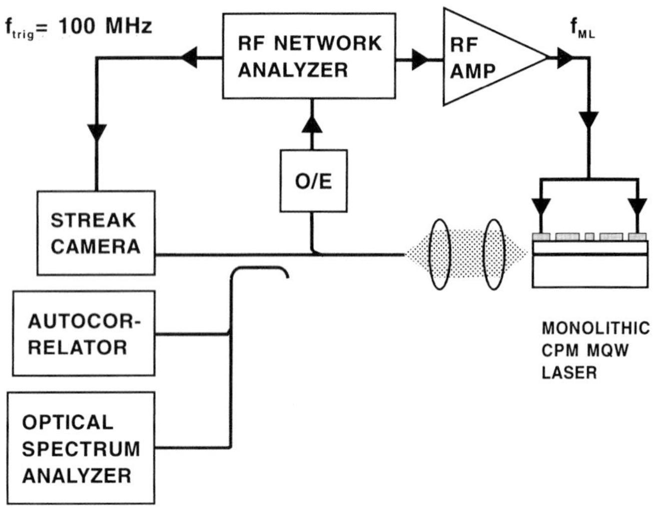

Figure 9.8. The experimental setup for characterizing a monolithic CPM laser.

to the two end modulators through the integrated microwave transmission lines. The time domain signals are characterized by a Hamamatsu synchro-scan streak camera and a noncollinear second harmonic autocorrelator. The optical spectrum is monitored simultaneously with an optical spectrum analyzer. The output light from the CPM laser is coupled into a single mode fiber so that it is more convenient to direct the output to various instruments. We also found it necessary to insert an optical isolator between the laser and the optical fiber. The reflected light would disturb the modelocking of the laser.

The small-signal frequency response of a 2.1-mm-long monolithic CPM laser is shown in Figure 9.9. When the laser is uniformly biased (Figure 9.9(a)), a relaxation oscillation-enhanced peak is observed around 3 GHz. The modulation response decreases rapidly beyond the relaxation oscillation until 19.2 GHz, at which point a strong resonance about 20 dB above the noise level is observed. This resonance corresponds to the photon round-trip time around the 2.1-mm-long cavity. The effect of the saturable absorber on the small-signal response is also investigated by applying a reverse bias of –0.4 V to the absorber, as shown in Figure 9.9(b). It is interesting to note that the cavity resonance-enhanced peak at 19.2 GHz is actually suppressed by the saturable absorber. This is because of dual-pulse nature of the CPM scheme. In colliding pulse operation, there are two counterpropagating pulses circu-

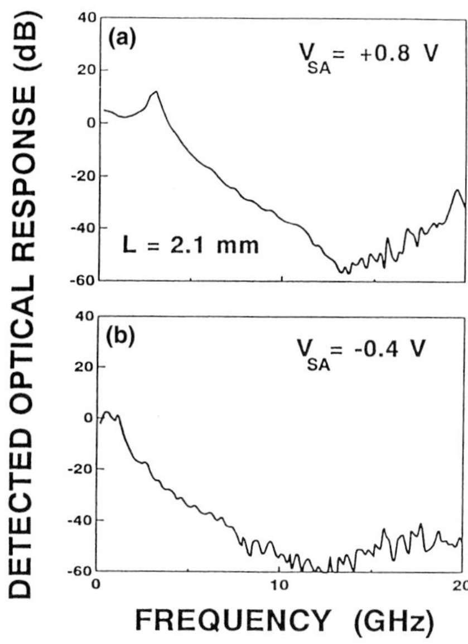

Figure 9.9. Small-signal frequency response of monolithic CPM laser when (a) the laser is uniformly biased and (b) the center saturable absorber is biased at −0.4 V. The cavity resonance peak in (b) moves to twice that of the peak in (a) because of the CPM operation.

lating in the cavity at any time. In ring cavity configuration, the two pulses are separated at the output coupler and only one of them is collected. In our linear cavity CPM configuration, both pulses go through the same path and are both collected from the same output mirror. Therefore, the fundamental repetition frequency of the linear CPM laser is actually twice that of the same laser operating in a single-pulse modelocking mode. The fundamental frequency of this CPM laser (38.5 GHz) is beyond the bandwidth of our photodetector and is not shown in Figure 9.9(b).

9.3.3 *Hybrid monolithic CPM laser*

The monolithic CPM laser is hybrid modelocked by fixing the input microwave frequency at the cavity resonance frequency of 38.5 GHz. As we gradually increase the microwave power, short optical pulses are formed and synchronized to the microwaves, which serves as an electrical clock. Figure 9.10 shows the optical pulse train measured by the synchro-scan streak camera. Stable optical pulses at 38.5 GHz are

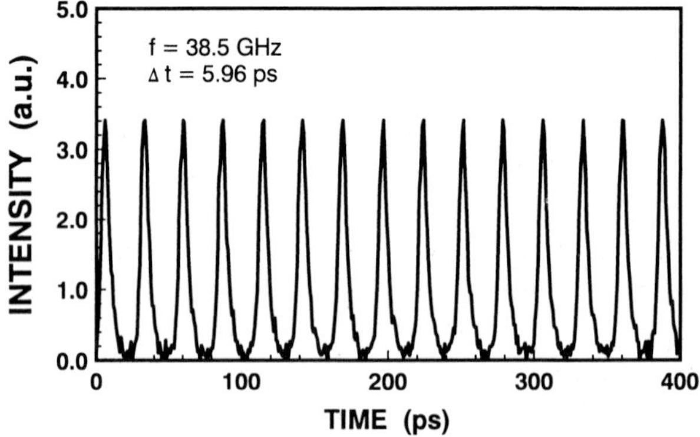

Figure 9.10. Optical pulse train of a 40 GHz hybrid monolithic CPM laser measured by synchro-scan streak camera. The measured pulse width is limited by the resolution of the streak camera.

observed. The pulse width of 6 ps measured here is limited by the resolution of the streak camera, which operates at a scan rate of 100 MHz.

As the RF power is gradually increased, the optical spectrum also changes as the laser becomes modelocked. Figure 9.11 shows the time-averaged optical spectra of the CPM laser with increasing RF power. When the RF power is small (−25 dBm for trace 00), there is a dominant single longitudinal mode. The spectrum starts to change at +5 dBm (trace 06). A significant change occurs at +10 dBm (trace 07) where

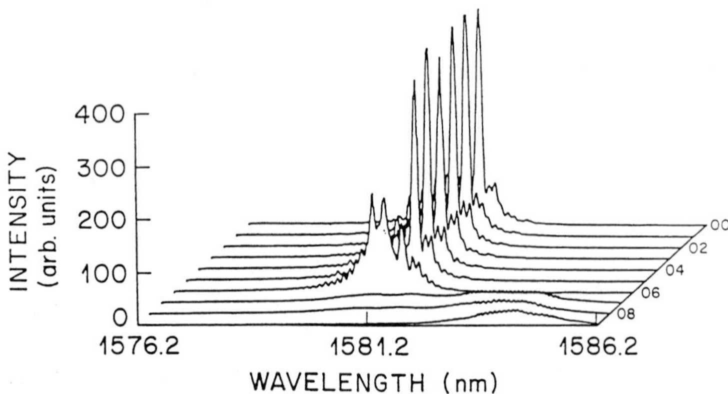

Figure 9.11. Time-averaged optical spectra of the hybrid monolithic CPM laser with increasing RF power (−25 dBm for trace 00 and increment step of 5 dBm). The laser is modelocked at 10 dBm of RF power (trace 07).

the dominant single mode is replaced by a series of longitudinal modes with peak heights suppressed by more than 10 dB. The spectral width also broadens to a few nanometers, and the peak lasing wavelength shifts to the longer wavelength side. This is the initiation point of modelocking, as confirmed by the simultaneous time-domain measurement with the streak camera.

The more precise measurement of the pulse width and the pulse shape are obtained by the noncollinear second harmonic generation (SHG) autocorrelator with a 5-mm-thick $LiNbO_3$ crystal. The SHG autocorrelation trace and time-averaged optical spectrum of the 2-mm-long CPM laser are shown in Figure 9.12(a) and (b), respectively. A clean autocorrelation trace with fully suppressed background level is observed. The autocorrelation trace agrees very well with that of a sech^2 pulse shape. The full width at half-maximum (FWHM) pulse width is 0.95 ps. With the measured optical spectral width of 2.4 nm, the time–bandwidth product is 0.32, which is very close to the transform-limited value of 0.31 for the sech^2 pulse shape.

The contribution of the saturable absorber is very well illustrated in Figure 9.13, which shows a series of autocorrelation traces with various absorber biases. The RF power is fixed at 25 dBm. When the saturable absorber is only slightly reverse biased at −0.131 V (the bottom trace), there is substantial energy in the pedestal region of the pulses. The DC component between pulses is also high (the notches on the right-hand side of the autocorrelation traces are the reference levels when the input light is blocked). As the reverse bias on the saturable absorber increases, the DC component decreases and the pedestal of the pulse gradually disappears. At a bias of −0.4 V, nearly transform-limited pulses are

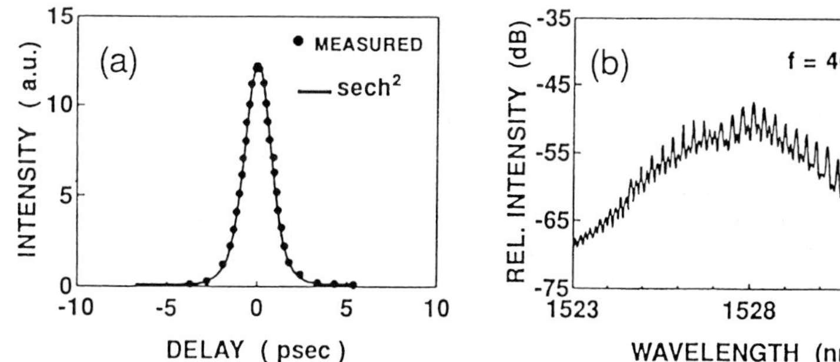

Figure 9.12. The second harmonic autocorrelation trace and optical spectrum of the 2-mm-long hybrid monolithic CPM laser operating at 40 GHz.

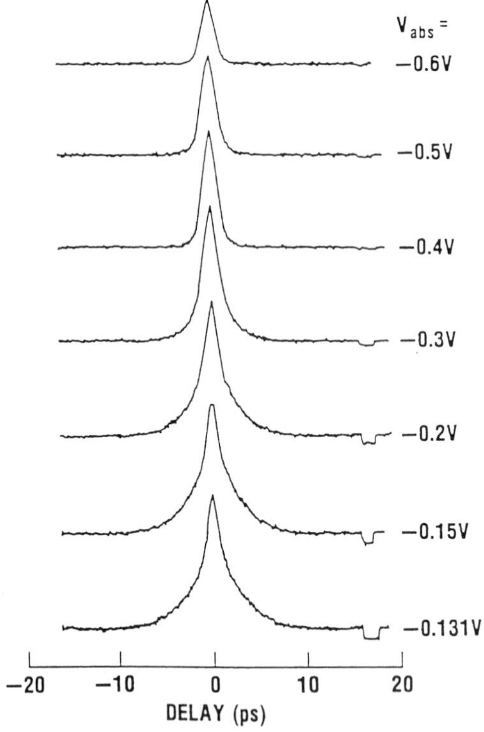

Figure 9.13. The autocorrelation traces of the 2-mm-long hybrid monolithic CPM laser with various saturable absorber biases.

obtained. With more reverse bias on the saturable absorber, good modelocking condition is still maintained, however, the peak power starts to decrease. The pulse width and the peak wavelength of the spectral envelope of the pulses versus the bias voltage of the saturable absorber are shown in Figure 9.14. As the pulse width decreases steadily with reverse bias on the absorber, the peak wavelength is red shifted.

In hybrid modelocking, the optical pulses are synchronized to the RF frequency by injecting the microwaves to the intracavity modulators. The RF frequency applied is equal to that of the cavity resonance frequency and is experimentally measured by the network analyzer as described earlier. For practical applications, the RF frequency (the electrical clock) is usually pre-determined by system requirement. On the other hand, the cavity length (and hence the cavity resonant frequency) is determined by the microfabrication process and cannot be changed after the device fabrication is finished. It is therefore necessary to evaluate the

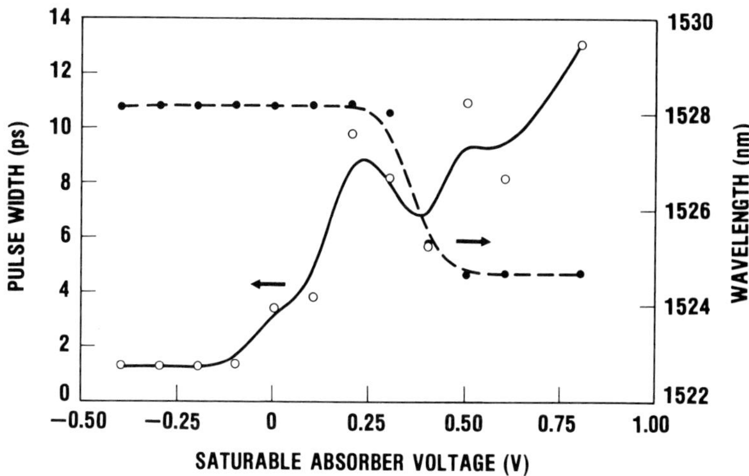

Figure 9.14. The pulse width and the peak wavelength of the 2-mm-long hybrid monolithic CPM laser versus saturable absorber bias.

precision required for the cavity length control. This question is answered by detuning the input RF frequency away from the cavity resonant frequency while monitoring the pulse width. The results are shown in Figure 9.15. The SHG autocorrelation width of the pulses remain relatively flat as the frequency varies from 37.2 GHz to 39.3 GHz. The acceptable detuning range is approximately 5% of the cavity resonance frequency. Compared with external cavity modelocked semiconductor lasers or monolithic active modelocked semiconductor lasers with integrated passive cavity, a much wider detuning range is achieved. This is attributed to the use of active waveguides throughout the cavity. The refractive index of the semiconductor is a very sensitive function of the carrier concentration at a wavelength corresponding to the bandgap energy. Thus the refractive index of the active waveguides in the monolithic CPM laser will be self-adjusted to keep up with the input RF frequency. This property greatly relaxes the fabrication tolerance of the cavity length.

9.3.4 *Passive monolithic CPM laser*

In hybrid modelocking of the monolithic CPM laser, the two pulses are seeded by the two integrated modulators and forced to collide in the central saturable absorber. In other words, we force the laser to operate in the CPM mode through microwave modulation of the two end modulators. It is interesting to see if the laser will be passive modelocked

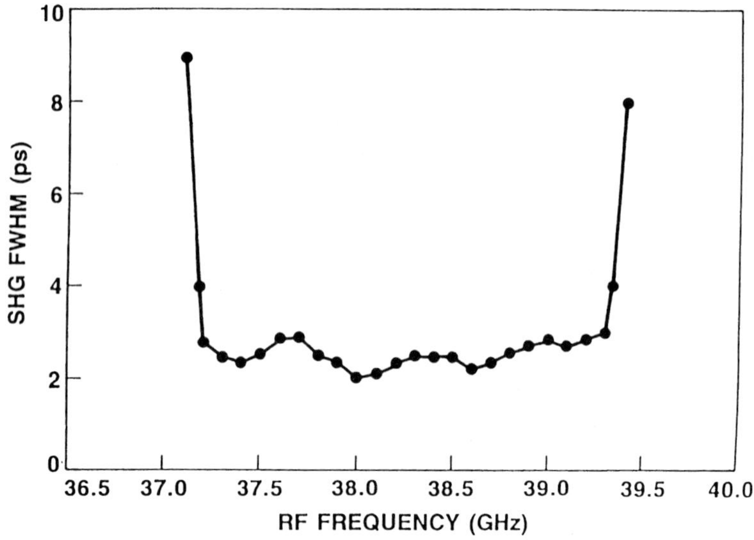

Figure 9.15. The width of autocorrelation traces versus the modulating RF frequency for the 2-mm-long hybrid monolithic CPM laser. The repetition frequency can be detuned by 5% without significantly changing the pulse width.

in the absence of microwave modulation, and whether the laser will still operate in the CPM mode.

Passive modelocking of the monolithic CPM laser is similar to that of hybrid modelocking except that the microwave modulation is removed and the modulator sections are now connected to the active waveguides (gain sections). The gain sections are forward biased with a current source and the central saturable absorber is reverse biased with a voltage source. It is confirmed that the monolithic CPM laser can also be passive modelocked. The pulse repetition frequency is still twice the round-trip frequency, confirming that CPM is indeed the preferred mode of modelocking for the monolithic CPM lasers. The SHG autocorrelation trace and the time-averaged optical spectrum of the same 2-mm-long CPM laser operating in passive modelocking mode are shown in Figure 9.16 (a) and (b), respectively. The autocorrelation trace is clean and fits very well with that of the sech^2 pulse shape. The FWHM pulse width is 1.1 ps, slightly longer than that of the hybrid CPM (0.95 ps). The time-averaged optical spectrum of the passive CPM laser shows a similar spectral width (2.37 nm) to that of the hybrid CPM mode. However, a much smoother envelope is observed for the passive CPM laser.

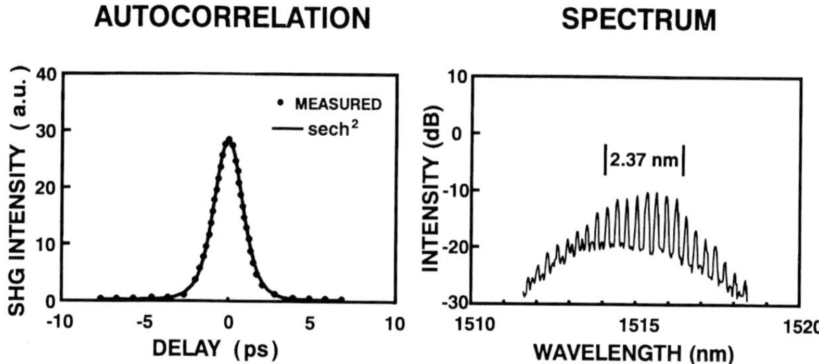

Figure 9.16. The autocorrelation trace and the optical spectrum of the 2-mm-long CPM laser under passive modelocking. The pulse width is 1.1 ps, and the time–bandwidth product is 0.34.

In passive CPM lasers, the pulse repetition frequency is no longer limited by the availability of microwave sources and amplifiers. Higher repetition frequency can be obtained by designing CPM lasers with shorter cavity lengths. Figure 9.17 shows the SHG autocorrelation traces and the time-averaged spectra of the passive monolithic CPM lasers with cavity lengths of 2 mm, 1mm and 534 μm, respectively. The repetition frequencies of these lasers as measured from the autocorrelation traces are 40 GHz, 80 GHz and 160 GHz, respectively. The length of the saturable absorber is 50 μm. Transform-limited optical pulses with sech^2 pulse shape are observed for all lasers, and the FWHM pulse widths are 1.1 ps, 0.83 ps and 1.0 ps, respectively.

It is noted that alternate longitudinal modes of the optical spectra in Figure 9.17 are suppressed, which means the mode spacing of the CPM laser is twice that of the same laser without modelocking. This is characteristic of the CPM lasers with linear cavities. Because there are two counterpropagating pulses in the cavity at any time, the linear CPM cavity can be considered as two symmetric halves with respect to the center of the cavity. If the optical phases of the two pulses are coherent, the optical pulse transmitting through the saturable absorber is equivalent to the reflection of the other pulse from the center of the cavity. Thus the effective cavity length of the CPM laser is equal to half of the total cavity length. The mode spacing is therefore doubled. The experimental observation here confirms that the optical phases of the two counterpropagating pulses are indeed coherent. This special feature actually serves as useful indicator for searching the biasing ranges for CPM opera-

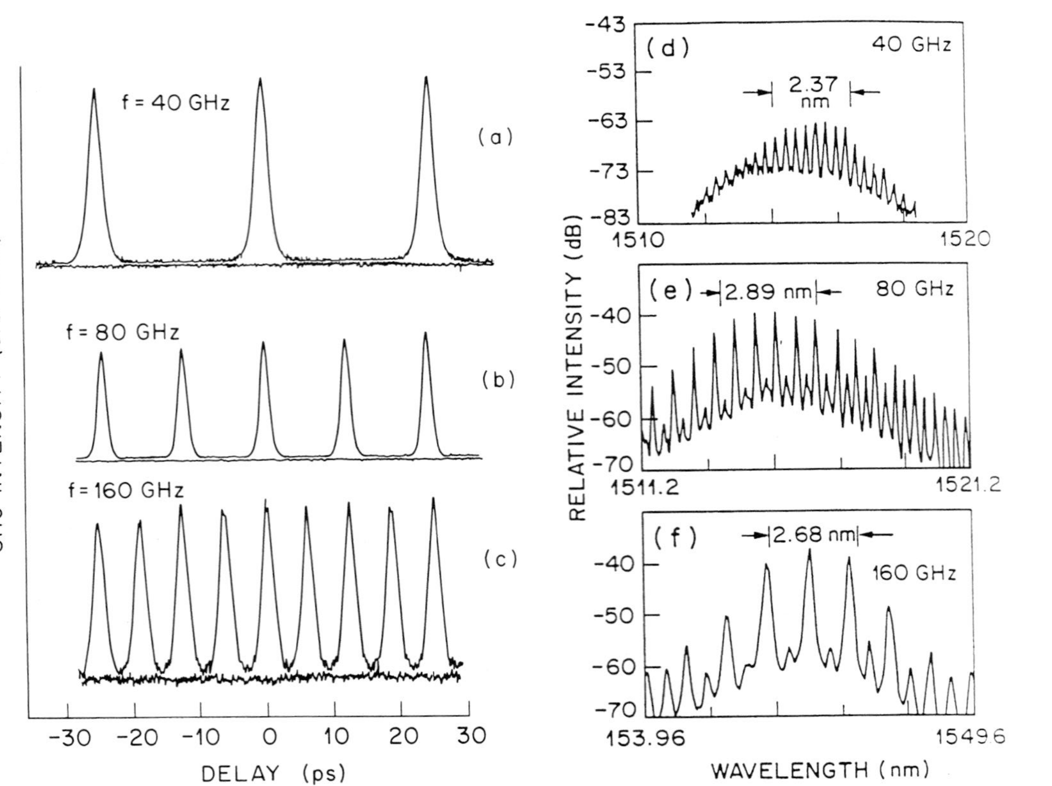

Figure 9.17. The autocorrelation traces (a, b, c) and the optical spectra (d, e, f) of passive CPM lasers of length 2 mm, 1 mm and 534 μm. The repetition rates are 40 GHz, 80 GHz and 160 GHz, respectively.

tion. The observation here also echoes the small-signal frequency response of the CPM laser as discussed in Figure 9.9.

Modelocked pulses with even higher repetition frequencies are obtained by designing the CPM laser with a shorter cavity. With a 250-μm-long cavity and a 15-μm-long saturable absorber, we have successfully generated an optical pulse train with a record high 350 GHz repetition rate. Figure 9.18 shows the SHG autocorrelation trace and the optical spectrum of the laser. Even at this high rate, the autocorrelation trace still fits very well to that of the sech^2 pulse shape. The pulse width of 640 fs is the shortest ever reported for monolithic modelocked semi-

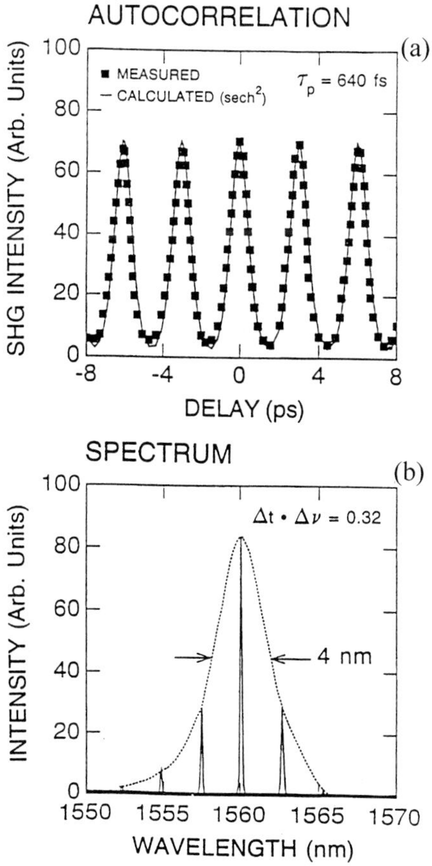

Figure 9.18. The autocorrelation trace and the optical spectrum of a 250-mm-long passive monolithic CPM laser. The repetition rate of the laser is 350 GHz, and the pulse width is 640 fs. The pulse shape agrees very well with sech^2.

conductor lasers without external compression. The measured spectral width in Figure 9.18(b) is 4 nm. The time–bandwidth product is 0.32, indicating that the optical pulses are nearly transform-limited. The optimum bias conditions for gain section current and the saturable absorber voltage are shown in Figure 9.19 for the 350 GHz passive CPM laser with 15 and 25-μm-long saturable absorbers. Two general trends are observed. First, shorter pulses are obtained from the CPM lasers with shorter saturable absorbers. For example, at a gain section current of 35 mA, the shortest pulse width is 700 fs for the laser with 25-μm-long saturable absorber, while it is 640 fs for the laser with 15-μm-long saturable absorber. Second, the pulse width is shorter when the laser is biased at higher gain current and more reverse bias voltage for the saturable absorber. For the laser with 15-μm-long saturable absorber, the shortest pulse width decreases from 700 fs to 610 fs when the gain section current increases from 30 mA to 40 mA.

9.3.5 *Parameter ranges for stable modelocking*

As discussed earlier, a necessary condition for passive modelocking is that the saturable absorber cross-section should be larger than the

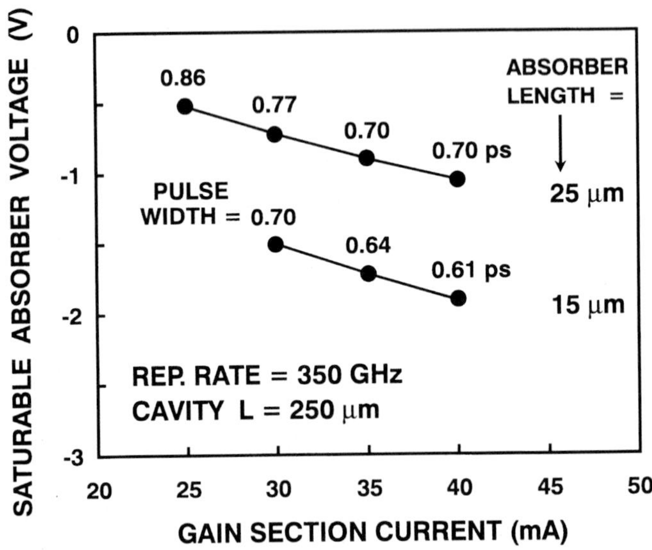

Figure 9.19. The optimum bias conditions and the pulse widths of two 350 GHz passive monolithic CPM lasers with saturable absorber lengths of 15 μm and 25 μm, respectively. The shortest pulse obtained is 610 fs long.

gain cross-section. However, the same condition is also a necessary condition for self-pulsation (Haus, 1976; Dixon and Joyce, 1979). The self-pulsation is a low frequency oscillation (from a few hundred MHz to a few GHz) resulting from the nonlinear interactions of electrons and photons in the presence of saturable absorbers. The occurrence of self-pulsation in nonuniformly pumped semiconductor lasers was first observed by Lee and Roldan (1970). The main difference between self-pulsation and modelocking is that (1) the self-pulsation frequency is much lower than that of modelocking in monolithic cavities, and (2) the photons are uniformly distributed along the cavity in self-pulsation, while in modelocking the photons are highly localized and their spatial distribution is much shorter than the cavity. The self-pulsation frequency does not depend on the length of the cavity, and changes with bias conditions. Since self-pulsation has long been observed for multi-segment lasers, it is not surprising that it is also observed in monolithic modelocked lasers with multiple contacts. However, if the bias ranges of self-pulsation overlap with the modelocking ranges, the modelocked pulses will be modulated by a self-pulsation envelope or even quenched by the self-pulsation.

In Lau and Paslaski (1992) it was concluded that it was difficult to generate very short pulses without the simultaneous self-pulsation envelope for single pulse tandem-contact modelocked semiconductor lasers. High reflection facet coatings were proposed to separate the modelocking region from the self-pulsation region (Paslaski and Lau, 1991). In the following we will show that the monolithic CPM configuration is particularly effective in discriminating against self-pulsation.

It has been shown that the effective cross-section of the saturable absorber is increased by a factor of three for the CPM configuration (Stix and Ippen, 1983). On the other hand, the saturable absorber cross-section for self-pulsation does not depend on the location of the absorber because the pulses generated by self-pulsation ($\sim$1 ns or 10 cm in spatial duration) are much longer than the cavity. Thus the ratio of absorber cross-section to gain cross-section is larger in CPM than in self-pulsation. This will effectively shift the modelocking region to lower bias levels and away from the self-pulsation region (Haus, 1976).

Because the autocorrelation measurement alone cannot detect the presence of self-pulsation, we also use a microwave spectrum analyzer and a sampling scope with a high speed photodetector (20 GHz bandwidth), together with the optical spectrum analyzer, to simultaneously characterize the optical pulses. The measurement result is shown in Figure 9.20 for a 1-mm-long monolithic CPM laser with 80 GHz repetition rate. The

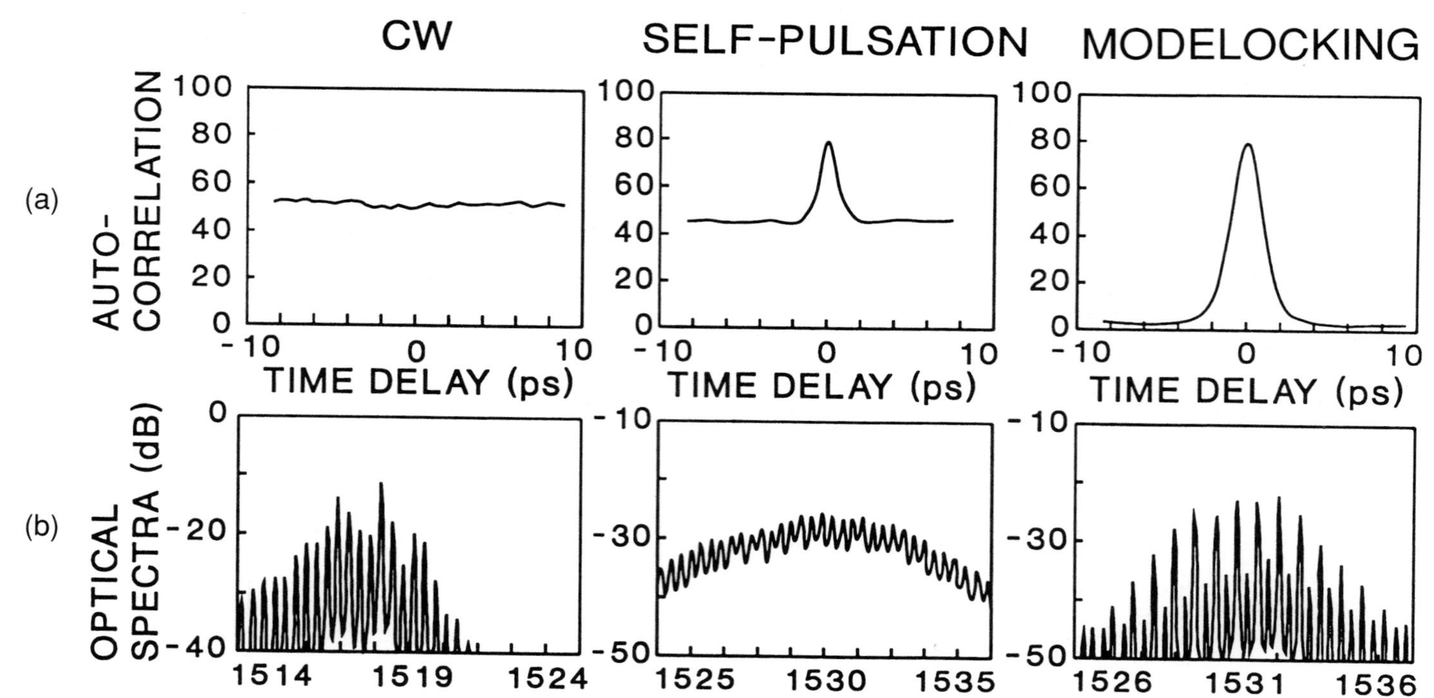
CW
SELF-PULSATION
MODELOCKING
(a)
AUTO-CORRELATION
100
80
60
40
20
0
-10
0
10
TIME DELAY (ps)
(b)
OPTICAL SPECTRA (dB)
0
-20
-40
1514
1519
1524
-10
-30
-50
1525
1530
1535
1526
1531
1536

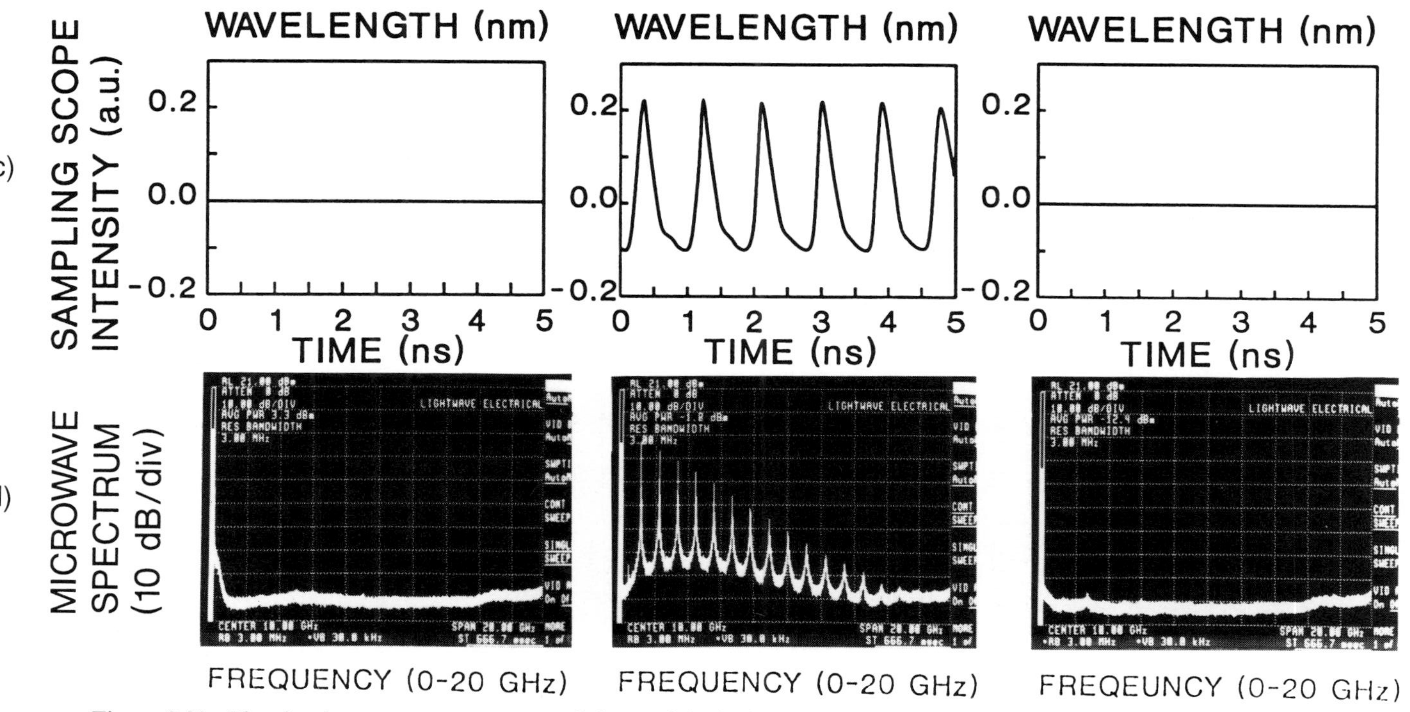

Figure 9.20. The simultaneous measurement of the modelocked pulses with (a) autocorrelator, (b) optical spectrum analyzer, (c) sampling scope with fast photodetector, and (d) microwave spectrum analyzer with fast photodetector. The laser is a 1-mm-long passive CPM laser with 80 GHz repetition rate. The bandwidth of (c) and (d) is 20 GHz. The measurement was done under three conditions: CW lasing (first column), self-pulsation (second column), and modelocking (third column).

third column of Figure 9.20 corresponds to the measurement results of the CPM pulses, while the first and second columns correspond to those of CW lasing and self-pulsation conditions, respectively. During modelocking, a clean autocorrelation trace and optical spectrum with double mode spacing as described above are observed. If the modelocking is free from self-pulsation, the sampling scope and the microwave spectrum analyzer should not record any signal because the fundamental frequency of modelocking (80 GHz) is much higher than the bandwidth of the measurement system (20 GHz). On the other hand, a strong oscillation appears in the sampling scope and a series of oscillation peaks corresponding the self-pulsation frequency and its harmonics are observed from the microwave spectrum analyzer when laser is in self-pulsation. The longitudinal modes in the optical spectrum are also broadened.

Using these measurements, we can distinguish whether the laser is in modelocking, self-pulsation, or both. The modelocking and self-pulsation regions (defined in the plane of gain current versus saturable absorber voltage) are mapped out in Figure 9.21(a) and (b) for the 1-mm-long

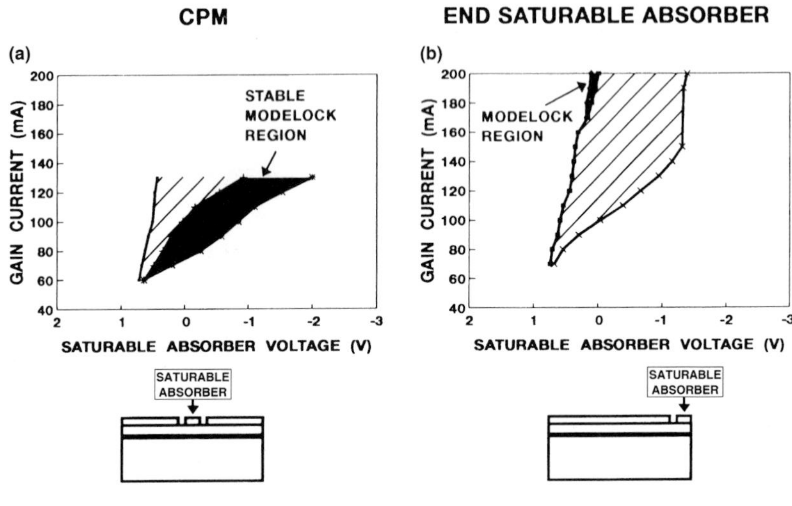

Figure 9.21. The modelocking and self-pulsation ranges for (a) 1-mm-long passive monolithic CPM laser, and (b) the adjacent monolithic modelocked laser with end saturable absorber. There is a wide clean modelocking region for the monolithic CPM laser, while it almost completely overlaps with the self-pulsation region for the laser with end saturable absorber.

CPM laser and a similar laser with saturable absorber adjacent to the facet, respectively. The two lasers are immediately next to each other on the same chip so that there is minimum material variation between them. Both lasers have the same cavity length and same saturable absorber size. A clean modelocked region is clearly observed for the monolithic CPM laser, while it almost completely overlaps with the self-pulsation region for the modelocked laser with end saturable absorber. In addition, the modelocking region is shifted towards the lower bias levels for the CPM lasers, which agrees with the qualitative argument given above. The corresponding autocorrelation traces of the pulses in the modelocked regions are compared in Figure 9.22(a) and (b) for the CPM laser and the laser with end saturable absorber, respectively. The pulses generated by the non-CPM laser have a wider pulse width. In addition, the autocorrelation trace shows a coherent spike on top of the modelocked pulses, indicating the simultaneous presence of the self-pulsation. The data presented here does not imply that it is impossible to produce clean modelocking for lasers with end saturable absorbers. Instead, it suggests that it is more difficult to generate clean modelocked pulses using end saturable absorbers, and the cavity needs to be optimized in a different manner (e.g., using different size saturable absorbers or facet coatings as suggested by Paslaski and Lau (1991)).

Another interesting comparison is the modelocking ranges of hybrid CPM and passive CPM for the same CPM laser. This is illustrated in Figure 9.23 for the 2-mm-long CPM laser. The power level of the input

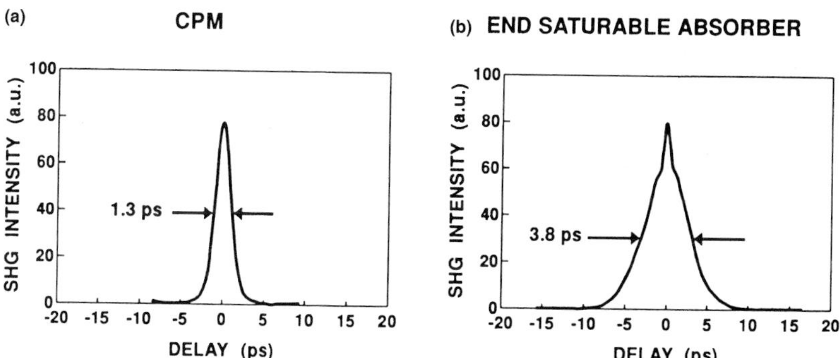

Figure 9.22. The autocorrelation traces of (a) the CPM laser and (b) the adjacent monolithic modelocked laser with end saturable absorber. Both lasers are biased in the modelocked regions, as shown in Figure 9.21. Shorter pulses are obtained with the monolithic CPM configuration.

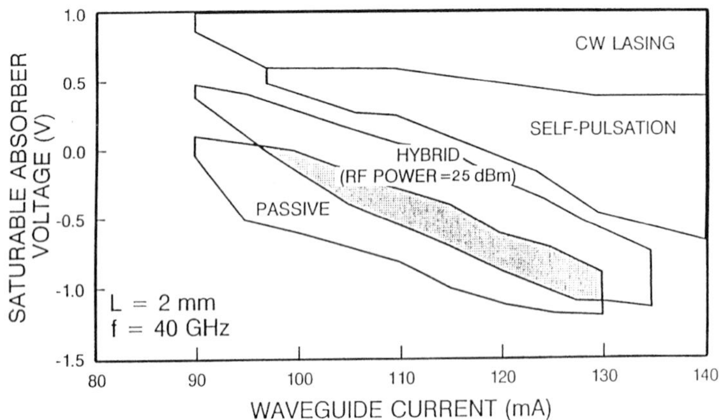

Figure 9.23. The bias ranges for hybrid CPM and passive CPM operations for a 2-mm-long monolithic CPM laser. The RF power level for hybrid CPM is 25 dBm.

microwave signal is 25 dBm for the hybrid CPM scheme. In hybrid modelocking, the optical pulses are also shaped by the microwave modulation and do not depend on the saturable absorber as much as in passive modelocking. Therefore, less reverse bias voltage is needed for the saturable absorber.

9.3.6 *Amplification and tuning of the monolithic CPM laser*

As shown in Figure 9.6 and Figure 9.23, when the monolithic CPM laser is tuned to the optimum modelocking condition, the average output power is around 1 mW. Higher output power can be obtained by increasing the pumping current to the gain section. However, that usually results in degradation of the pulse quality (higher frequency chirp and larger DC background). To maintain the quality of the pulses, it is therefore more desirable to amplify the pulse energy using external amplifiers. Erbium-doped fiber amplifiers (EDFA) are ideal candidates for amplifying ultrashort optical pulses at high repetition rates without distortion, because the relaxation time of the EDFA ($\sim$ 1 ms) is much longer than both the pulse width and the period (for a review, see Zyskind *et al.*, 1993). An experiment has been conducted to compare the pulse shapes of the original and amplified pulses. The experimental setup is shown in Figure 9.24. A CPM laser operating at 80 GHz repetition rate is used as the optical source. The light pulses are coupled into an optical fiber and then amplified by a 1480 nm diode-pumped EDFA. Optical isolators

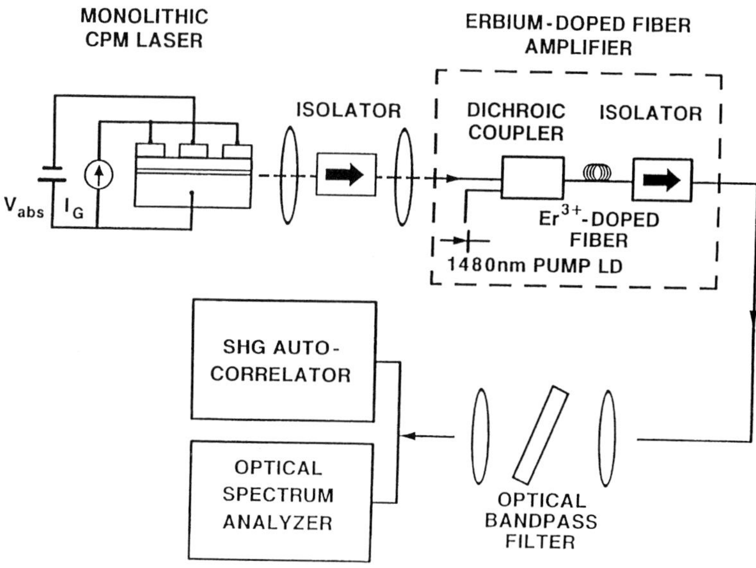

Figure 9.24. Experimental setup for pulse amplification with an erbium-doped fiber amplifier (EDFA).

are inserted on both sides of the EDFA to prevent reflections back to the CPM laser and suppress lasing in the EDFA. The wavelength of the CPM laser is designed to be 1.53 mm, matching that of EDFA gain peak. Under normal passive modelocking conditions, the optical pulses have a duration of 1.28 ps, a time–bandwidth product of 0.34, and a peak power of 5 mW. The optical pulses are amplified by 20 dB when the average input power entering the EDFA is −8 dBm. The peak power in the fiber is boosted to 160 mW. The autocorrelation trace is almost unchanged from that of the input pulse. Higher peak power ($\sim$ 1 W) could be obtained by using EDFAs with higher pump power.

For some applications, such as optical soliton transmission, multicolor solitons or nonlinear optical logic devices, fine tuning of the optical wavelength and pulse width is desired. In external cavity modelocked lasers, the wavelength can be mechanically tuned by etalons or gratings. Such a scheme is more difficult to apply for monolithic modelocked lasers because the optical elements and the cavity configuration are already fixed during the fabrication stage (which is precisely the advantage of monolithic lasers). On the other hand, the modelocking in monolithic lasers is very stable with respect to temperature changes. Thus we can use temperature variation to tune the wavelength of the modelocked pulses. The experimental setup is similar to that in Figure 9.24. The CPM laser is mounted on a thermoelectric cooler. Figure 9.25 shows the autocorrelation traces and the optical spectra of the modelocked pulses for heat sink temperatures of (a) 25°C, (b) 18°C, and (c) 11°C. The spectral envelope shifts continuously towards shorter wavelength with a tuning rate of 0.63 nm/°C. The peak wavelength moves from 1537.2 nm to 1528.4 nm. The variation of the pulse width versus the peak wavelength is shown in Figure 9.26. The pulse width is slighter shorter (1.3 ps) when the peak wavelength coincides with the EDFA gain peak (the amplified spontaneous emission spectra of the EDFA is shown in the inset of Figure 9.26). When the peak wavelength deviates from the gain peak, the pulse width becomes slightly longer (1.6 ps) due to gain dispersion. A

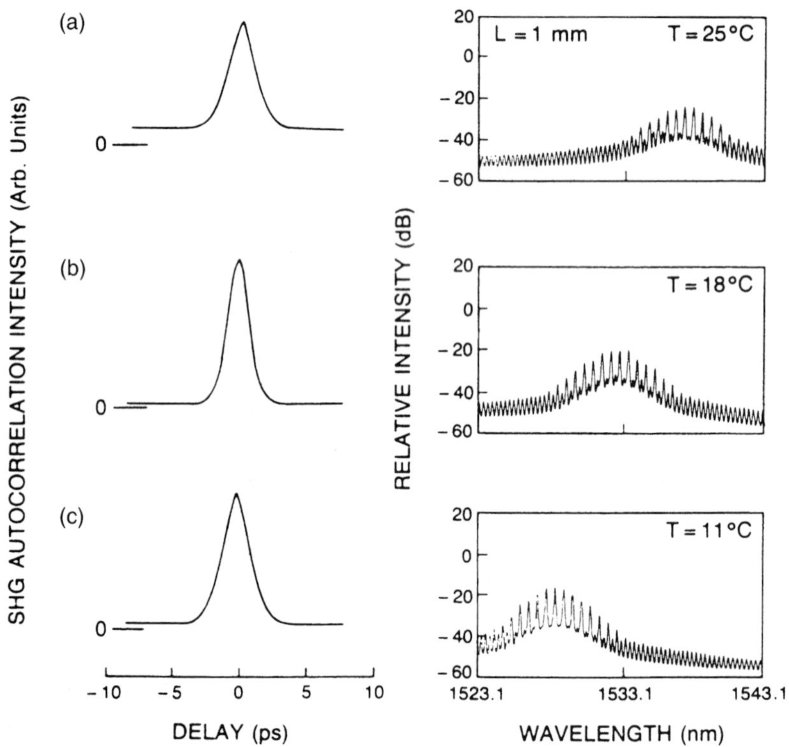

Figure 9.25. The autocorrelation traces and optical spectra of a 80 GHz monolithic CPM laser at heat sink temperatures of (a) 25°C, (b) 18°C, and (c) 11°C. The center wavelength of the modelocked pulses shifts continuously from 1537.2 nm to 1528.4 nm without significant change of pulse width.

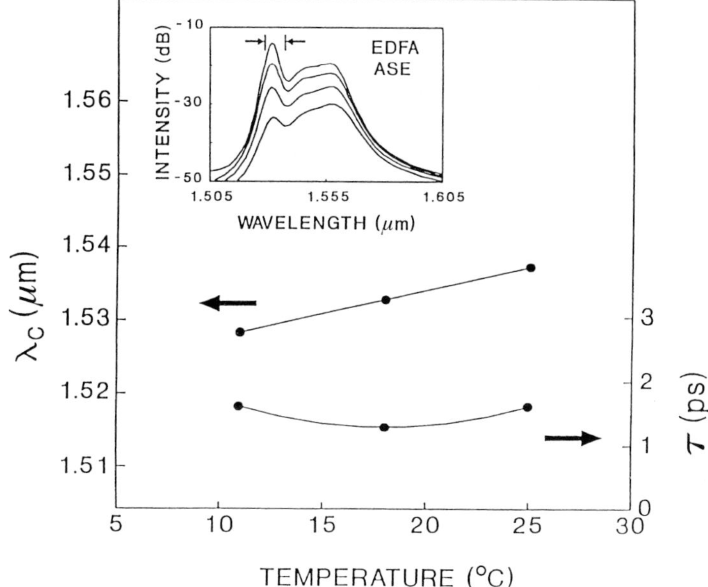

Figure 9.26. The variation of the pulse width versus center wavelength for the temperature tuned passive monolithic CPM laser. The inset shows the amplified spontaneous emission (ASE) spectra of the EDFA used in this experiment. Wavelength tuning across the 1530 nm peak (8.8 nm) is accomplished.

total tuning range of 8.8 nm is achieved. The current tuning range is limited by the EDFA gain bandwidth on the short wavelength side and by the high temperature performance of the CPM laser on the long wavelength side. Wider tuning range can be achieved by designing longer peak wavelength for the CPM laser at room temperature.

The advantage of temperature tuning is that the pulse width is not sacrificed. The wavelength can also be tuned using some built-in tunable wavelength filters at the expense of pulse width (due to narrower bandwidth limited by the filters). A tuning range of 20.2 nm has been demonstrated for an actively modelocked laser with integrated vertical-coupler filter (Raybon *et al.*, 1993). The pulse width is between 10 and 12 ps for a repetition frequency of 15.8 GHz. A passive modelocked DBR laser is also demonstrated with a pulse width of 5.4 ps and repetition frequency of 80 GHz. The wavelength tuning has not yet been reported, though in principle it is possible to tune the Bragg wavelength electrically. It is noted that temperature tuning can be applied to these lasers to extend the wavelength tuning range.

For applications in solitons, sometimes longer pulses are desired because of the finite available optical power in the fiber. The required peak power for an optical soliton is inversely proportional to the square of the pulse width. For example, the power required for a 2 ps pulse is 16 times less than that of 0.5 ps pulse. Again we find it more advantageous to adjust the pulse using external tuning elements. By reducing the optical bandwidth with various etalons outside the cavity, pulses can be adjusted to have longer pulse width. Figure 9.27 shows the autocorrelation traces and the optical spectra of the filtered CPM pulses with filter bandwidths

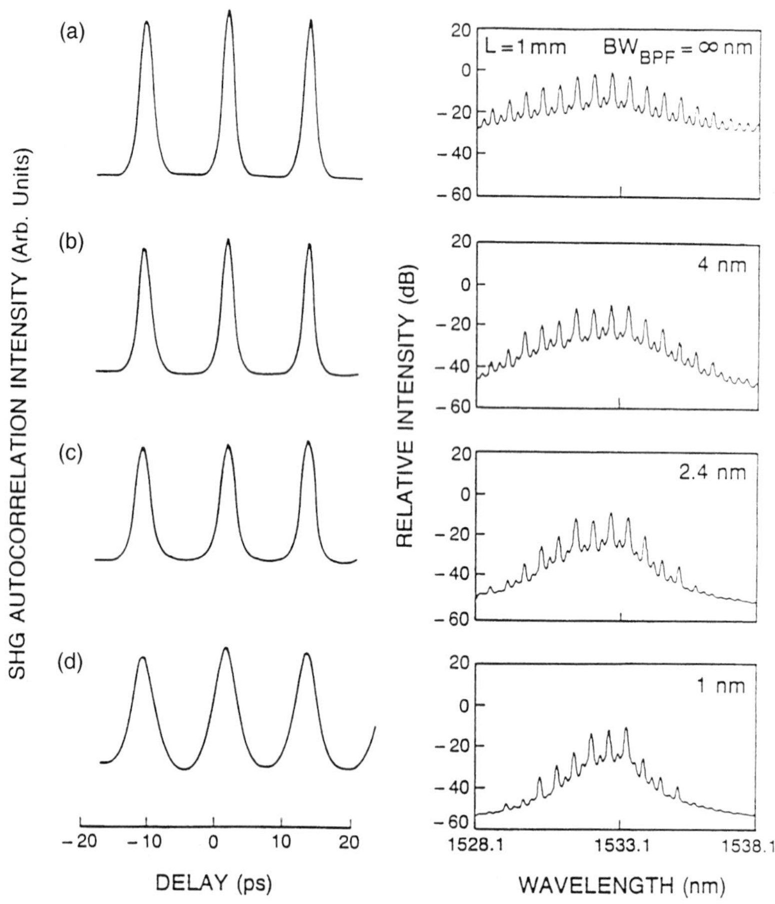

Figure 9.27. The autocorrelation traces and the optical spectra of the filtered CPM pulses with a filter bandwidth of (a) (no filter), (b) 4 nm, (c) 2.4 nm, and (d) 1 nm. Transform-limited pulses with durations of (a) 1.3 ps, (b) 1.6 ps, (c) 1.8 ps, and (d) 2.9 ps are obtained.

of ∞, 4 nm, 2.4 nm, and 1.0 nm, respectively. The pulse widths are 1.3 ps, 1.6 ps, 1.8 ps, and 2.9 ps, respectively. One advantage of this approach is that the high quality of the pulses is preserved. The filtered pulses remained transform-limited with time–bandwidth products between 0.31 and 0.34.

9.4 Applications of monolithic CPM lasers

9.4.1 *Millimeter-wave generation*

Traditionally, microwaves and millimeter-waves are generated by two terminal solid state devices such as GUNN and IMPATT diodes. However, the efficiency and power of these devices decrease dramatically when the frequency is higher than 100 GHz (Yngvesson, 1991). Though resonant tunneling diodes have been demonstrated to oscillate up to several hundred GHz (Brown *et al.*, 1991), their output power is very low. With the development of ultrafast optical sources and photodetectors, opto-electronic generation of millimeter-waves becomes more attractive for frequencies above 100 GHz. Optical fiber, which has an extremely wide bandwidth, is also very suitable for distributing the millimeter-wave modulated optical signals.

There are several methods to generate microwaves and millimeter-waves using opto-electronic techniques. Heterodyning of two single frequency lasers in a high speed photodetector is capable of generating microwave signals with variable frequency. However, in order to reduce the phase noise of the microwave signals, the two lasers need to be phase locked, which is difficult in that frequency range (Seeds, 1993). Using photoconducting antennas, bursts of broadband (a few GHz to 4 THz) pulses have been generated (Froberg *et al.*, 1992). These broadband pulses are useful for characterizing materials at millimeter-wave frequencies.

Since the fundamental repetition frequencies of monolithic modelocked lasers are between a few GHz and several hundred GHz, CW microwaves and millimeter-waves can be generated by converting the optical signals into electrical signals through high speed photodetectors (Scott *et al.*, 1992; Helkey *et al.*, 1993). The monolithic CPM lasers are particularly attractive for opto-electronic generation of ultra-high frequency millimeter-waves because the repetition frequency is doubled. CW modelocked optical signals at 350 GHz have been demonstrated

(Chen, Y. K. *et al.*, 1991). It is also shown that these high frequency pulses can be amplified by EDFAs without much distortion (Wu *et al.*, 1991b). Photodetectors with bandwidths of 510 GHz and 375 GHz have also been reported (Chou and Liu, 1992; Chen, Y. *et al.*, 1991). Thus it could be more efficient to generate and amplify millimeter-waves in the optical domain before converting them to electrical signals.

9.4.2 *Source for ultra-high bit-rate systems*

Recent development of ultrafast all-optical and electro-optic multiplexing and demultiplexing techniques has made it attractive to realize ultra-high bit-rate (50 Gbit/s to 100 Gbit/s) optical communication systems using ultrashort optical pulses. Using two cascaded Mach–Zehnder (MZ) modulators, a 49.6 Gbit/s optical signal is demultiplexed into two 24.8 Gbit/s channels (Jinno, 1992). Using an ultrafast all-optical nonlinear optical loop mirror (NOLM), Andrekson *et al.* have demultiplexed a 64 Gbit/s time-division multiplexed data stream into 4 Gbit/s channels (Andrekson *et al.*, 1992). The experimental setup and the demultiplexed data of the latter experiment are shown in Figure 9.28(a) and (b), respectively. The optical source used in that experiment is an active modelocked external cavity laser operating at 4 GHz. The output pulses were linearly compressed to 7 ps and then multiplexed to 64 Gbit/s using a series of fiber couplers and fiber loops. Though excellent experimental results were obtained, such optical sources are very sensitive to the fiber length variation caused by mechanical or thermal disturbance. The monolithic CPM lasers are ideal candidates for replacing the ultra-high bit-rate optical sources. The optical pulses can be directly generated by the laser and do not rely on the mechanical delay lines, which are also a source of timing jitters. In addition, transform-limited picosecond optical pulses are directly generated, thus eliminating the pulse compression process and also permitting the system to operate to 100 Gbit/s.

9.5 Other monolithic modelocked semiconductor lasers

The monolithic modelocked semiconductor laser is an active research area. In addition to the monolithic CPM quantum well laser described in detail here, many other modelocking schemes have also been demonstrated, including active modelocking, passive modelocking

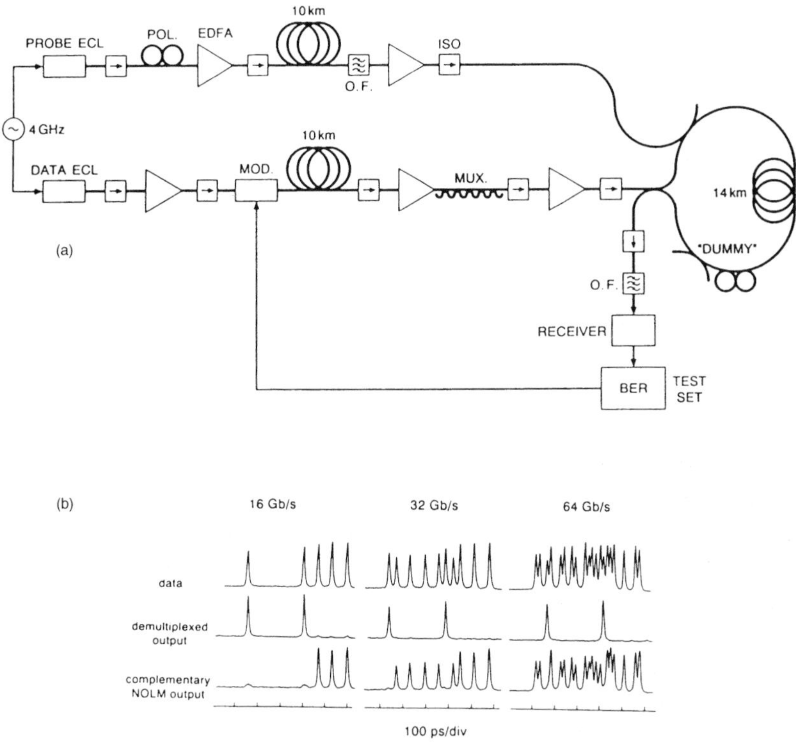

Figure 9.28. (a) The experimental setup and (b) the input and output waveforms of the 64 Gbit/s demultiplexing experiment using an all-optical nonlinear optical loop mirror as ultrafast demultiplexer (from Andrekson *et al.*, 1992).

and hybrid modelocking. The cavity configurations range from the linear Fabry–Perot cavity to the linear DBR cavity and ring cavity. To compare different modelocking schemes, we have summarized the performances (pulse width, repetition frequency, and time–bandwidth product) of various monolithic modelocked lasers reported to date in Table 9.1. It is understood that comparison of different structures reported by different groups could be very difficult. Care must be taken to interpret the comparison results. Nevertheless, some insights could be gained by observing the general trends.

From Table 9.1, first we observe that the pulse widths are in general shorter for hybrid modelocking and passive modelocking schemes than for active modelocking schemes. The microwave power required to achieve pure active modelocking is usually very high. As a result, active modelocking has higher frequency chirp. The time–bandwidth products

Table 9.1.

Modelocking schemes	Pulse width	Repetition frequency	$\Delta t \cdot \Delta \nu$	Active layer materials	Reference
Active modelocking with extended cavity	4.0 ps	40 GHz	1.4	Bulk InGaAsP	Tucker *et al.*, 1989
	9.0 ps	4.4 GHz	8.8	InGaAs/ InGaAsP MQW	Raybon *et al.*, 1992
	12 ps	2.2 GHz	10.6	InGaAs/ InGaAsP MQW	Hansen *et al.*, 1993a
Passive modelocking with tandem-contact lasers	1.75 ps	65 GHz	–	GaAs/ AlGaAs MQW	Lau and Paslaski, 1992
	2.3 ps	100 GHz	0.7	GaAs/ AlGaAs DH	Vasil'ev and Sergeev, 1989
	2.4 ps	108 GHz	1.1	GaAs/ AlGaAs MQW	Sanders *et al.*, 1990a
	6 ps	42 GHz	2.4	InGaAs/ GaAs MQW	Sanders *et al.*, 1990b
	2.5 ps 4.8 ps	15 GHz 5.9 GHz	– –	GaAs/ AlGaAs SQW	Brovelli *et al.*, 1992
Hybrid modelocked laser	1.4 ps	32 GHz	–	Bulk InGaAsP	Morton *et al.*, 1990
	2.0 ps	17.6 GHz	0.59	GaAs/ AlGaAs	Brovelli and Jackel, 1991
Hybrid monolithic CPM laser	1.4 ps	32 GHz	0.30	InGaAs/ InGaAsP MQW	Wu *et al.*, 1990
	0.95 ps	40 GHz	0.32	InGaAs/ InGaAsP MQW	Wu *et al.*, 1991a
Passive monolithic CPM laser	1.1 ps 0.83 ps 1.0 ps 0.64 ps	40 GHz 80 GHz 160 GHz 350 GHz	0.34 0.31 0.34 0.32	InGaAs/ InGaAsP MQW	Chen, Y. K. *et al.*, 1991
Passive self CPM laser	1.75 ps	40 GHz	0.77	InGaAs/ InGaAsP MQW	Derickson *et al.*, 1992

Table 9.1 *(continued)*

Modelocking schemes	Pulse width	Repetition frequency	$\Delta t \cdot \Delta \nu$	Active layer materials	Reference
Active modelocked ring cavity	27 ps	9 GHz	0.47	InGaAs/ InGaAsP MQW	Hansen *et al.*, 1992
Passive modelocked ring laser	1.3 ps	86 GHz	0.43	GaAs/ AlGaAs SQW	Hohimer and Vawter, 1993
Active modelocked DBR laser	15 ps	8.3 GHz	0.41	InGaAs/ InGaAsP MQW	Raybon *et al.*, 1991
Passive modelocked DBR laser	5.4 ps	80 GHz	0.65	InGaAs/ InGaAsP MQW	Arahira *et al.*, 1993
Active modelocked laser with vertical coupler filter	10–12 ps	15.8 GHz	0.58–1.4	InGaAs/ InGaAsP MQW	Raybon *et al.*, 1993
Active, hybrid, and passive modelocking with extended cavity	6.2 ps (A) 4.4 ps (H) 5.5 ps (P)	8.67 GHz	7.0	InGaAs/ InGaAsP MQW	Hansen *et al.*, 1993b
	13 ps (A) 6.5 ps (H) 10 ps (P)	5.5 GHz	4.3 3.5 4.0	GaAs/ AlGaAs MQW	Derickson *et al.*, 1991

are usually much greater than units in the absence of bandwidth limiting elements (such as DBRs or filters) in the cavity. On the other hand, active modelocking is perhaps the most straight forward modelocking scheme for monolithic cavities. Repetition frequency as low as 2.2 GHz has been demonstrated for a 2-cm-long cavity. The highest frequency reported is 40 GHz, limited by the available high frequency synthesizers and amplifiers. Passive modelocking generates shorter optical pulses with much less frequency chirp. The time–bandwidth products are usually within a factor of two of the transform-limited value. In addition, much higher repetition frequency can be achieved since no RF modulation is required. However, it requires more delicate balance between gain media and saturable absorbers. The bias ranges for passive modelocking are usually narrower. Most of the passive modelocked lasers in Table 9.1 use quantum wells as active media (except Vasil'ev and Sergeev, 1989) because the ratio of absorber cross-section to gain cross-section is larger than that of bulk active media (Lau, 1990).

To date, monolithic CPM lasers are the only monolithic modelocked lasers reported that generate subpicosecond optical pulses. Since most of the saturable absorbers in monolithic modelocked lasers consist of the same quantum well materials as the gain media, the enhancement of the saturable absorber cross-section by CPM is particular valuable. In addition, the discrimination against self-pulsation also allows a wider bias range of clean pulse generation. Another interesting trend is that the pulse width is shorter for modelocked lasers with higher repetition frequency. This could suggest that the current pulse width obtained is limited by the dispersion of the cavity. Higher frequency lasers have shorter cavities and therefore less dispersion. To further reduce the pulse width, some integrable dispersion compensation scheme needs to be developed.

9.6 Conclusion and future direction

In conclusion, we have described a novel monolithic colliding pulse modelocked semiconductor quantum well laser that generates picosecond and subpicosecond optical pulses. With a monolithic cavity, it is possible to scale down the cavity length to generate optical pulses at hundreds of GHz. With a 250-μm-long cavity, a record high repetition frequency of 350 GHz has been experimentally demonstrated. The transform-limited pulse width of 640 femtoseconds is also the shortest ever reported for passive modelocked semiconductor lasers. Passive modelocking of the monolithic CPM laser has also been demonstrated at 40 GHz, 80 GHz, and 160 GHz with a pulse width of 1 ps or shorter. The CPM laser is also hybrid modelocked at 32 GHz and 40 GHz with pulse duration of 1 to 1.4 ps. All the pulses obtained have pulse shapes of sech^2 and the nearly transform-limited time–bandwidth products of 0.31. The monolithic cavity does not require optical alignment, and is very stable with respect to mechanical or thermal perturbations. The monolithic CPM design is also advantageous to suppress the low frequency self-pulsation typically present in multi-segment semiconductor lasers.

The technique presented here can be extended to monolithic CPM lasers with lower (< 10 GHz) or higher frequencies. Better understanding of the ultrafast physical mechanisms of the recovery time of saturable absorbers is needed. The 350 GHz laser we demonstrated shows that there is at least one ultrafast process with response time faster than 2.86 ps (period of the pulse train) since the absorber must recover before the next pulse arrives. These ultrafast picosecond optical sources have many applications, such as opto-electronic generation of millimeter-waves at frequencies above 100 GHz, and optical sources for ultra-high bit-rate time-division multiplexed systems. The advancement of ultrafast optical sources will also inspire the research of ultrafast photodetectors and ultrafast all-optical logic devices.

Acknowledgments

The work on monolithic colliding pulse modelocked diode lasers was done at AT&T Bell Laboratories with collaboration with Drs. T. Tanbun-Ek, R. A. Logan and J. R. Simpson.

References

Andrekson, P. A., Olsson, N. A., Simpson, J. R., DiGiovanni, D. J., Morton, P. A., Tanbun-Ek, T., Logan, R. A. and Wecht, K. W. (1992) *IEEE Photon. Technol. Lett.*, **4** , 644–7.

Arahira, S., Matsui, Y., Kunii, T., Oshiba, S. and Ogawa, Y. (1993) *Electron. Lett.*, **29**, 1013–15.

Arakawa, Y. and Yariv, A. (1986) *IEEE J. Quantum Electron.*, **22**, 1887–99.

Bowers, J. E., Morton, P. A., Mar, A. and Corzine, S. W. (1989) *IEEE J. Quantum Electron.*, **25**, 1426–39.

Brovelli, L. R. and Jackel, H. (1991) *Electron. Lett.*, **27**, 1104–6.

Brovelli, L. R., Jackel, H. and Melchior, L. H. (1992) *Proc. Conf. Lasers and Electro-Optics*, paper JThB3 (Optical Society of America: Washington, DC).

Brown, E. R., Soderstrom, J. R., Parker, C. D., Mahoney, L. J., Molvar, K. M. and McGill, T. C. (1991) *Appl. Phys. Lett.*, **58**, 2291–3.

Chen, Y. K. and Wu, M. C. (1992) *IEEE J. Quantum Electron.*, **28**, 2176–85.

Chen, Y. K., Wu, M. C., Tanbun-Ek, T., Logan, R. A. and Chin, M. A. (1991) *Appl. Phys. Lett.*, **58**, 1253–55.

Chen, Y., Williamson, S., Brock, T., Smith, F. W. and Calawa, A. R. (1991) *Appl. Phys. Lett.*, **59**, 1984–6.

Choi, H. K. and Wang, C. A. (1990) *Appl. Phys. Lett.*, **57**, 321–3.

Chou, S. Y. and Liu, M. Y. (1992) *IEEE J. Quantum Electron.*, **28**, 2358–68.

Derickson, D. J., Morton, P. A. and Bowers, J. E. (1991) *Appl. Phys. Lett.*, **59**, 3372–4.

Derickson, D. J., Helkey, R. J., Mar, A., Bowers, J. E., Tanbun-Ek, T., Coblentz, D. L. and Logan, R. A. (1992) *Proc. Optical Fiber Communications Conf.*, paper ThB3 (Optical Society of America: Washington, DC).

Dixon, R. W. and Joyce, W. B. (1979) *IEEE J. Quantum Electron.*, **15**, 470–4.

Downey, P. M., Bowers, J. E., Tucker, R. S., and Agyekum, E. (1987) *IEEE J. Quantum Electron.*, **23**, 1039–47.

Fork, R. L., Greene, B. I. and Shank, C. V. (1981) *Appl. Phys. Lett.*, **38**, 671–2.

Froberg, N. M., Hu, B. B., Zhang, X.-C. and Auston, D. H. (1992) *IEEE J. Quantum Electron.*, **28**, 2291–301.

Garmire, E. M. and Yariv, A. (1967) *IEEE J. Quantum Electron.*, **3**, 222.

Hansen, P. B., Raybon, G., Chien, M. D., Koren, U., Miller, B. I., Young, M. G., Verdiell, J. M. and Burrus, C. A. (1992) *IEEE Photon. Technol. Lett.,* **4**, 411–13.

Hansen, P. B., Raybon, G., Koren, U., Miller, B. I., Young, M. G., Newkirk, M., Chien, M. D., Tell, B. and Burrus, C. A. (1993a) *Electron. Lett.*, **29**, 739–41.

Hansen, P. B., Raybon, G., Koren, U., Iannone, P. P. U., Miller, B. I., Young, M. G., Newkirk, M. A. and Burrus, C. A. (1993b) *Appl. Phys. Lett.*, **62**, 1445–7.

Haus, H. A. (1975) *IEEE J. Quantum Electron.*, **11**, 736–46.

Haus, H. A. (1976) *IEEE J. Quantum Electron.*, **12**, 169–76.

Helkey, R. J., Derickson, D. J., Mar, A., Wasserbauer, J. G. and Bowers, J. E. (1993) *Microwave and Opt. Technol. Lett.*, **6**, 1–5.

Ho, P. T., Glasser, L. A., Ippen, E. P. and Haus, H. A. (1978) *Appl. Phys. Lett.*, **33**, 241–2.

Hohimer, J. P. and Vawter, G. A. (1993) *Appl. Phys. Lett.*, **63**, 1598–600.

Ippen, E. P., Eilenberger, D. J. and Dixon, R. W. (1980) *Appl. Phys. Lett.*, **37**, 2679.

Ippen, E. P., Haus, H. A. and Liu, L. Y. (1989) *J. Opt. Soc. Amer. B*, **6**, 1736–45.

Jinno, M. (1992) *IEEE Photon. Technol. Lett.*, **4**, 641–4.

Kawanishi, S., Takara, H., Uchiyama, K., Kitoh, T. and Saruwatari, M. (1993) *Proc. Optical Fiber Communications Conf.*, paper PD2 (Optical Society of America: Washington, DC).

Koch, T. and Bowers, J. (1984) *Electron. Lett.*, **20**, 1038–40.

Krausz, F., Fermann, M. E., Brabec, T., Curley, P. F., Hofer, M., Ober, M. H., Spielmann, C., Wintner, E. and Schmidt, A. J. (1992) *IEEE J. Quantum Electron.*, **28**, 2097–2122.

Lau, K. Y. (1988a) *Appl. Phys. Lett.*, **52**, 257–9.

Lau, K. Y. (1988b) *Appl. Phys. Lett.*, **52**, 2214–6.

Lau, K. Y. (1990) *IEEE J. Quantum Electron.*, **26**, 250–61.

Lau, K. Y. and Paslaski, J. (1992) *IEEE Photon. Technol. Lett.*, **3** , 975–6.

Lee, T. P. and Roldan, R. H. (1970) *IEEE J. Quantum Electron.*, **6**, 339.

Leep, D. A. and Holm, D. A. (1992) *Appl. Phys. Lett.*, **60**, 2451–3.

Martins-Filho, J. F., Ironside, C. N. and Roberts, J. S. (1993) *Electron. Lett.*, **29**, 1135–6.

Mollenauer, L. F., Evanglides, S. and Haus, H. A. (1991) *J. Lightwave Technol.*, **9**, 194–7.

Mollenauer, L. F., Gordon, J. P. and Evanglides, S. (1992) *Opt. Lett.*, **17**, 1575–7.

Morton, P. A., Bowers, J. E., Koszi, L. A., Soler, M., Lopata, J. and Wilt, D. P. (1990) *Appl. Phys. Lett.*, **56**, 111–13.

Nakazawa, M., Yamada, E., Kubota, H. and Suzuki, K. (1991) *Electron. Lett.*, **27**, 1270–2.

Nelson, B. P., Blow, K. J., Constantine, P. D., Doran, N. J., Lucek, J. K., Marshall, I. W. and Smith, K. (1991) *Electron. Lett.*, **27**, 704–5.

Paslaski, J. and Lau, K. Y. (1991) *Appl. Phys. Lett.*, **59**, 7–10.

Raybon, G., Hansen, P. B., Koren, U., Miller, B. I., Young, M. G., Chen, M., Burrus, C. A. and Alferness, R. C. (1991) *Proc. 17th European Conference on Optical Communication (ECOC)* pp. 40–3.

Raybon, G., Hansen, P. B., Koren, U., Miller, B. I., Young, M. G., Newkirk, M., Iannone, P. P., Burrus, C. A., Centanni, J. C. and Zirngibl, M. (1992) *Electron. Lett.*, **28**, 2220–1.

Raybon, G., Hansen, P. B., Alferness, R. C., Buhl, L. L., Koren, U., Miller, B. I., Young, M. G., Koch, T., Verdiell, J. M. and Burrus, C. A. (1993) *Opt. Lett.*, **18**, 1335–6.

Sanders, S., Eng, L., Paslaski, J. and Yariv, A. (1990a) *Appl. Phys. Lett.*, **56**, 310–11.

Sanders, S., Eng, L. and Yariv, A. (1990b) *Electron. Lett.*, **26**, 1087–8.

Scott, D. C., Plant, D. V. and Fetterman, H. R. (1992) *Appl. Phys. Lett.*, **61**, 1–3.

Seeds, A. J. (1993) *Optical and Quantum Electron.*, **25**, 219–29.

Shank, C. V. (1988) In W. Kaiser (ed.) *Ultrashort Laser Pulses and Applications*, pp. 5–34 (Springer-Verlag: Berlin).

Silberberg, Y., Smith, P. W., Eilenberger, D. J., Miller, D. A. B., Gossard, A. C. and Wiegmann, W. (1984) *Optics Lett.*, **9**, 507–9.

Sipe, J. E. (1993) personal communication.

Stix, M. S. and Ippen, E. P. (1983) *IEEE J. Quantum Electron.*, **27**, 1426–39.

Tanbun-Ek, T, Logan, R. A., Temkin, H., Berthold, K., Levi, A. F. J. and Chu, S. N. G. (1989) *Appl. Phys. Lett.*, **55**, 2283–5.

Thijs, P. J. A., Fiemeijer, L. F., Kuindersma, P. I., Binsma, J. J. M. and Dongen, T. V. (1991) *IEEE J. Quantum Electron.*, **27**, 1426–39.

Tsang, W. T. (1987) In Dingle, R. (ed.), *Semiconductors and Semimetals*, **24**, pp. 397–458 (Academic Press: San Diego).

Tucker, R. S., Koren, U., Raybon, G., Burrus, C. A., Miller, B. I., Koch, T. L. and Eisenstein, G. (1989) *Electron. Lett.*, **25**, 621–2.

Valdmanis, J. A. and Mourou, B. (1986) *IEEE J. Quantum Electron.*, **22**, 69–78.

van der Ziel, J. P. (1985) In Tsang, W. T. (ed.), *Semiconductors and Semimetals*, **22**, Part B, pp. 1–68 (Academic Press: Orlando).

van der Ziel, J. P., Tsang, W. T., Logan, R. A., Mikulyak, R. M. and Angustyniak, W. M. (1981) *Appl. Phys. Lett.*, **39**, 525–7.

Vasil'ev, P. P. and Sergeev, A. B. (1989) *Electron. Lett.*, **25**, 1049–50.

Vasil'ev, P. P., Morozov, V. N., Popov, Y. M. and Sergeev, A. B. (1986) *IEEE J. Quantum Electron.*, **22**, 149–52.

Wu, M. C., Chen, Y. K., Tanbun-Ek, T., Logan, R. A., Chin, M. A. and Raybon, G. (1990) *Appl. Phys. Lett.*, **57**, 759–61.

Wu, M. C., Chen, Y. K., Tanbun-Ek, T. and Logan, R. A. (1991a) *Proc. Picosecond Electronics and Optoelectronics*, **9**, pp. 176–80 (Optical Society of America: Washington, DC).

Wu, M. C., Chen, Y. K., Tanbun-Ek, T., Logan, R. A., and Chin, M. A. (1991b) *IEEE Photon. Technol. Lett.*, **3**, 874–6.

Wu, M. C., Chen, Y. K., Tanbun-Ek, T. and Logan, R. A. (1993) In Martin, J.-L., Migus, A., Mourou, G. A. and Zewail, A. H. (eds.) *Ultrafast Phenomena VIII*, pp. 211–16 (Springer-Verlag: Berlin).

Yngvesson, S. (1991) *Microwave Semiconductor Devices*, p. 137 (Kluwer Academic Publisher: Boston).

Zyskind, J. L., Giles, R. R., Simpson, J. R. and DiGiovanni, D. J. (1993) *Trends in Telecommunications*, **8**, 37–42.

Index

absorption
  acoustic, 114
  broadening, 296
  defect, 339
  electron-heavy-hole, 226
  excitonic, 285, 286, 316
  linear, 171
  non-bleachable, 170
  optical, 114
  pump light, 95, 230, 233
  recovery of, 125, 286, 296
  saturable, 170, 172, 316, 384, 386
  saturation, 170, 299
  two photon, 217, 253
absorption band, 94, 114, 131, 132, 141, 339
absorption coefficient, 9, 247
absorption cross-section, 60, 170
absorption length, 95, 110
absorption media, 1
acousto-optic modulator, 100, 105, 141, 142, 192, 195
amplification, 100, 129, 130, 211, 300, 410, 282
  and tunability, 129
  chirped pulse, 324
  distributed, equivalence with decreasing dispersion, 173
  external, 156, 213, 291, 300
  lack of gain saturation, 99
  periodic, 160
  regenerative, 131
  theory, 99
amplified spontaneous emission, 142, 412
amplifier
  conduction band of, 308
  dephasing time of, 60
  diode, 118, 119, 144, 284, 305
  dispersion, 168
  erbium doped fiber (EDFA), 35, 42, 143, 146, 152–7, 181, 275, 329, 336, 369, 370, 372, 373, 378, 410–13
  external, 320, 324, 410
  fiber, 35, 143, 154, 205, 275, 329, 336, 341, 369, 370, 373, 378, 410
  for fanout, 311
  Nd
    glass, 106, 133
    YLF, 131
  noise, 67
  regenerative, 131
  saturable, 121
  saturable diode (SDA), 118
  semiconductor, 275, 277, 388
  solid state, 130, 131
  spontaneous emission, 300
  traveling wave (TWA), 210, 211, 275–82, 285, 286, 290, 293–304, 316, 322, 324
amplifier bandwidth, 100
amplifier power gain, 230
amplitude modulated, 96, 102–7, 112, 114, 128
amplitude modulator, 60, 113, 116, 117, 120, 141, 143
anomalous dispersion, 143, 150
antireflection coating, 110, 201, 209, 210, 217, 225–8, 256, 276, 278, 281, 334, 340, 355, 364, 365
antiresonant Fabry–Perot saturable absorber, 125
artificial fast saturable absorber (AFSA) 2, 3, 26, 140, 153, 154, 174
astigmatism compensation, 108
Auger recombination, 226

bandwidth
  communication, 309, 310
  detector, 335, 393, 395, 405
  fiber, 415
  filter, 46, 244, 248, 252, 413, 414
  FM operation, 111
  frequency comb, 127, 128
  gain, 10, 15, 18, 42, 47, 57, 100, 114, 180, 211, 215, 243, 248, 249, 324, 413, *see also* gain bandwidth
  GVD broadening, 299
  laser, 48, 72
  laser oscillation, 67, 68
  lasing, 142, 297
  modulation, 190, 361
  modulator, 142
  optical, 299, 329, 333, 336, 348, 363, 370, 414
  phase noise sideband, 289
  relative, 1
  switch tunability, 58
  table of modelocked lasers, 107
  XPM induced, 222, 244
bandwidth-limited pulses, 80, 81, 195, 199, 209
bandwidth limiting, 63, 186, 245, 385
  by gain, 13, 18, 44, 46
  general, 77, 85, 154, 160, 209, 419
beat-note frequency, 162
beat-note line, 65, 66
beat-note linewidth, 65, 66, 68, 70, 89
  broadening, 66–70, 89
beat-note signal, 66
beat-note sources, 173

birefringence
  cavity, 162
  circular, 180, 181
  elliptical, 180
  extrinsic, 181
  form, 181, 182
  high, 181, 182, 189, 190, 198, 200, 201, 203
  internal, 181
  linear, 148, 149, 181, 185, 189
  low, 179, 181, 200, 201, 203
  natural, 155
  nonlinear, 145
  of the cavity, 144
  thermal, 94, 106
  zero, 197, 198
birefringence axes, 187, 200
bit error rate, 379
Bragg reflector, 338, 359, 372
  bandwidth, 255, 330, 334, 336, 339, 355, 360, 361, 364, 365, 373
  chirped, 346–54, 357–9, 364, 367, 373, 375, 379
    wide bandwidth, 364
  diffractive, 213
  distributed (DBR), 213–17, 225, 230, 241, 243, 255, 256, 261, 341, 355, 356, 385, 413, 417, 419
  fiber, 329, 339, 344
  first order, 335
  integrated, 339, 340, 341, 365
  linear, 343, 345, 346, 348
Bragg wavelength, 337, 339, 343–7, 366–9, 413
Brewster surface, 108, 110
broadening
  homogeneous, 96, 99, 104
  inhomogeneous, 115
building integrated timing supply, 320
buried heterostructure, 391

carrier
  distribution, non-uniform, 244
carrier concentration, 305, 399
carrier cooling, 304, 305, 306
carrier density, 224, 231, 233, 235, 246, 247, 343, 349, 355, 361, 363, 364, 369, 384
carrier depletion, 293, 304, 343, 361
carrier diffusion, 355
carrier dynamics, 246
carrier frequency, 5, 19, 24, 60, 172, 185, 289, 295, 305–7, 310
carrier generation, 170
carrier heating, 226, 253, 260, 300, 301
carrier injection, 360, 386
carrier lifetime, 247
carrier recombination, 170
carrier recovery, 304
carrier thermalization, 170, 301, 305, 307, 308
carrier transport, 224–6, 244–54, 260
carrier wavelength, 310
carriers, 171, 255
chirp, 45, 80, 144, 233, 278, 290, 293, 317, 320
  by FM modelocking, 106, 174
  down-, 186, 235, 237, 239, 244, 245
  due to SPM, 242
  frequency, 104, 275, 294, 324, 384, 386, 410, 417, 419
  linear, 97, 291, 317, 346, 354, 355, 365, 373
  nonlinear, 241, 293
  phase, 44
  positive, 245, *see also* chirp, up-
  up-, 186, 235, 237, 244, 245
  zero, 27, 44, 97, 99, 102, 232, 244, 245, 343–6, 359
chirp compensation, 211, 260, 278, 284, 299, 324
chirp parameter, 15
chirp rate, 364, 369, 375
chirped Bragg reflector, 346–51, 357, 359, 364, 367, 373, 375, 379
chirped device, 349
chirped grating, 348, 351, 369
chirped pulse, 196, 215, 230
  amplification (CPA), 100, 196, 300
  as solution to the master equation, 48
  compensation of, 240
  filtering of, 17
  from nonlinear effects, 11
  from non-optimal compression, 303
  from XPM, 239
  in APM, 23, 27
  in FM modelocking, 106
  in hybrid modelocking, 297
  in modelocked VCSEL, 221
  in VCSEL, 222, 229, 235, 243, 260
  nonlinear chirp from VCSEL, 221
  primary mechanisms for, 230
  scaling with energy, 223
  sign of, 244
  up-chirp, 245
clock, 171, 189, 211, 309–16, 323, 324, 330, 332, 395, 398
coherence spike, 219, 276, 281, 282, 409
colliding pulse modelocking, 208, 210, 211, 241, 385–416, 420
commutator, 74, 79
compressor, 210, 221, 241, 290, 291, 292, 293, 294, 295, 299, 300
coupled cavity modelocking, 117, 119
cross-phase modulation (XPM), 186, 222, 230, 233, 239, 242–4, 260
cubic phase compensation, 293
current versus voltage, 256, 393
CW operation, 94, 106, 123, 125, 193, 217, 257, 330, 348, 359, 372

diode amplifier, 118, 119, 144, 284
direct current, 257, 258, 259, 392
dispersion
  anomalous, 150, *see also* dispersion, negative
  compensation, 168
  compensation, cubic, 293, 295, 323
  compensation, external, 219, 240
  compensation, internal, 275, 296, 297
  decreasing, 173
  gain, 77, 241
  group delay (GDD), 59, 67, 71–8, 81, 82, 84, 87, 88
  group velocity (GVD), 3, 11, 13, 18, 219, 240–2
  in the master equation, 14, 48
  in the nonlinear Schrödinger equation, 19
  measurement of, 162
  negative, 4, 12, 17, 18, 23, 40, 49, 71, 117, 169, 196–8, 221, 291
  positive, 17, 119, 221
  pulse broadening from, 17
  pulse-width dependence, 15
  second-order, 12, 159–62
  third-order, 44, 45, 46, 49, 77, 78, 82–4, 87, 157–61, 196
dispersion decreasing fiber, 157, 173, 174
dispersive delay line, 75
dispersive waves, 83–7, 158, 163–8, 170, 175
distributed feedback, 275, 355
double heterojunction, 209, 210
dynamics, 296, 300
  amplification, 300
  gain, 293, 299, 302–6, 324
  intracavity, 296
  modelocking, 72, 75, 79, 297

photon, 246
pulse, 72, 80
pulse shaping, 59, 72, 74, 76, 79, 84, 296
pulse shaping, periodic, 80
start-up, 171
system, 292
temporal, 301, 343

electron-heavy-hole, 226
electro-optic tuner, 209, 210
emission cross-section, 142
emitter coupled logic, 311
energy quantization, 154, 386
erbium, 42, 143, 175, 181, 198, 201, 205, 275, 329, 336, 355, 369–73, 378
etalon effect, 112, 228
external cavity, 117, 119, 120, 122, 125, 278, 293, 316
air, 333
coupling to, 355, 375
CPM laser, 217, 388
diode laser, 208–11, 276, 280, 281, 284, 296, 301, 324, 384, 386, 388, 399, 416
fiber, 339, 341, 348, 349
length, 122, 375
linear, 172
mode spacing, 287, 296, 316, 358
multiple pulses, 385
TWA laser, 281, 286
VCSEL, 255
waveguide, 334
external cavity laser, 256, 281, 284, 285, 388, 416
external cavity mirror, 210
external cavity modelocking, 282, 293, 301, 316

Fabry–Perot
antiresonant saturable absorber (A-FPSA), 125, 171
suppression of modes, 281, 340
Fabry–Perot effect, 225, 278
Fabry–Perot interferometer, 336, 338, 349, 356, 361, 363, 370, 377
Fabry–Perot lasers, 361
Fabry–Perot modes, 356, 358, 363, 370
Fabry–Perot resonator, 121, 172, 183, 187, 192, 417
Faraday mirror, 184, 190, 200
Faraday rotator, 190, 200
feedback
cavity length control, 22, 27, 144, 145
from a Bragg grating, 343, 344, 345, 349
from a chirped Bragg grating, 359
in nonlinear coupler modelocking, 170
jitter control, 211
loop, F8L, 146, 153, 155, 356
repetition rate control, 156, 174
to reduce acoustic pickup, 358
see also frequency shifted feedback
fiber loop, 146, 147, 150, 184, 416, *see also* loop mirror
figure eight laser, 140, 146–68, 189, 198
filter, 17, 27, 125, 150, 152, 155, 162, 210, 225, 242–5, 252–3, 286, 385, 413, 415
bandwidth, 314
birefringent, 222, 223, 252, 260
sliding-frequency guiding, 331–3, 378, 379
spectral, 101, 244, 260, 281, 291, 300, 301, 303, 308
filter bandwidth, 46, 244, 248, 252, 414
Fourier components, 5, 6, 159, 165
Fourier transform, 7, 8, 97, 159, 243
Fourier transform limit, 243–4
frequency modulated, 96, 104–7, 111–15, 128, 130

frequency shifted feedback, 126, 127, 133

gain bandwidth, 5, 7, 15, 18, 44, 46, 93, 96, 99, 102, 107, 116, 133, 211, 248–9, 413
gain compression, 283, 299
gain cross-sectional area, 213
gain dispersion, 40, 63, 72,77, 242–3, 248, 412
gain relaxation time, 10, 60, 61, 64
gain saturation, 69, 85, 131, 211, 213–14, 225, 247, 259, 283, 298, 300, 324, 357
dynamic, 60, 64, 117
in pulse formation, 57, 64, 72, 99, 127, 222–4
induced chirp, 144, 222, 230–41, 243, 244, 260
single pass, 119, 142
weak, 65
gain saturation energy, 211, 213, 214, 225, 247, 259
gain switching, 214–16, 274, 275, 383, 384
gain window, 10, 60, 64, 65
Gaussian pulse, 7, 96, 97, 99, 100, 101, 102, 103, 104, 116, 219
graded index separate confinement heterostructure, 391
grating, 67, 69, 330, 339–46, 349, 351, 354, 355, 357, 360, 361, 365–9, 373–80
bandwidth, 330
complex refractive index, 67–9, 70
for dispersion compensation, 210, 219, 221, 228, 241, 260, 275, 290–5, 299, 300, 316, 322, 324
for tuning, 210, 278, 281, 294, 333, 411
linearly chirped, 348
population, 70
transient, 388, 390
grating pulse compressor, 290

high birefringence, 181, 182, 189, 190, 198, 200, 201, 203
high reflector, 125, 278
homogeneous, 66
hot carrier thermalization, 301
hybrid soliton pulse source (HSPS) 329–80
chirped Bragg reflector, 348
high power output, 368
packaging, 373
relative intensity noise spectrum, 345, 359
spectral instabilities, 329
temperature tuning of wavelength, 365
wavelength self-tuning, 349

impact ionization avalanche transit time, 415
infrared, 2, 72, 208, 212, 216, 218
inhomogeneous broadening, 66

Jones matrix, 182–5, 188, 191
Jones vector, 182

Kerr, 133, 145, 171, 200, 203
Kerr coefficient, 19, 25, 31, 52, 54
Kerr effect, 2, 13, 14, 18, 19, 30, 32, 48, 50, 71, 117, 124, 145
Kerr index, 32
Kerr induced SPM, 58, 63, 71, 72, 74
Kerr lens modelocking, 3, 32, 35, 39, 42, 43, 48, 49, 58, 59, 118, 123, 124, 125, 208
Kerr medium, 14, 26, 27, 28, 30, 33, 43, 49, 50, 53, 54, 75, 80, 117, 124
Kerr modulation, 40
Kerr nonlinearity, 2, 37, 49, 58, 60, 71, 81, 89, 119, 132, 163, 179
Kerr phase, 25, 28, 31, 48, 49
Kerr shift modelocking, 125
Kerr tensor, 50
Kerr-type modelocking, 171, 200, 203

Kuizenga–Siegman theory, 112–15, 142

laser
  all-fiber, 3, 4, 32, 35, 37, 40, 41, 42, 43, 140
  color center, 1
  diode pumped solid state laser (LDPSSL), 94, 95, 96, 114, 131
  diode pumping of, 93, 96, 106, 107, 112–17, 123, 124, 126, 129–31
  end pumping, 110–11
  frequency shifted feedback, 126–8, 133
  hybrid, 348, 354, 356, 359
  multi-section, 353, 364, 368
  optical parametric oscillators, 129–30
  planar waveguide, 128–9
  resonators, 108–10
  ring, 3, 36, 39, 43, 115, 146, 153, 172, 183, 189, 385, 389, 419
  semiconductor diode, 1, 10, 11, 102, 111–19, 133, 141, 208–61, 274–324, 330–80, 383–420
  side pumping, 110–11
  solitary, 59, 67, 76, 82, 84, 85, 86
  soliton, 2, 3, 22, 49, 58, 117, 119, 154, 164
  synchronously modelocked, 216
  Ti:sapphire, 2
laser diode arrays, 93, 94
laser resonator, 28, 32, 33, 66, 73, 81, 108, 126
light versus current, 392
line broadening, 66, 67, 70
linewidth, 65, 66, 68, 77, 89, 99, 106, 108, 110, 112, 128, 230, 330, 354, 358, 359
linewidth enhancement factor, 230, 242, 245
lithium triborate, 129, 130
locking range, 404, 405, 408
loop mirror, *see also* nonlinear optical loop mirror *and* nonlinear amplifying loop mirror
  nonlinear, 140, 144, 146, 150, 379
  optical, 144, 146, 147, 149, 174, 416
low birefringence, 179, 200, 201

Mach–Zehnder, 3, 27, 29, 32, 36, 142, 416
Mach–Zehnder interferometer, 3
master equation, 3, 9, 10, 12, 14, 17, 18, 25, 37, 40, 47, 48, 49
master oscillator-power amplifier, 276, 370, 372
millimeter wave generation, 415
mode-beating fluctuations, 59, 61–3, 65
mode-hopping, 356
modelocked gain, 7
modelocked lasers
  all-fiber, 18, 35
  dye, 208
  FM, 106
  monolithic semiconductor, 385, 386
  Nd:YAG, 22, 42, 112, 168
  semiconductor, 208–14, 255, 260, 333, 348, 349, 370, 384–6, 399, 403, 405, 416, 420
  solid state, 66, 68, 93, 106, 107, 111, 133
  Ti:sapphire, 47
modelocked operation, 74, 343
modelocked pulse, 58–63, 71, 78, 82, 85, 86, 142, 144, 156, 192, 210, 257, 259, 280, 290, 296, 299, 336, 368, 385, 390, 405, 409, 412
modelocked spectrum, 78, 83, 85, 87, 198, 296, 297
modelocked systems, 18, 49, 70, 102, 114, 171, 286, 287, 324
modelocking
  active, 2, 4, 10, 95, 96–116, 140–5, 169, 174, 275, 276, 278, 287, 290, 323, 336
    frequency domain, 4
    solid state lasers, 96
    techniques, 95
    theory, 2
    time domain, 7
  additive pulse (APM), 3, 23, 27–30, 35–44, 48, 49, 58, 59, 117–25, 129, 172, 208, 229
  carrier type (CTM), 170–2
  colliding pulse (CPM), 387
  coupled cavity, 3, 23
  harmonic, 144, 153, 155, 172, 198
  hybrid, 145, 278, 284, 287, 288, 290, 296, 299, 323, 385, 395, 418, 419
  Kerr lens (KLM), 3, 32, 49, 118, 123–5
  passive, 7, 37, 48, 49, 57, 140–5, 171, 174, 179, 187, 189, 192–5, 198, 199, 285, 288
  passive mechanisms, 3, 26, 95, 96, 115–25, 131–3, 385, 399, 418, 419
  passive, self-starting, 68, 204
  second harmonic generation, 113
  synchronous pumping, 114, 129
  VCSEL, 214, 217
modes
  axial, 4, 5, 6, 66
modulation frequency range, 339
modulation instability, 71, 72, 158
modulator, 2, 4, 6, 101, 103, 115, 116, 132, 156, 331, 390, 400
  acousto-optic, 100, 105, 141, 142, 192, 195
  active, 96, 145, 153
  amplitude, 95, 100, 105, 106, 114
  bulk, 141
  frequency, 100, 104, 113
  intracavity, 95, 398
  linear, 119
  nonlinear, 96, 117
  optical, 60, 389
  phase, 60, 95, 104, 105, 106, 108, 111, 120, 121, 128, 143, 144, 145
modulator resonance, 105
modulator transmission function, 100, 104
multiple quantum well, 118, 125, 171, 210, 217, 222–9, 237, 251, 252, 255, 260, 274, 275, 284–7, 299, 316, 361, 365, 368, 386, 387, 392
multiple quantum well saturable absorber, 117, 284, 296, 297
multiplexer, 143

Nd:glass, 1, 65, 70, 94, 95, 102, 106–8, 112, 113, 118, 125, 128, 131, 133
neodymium, 141, 180, 192
noise generator, 249
nonlinear amplifying loop mirror, 146, 147, 152, 153, 163, 167
nonlinear optical loop mirror, 146, 147, 149, 150, 174, 416
nonlinear scale length, 150
nonlinear Schrödinger equation (NLSE), 3, 18, 19, 49, 150, 184, 357
optical clock distribution, 211, 309, 324
optical clock recovery, 309, 324
optical parametric oscillator, 129, 130
optical spectrum analyzer 336, 358, 361, 363, 394, 405
opto-electronic, 212, 311, 330, 383, 415, 421
opto-electronic integrated circuit, 212, 330
organometallic vapor phase epitaxy, 391, 392
output coupler, 68, 102, 143, 146, 153–7, 162, 168, 170, 196, 201–4, 217, 218, 222, 225, 226, 248–51, 256–9, 278, 296–8, 355, 395
output power saturation, 219

packaging, 212, 324, 333, 339, 348, 353, 368, 373, 378
penetration depth, 343, 346, 347, 364
periodic perturbations, 76, 77, 83, 84, 89, 157
phase modulator, 60, 104, 105, 108, 111, 120, 121, 128, 143, 144, 145

photodetectors, 393, 405
piezoelectric transducer, 144, 229
planar waveguide laser, 128
Poincaré sphere, 191
polarization beat length, 180–2, 185, 187, 190, 198, 201
polarization controllers, 148, 154, 161, 189, 191, 192, 199
polarization evolution
  linear, 181, 189
  nonlinear, 140, 179, 180, 185, 187, 189, 191, 198, 204
polarization maintaining, 144, 145, 148, 149, 152, 170, 174, 179, 200, 205
polarization maintaining fiber, 144, 145, 149, 170, 200
polarization switching, 146, 187, 190, 197
praesedymium, 147
pulse
  bandwidth, 99, 101, 102, 143, 153, 164, 168, 174
  broadening, 81, 242, 243–4, 299, 303, 308, 386
  broadening, GVD induced, 81, 299
  chirping, carrier induced, 211
  compression, 1, 221, 222, 228, 237, 260, 283, 291–4, 299–303, 321–4, 416
  compression, cubic, 295
  compression, extracavity, 15, 156, 208, 211, 404
  compression, fiber, 1, 156
  compression, in dispersion decreasing fiber, 173
  compression, intracavity, 88, 119, 196
  compression, soliton, 157, 275
  shaping, discrete, 73, 74, 78

Q-switch, 1, 96, 112, 113, 129, 140, 172

radio frequency, 96, 100, 105, 111, 128, 143, 210, 216, 257, 287, 290, 297
Raman, 113, 168, 169, 173, 175
Raman self-scattering, 173
regenerative amplifier, 131
relative intensity noise, 345, 346, 359
relaxation oscillation, 94, 95, 215–16, 394
resolution bandwidth, 289
response time, 2, 58, 145, 421
ring resonators, 5
rotating wave approximation, 185, 193

Sagnac interferometer, 35, 140, 146, 147, 183
saturable absorber, 15, 46, 59, 60, 179, 235, 245, 287, 291, 316, 419, 420
  antiresonant Fabry–Perot, 125
  artificial fast (AFSA), 2, 3, 26, 140, 149, 153, 154, 174
  dye, 1, 57, 58
  Er-doped fiber, 172, 175
  fast, 8, 9, 10, 11, 29, 48, 58, 63, 85, 117, 120, 121, 124, 125, 149
  mirror, 299
  multiple quantum well, 117, 125, 171, 275, 284–6, 296, 297, 299, 385–410
  proton bombarded, 384, 386
  semiconductor, 145, 170, 175, 204, 205, 209–11, 216, 261, 353
  slow, 11, 386
saturable absorber action, 11, 18, 29, 48, 63, 85
saturable absorber medium, 9
saturable diode amplifier, 118
saturation, 86, 152, 194, 219, 231, 242, 282, 297–9, 388, *see also* gain saturation
  of dyes, 57, 96
  of SAM, 44
saturation energy, 99, 170, 171, 211, 213, 214, 222, 225, 247, 251, 259, 388
saturation formula, 14
saturation intensity, 120, 285
saturation power, 9, 14, 64, 69, 154, 277, 278
second harmonic, 394
second harmonic generation, 117, 118, 128, 133, 397, 399, 400, 401, 403
self amplitude modulation (SAM), 25, 37, 39, 44, 62–4, 74, 75, 79–81, 85, 86, 89
  coefficient, 9, 15, 27, 29, 31, 32, 35, 40, 41, 49, 59, 72
self focusing, 117, 124, 125
self starting, 27, 59, 62, 65, 89, 116, 119, 152, 153, 170, 171
self-phase modulation (SPM), 117, 167, 170, 222, 230, 233, 235, 239, 242–5, 260, 304
  and negative GDD, 58, 59, 71–6, 81, 82, 84, 88
  coefficient, 15, 19, 25, 29, 31, 32, 40
  in the nonlinear Schrödinger equation, 11, 13, 14
  induced chirp, 260
  modulation, 39
self-pulsation, 405–9
self-tuning
  wavelength, 330, 347, 351, 365, 370, 373, 377–9
sidebands, 6, 39, 83, 84, 140, 153, 157–68, 171, 289, 322, 323
side-mode suppression ratio, 358
silicon optical bench, 330, 333, 334, 339, 347, 373
soliton, 19, 20, 37, 38, 41–5, 89, 140, 150–70, 174, 203, 211, 336, 360, 383, 414
  average, 18, 42, 59, 75, 79, 82, 84, 158
  fundamental, 3, 21, 81, 150, 152, 187, 197
  higher order, 49
  perturbed, 3
  phase velocity of, 46
  second order, 3, 21, 22
soliton behavior, 3, 18
soliton compression, 157, 228, 275
soliton energy, 150, 165
soliton evolution, 46
soliton filter, *see also* nonlinear optical loop mirror, 150, 152
soliton formation, 23, 49, 152, 173, 185
soliton interaction, 164
soliton peak power, 150
soliton period, 22, 23, 76, 168, 185, 186, 187, 197, 201
soliton power, 185, 187, 197
soliton propagation, 145, 158, 159, 332
soliton pulse width, 75, 81, 88
soliton transmission, 180, 226, 352, 354, 375, 378, 379, 380, 411
soliton transmission system, 208, 339–33, 342, 348, 368, 373
soliton trapping, 179, 180, 186, 187, 199, 205
solitonic sidebands, 157
spatial hole-burning, 5, 63, 69, 190, 192, 224, 246
spectra
  time resolved, 302, 324
spectral broadening, 243–5, 286
spectral filtering, 101, 281, 291, 300, 301, 303, 308
spectral hole-burning, 63, 253, 370
spectral instabilities, 342, 348
spectral intensity
  bandwidth, 97
spectral narrowing, 221, 243–5
spectral resonances, 83–7, 157, 202
spontaneous emission noise, 249
split step Fourier method, 150
spurious reflections, 39, 67, 68, 201

stability, 10, 17, 85, 86, 89, 94, 128, 187, 199, 288, 338, 340, 341
  environmental, 180, 200, 204
standing wave, 5, 10, 40, 89, 100, 105, 114, 224–5, 246
steady state, 6, 8, 10, 14, 15, 25, 45, 59, 72, 74, 79, 80, 81, 89, 96, 102–4, 116, 129, 250, 314, 390
step recovery diode (SRD), 143, 257, 340, 341, 361, 364
stretched pulse-additive pulse modelocking, 40, 41
substantial broadening of the beat-note linewidth, 68
synchronization, 141, 156, 171, 309, 316, 320, 324, 419
synchronous pumped, 40, 41, 107, 130
synchronous pumping, 107, 175, 216–17

temperature
  insensitive, 182, 190
  room, 212, 213, 215, 217, 218, 222, 226, 260, 261, 413
  sensitive, 189
  tuning, 411–13
thermionic emission, 224, 246
time division multiplexing (TDM), 320, 383, 416
time–bandwidth product, 195, 202, 216, 243, 332, 339, 341, 363, 368, 373, 397, 404, 411, 415–20
  control of, 222
  diode in external cavity, 209
  effect of filtering on, 223, 260, 292
  Gaussian pulse, 97, 99, 102
  in solitary lasers, 81
  in soliton transmission, 352
  of a chirped pulse, 219, 290, 297
  of a modelocked laser diode, 336
  of a VCSEL, 221, 230
  of HSPS, 370
  of synchronously pumped VCSEL, 228
  plotted vs RF power, 377
  use with autocorrelation, 225
  vs RF power, 353
  with APM, 229
timing jitter
  between synchronized sources, 289, 290, 316–21
  estimated from phase noise, 288, 289
  from delay lines, 311, 416
  from RF drive, 320
  in gain switched lasers, 384
  in hybrid modelocking, 211, 289
  in passive modelocking, 289, 322, 323
  in pulse width measurements, 258
  in VCSEL, 258
  low, 309, 311, 330, 384, 419
  measured with correlator, 312, 319
transverse mode confinement factor, 213, 231
transmission line, 389
traveling wave optical amplifier (TWA), 276
two-photon absorption, 253

ultraviolet, 140, 339
unchirped Bragg reflector, 344

VCSEL modelocking, 213–14
vertical cavity laser, 216
vertical cavity surface emitting laser (VCSEL), 208, 320
vertical coupler filter, 385, 419

wave plate, 29, 149
waveguide, 5
wavelength division multiplexing, 19, 143, 201, 320, 332, 333, 339
wavelength tuning, 385, 412, 413
weak pulse shaping approximation, 72, 75, 79

zero birefringence, 197, 198